The BIRDS of SONORA

The BIRDS *of* SONORA

STEPHEN M. RUSSELL

GALE MONSON

Illustrations by Ray Harm

THE UNIVERSITY OF ARIZONA PRESS TUCSON

The University of Arizona Press

First printing

∞ This book is printed on acid-free, archival-quality paper.

Manufactured in the United States of America

Library of Congress Cataloging-in-Publication Data

Russell, Stephen Mims, 1931-

The birds of Sonora / Stephen M. Russell, Gale Monson;

illustrations by Ray Harm.

p. cm.

Includes bibliographical references (p. 341) and index.

ISBN 0-8165-1635-9 (cloth: acid-free paper)

1. Birds–Mexico–Sonora (State). I. Monson, Gale. II. Title.

QL868.R87 1998

598´.0972´17–dc21 97-33938

CIP

British Library Cataloguing-in-Publication Data

A catalogue record for this book is available from the British Library.

Frontispiece: Black-throated Magpie-Jay

To the late Donald Lamm and to Arnold Moorhouse and Ruth Russell, who accompanied us on so many trips to Sonora and contributed so significantly to this work.

Contents

Illustrations

OVERVIEW MAPS

ILLUSTRATIONS BY RAY HARM

Acknowledgments

We gratefully acknowledge the contributions of many persons who provided assistance in one way or another, especially those who accompanied us on field trips and camped with us in all parts of Sonora. In this category we wish especially to mention Arnold Moorhouse of Douglas, Arizona, who often traveled with us on our trips and whose skills in finding birds and then showing them to us and others were, to say the least, remarkable; Donald Lamm of Tucson, who participated in many trips and was Russell's colleague in trapping and banding birds in the coastal mangroves; and Ruth Russell, supreme nest-finder and bird-bander, who was always there to provide assistance in so many ways.

We were fortunate in all our expeditions to experience universal hospitality on the part of those Mexicans who owned or otherwise were in charge of the places we wished to explore and to make our camps. Such hospitality made our trips pleasant and memorable, and we are much indebted to all of our many hosts.

We also enjoyed the cooperation and friendship of many others who camped with us, some on many trips, and helped find birds and, in some cases, capture and band them. The length of the following list of participants will indicate the amount of assistance, in total, given us: George Austin, Harriette Barker, Jon Barlow, Marty and Frank Barry, Gary Bateman, John Bates, Paula Becker, Jerry Bock, Nancy Bock, Rick Bowers, Walter Bulmer, William Calder, Ross and Bette Chapin, Larry Christensen, William Clark, Kristin and Steve Coleman, Mark Courtney, Richard

Crossin, Douglas and Natalie Danforth, Stephen Deatrick, Carol DeWaard, Rick Dorrance, Bruce Duell, John Dunning Jr., Dan Fischer, Mary Foster, Sharon Goldwasser, Cyrena Goodrich, Patrick Gould, Clive Green, Grace Gregg, Kathy Groschupf, Alan Gubanich, William Gunn, Mary Jean Hage, Susan Ogden Hampton, Billie Hardy, John Haugh, Holly Hobart, John Hoffman, Jack Holloway, Kirk Horn, Lawrence Huber, Tom Huels, Terry Johnson, Michael Kaspari, Jeff Kingery, Mark Kot, Mark Larson, Darrel Lee, Diana Lett, Susan Lohr, Jan and Gordon Luepke, John Luepke, Don and Harriet Marble, Paul Martin, Nora Mays, Marie McGee, Kip and Joan Metzger, Nelson Moore, Mary Ellen Morbeck, Robert Morse, Karen Nickey, Robert Ohmart, Steve Olson, Carol O'Neil, Francisco Ornelas, Steve Prchal, Tom Rehm, Vince Roth, Harold Russell, Jennifer Shopland, David Sibley, Ronnie Sidner, Jim Siegel, Peter Siminski, E. Linwood Smith, Douglas Stotz, Peter Sundt, Sherman Suter, Bonnie Swarbrick, Steve Trimble, Ray Turner, Genevieve Tvrdik, Sarah Vetault, Diana Warr, Richard Weatherly, Steve Wilson, and Eric Yensen.

The following individuals provided notes and lists of birds observed on personal journeys to Sonora and graciously permitted us to use them in the preparation of this book: Daniel Anderson, Richard Bailowitz, Thomas Bancroft, John Bates, Rick Bowers, N. Boyajian, David Brewer, Judie Bronstein, David Brown, José Campoy F., Edward Chalif, J. Coons, Richard Cunningham, Chris Cutler, Douglas Danforth, Tom Deecken, Salome Demaree, Steve Dobrott, Louis Dombrowski, John Dunning Jr., Steve Ganley, John Goodwin, Clive Green, Grace Gregg, Kathy Groschupf, Jonathan and Roseann Hanson, Bill Harrison, Susanna Henry, Holly Hobart, Jack Holloway, Greg Homel, Steve Howell, Doug Hulmes, William Hunter, Phil Jenkins, Roy Johnson, Kyle Jones, Kenn Kaufman, Merritt Keasey, Lloyd Kiff, Donald Lamm, Wade Leitner, John Levy, Seymour Levy, Ruth Loveless, John Luepke, Joe Marshall, Paul Martin, Matt Matthiessen, Arnold McGill, Barbara McKnight, Eric Mellink, Terry Meyer, Nelson Moore, Arnold Moorhouse, Karen Nickey, Francisco Ornelas, Eduardo Palacios, David Pearson, Robert Perrill, Allan Phillips, Michael Plagens, Peter Pyle, Amadeo Rea, James Rising, Gary Rosenberg, Will Russell, Kenneth Stager, David Stejskal, Richard Taylor, Scott Terrill, Dick Todd, Robert Tweit, Tom Van Devender, Betsy Webb, Janet and Robert Witzeman, and Eric Yensen.

We are grateful for special assistance provided by Joe Marshall and the late Allan Phillips, for the use of their voluminous notes and for valuable comments on various bird species. We would also like to extend our gratitude to David Stejskal, an expert on the birds of Puerto Peñasco and the compiler of Christmas Bird Counts from there as well as from Alamos; to Bryan Brown and his associates, who made special contributions of their records obtained on their canoe trips down major streams of Sonora in search of Bald Eagles; to James Voelzer, who sent us copies of U.S. Fish and Wildlife Service surveys of wintering waterfowl; to Kathleen Klimkiewicz of the National Biological Service, who sent us lists of banded birds recovered in Sonora; to Michael Kunzmann, who provided essential assistance with the maps; to Dave Barker, Hal Coss, John Hafner, Joe Jehl, Phil Jenkins, Lloyd Kiff, Gary Kramer, Charles Mason, Guy McCaskie, Raymond Paynter Jr., Andres Sada, Paul Springer, Rebecca Van Devender, Tom Van Devender, and David Yetman, who clarified various problems of bird and plant identification for us or rendered assistance with other technical matters; and to Steve Howell and Kenn Kaufman for their comments on the manuscript. We extend our thanks to Sally Bennett for careful and constructive editing. We extend our appreciation to the curators of several institutions (listed on pages 13 and 14) for providing access to collections in their care.

We are indebted to artist-naturalist Ray Harm of Sonoita, Arizona, for the artwork that he created for the pages and the dust jacket of this book. He graciously contributed the use of these illustrations.

The BIRDS *of* SONORA

Introduction

The state of Sonora has features that result in a high diversity of bird species (we report on 525 species, including 35 that we consider hypothetical), making it of interest to biologists and other naturalists. It has an extraordinary geography. Here the northern fringe of the tropics meets the southern end of the temperate region, bringing plants and animals from both sources into a single state. Sonora has varied deserts and grassy plains. Rugged mountains may be arid and barren or forest clad and split by great barrancas. The long coastline of El Golfo de California has beaches of sand, mud, or rock. Mangrove thickets characterize some estuarine situations. River valleys are bordered by gallery woodlands or by irrigated fields. Several large dams create extensive impoundments to supply water for immense agricultural developments.

Sonora's climate varies seasonally from hot and dry to cool and moist. The elevational range is considerable, from sea level to 2630 m (map 1). The annual precipitation may be as low as 5 cm or may exceed 100 cm.

These characteristics produce a rich, diversified, and intensely interesting biota, which still awaits full study and appreciation. The opportunity to study Sonora's fascinating natural history must be taken soon, for it is slipping away as a consequence of human impact. We hope this volume will contribute to an understanding of the changing natural history of Sonora.

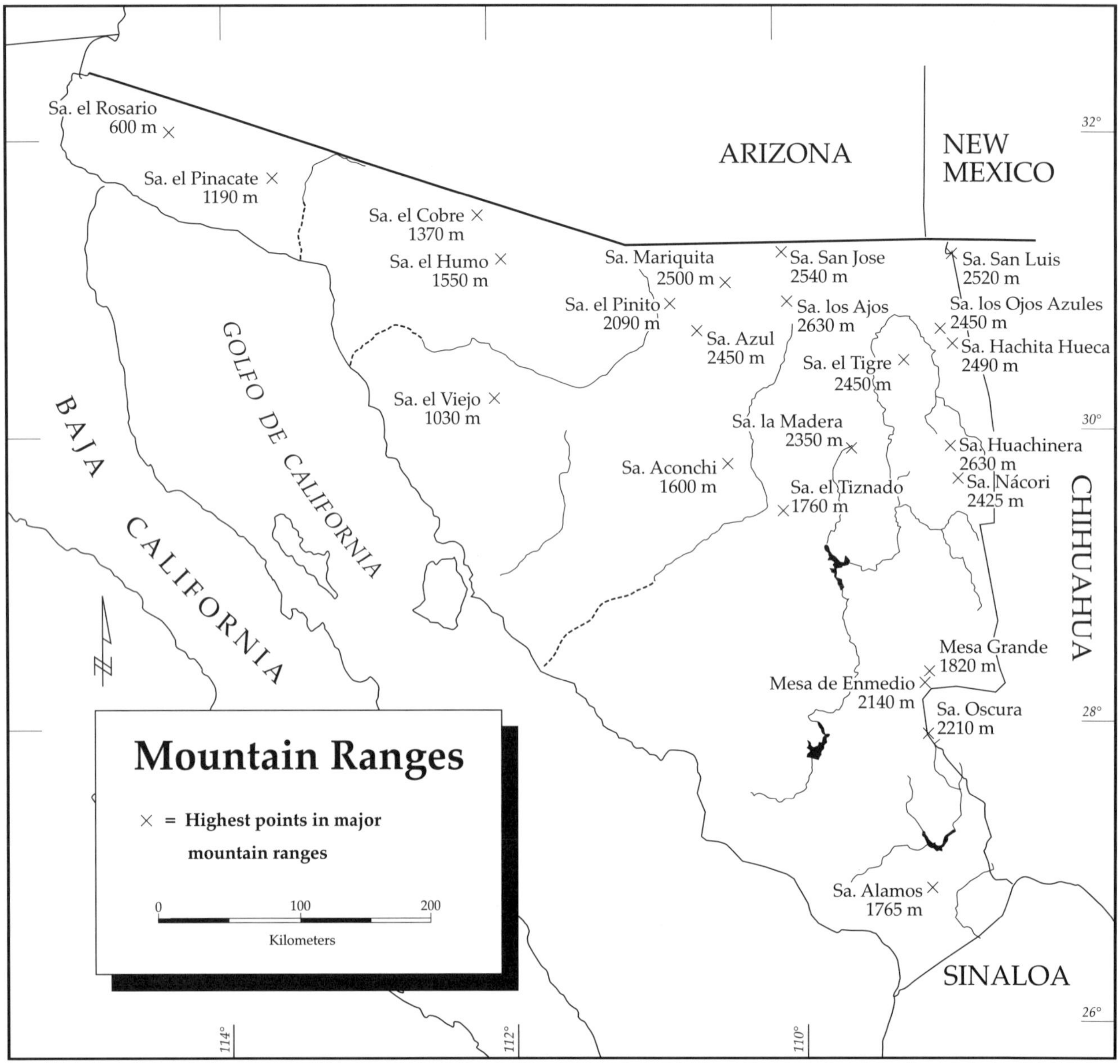

Map 1

GEOGRAPHY

Sonora is the most northwestern state of mainland Mexico. It is bounded on the north by Arizona (and a bit of New Mexico); on the east by the largest Mexican state, Chihuahua; on the south, where it is most narrow, by the state of Sinaloa; and on the west by El Golfo de California and, for about 80 km, the Río Colorado. It spans about 600 km from north to south and about 400 km from west to east at its midsection, for a total land area of 182 491 km^2. Hermosillo (see map 2), the capital city, lies near the geographical center of the state and is at nearly the same latitude as San Antonio, Texas; Cairo, Egypt; Delhi, India; the Midway Islands; and Brisbane, Australia.

Several major river systems traverse the state (see map 3), draining chiefly southward. These (from north to south), at their outlets into El Golfo de California, are named as follows: Río Concepción (formed by Ríos Magdalena and Altar), Río Bacoachi, Río Sonora (whose principal tributary is the Río San Miguel), Río Yaqui (whose main headwaters are the Ríos Bavispe, Moctezuma, and Aros), Río Matape, and Río Mayo. The first two tend to be dry much of the year, while

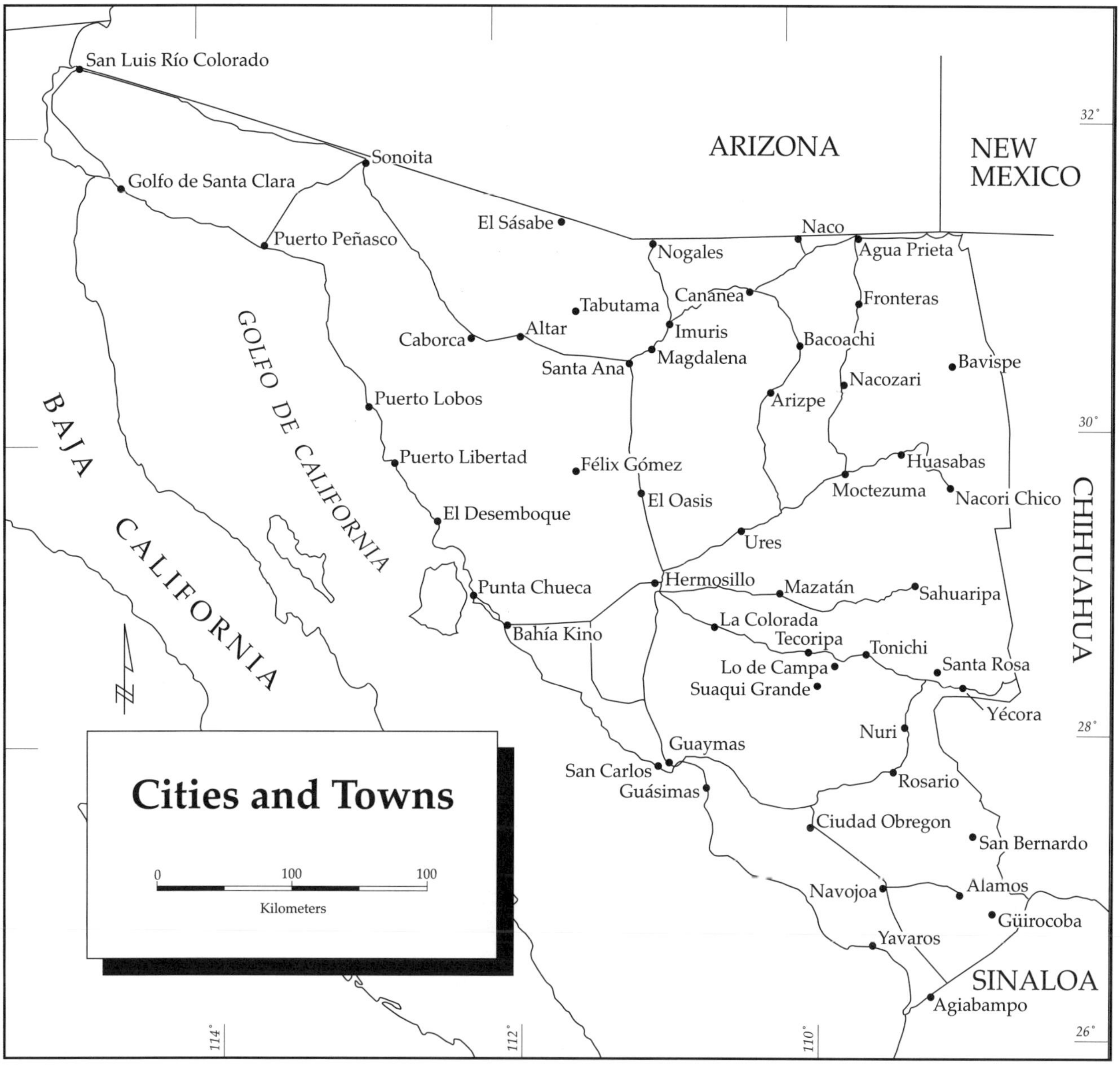

Map 2

the remainder, although having low flows at times, are considered major permanent streams in a desert region such as Sonora. We mention also the now almost dry Río Colorado, which forms the boundary with Baja California Norte; the Río Sonoita, a short stream in the far northwest; and the Río Fuerte, which for a short distance forms the boundary between Sonora and Sinaloa and drains extreme southeastern Sonora via the Río Cuchujaqui (or El Cuchujaqui). The well-known Alamos locality is in the Río Fuerte watershed. No river system in Sonora has a flow that reaches the coast throughout the year.

Major dams and their reservoirs are referred to as *presas.* In Sonora these are Presa Abelardo L. Rodriguez on the Río Sonora (at Hermosillo); Presa la Angostura, Presa Plutarco Elias Calles (popularly known as Presa el Novillo), and Presa Alvaro Obregon in the Río Yaqui drainage; Presa Adolfo Ruiz Cortinez (popularly known as Presa Mocuzari) on the Río Mayo; and Presa Francisco L. Alatorre on the Río Matape. To these might be added three lesser presas: Presa Cuauhtémoc on the Río Altar about 40 km northeast of Altar; Presa Jacinto Lopez on the Río Cuquiarichi about 12 km west of Fronteras; and Presa Agua Caliente, a part of the

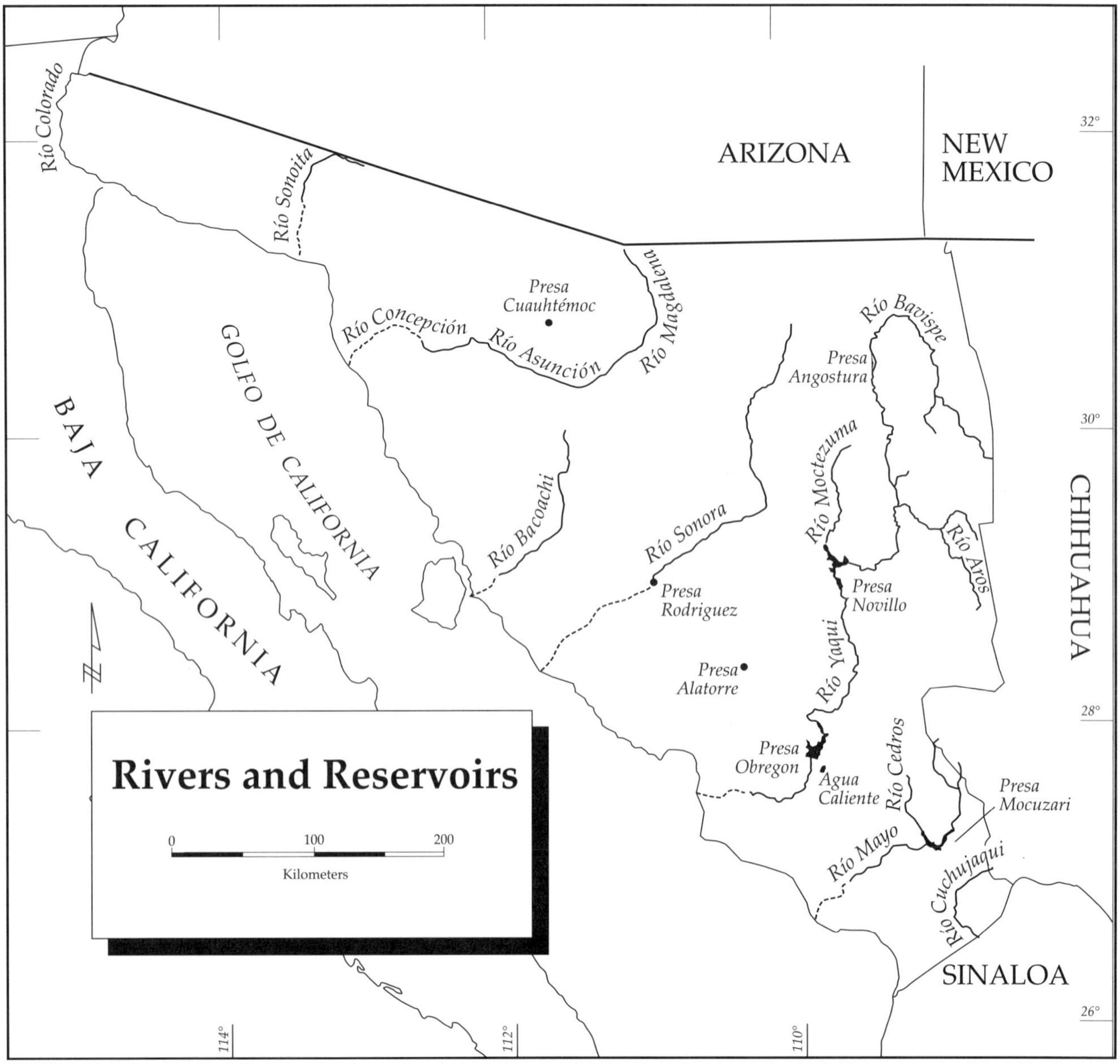

Map 3

water distribution system below Presa Obregon. These lesser presas are not always shown on maps. The reservoirs behind these dams are sometimes called *lagos* (lakes) or *embalses.*

PLANT COMMUNITIES OF SONORA

It is difficult to describe plant communities because they blend with one another over distances spanning many kilometers. We recognize major plant communities in a nontechnical approach, identifying them only enough to define general categories. Plant names used in the text are listed in appendix A, where the family is also identified. For a more detailed review see D. Brown 1994. Many species of birds have rather specific habitat requirements, such that alteration of some single habitat component may result in a change of their population numbers or even their distribution. Our studies of the birds of Sonora have been inadequate to determine the exact requirements of any species, and thus we describe habitat preferences in general terms. We recognize major plant communities as follows (map 4).

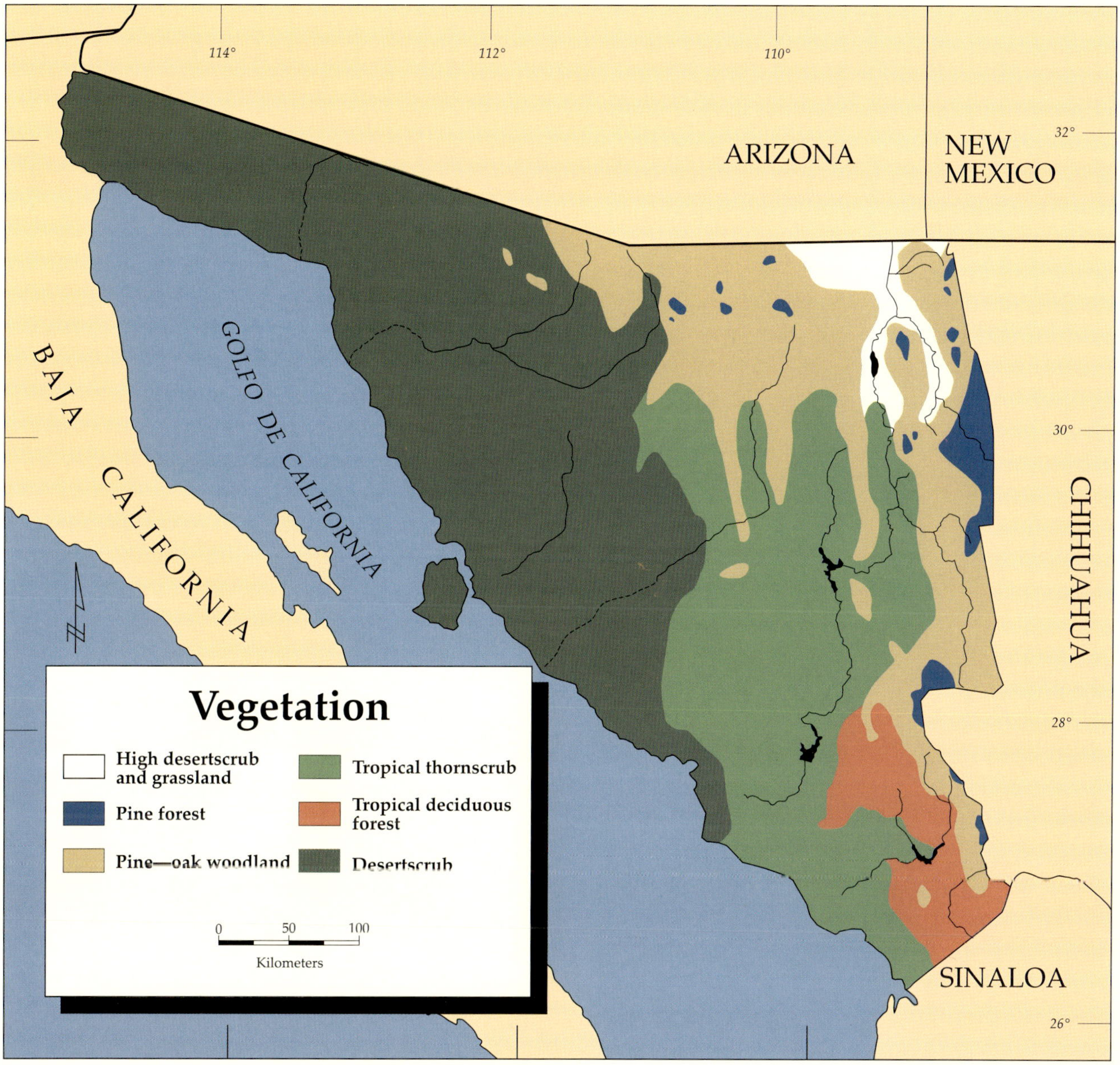

Map 4

High Desertscrub and Grassland

This plant community is limited to a small area in the extreme northeast. Shrubs are mainly creosote bush *(Larrea tridentata)*, tarbush *(Flourensia cernua)*, whitethorn acacia *(Acacia neovernicosa)*, and scrub mesquite *(Prosopis glandulosa* var. *torreyana)*. Grasses are chiefly gramas (*Bouteloua* spp.). This community is referred to as Chihuahuan Desertscrub and Plains and Great Basin Grasslands by D. Brown (1994).

Pine Forest

Found only in the upper parts of the higher mountains, this plant community occurs mostly along or near the northeastern and eastern borders. Ponderosa pine *(Pinus ponderosa)*, Mexican white pine *(Pinus ayacahuite)*, Douglas-fir *(Pseudotsuga menziesii)*, oaks, and grasses are the predominant vegetation.

Pine-Oak Woodland

Typically this plant community occurs below the pine forest, but it is also found in the higher mountains along the northeastern and eastern borders. Plant species include Chihuahua pine *(Pinus leiophylla)*, Apache pine *(Pinus engelmannii)*, Arizona white oak *(Quercus arizonica)*, silver-leaf oak *(Quercus hypoleucoides)*, Chihuahuan oak *(Quercus chihuahuensis)*, Arizona cypress *(Cupressus arizonica)*, alligator bark juniper *(Juniperus deppeana)*, manzanita *(Arctostaphylos pungens)*, and grasses. A number of additional species of Mexican pines and oaks of Sierra Madrean origin are present. Oaks are often dominant over large areas, forming oak woodlands that in heavy stands are known as *encinal*.

Desertscrub

This plant community has an extensive distribution. It is found below pine-oak woodland, in central and western Sonora west to the coast and east to Mexico Highway 15. The species composition is multitudinous but generally includes velvet mesquite *(Prosopis velutina)*, ironwood *(Olneya tesota)*, palo verde (*Cercidium* spp.), saguaro *(Carnegiea gigantea)*, chollas (*Opuntia* spp.), bur sages (*Ambrosia* spp.), and desert hackberry *(Celtis pallida)*. This community is called Sonoran Desertscrub by D. Brown (1994).

Tropical Thornscrub

This plant community is found in foothills and valleys south of about latitude 30° N, east of Mexico Highway 15 and south to the coast, below pine-oak woodland but excluding considerable areas south of the Rosario region that are in the tropical deciduous forest community. The community comprises tall, thorny, spring-deciduous shrubs that are not as tall as the accompanying hecho or hairbrush cactus *(Pachycereus pecten-aboriginum)*. Some of the characteristic species are hecho, organ pipe cactus *(Stenocereus thurberi)*, boatspine acacia *(Acacia cymbispina)*, canyon ragweed *(Ambrosia ambrosioides)*, kapok *(Ceiba acuminata)*, palo de asta *(Cordia sonorae)*, tree ocotillo *(Fouquieria macdougalii)*, vara blanca *(Croton fragilis)*, tree morningglory *(Ipomoea arborescens)*, palo blanco *(Piscidia mollis)*, *Pithecellobium* spp., and sitavaro *(Vallesia glabra)*. This community is called Sinaloan Thornscrub by D. Brown (1994).

Tropical Deciduous Forest

Often disjunct with the foregoing community and sometimes restricted to riparian sites within the foregoing community, this plant community is distributed throughout southern Sonora. It is most consolidated toward the border with Sinaloa, and it represents the northernmost portion of a plant community that extends southward on the Pacific coast to northwestern Costa Rica. The vegetation is characterized by spring-deciduous trees, taller than hechos. Some of the most characteristic plants are hecho, canyon ragweed, *Bursera* spp., twin-flower cassia *(Cassia biflora)*, kapok, guácima or prickle-nut tree *(Guazuma ulmifolia)*, tree morningglory, feather tree (*Lysiloma* spp.), palo blanco, *Pithecellobium* spp., papache *(Randia echinocarpa)*, organ pipe cactus, amapa (*Tabebuia* spp.), Montezuma baldcypress or sabino *(Taxodium mucronatum)*, and uvulama *(Vitex mollis)*. This community is referred to as Sinaloan Deciduous Forest by D. Brown (1994).

Other Vegetation Types

In some areas of central Sonora are relics of grasslands and savannas that once stretched over considerable areas but were overgrazed for many years and now are supplanted by desertscrub and tropical thornscrub and by the alien buffel grass *(Pennisetum ciliare)*. Buffel grass (almost useless to birds) has been planted after the clearing of many thousands of hectares of tropical deciduous forest and tropical desertscrub. The resulting near-monoculture is praised by the cattle industry, but the practice threatens the level of diversity of plants and animals in the entire region.

In addition to the vegetative communities listed above are gallery woodlands along the streams of the northern half of Sonora and cottonwood (*Populus* spp.)

stands, now almost relics, along the lower Ríos Yaqui and Mayo. Freshwater wetlands are scarce and for the most part man-made. They may be the inundated lands behind simple *represos* or *charcos* for livestock water or lands fringing the reservoirs behind major dams. Saltwater marshes, estuaries, *esteros* (bays or inlets without a regular influx of fresh water), and mangrove swamps are found along El Golfo de California. Freshwater marshes, or *ciénegas,* are rare. Moreover, there are farmlands with their irrigation canals and drains, villages *(pueblos),* and cities, as well as the open sea with sandy beaches and rocky shorelines.

In noting the changes in plant life as one travels from north to south along Mexico Highway 15, the difficulty of detecting the change from desertscrub to thornscrub is apparent. Tropical plant species when first met do not appear en masse, but gradually and in small numbers. A species that might be considered typical of much of Sonora is organ pipe cactus, which reaches almost to the northern border in the upper Río Yaqui watershed and actually crosses it in the Sonoita region. Somewhat farther south one encounters chino *(Pithecellobium mexicanum),* a large leguminous tree growing along washes. Next noted as new plants are tree ocotillo and tesajo cactus *(Stenocereus alamosensis).* Approaching Hermosillo, tree morningglory and the deep-blue-flowered guayacán *(Guaiacum coulteri)* are found. From Hermosillo to Guaymas more new species put in their appearance, including the extremely graceful *Acacia willardiana* of rocky hillsides and the dense dark *Forchammeria watsoni,* but even this far south the vegetation remains more desertscrub than thornscrub. Traveling east or southeast from Hermosillo, typical tropical thornscrub is soon encountered, increasingly so as hills in the drainage of the Río Yaqui are reached. Southward from Hermosillo to Guaymas and as one nears Cd. Obregon, this community is met as the foothills and mountains are entered, away from the coast. In the Cd. Obregon region and southerly, thornscrub becomes dominant.

Desertscrub and thornscrub of the coastal plain each contain some species not observed as one progresses inland. Along the coast south of the Río Mayo delta, thornscrub becomes dominated by hecho and organ pipe cactus.

The tropical deciduous forest of southern Sonora is found usually on hills and steep slopes. In July, it appears as a blanket of varied green forming a canopy 10–18 m above the ground. Until late October in Sonora the region appears green and leafy, but then the drought begins that may persist until the following June. Leaves begin falling, and by December the forest is almost bare. Spring and early summer may be a period of extreme stress for both plants and animals, for there is little water, temperatures are high, and the sunshine is intense.

SEASONAL CHANGES

Seasonal changes in the plant communities of Sonora correspond to the rainfall seasons. The summer rainy season produces most of the annual precipitation; depending on the altitude it lasts roughly from mid-June to mid-September. Then comes the fall drought, which in some years is interrupted by heavy rain associated with hurricanes, followed by occasional light rain falling periodically from December through February. A long spring drought characterized by increasing temperature persists until the summer rains begin.

The mean annual rainfall varies from as little as 5 cm along the northwestern coast to as much as 100 cm in the high sierras of the east. It is not predictable, for in some years drought conditions prevail but in other years it will be much wetter than normal. It is possible for an unusually dry winter to be followed by an unusually dry summer, or vice versa, producing conditions extremely stressful to plants and animals. Another departure from normal is the delay of summer rains well into July, when they may be comparatively light, normal, or heavy.

The breaking of the spring drought in the lowlands and foothills is quite dramatic. Away from permanent streams, most plants spring to life and a lush green mantle spreads over the countryside where only a few days earlier all was sere and quite leafless. Annual plants sprout and grow prodigiously to form near-impenetrable thickets. Flowers put in their appearance, followed sooner or later by fruits of many kinds. Insects multiply, and their variety becomes more visible.

Nowhere is this contrast between seasons more striking than in the tropical thornscrub and the tropical deciduous forest. Let Howard Scott Gentry in his *Río Mayo Plants* (1942:33) describe it:

> In the long spring dry season the forest is a dreary scene: a naked infinite host of trunks and branches, spreading interminably over the volcanic hills and mesas, bared to the fiery sun, under which the last leaf seems to have withered and died. He who walks this land in the month of May walks with a parched throat. The plants are waiting for the rains and their union with the soil, when like a piece of magic they turn the look of the dead into a fiesta of voluptuous growth, all in a brief few days. The sudden starting of new leaves with the first summer rain is aided greatly in many plants by the advanced development they have attained in the bud. Ten days is sufficient to bring the bare forest into green leaf, though it requires several weeks to bring the leaves to maturity.

This phenomenon of abrupt change in the moisture supply, although it varies somewhat from year to year and is of lesser scope at higher elevations, over the breadth of evolutionary time certainly has had a pronounced effect on the annual life cycle of many of the breeding birds of Sonora. Much remains to be learned about the relationship of climate to breeding behavior. We touch on this in some of the species accounts. Generally, most species nest as the spring warms, but some species do not nest until the summer rains provide ample food and cover.

MIGRATIONS

Many Sonoran birds are year-round residents, keeping to a local territory. Others are termed summer residents; these nest after returning from a winter stay to the south. Following nesting, the summer residents migrate southward. Some mountain species move to adjacent lower elevations before they move south for the winter; a few migrate only altitudinally and spend the winter in the lowlands.

Another group is composed of birds that breed in the United States and Canada, then migrate south for varying distances to spend the colder months—some not farther south than Sonora, others passing through Sonora to continue as far as southern South America. Some migrants from the north barely enter the confines of Sonora. A final group consists of strays of species otherwise not found in Sonora.

It is of interest that although most migrants cross or enter Sonora along a northwest-southeast axis, there are some that enter on a northeast-southwest axis. Some species that breed in eastern or central North America are represented by a scattering of individuals over the Sierra Madre Occidental to Sonora, especially to the southern part. These are chiefly wintering birds, not strays. These seeming eccentricities result because Mexico literally is a funnel receiving migrants from the very broad North American mainland.

The length of the migration period may vary considerably among different species. Some southbound fall migrants reach Sonora in early July or even late June. In the case of waterbirds, some individuals that would be expected to migrate north in the spring will linger into, or even remain for, the summer.

Some waterbirds, notably seabirds, nest in colonies. Most colonies are on islands in El Golfo de California, but nonbreeding individuals may be found along the Sonora coast during the nesting period.

HUMAN EFFECTS ON THE ENVIRONMENT

We have been very aware of the influence of man on the countryside. Ranchers and their livestock have been a dominant feature for 300 years, and they have had a profound impact on natural habitats. All too frequently, the effects have been harmful. But the impact of past centuries of land use are insignificant in comparison to changes now taking place. In recent years vast areas have been cleared of all existing plant life and planted with African buffel grass, creating a monoculture of little value to wildlife (Bowden 1993). Even more of Sonora is destined to be altered this way.

The original desertscrub and thornscrub of much of the coastal plain has been removed to make way for irrigated districts: El Distrito de Riego de Hermosillo southwest of that city, El Distrito de Riego Yaqui, and El Distrito de Riego Río Mayo. Water for these farms comes from large presas—reservoirs impounded by dams on major streams. Most of the great cottonwood groves that once marked the routes of permanent rivers are gone, replaced by an agribusiness that has no place for trees. The cottonwood groves supported a diversity of birds among the highest of any plant community in Sonora. Fortunately, groves of Montezuma baldcypress, or sabino, are being protected.

The pine forests of Sonora have been logged except in small areas where it would be economically infeasible to do so. This logging began with the first settlement of the mountains by the Spanish, but it was carried out on a large scale principally in the 1940s to 1970s. Logging continues to the present day, but on a reduced scale involving chiefly smaller trees not taken on the first cutting. It is difficult to find any high mountain in Sonora that is not crisscrossed by bulldozed logging trails, although some of these are already becoming dim with time. This lumbering has had a great effect on wildlife, especially snag-inhabiting birds and mammals.

The delta of the Río Colorado, partly in Sonora but largely in Baja California Norte, was once dominated by riparian plants supporting a vast amount of wildlife at all seasons. Diversion of Colorado River water in the United States to serve the demands of agriculture and hydroelectric power permanently altered habitats along the lower portion. With the completion of Hoover Dam in 1934, and the earlier building of dams on the Gila-Salt River systems, the flow of water to the delta all but ceased. Now, the former riparian thickets, where not replaced by farmlands, have become a great mudflat, scoured twice daily by a tide that fluctuates as much as seven meters. Now and then, rare periods of unusually heavy runoff in the United States will spill surplus water into the delta. The Ciénega de Santa Clara in the Sonoran portion of the delta has been maintained by irrigation drainage water and by water released by a huge desalinization plant (now in disuse). The ciénega's future is uncertain.

RECENT CHANGES IN BIRD DISTRIBUTION AND NUMBERS

Although it is an easy supposition that the ranges and numbers of breeding birds do not materially change, the opposite may be true. A species' range and population size may contract or expand for various reasons, not the least of which is human disturbance. Other causes include meteorological conditions, parasitism, predation, disease, and interaction with other species.

A striking phenomenon is the recent northward extension of the breeding range of some species (e.g., Buff-collared Nightjar, White-eared Hummingbird, Thick-billed Kingbird, Five-striped Sparrow) from Sonora into Arizona, coupled with the casual occurrence without nesting of other Sonoran breeding species (e.g., Aztec Thrush, Tropical Parula, Fan-tailed Warbler, Yellow Grosbeak) in Arizona. Whether such changes represent steady progression or are simply sporadic irruptions remains to be understood. Comments are made in the species accounts on recent distributional changes.

Plan of the Species Accounts

THE PROBLEM OF SIGHT RECORDS

Most of this book is based upon our own observations and those of many contributing observers. We have not used sight records about which we have reservations. We have regarded all records critically and made decisions based on the available details, the known expertise of the observer, the relative difficulty in identifying the bird, and the likelihood of the bird's occurrence. Most reports of unusual observations have been discussed with the observer. For a detailed in-depth treatment of the problems involved, see Phillips's (1986:xxiii–xxxii) "Why we can *not* trust our eyes in science."

When the occurrence of a species in Sonora is represented by few records, we cite all of them. If there are few records and a specimen exists, we may name the collection in which it has been reposited, or we may instead cite a reference to the specimen. The following abbreviations for specimen collections are used:

AMNH	American Museum of Natural History
CMNH	Carnegie Museum of Natural History
DVM	Delaware Valley Museum
LACM	Los Angeles County Museum
LSUMZ	Louisiana State University Museum of Zoology

MCZ	Museum of Comparative Zoology, Harvard College
MVZ	Museum of Vertebrate Zoology, University of California, Berkeley
OC	Occidental College
SDMNH	San Diego Museum of Natural History
SWC	Southwestern College, Winfield, Kansas
UA	University of Arizona
UCLA	University of California, Los Angeles
UK	University of Kansas Museum of Natural History
USNM	U.S. National Museum, Washington, D.C.
WFVZ	Western Foundation of Vertebrate Zoology
YU	Yale University, Peabody Museum of Natural History

Our knowledge of birds collected by Allan R. Phillips has come primarily from his field catalog and notes. His specimens have been deposited mainly in the Delaware Valley Museum, but some of his specimens are in places unknown to us, so we generally do not ascribe Phillips specimens to any one collection.

HYPOTHETICAL SPECIES

Hypothetical species are included in the text in their taxonomic order (American Ornithologists' Union) but are placed in brackets. Hypothetical species are those that have been reported with substantial documentation yet are not supported by a specimen, a clearly diagnostic photograph, or extensive detailed information from multiple observers. We include a species in this book without brackets if it has been seen and well documented in at least two observations by experienced and reliable observers and we have studied the documentation and are convinced of the accuracy of the report. In the case of a few well marked and distinctive species, we have accepted a record if there has been only one observation by three or more competent observers. The reader may wish to apply different rules for judging the acceptability of records.

SUBSPECIES

We do not treat birds at the subspecies level unless obvious morphological differences exist that are readily discernible in the field (e.g., Mangrove versus Yellow Warblers). In those few cases, we discuss such subspecies in their Sonoran context. We are aware of the importance of subspecific studies, but our objective in this book is to describe distributional, behavioral, and ecological information at the species level.

NAMES OF BIRDS

Because systematic and taxonomic treatment is not one of our goals, we follow the presentation of the sixth edition of the *Check-list of North American Birds* (AOU 1983) and its supplements (AOU 1985, 1987, 1989, 1991, 1993, 1995) for technical and English-language names. The most recent supplement appeared in 1997 (AOU 1997) and included announcements of major changes to appear in the seventh edition of the *Check-list of North American Birds* (AOU in press). The supplement was published after the deadline for major revisions in *The Birds of Sonora.* Many of the changes pertain to the sequence in which groups are listed and to the ranking of taxonomic categories. Some changes apply at the species level and involve splits or name changes. Where appropriate in this book, we do mention revisions at the species level, although we could not modify the accounts accordingly. For Spanish-language vernacular names we follow *Listado de nombres comunes de las aves de México* (Escalante et al. 1996). Where we are aware of a differing prevalent vernacular name in Sonora, we provide it.

HOW RECORDS ARE CITED

A. J. van Rossem's thorough and critical work titled *A Distributional Survey of the Birds of Sonora, Mexico,* published in 1945, has been the major ornithological publication on Sonora. We cite many records (both specimen and sight) from that work and utilize several other sources. Published records are acknowledged in the "Literature Cited" section. Records in the text that are not accompanied by an observer's name are attributable to Monson or Russell or to someone accompanying one or both of them in the field (in exceptional instances we do cite the name of the observer). The surname and initial of the observer are given for other unpublished records; the records have been provided to us in the form of field notes, lists, or correspondence.

TERMS FOR RELATIVE ABUNDANCE OF SPECIES

In employing terms for relative abundance of species, it should be remembered that any species will vary in numbers in different parts of its range; some species occupy a specialized habitat of limited extent and are therefore local; and birds vary in abundance according to season. It is apparent that more than one term may be used to describe the relative abundance of a given species.

The following subjective designations of abundance are used to represent the probable results of a morning's effort in the field in appropriate habitat by a skilled observer familiar with the area:

abundant	invariably seen, often in conspicuous flocks
common	frequently encountered, as individuals or in pairs or small groups
fairly common	a few individuals or pairs may be seen
uncommon	present in area (region, habitat) but may not be found in a day or two of field observations
rare	present in an area but infrequently seen (several days could be needed to find it)
casual	out of usual range but could be expected every year or two
accidental	far from normal range and not to be expected again
irregular	within its range but not present every year

PLACE NAMES

Appendix B is a gazetteer of all localities mentioned in the text. All place names follow the usage in the excellent topographic charts (scale 1:50000) produced by the Mexican government (Mexico 1971–1982). We have preferred to use Mexican words for physical features (i.e., *sierra* for mountain range, *río* for river, *charco* or *represo* for stock tank or water hole, *ciénega* for marsh, *presa* for dam and accompanying reservoir, *isla* for island, *estero* for estuary, *bahía* for bay, *laguna* for lake or pond, *arroyo* for gully or wash, etc.). We use "El Golfo de California" rather than "The Gulf of California."

THE SONORAN ISLANDS

We do not treat any of the islands of El Golfo de California as a part of this volume. We limited our field studies to mainland Sonora.

THE MAPS

A map for each species illustrates all localities for which we have records. Some localities lie at the poorly defined boundary of Sonora with Chihuahua; symbols at these localities may thus appear outside the boundary line.

Symbols reflect our knowledge of the breeding status. "Breeding confirmed" is based upon the observation of a nest containing eggs or young or the presence of young birds incapable of sustained flight. We have used the expression "breeding presumed" to reflect strong circumstantial evidence, such as the presence of paired and territorial birds, old nests, or nests under construction. The use of the expression "transient" indicates that individuals move through the locality during their migration. "Transient and winter" implies that some migrants may also be present in winter. Symbols indicating breeding status receive precedence over other symbols. Thus, at a locality where a species is thought to breed, there will be only one symbol, even if the species is also present as a transient or as a transient and in winter. We use "locality record" to indicate that a species has been found at least once, but we lack evidence that would enable us to place it in one of the other categories.

BRIEF HISTORY OF ORNITHOLOGY IN SONORA

The scant early ornithology of Sonora was summarized by van Rossem (1945). The collections of van Rossem and his colleagues, notably Seth Benson, W. W. Brown, Chester Lamb, Robert T. Moore, William J. Sheffler, and John T. Wright, form much of the foundation for van Rossem's book. His work is the major landmark in the study of Sonora's birds, especially because of his critical examination of many specimens, as well as a thorough review of the existing literature.

Since van Rossem's time, the decade of the 1950s ushered in a period of increased activity. Allan R. Phillips made important collections in several localities throughout Sonora from Cd. Obregon north, essentially ending in 1958 when he moved from Tucson, Arizona, to the city of Mexico. Joe T. Marshall (1957) published an exceptionally fine analysis of the distribution of the breeding birds of the mountains of northeastern Sonora and southeastern Arizona. Peter Alden led many bird tours to western Mexico; his resultant volume (1969) is almost wholly based on sight records and contains some accounts that are vague. It furnished detailed information (now somewhat dated) on some of the best places to observe birds in Sonora, chiefly coastal, and served to develop an interest in Sonoran birds on the part of birders in general.

The published works mentioned above, a few additional miscellaneous papers, the accretion of unpublished records of specimens and of sight observations made available to us, plus our own studies, form the basis for this book. We have been extremely fortunate to have detailed field lists and notes from most of the individuals who have observed Sonoran birds in recent years. Our coverage of the Puerto Peñasco and Alamos areas has been much enhanced by these contributions.

FIELDWORK BY THE AUTHORS

Russell began his ornithological studies in Sonora in 1964. His 119 trips have involved 565 days in the field, with campsites at 188 different localities. Russell was accompanied on many of these trips by experienced observers who contributed immensely in gaining information about the birds encountered. From 1970 to 1978, with Donald W. Lamm and others, he banded extensively in mangrove swamps at Punta Santa Rosa and Punta Chueca. Some collecting was done until 1983, and recognition is made of the Mexican government for issuing permits. The specimens were deposited at the University of Arizona and the National Autonomous University of Mexico (Universidad Autónoma de Mexico). Monson has made 55 trips to Sonora. Of these, 31 trips totaling 127 days in the field were mostly to Alamos, San Carlos, and Bahía Kino but also to Arroyo Cajón Bonito, the Sierra San Luis, and several other areas. Another 24 trips were made with Russell.

Sonora covers a vast expanse, and travel is slow in remote areas. Most places have been visited very few times. Lo de Campa was selected in January 1983 as a locality for repeat visits. This site on Arroyo San Javier southeast of Tecoripa and not far west of the Río Yaqui is in an exceptionally diversified habitat, with water in the arroyo year-round. It is distinguished by having the most northerly grove of Montezuma baldcypresses in North America, to our knowledge. Many species characteristic of the tropical deciduous forest are present. Many birds were banded there, providing information on residency status and site fidelity.

Species Accounts

FAMILY GAVIIDAE

Loons

RED-THROATED LOON

Colimbo Menor

Gavia stellata

This loon has been found about ten times in Sonora in winter just offshore in El Golfo de California, south to Huatabampito, from 27 November to 9 April. No specimens have been taken. We have not included a number of reports of the Red-throated Loon because it is often confused with Common and Pacific Loons and the records are not qualified by details.

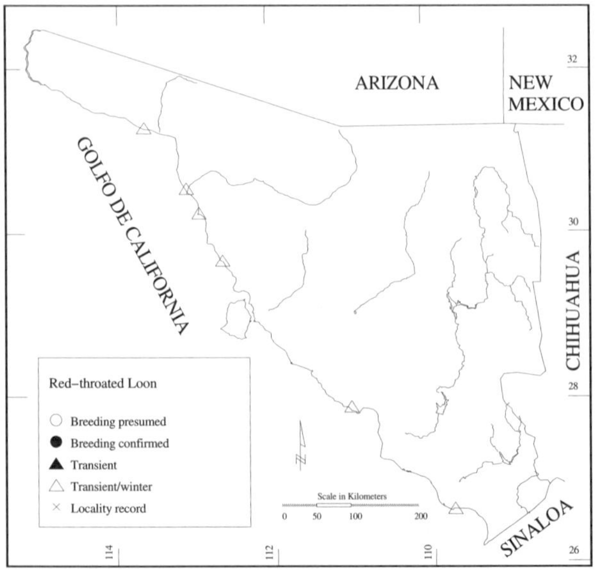

PACIFIC LOON

Colimbo Pacífico

Gavia pacifica

The Pacific Loon is a regular and fairly common winter visitant along the shore of Sonora, less common south of Guaymas. That sizable numbers can occur is suggested by the total of 1746 reported on a Christmas

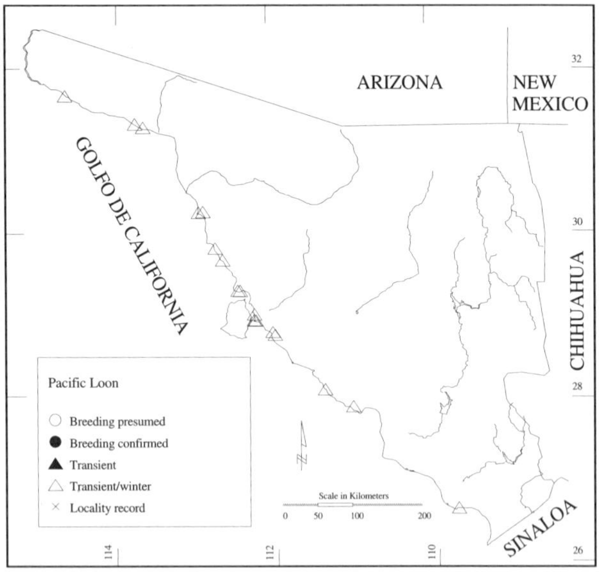

Bird Count (3 January 1981) at Puerto Peñasco, including some counted from a boat. The earliest fall arrival date is 8 November; the latest in spring is 3 May. Zimmerman and Boettcher (1967) stated that large numbers of Pacific Loons were seen offshore at Puerto Peñasco on 24 November. We feel a summer sight record (Gallucci 1981) from Puerto Peñasco needs documentation, as does an undated record (Alden 1969) for Mazatlan, Sinaloa.

COMMON LOON

Colimbo Mayor

Gavia immer

As befits its name, the Common Loon is the loon most frequently noted along the coast of Sonora. This is contrary to past literature, at least until Zimmerman and Boettcher (1967) found a dozen at Puerto Peñasco in 1965. The Common Loon is readily seen from the shore during winter at almost any locality north of Guaymas. The recorded wintering span is 23 September to 28 April. There are several summer records. Russell observed a single loon off Punta Cirio on 20 June 1978. D. Stejskal studied one off Puerto Peñasco from 2 to 5 July 1976, where an astonishing six or more, four in breeding plumage, were reported on 25 and 26 July 1979 (Gallucci 1981).

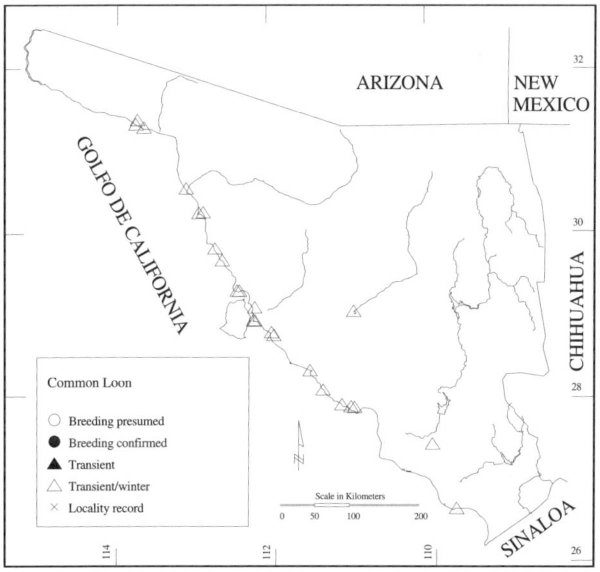

Inland, one was found on a freshwater pond 10 km southeast of Cd. Obregon on 18 January 1982 (Monson), and another was at Hermosillo on 15 November 1986 (J. Dunning). It is likely that Common Loons appear more frequently on inland ponds and reservoirs than these two records indicate.

R. Taylor has told us of finding about ten dead Common Loons washed up on a 1.5 km stretch of beach at Bahía Kino on 24 February 1987 following two days of violent winds. The same storm caused injury or death to one each of Blue-footed and Brown Boobies and about six Brown Pelicans. Possibly some summer records are of birds disabled by storms.

FAMILY PODICIPEDIDAE

Grebes

LEAST GREBE
Zambullidor Menor
Tachybaptus dominicus

The Least Grebe is an uncommon resident of southern Sonora, principally on ephemeral flooded shallow basins of rainwater or irrigation water. Bordering cattails or other emergent aquatic vegetation is a requisite. Saltwater records are lacking.

We find few breeding records in Sonora. R. Taylor saw three adults accompanied by two young riding on a parent's back at a roadside pond about

Least Grebe

16 km southeast of Empalme in early April 1983. C. Schwalbe noted an adult incubating five eggs near Alamos on 20 August 1982; the nest was on a mound of vegetation in a small stock pond and about 1.5 m from shore. We watched a pair building a nest in the floating tip of a canyon ragweed in a ranch pond at Lo de Campa on 20 August 1984. Russell found two juveniles attended by two adults on the outskirts of Navojoa on 28 November 1965. D. Stejskal observed ten juveniles with two adults on a pond near Alamos on 26 November 1983.

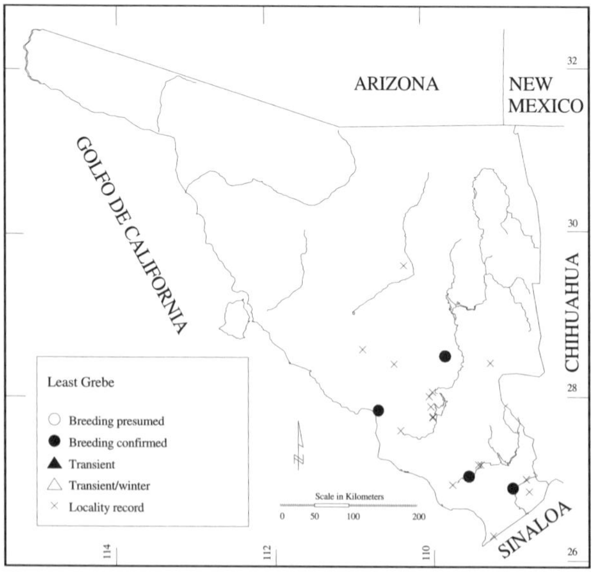

Least Grebes have not been found in northern Sonora, although there are a dozen records from southern Arizona and the species nested once in extreme southeastern California (McMurry and Monson 1947).

PIED-BILLED GREBE

Zambullidor Pico Grueso

Podilymbus podiceps

Van Rossem (1945:29) said of this grebe, "Detected only as a spring migrant . . . two known specimens were taken." Since his time, the bird has been seen infrequently, but statewide, throughout the year. We noted courtship at Presa Cuauhtémoc on the Río Altar and at Presa Lopez near Cuquiarichi west of Fronteras. We saw striped young with parents in cattail-bordered backwaters of the Río Yaqui below Presa Obregon on 10 July 1987. Pied-billed Grebes may nest on other waters, including Presa Alatorre on the Río Matape, and formerly they may have bred in the Río Colorado delta. We do not know if the breeding population in Sonora is permanently resident.

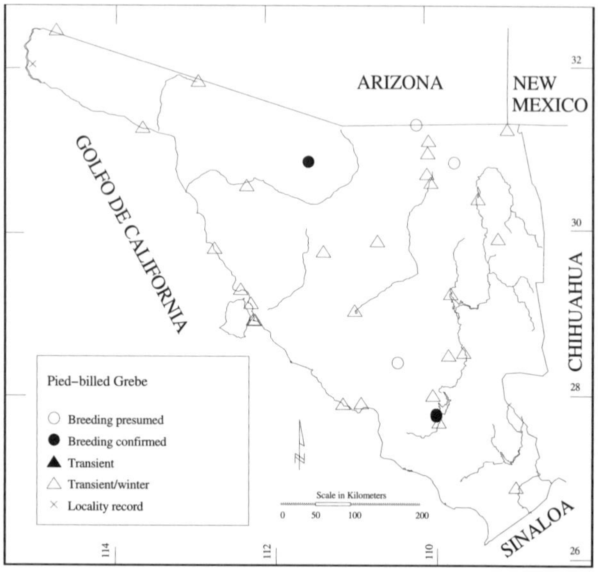

From October to April it is an uncommon transient and winter resident on waters throughout the state, including large ponds on ranches, as well as coastal lagoons and estuaries. It is sometimes seen on pools in the Río Cuchujaqui near Alamos.

HORNED GREBE
Zambullidor Cornudo
Podiceps auritus

This species is a casual visitant in Sonora (Russell and Lamm 1978): three were observed at Puerto Peñasco on 23 October 1964 (W. Bulmer, Russell); one at Bahía Kino on 21–23 February 1969 (E. Chalif); two at Puerto Peñasco on 31 October 1970 (R. Witzeman); and three at Punta Santa Rosa north of Bahía Kino on 29–30 November 1974 (Russell).

In Arizona, principally along the Colorado River, it is considered an uncommon fall transient, an uncommon to sparse winter resident, and a sparse spring transient (Monson and Phillips 1981).

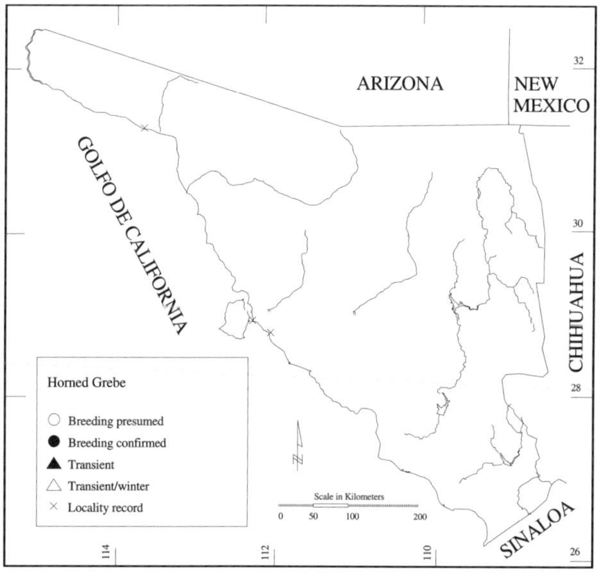

EARED GREBE
Zambullidor Orejudo
Podiceps nigricollis

The Eared Grebe is a common winter visitant and migrant in Sonora. It is most numerous along the coast but also may be present in numbers on large inland reservoirs. We have seen a few on stock ponds. Permanent streams rarely attract it.

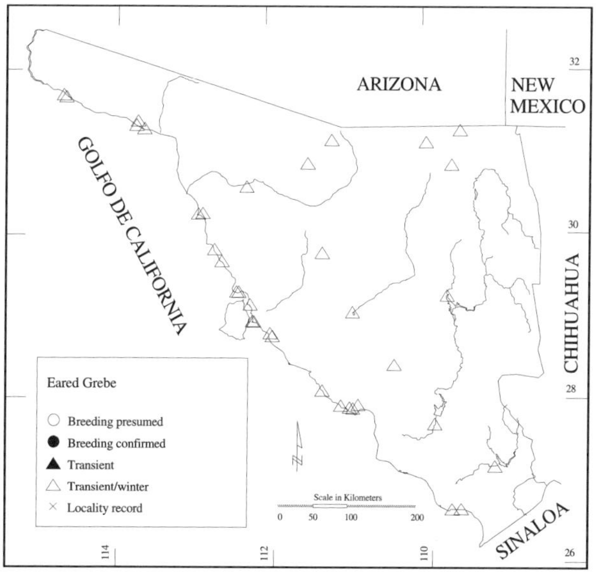

The species has been found many times from 22 August (two on a pond at El Arpa west of El Oasis) to 3 May (one at Punta San Nicolás south of Bahía Kino; R. Taylor). Most birds arrive in November and depart in the first half of April. The maximum number reported is 850, at Puerto Peñasco on 16 December 1989 by D. Stejskal and S. Ganley. Stejskal informs us that he saw one once in summer, in July 1976, at Puerto Peñasco. Russell saw singles at Punta Cirio on 26 May 1975 and on 29 May 1971, and A. Moorhouse observed one 22 km south of Naco on 10 June 1986.

WESTERN GREBE
Achichilique Pico Amarillo
Aechmophorus occidentalis

The bird long known as Western Grebe was split by the American Ornithologists' Union in 1985 to form two species, the Western and Clark's Grebes. Both are winter visitants in Sonora. Only since 1985 have *Aechmophorus* grebes in Sonora been studied carefully to determine the species. On the basis of observations accrued since then, the Western has been more frequently reported, and we consider it uncommon.

Records range from mid-November to late April, principally along the coast of El Golfo de California.

Van Rossem and Hachisuka (1937) reported an *Aechmophorus* specimen (MCZ) from Sonora; the specimen cannot now be located for species identification.

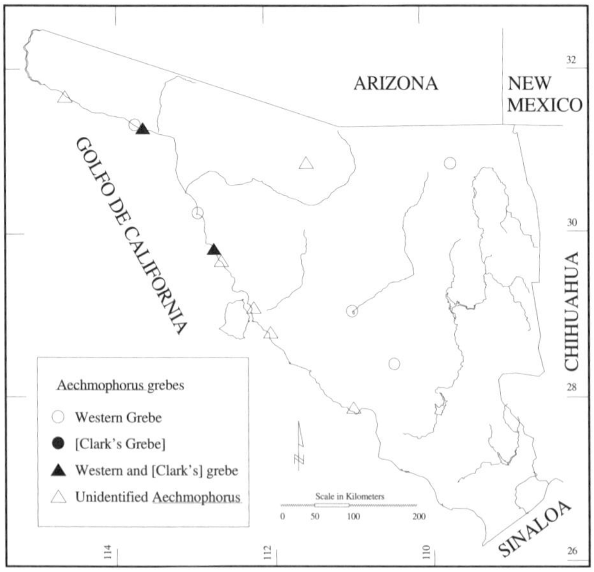

CLARK'S GREBE
Achichilique Pico Naranja
Aechmophorus clarkii

We have only two Sonoran records of this "new" (refer to the preceding account) species: five seen at Punta Cirio on 6 January 1988 (Russell) and one seen at Puerto Peñasco with Western Grebes on 16 December 1989 (D. Stejskal, S. Ganley).

FAMILY PROCELLARIIDAE
Shearwaters and Petrels

[NORTHERN FULMAR
Fulmar Norteño
Fulmarus glacialis

The Northern Fulmar is accidental on the coast of Sonora. Two in the dark phase were carefully identified at Puerto Peñasco on 24 March 1968 (W. Bolte, R. Cunningham *in* Russell and Lamm 1978). A skull (UK) found on Isla Tiburón on 22 June 1963 by J. D. Smith is the only other record from El Golfo de California.]

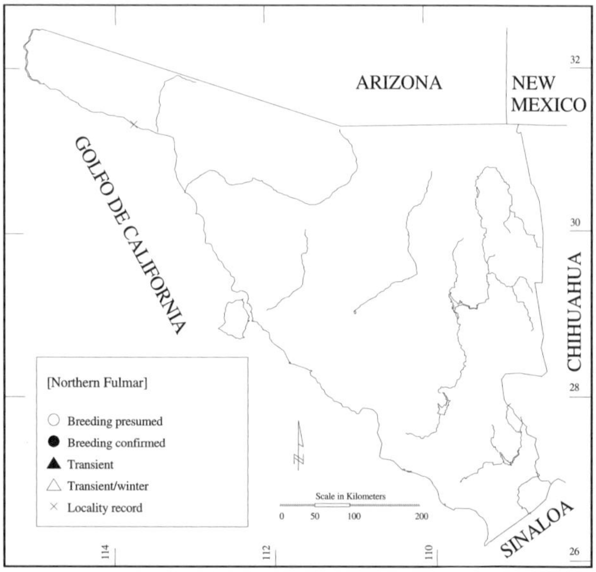

SOOTY SHEARWATER
Pardela Gris
Puffinus griseus

There are few records of this species from the shore of Sonora. Gallucci (1981) stated that he saw 18 off Puerto

Peñasco, 25–27 July 1979, but convincing details are lacking. D. Stejskal has seen a bird of this species three times in the Puerto Peñasco area. A skull (UA) found by Russell on the beach near Punta Chueca on 22 December 1974 was identified as this species by J. Jehl. The Sooty Shearwater, which does not nest in the northern hemisphere, ranges post-breeding throughout the Pacific Ocean.

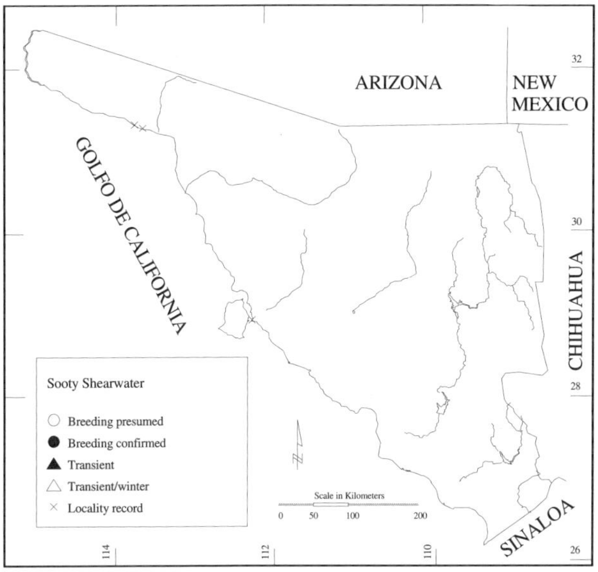

BLACK-VENTED SHEARWATER

Pardela Mexicana

Puffinus opisthomelas

This tubenose nests on islands off the outer coast of Baja California. L. Huber collected two females (UA) with tiny ova a kilometer southwest of the mouth of Bahía la Choya near Puerto Peñasco on 1 or 2 August 1970. D. Stejskal has observed these birds twice at Puerto Peñasco: two on 3 September 1979 and two on 16 December 1989. Stejskal anticipates that careful, extensive observation from the shore will provide records nearly year-round, as the species regularly feeds in inshore waters in other parts of its range.

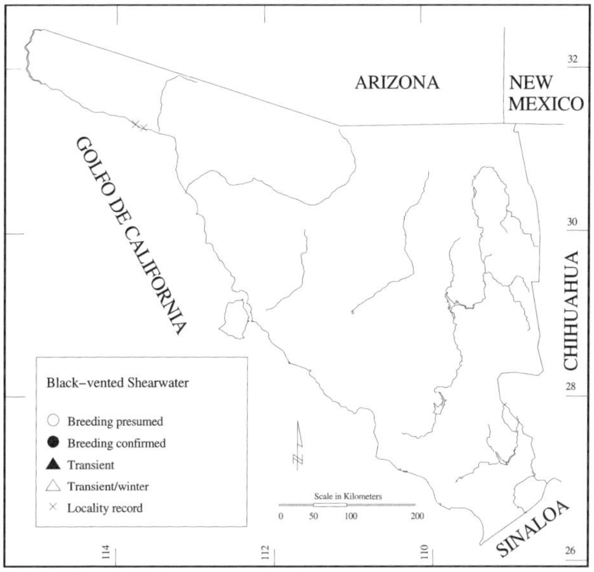

FAMILY HYDROBATIDAE

Storm-Petrels

BLACK STORM-PETREL

Paíño Negro

Oceanodroma melania

The Black Storm-Petrel breeds in considerable numbers on islands in the northern third of El Golfo de California. During the spring and early fall many have been reported in the gulf and along the coast. Van Rossem (1945) told of great numbers flying north between 17 April and 5 May, presumably in the areas of Guaymas and Isla San Esteban. On a late March boat trip out of Guaymas, Alden (1969) saw thousands. Between 50 and 60 followed boats into the Puerto Peñasco harbor on 26 April 1983 (Mark Larson). About 4500 were observed from shore near Puerto Peñasco on

2 September 1991 (D. Pearson). D. Stejskal describes the Black Storm-Petrel as common in some years close to shore at Puerto Peñasco during the summer months.

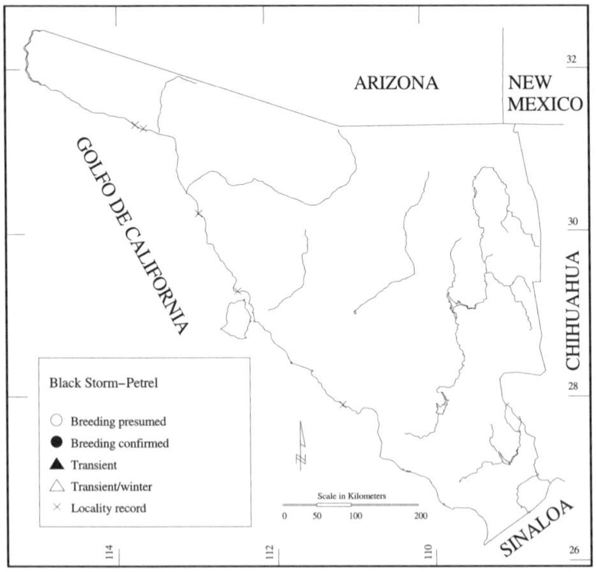

There are only three coastal records for the period of October through March. H. Hobart saw one in Puerto Peñasco on 2 January 1984. C. Cutler and M. Walsh studied three at Puerto Peñasco on 15 February 1992. L. Dombrowski and K. Kaufman saw at least 15 at Puerto Peñasco on 20–21 February 1993. Specimens were taken in the northern part of the gulf near Roca Consag on the Baja California side in the night of 31 January 1940 (van Rossem 1945).

LEAST STORM-PETREL

Paíño Mínimo

Oceanodroma microsoma

This storm-petrel nests on several islands in El Golfo de California. In spring migration, van Rossem (1945:31) stated, "Between April 17 and 21, 1925, in the vicinity of Guaymas, San Esteban and Tiburón Islands, a tremendous migration flight was in progress and some of the resting rafts certainly contained several hundred individuals." Alden (1969) reported thousands off Guaymas in late March. Three were observed at Puerto Peñasco on 23 March 1974 by J. and R. Witzeman and G. Bauer. D. Stejskal has seen it from the shore at Puerto Peñasco only in spring, and then uncommonly. Reports from along the coast of Sonora are lacking for summer and winter; the only fall record from shore is of about 500 seen with a much larger gathering of Black Storm-Petrels near Puerto Peñasco on 2 September 1991 (D. Pearson).

Observers on the coast of Sonora frequently reported petrels "far from shore." These birds were so distant that they could not be identified with any certainty.

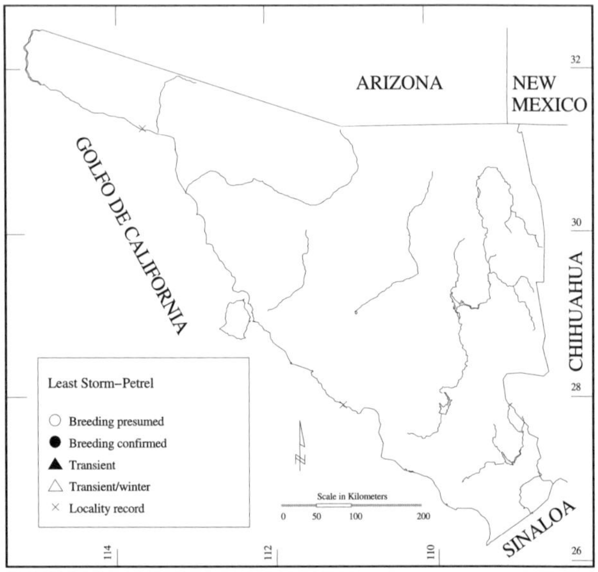

FAMILY PHAETHONTIDAE

Tropicbirds

RED-BILLED TROPICBIRD

Rabijunco Pico Rojo

Phaethon aethereus

The species nests regularly on Isla San Jorge, about 43 km southeast of Puerto Peñasco, yet there are only three recent observations from shore. Two were seen on a Christmas Bird Count by S. Terrill and colleagues on 3 January 1981, and one was observed by T. Gatz on 20 February 1984, both at Puerto Peñasco. Another was observed at Puerto Lobos on 20 February 1993 by C. Cutler and W. Leitner. Van Rossem and Hachisuka (1937) reported the species at the mouth of the Río Colorado on 25 April 1925.

There is evidence that Red-billed Tropicbirds may have once nested on the mainland. Records in the Western Foundation of Vertebrate Zoology indicate

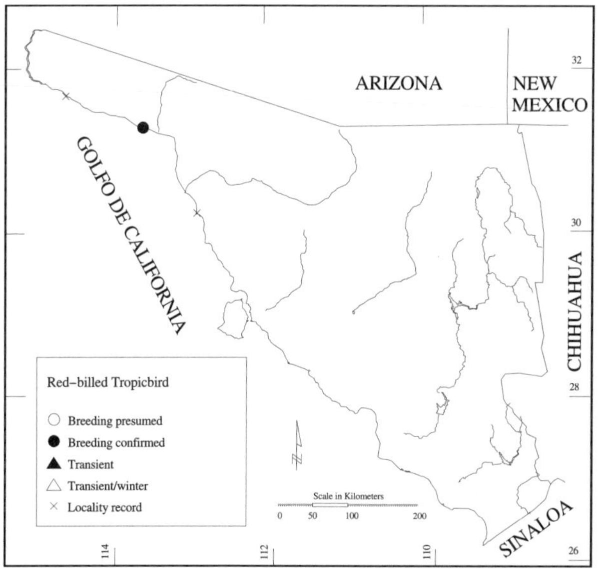

that an egg initially in the collection had been taken from the back of a small cavity under a rock at Puerto Peñasco on 8 April 1958. The collector, Bill Striebling, stated that an adult bird was taken off the nest.

FAMILY SULIDAE

Boobies and Gannets

BLUE-FOOTED BOOBY

Bobo Pata Azul

Sula nebouxii

The Blue-footed Booby breeds on several islands in El Golfo de California. Nelson (1978:520) wrote that Isla

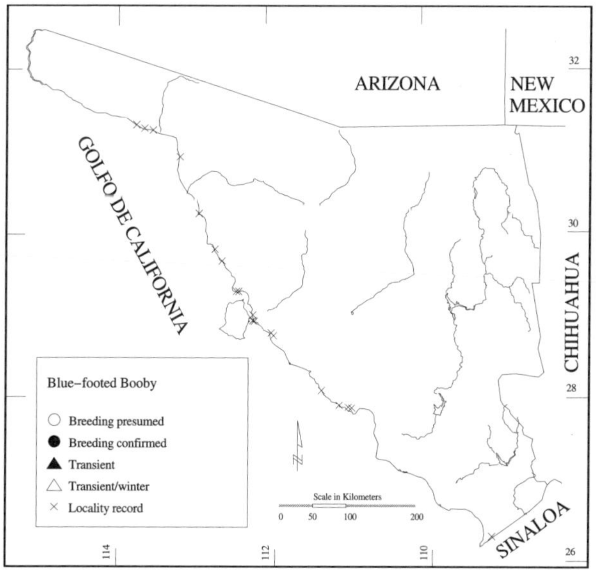

San Pedro Mártir is "one of the world's main stations for this species, possibly exceeded only by Lobos de Tierra (Peru)."

Although the literature indicates that the Blue-footed Booby is rarely seen from shore north of Bahía Kino in summer, we have found it at Puntas Santa Rosa and Chueca from May to September, albeit in rather small numbers. Many may be seen from shore from

Blue-footed Booby

September through March, when hundreds are fishing at Bahía Kino and San Carlos, for instance. By early April most have departed, presumably to nesting islands. One found dead by R. Ziegler at Bahía la Choya near Puerto Peñasco on 5 January 1992 had been banded as a fledgling by H. Drummond on Isla Isabel, Nayarit, on 28 May 1991.

BROWN BOOBY
Bobo Café
Sula leucogaster

The Brown Booby is abundant in El Golfo de California, where it is perhaps more numerous than the Blue-footed. Large numbers breed on several islands in March, April, and May. From late September until April the species often occurs inshore in groups of tens to hundreds. Although uncommon near shore in summer, some are regularly found then at Puerto Peñasco and Punta Santa Rosa.

A remarkable manifestation of fishing Brown Boobies was observed on 27 February 1974 by B. and D. McKnight at Bahía Kino. In the late afternoon they heard a repeated booming sound, which aroused their curiosity. As the former wrote (pers. comm.): "We had a magnificent display of Brown Booby flight and diving: looking to the north up the coast we saw uncountable *thousands* of these birds flying at a moderate height over the water—not above the level of the hills beyond—in an enormous counter-clockwise circle; evidently they

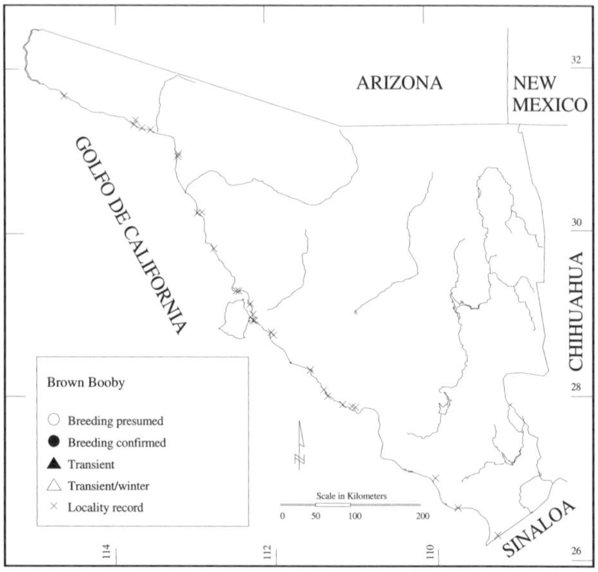

were over a large school of fish, for periodically perhaps half of the enormous flight would simply *drop* into the water with a kind of formation dive. And within 10–12 seconds, we would hear the boom made by their drop. We watched them again and again in complete fascination."

FAMILY PELECANIDAE

Pelicans

AMERICAN WHITE PELICAN
Pelícano Blanco
Pelecanus erythrorhynchos

White Pelicans are now rather uncommon winter and transient birds along the coast of El Golfo de California, although formerly they were much more plentiful (van Rossem 1945). The greatest number of coastal birds noted in the past 25 years is 6290 between the Río Colorado delta and Isla Tiburón in 1981, counted during an aerial winter survey of Mexican waterfowl by the U.S. Fish and Wildlife Service (1981). Inland, the most reliable place to find them, year-round, is Presa

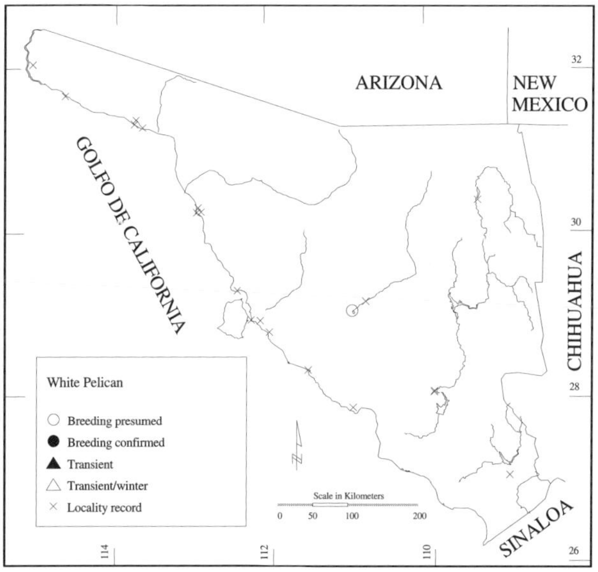

Rodriguez at Hermosillo, where the largest number on record is 600 on 5 April 1990. As many as 350 may be present in summer, and we have undocumented reports of nesting. Sometimes they may be seen soaring over the highway on the northeastern outskirts of Hermosillo. The only other inland localities of record are Presa Obregon, where large numbers have been found (e.g., about 500 on 18 April 1989; J. Levy); Presa Angostura (two records, 6 April 1980 and 26 March 1989); and near Alamos (15 flying birds, 31 October 1993; R. Russell). One additional summer observation is of 46 individuals at Punta Santa Rosa on 20 June 1974. Extreme dates, excluding the resident possibly breeding birds at Presa Rodriguez and the June birds at Punta Santa Rosa, are 6 October (1979, Puerto Peñasco) and 25 April (1989, Estero Tastiota).

BROWN PELICAN
Pelícano Pardo
Pelecanus occidentalis

This is an abundant and ubiquitous bird of El Golfo de California, flying in stately fashion low over the waves or plunging into the salt water for its scaly prey. Often it can be seen resting in groups on rocky shores and on beaches or perching on pilings or on the sides of boats. Many hundreds reside on Isla Alcatraz at the entrance to Bahía Kino. Hundreds can usually be seen off Puerto Peñasco, Puerto Lobos, Bahía Kino, and Bahía San

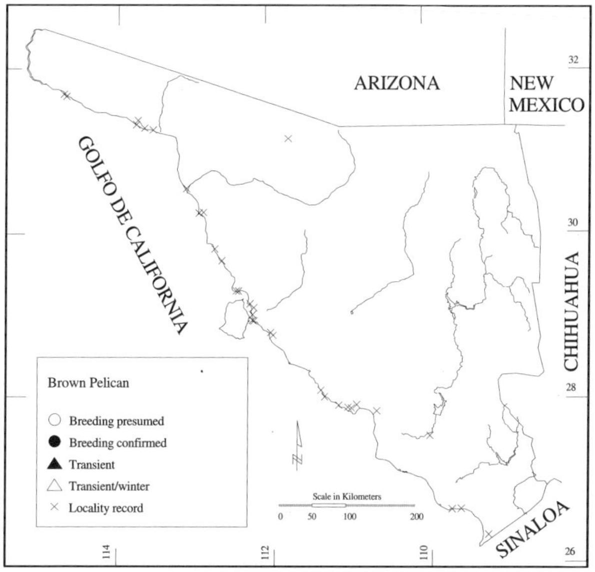

Carlos, especially in winter. The pelican nests commonly on most gulf islands.

According to Anderson (1983), after nesting, many Brown Pelicans disperse across Baja California into the Pacific Ocean and southward into more tropical waters. Almost every year a few stray into southern Arizona in the summer, so it is not too surprising that we saw one flying high, headed north, about 40 km south of El Sasabe on 18 July 1975.

We are not aware of any studies that have established that Brown Pelicans in El Golfo de California are affected by pesticides. Perhaps more threatening to them is invasion of their nesting colonies by tourists and even well-meaning naturalists (Anderson et al. 1976).

FAMILY PHALACROCORACIDAE

Cormorants

DOUBLE-CRESTED CORMORANT

Cormorán Orejudo

Phalacrocorax auritus

Double-crested Cormorants are widespread and common residents of Sonora. Along the coast of El Golfo de California they may be abundant. We estimated that 8000 were fishing and flying offshore at Bahía Kino on 7 April 1976, and we counted 7500 flying south at Punta Santa Rosa in a 2.5 hour period on 11 October 1973. Their population dwindles somewhat in summer because part of the wintering numbers are birds that breed north of Mexico. Anderson (1983) stated that Double-crested Cormorants are not abundant in El Golfo de California, but he listed them as common breeders on the islands. At least 1000 were nesting on Isla Alcatraz, in Bahía Kino, on 7 November 1986, when both eggs and small young were found (C. Morgan, R. Taylor).

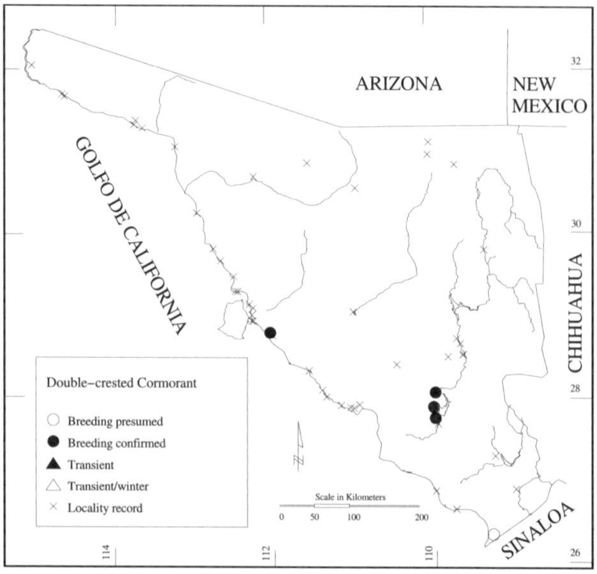

Inland, the principal nesting takes place along the Río Yaqui, wherever there are dead trees or brush to support their nests. We counted about 400, some sitting on nests, in water-killed shrubs on the Agua Caliente arm of Presa Obregon on 22 September 1985. We saw at least 450 adults sitting on nests in flooded dead willows in the Suaqui arm of the same reservoir on 10 April 1986. No young were seen, so we assumed that the birds were mostly on eggs. Besides Double-crested Cormorants, lesser numbers of Neotropic Cormorants, Anhingas, Great Blue Herons, Great Egrets, and Black-crowned Night-Herons were nesting in the same rookery, which was about 300 x 400 m. Double-crested Cormorants nested higher, and more toward the edge of the colony, than did the Neotropic Cormorants.

The level of Presa Obregon seems to have fallen drastically in some successive years, allowing cottonwoods, willows, and leguminous shrubs to grow to a height of 10 m or more on the exposed ground; then, when the water rose again, the plants were flooded and died, as was the case in the mid-1980s. This cycle may have occurred several times since the dam was first closed in 1947. Without these fluctuations, there would be no nesting trees or brush for the cormorants and their companion nesting species. The reservoir supports a tremendous population of tilapia (*Tilapia zillii*) plus other fishes.

At least 30 active nests were found in tall leafy cottonwoods along the Río Yaqui not far below the Obregon dam on 11 April 1986. Many cormorants were flying up and down the river at this site. Some 30 to 40 nests, some of them with birds presumably sitting on eggs or small young, were found in dead cottonwoods near the head of the new Presa Lopez, on the Río Cuquiarichi west of Fronteras at an elevation of about 1200 m, on 10 March 1987.

This species may have increased in numbers and range since the publication of van Rossem's survey (1945). He cited no inland records, nor have we found any evidence of nesting inland prior to the 1980s. Although many species of colonially nesting Pelecaniform and Ciconiiform birds are represented in eggs from Sonora in the collection of the Western Foundation of Vertebrate Zoology, no Double-crested Cormorant eggs are present.

NEOTROPIC CORMORANT

Cormorán Oliváceo

Phalacrocorax brasilianus

In Sonora, the Neotropic Cormorant is at the northwestern edge of its North American breeding range. It is, unlike the Double-crested Cormorant, predominantly a freshwater inhabitant in this part of its range. It occasionally has been found on the coast. It is common on the lower Ríos Yaqui and Mayo and is regularly found along smaller streams, such as the Río Cuchujaqui near Alamos. It favors reservoirs, particularly the larger ones. There it is an associate of the Double-crested Cormorant, sometimes outnumbering it. It is always present at Hermosillo's Presa Rodriguez.

We found a nesting colony of about 350 birds on the Suaqui arm of Presa Obregon on 10 April 1986, at which time they were sitting on eggs or small young, although with even larger numbers of Double-crested Cormorants and other colonial nesters. A small colony was observed well to the north on a reservoir at Villa Pesqueira, on the road between Mazatán and San Pedro de La Cueva on 19 August 1974 (S. and J. Levy).

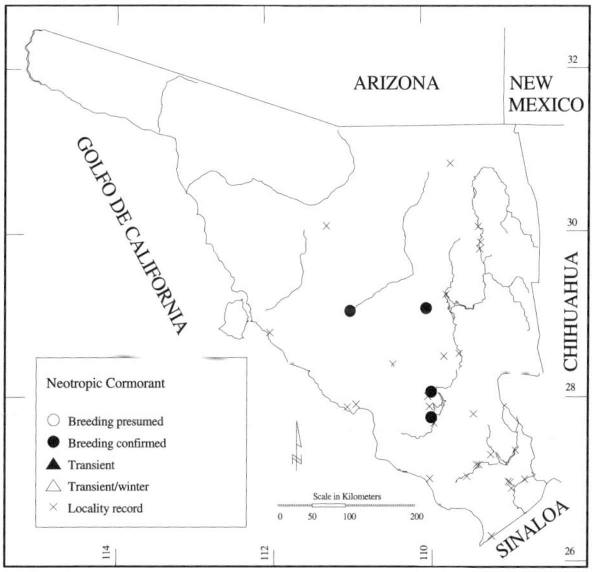

The northernmost sightings in Sonora appear to be of two at a charco, or stock pond, south of El Carrizo, which is about 35 km south-southwest of Benjamin Hill, on 11 August 1985 (S. Levy) and a single bird at Presa Jacinto Lopez on 16 October 1993 (S. Russell and others). Within the past 25 years, small numbers have reached extreme southern Arizona, including the San Bernardino National Wildlife Refuge east of Douglas just north of the Arizona-Sonora boundary.

BRANDT'S CORMORANT
Cormorán de Brandt
Phalacrocorax penicillatus

Although this cormorant breeds on islands in El Golfo de California, it is rarely seen on the coast of Sonora. Van Rossem noted it at Guaymas on 7–11 February 1946. Summer records are lacking; the span of our six records is from early November to 21 April. Brandt's Cormorant is probably more common and found over a greater spread of seasons in our area than these records indicate, because it is difficult to distinguish from the Double-crested Cormorant.

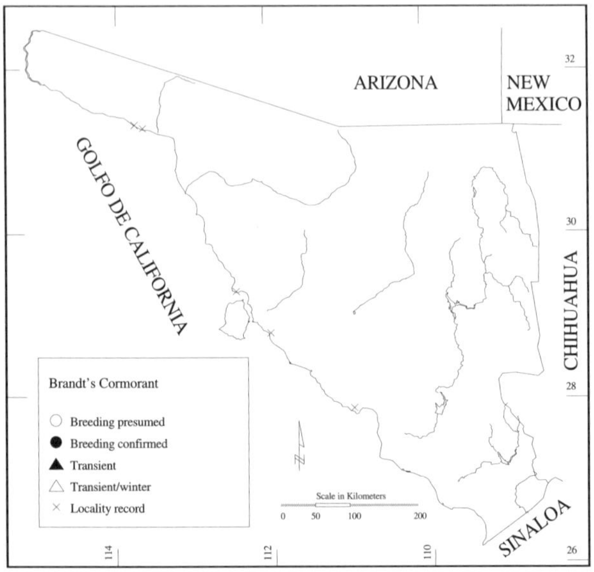

FAMILY ANHINGIDAE

Darters

ANHINGA
Anhinga Americana
Anhinga anhinga

The Anhinga, a nesting species in adjacent Sinaloa, was not found in Sonora until 1975, when D. Lamm observed two circling over the mangroves at Punta Santa Rosa on 29 May. K. Kaufman saw one on the Río Cuchujaqui near Alamos on 23–24 November 1977. All recent records have been from the lower Río Yaqui drainage in the mid-1980s, although the birds may have been there for some time before we discovered them.

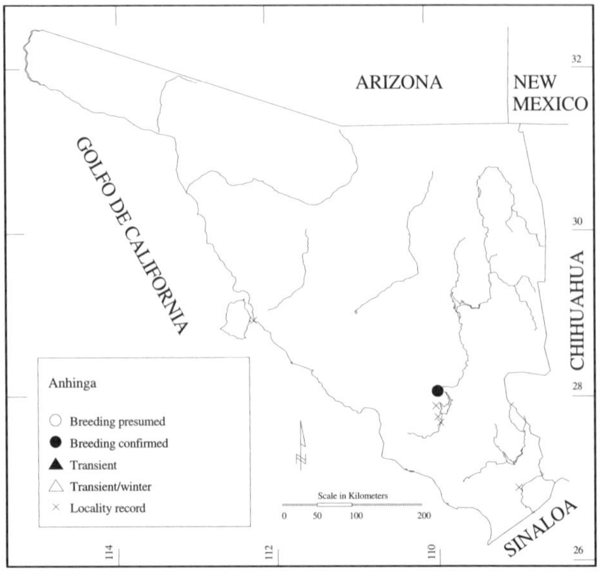

We counted 44 Anhingas on 21 September 1985 sailing above the Suaqui arm of Presa Obregon near Piojillo. The next day, four more were seen on the Agua Caliente arm of Presa Obregon. We observed at least 40 at Presa Agua Caliente southeast of the Obregon dam on 3 January 1986 and then found more than 300 helping to make up a large heronry on the Suaqui arm near Piojillo on 10 April 1986. At least 20 of this number were apparently on eggs. Commercial fishermen from the nearby village of Piojillo caught many Anhingas in their gill nets, whereas the nets took only a few cormorants. A rancher close to the boat landing fed the netted birds to his pigs, resulting in piles of wing and tail feathers littering his pasture. It seemed strange that any Anhingas were still left alive.

FAMILY FREGATIDAE

Frigatebirds

MAGNIFICENT FRIGATEBIRD

Fragata Magnífica

Fregata magnificens

Although the Magnificent Frigatebird does not nest in El Golfo de California, it is often abundant along the coast of Sonora in summer and fall. It is uncommon north of Puerto Libertad in winter and spring. Apparently its numbers vary considerably from year to year. The largest aggregations have been observed in September and October. We once counted 1000 flying north, 3 km inland from the coast just north of Bahía Kino (7 September 1973). Of these, 90% were adult males. At one time during this flight 200 milled in a

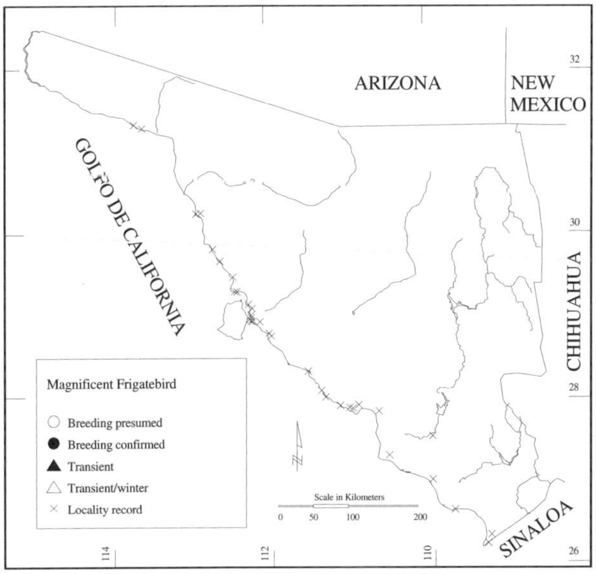

thermal as they gained altitude and then streamed northward in small flocks. On 24 September 1977, more than 2000 passed over Punta Chueca in a few hours.

Frigatebirds appear to soar effortlessly as they take advantage of air currents over the gulf and adjacent inland hills. One September midnight as we camped on dunes near Punta Santa Rosa, a nearby storm created winds of about 45 km/hr. Twenty-five frigatebirds silhouetted against gathering clouds hovered as if suspended low overhead. They faced the wind, scarcely moving their wings, but occasionally shifting positions.

The observation of these fabulous birds soaring over rocky islands and promontories clad in giant cacti is surely one of the premier sights in Sonora. Close views of them may be obtained when they mix with gulls low over the beach as offal is thrown to the birds by fishermen. As their name suggests, frigatebirds harass other flying birds, causing them to drop or disgorge their prey. One such frigatebird, chased by another, regurgitated a small fish that fell to the beach at Punta Santa Rosa. The fish, identified by D. Thomson, was a Spotted Sand Bass (*Paralabrax maculatofasciatus*).

Magnificent Frigatebird

FAMILY ARDEIDAE

Bitterns and Herons

AMERICAN BITTERN

Avetoro Norteño

Botaurus lentiginosus

The secretive American Bittern is a rare transient and winter visitant in Sonora. There are seven records involving eight individuals, only three in the past 40 years. Two observations along the coast were in November and January, one from the Río Colorado delta was in late August, and the remaining three were from March and April. A bird at Bahía Kino was flushed from a dense stand of mixed salicornia and small mangroves. This species may be more common than our records indicate, but it has also become a rare bird in Arizona in recent years.

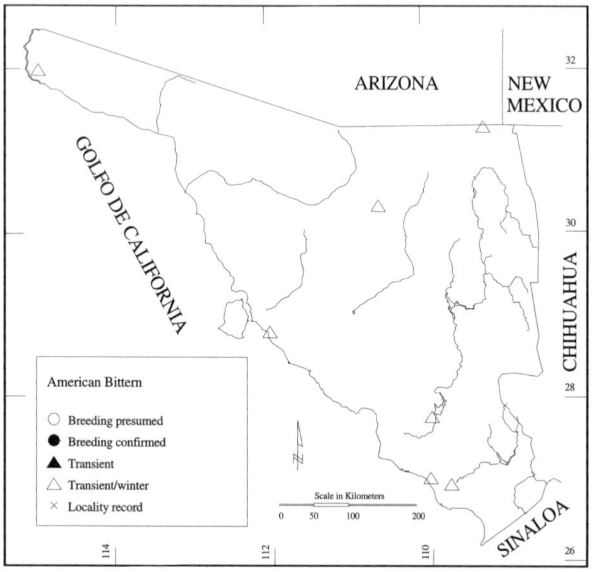

LEAST BITTERN

Avetoro Mínimo

Ixobrychus exilis

We consider the Least Bittern an uncommon summer resident on the basis of a dozen records from 28 April through 17 September. It lives in dense emergent aquatic vegetation, including mangroves, and is not easily or often seen. Van Rossem (1945) thought it fairly common and a breeding species from Bahía Kino southward along the coast of El Golfo de California.

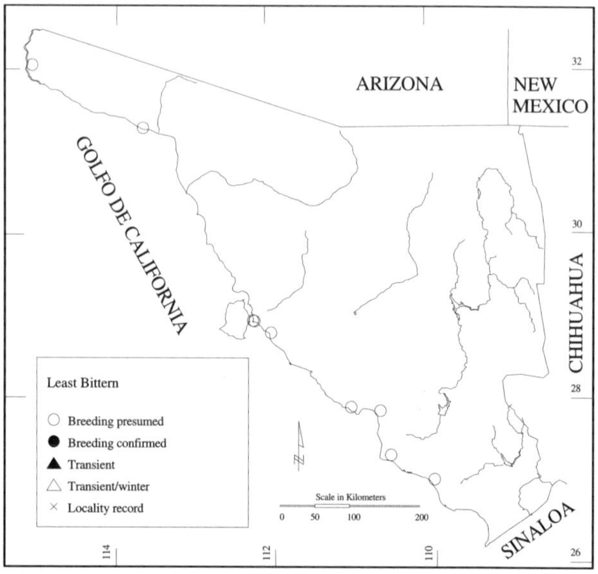

His only reports were in late April and May. R. Taylor saw five, including three in one group, in mangroves at Bahía Kino between 29 May and 2 June 1987. These dates suggest breeding. Russell and Lamm (1978) mist-netted individuals at Punta Santa Rosa on 18 August 1975, 8 September 1973, and 17 September 1974, in mangroves.

It is probable that the Least Bittern nested in the Río Colorado delta before the construction of Hoover Dam stopped the flow of the river and eliminated nesting habitat; it may nest there now in cattail-lined irrigation drains or in the Ciénega de Santa Clara (Todd 1986; Eddleman 1989). The extent to which the Least Bittern migrates into Sonora is unknown. There are no inland records.

BARE-THROATED TIGER-HERON

Garza-tigre Mexicana

Tigrisoma mexicanum

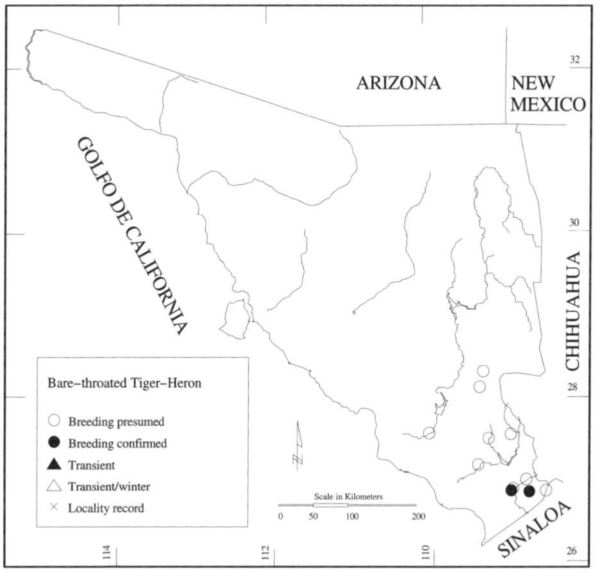

A strictly tropical and subtropical bird, the Bare-throated Tiger-Heron is an uncommon resident along streams in southeastern Sonora. Van Rossem's (1945) records are from south of Alamos, but more recently the species has been found at a number of more northerly localities.

The Bare-throated Tiger-Heron is usually solitary, infrequently observed as it stalks small fish in shallow streams or flies from a perch high in a Montezuma baldcypress. Along the Río Cuchujaqui it is usually seen as it flushes from a tree or rock bluff.

Bare-throated Tiger-Heron

GREAT BLUE HERON

Garza Morena

Ardea herodias

The Great Blue Heron is a fairly common resident throughout Sonora, although it is not known to nest in every locality where it is regularly found. Possibly there is some movement of birds into Sonora from the north outside of the breeding season. It is a successful and adaptable bird, able to feed on a variety of animal life from fish and crustaceans to amphibians, reptiles, and mice.

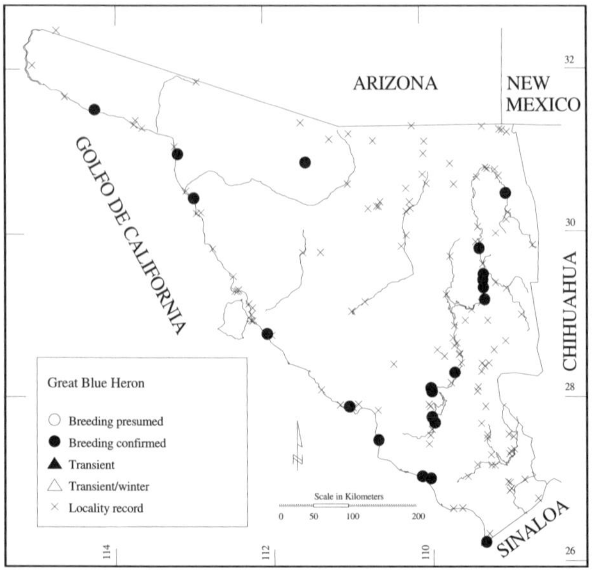

Great Blue Herons nest in small heronries in the tops of cottonwoods or other tall trees along or near permanent streams. We are aware of 17 such heronries in the Río Yaqui drainage alone, and more may be undetected. Additionally, Great Blue Herons nest in heronries associated with cormorants and other herons in drowned trees and brush of major reservoirs. We found at least 40 on nests at the head of Presa Obregon, many of the nests with young more than a quarter grown, on 10 April 1986. Adults were copulating in a small heronry of nine nests in a solitary cottonwood in the center of an alfalfa field near Granados on 19 March 1985. Eight nests, one with chicks, were found on Isla Huivulai in Estero Tóbari on 14 May 1994 (Palacios and Mellink 1995). Eggs have been found in Sonoran nests during March, April, and May.

This species forages along streams and ponds up to an elevation of 1500 m. Solitary birds are sometimes seen flying quite far overhead, en route to feeding sites or roosts. It is not unusual to find one at a small ranch pond.

GREAT EGRET

Garza Blanca

Ardea alba

The Great Egret is uncommon in Sonora but may be present along the coast of El Golfo de California in estuaries and lagoons and inland on reservoirs, rivers, and streams, chiefly during the winter. It is most often seen singly or in pairs but sometimes in groups of fewer than 15. An exception to this pattern was a throng of about 650 on 7 January 1986 in the upper end of the Cedros arm of Presa Mocuzari, where large numbers of cormorants were also present. The attraction was likely to have been numerous fish in the receding water. It is unknown to what extent birds from farther north contribute to the wintering population. The origin of the concentration at Mocuzari is dubious.

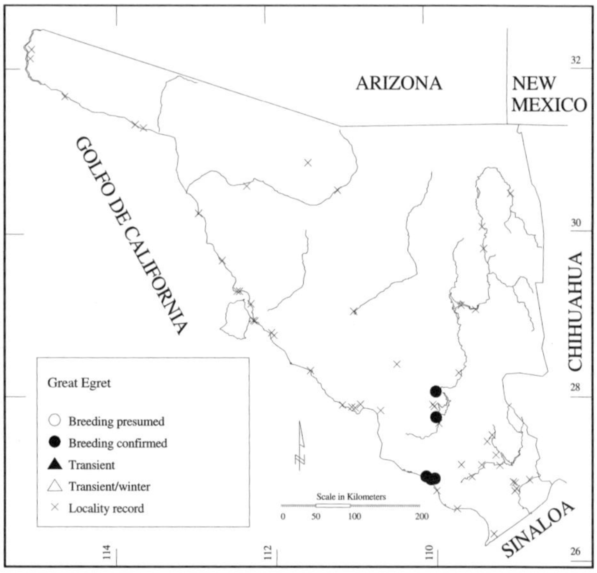

Van Rossem (1945:39) stated: "Recorded as breeding at Tóbari Bay, and undoubtedly does so at other points." Palacios and Mellink (1995) found two colonies totaling about 30 pairs nesting on Isla Huivulai on 14 May 1994. But we have found only two additional breeding colonies. A colony of 24 pairs occupied nests in two tall leafy cottonwoods below the Obregon dam on 11 April and 15 May 1986. About 65 pairs were interspersed in a large heronry used mainly by cormorants near Piojillo on the head of Presa Obregon on 10 April 1986. One bird was standing on a nest, others were carrying sticks, and some were courting.

SNOWY EGRET

Garceta Pie-dorado

Egretta thula

The Snowy Egret may be found in small numbers throughout the state and at any season of the year. Nesting may have occurred in Sonora in May 1994, when about 40 pairs were found in heronries on Isla Huivulai (Palacios and Mellink 1995). There are few summer records. Sonoran birds are wanderers or nonbreeders, and many are doubtless migrants. Occasionally groups number more than 20 or 25; the largest gatherings we have recorded have been of about 100

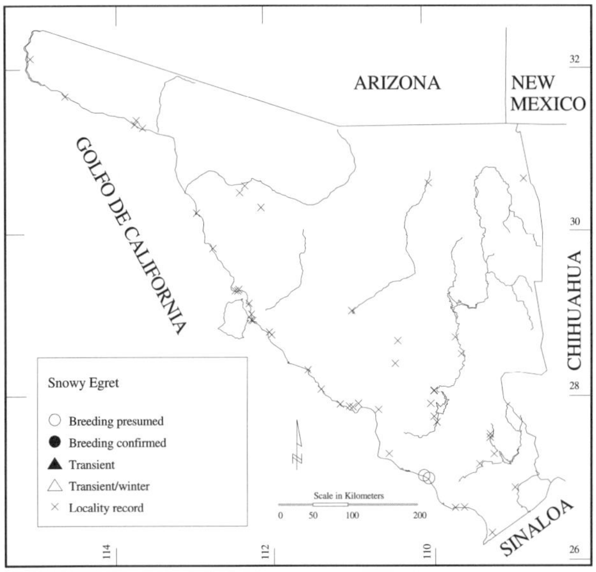

birds (Presa Mocuzari in January, Bahía Kino in September, and Puerto Peñasco in February).

Wilbur (1987) mentions that the Snowy Egret breeds, or has bred, in the Río Colorado delta on the Baja California side. The species does not nest on the Sonora side of the delta, although it may have done so before the construction of Hoover Dam.

LITTLE BLUE HERON

Garceta Azul

Egretta caerulea

The rare Little Blue Heron is found year-round in very small numbers, not exceeding five in any instance, regularly as far north as the Bahía Kino region (including Punta Santa Rosa). Its habitat is chiefly mangrove-edged lagoons and bays. We have one definite record away from the coast: an adult seen below a dam on the Río Babasac near Cocóspera northeast of Imuris on 13 June 1988 (J. Levy). We are at a loss to understand why Alden (1969) characterized it as abundant at Guaymas.

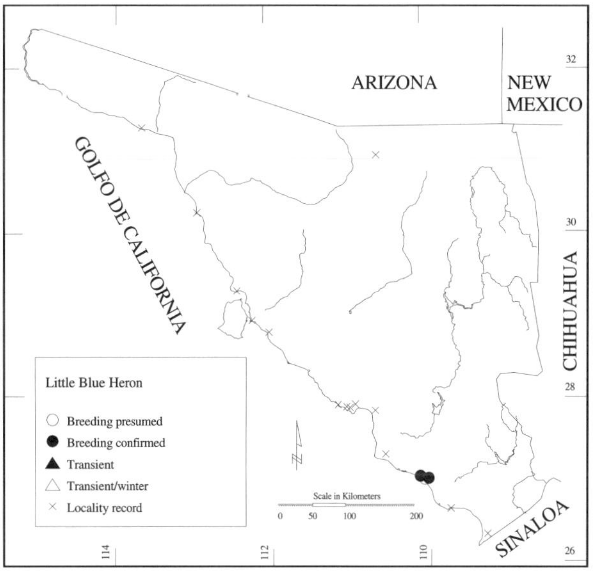

One of only two Sonoran nesting records is from Bahía Tóbari on 11 June 1930, when G. Bancroft took four clutches of fresh eggs (WFVZ). The colony, which was in mangroves, included many ibises, egrets, and herons. Six pairs with nests were noted in heronries on Isla Huivulai on 14 May 1994 (Palacios and Mellink

1995). We have seen only three all-white individuals (probably less than a year old); they may have wandered north from more southerly breeding sites.

TRICOLORED HERON
Garceta Tricolor
Egretta tricolor

Along the coast of El Golfo de California, in lagoons and shallow water, Tricolored Herons are rare. The birds are scattered, usually found one or two at a time; the largest aggregation we have recorded was six birds at Bahía Kino on 28 April 1972. They have been found in all months, except August, from Punta Chueca south on the coast. The species has been reported from Puerto Peñasco by Alden (1969) without details and from La Ciénega de Santa Clara by us.

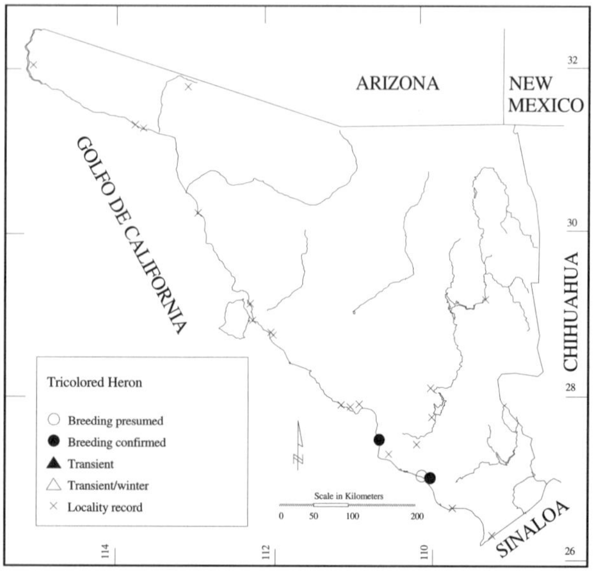

In 1930, nesting Tricolored Herons were found by G. Bancroft in mangroves at Bahía Tóbari and Estero la Luna in late May and early June. Fourteen clutches (WFVZ) consisted of three or four eggs, and incubation was just under way. There are no other nesting records. Palacios and Mellink (1995) found five feeding on Isla Huivulai in Estero Tóbari on 14 May 1994 and presumed they were nesting in an adjacent heronry.

REDDISH EGRET
Garceta Rojiza
Egretta rufescens

Reddish Egrets in Sonora are uncommon residents of coastal lagoons and tidal flats, where they are seen chiefly as single birds throughout the year. They are more frequent from Bahía Sargento southward, being rare at Puerto Lobos and Puerto Peñasco. The maximum number we have seen on any one day is nine, at Huatabampito on 10 December 1983. As far as known, no records of the white morph exist for Sonora, but many observers did not comment on the color of birds seen.

The several breeding records, with one exception, are prior to 1947. Nests containing eggs were found

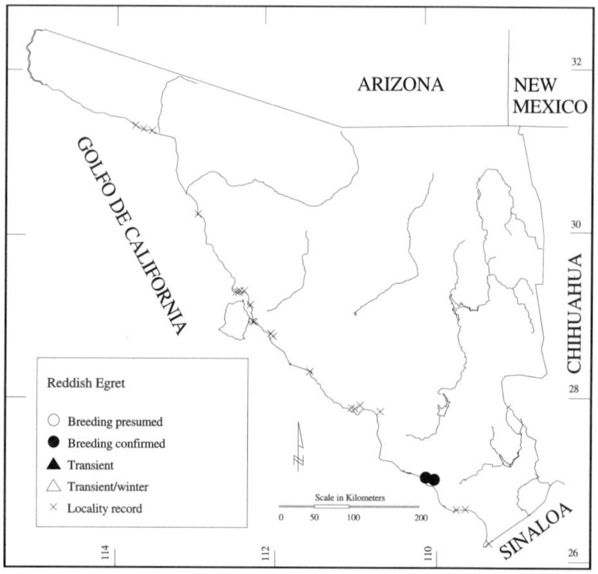

from late April to late May, mostly in mangroves. One nest, in Bahía Guásimas, was situated "well above the ground in a giant cactus" (G. Bancroft notes, WFVZ). Since 1947, only one possible nesting has been noted, when six pairs with nests were seen in heronries on Isla Huivulai in Estero Tóbari on 14 May 1994 by Palacios and Mellink (1995).

CATTLE EGRET
Garza Ganadera
Bubulcus ibis

The Cattle Egret had not been recorded in Mexico at the time of the writing of the *Distributional Check-list of the Birds of Mexico* (Friedmann et al. 1950). The species apparently arrived in South America from Africa in the 1870s and reached Florida in the 1970s. Alden did not find it in Sonora during his west Mexican trips in the 1960s. The first detection was of two in mangroves at Punta Sargento on 11 April 1970 (Russell and Lamm 1978). It was next seen in 1977 in the Bahía Kino area; since then it has become more common. Now we find it in most agricultural areas except in the northeast. It has seldom been found along the coast, and then usually in small numbers. Mellink and Palacios (1993) reported at least three nests at La Salina in May 1992, with broken eggs on the ground. Palacios and Mellink (1995) found about 200 pairs nesting in heronries on Isla Huivulai in Estero Tóbari in May 1995.

Usually Cattle Egrets associate with livestock in pastures, especially when flooded fields or wetlands are adjacent. The areas most frequented in Sonora are farming districts around Cd. Obregon, Navojoa, and Hermosillo and south of San Luis Río Colorado in the Río Colorado delta. Most observations are from November through April. The only reports for June through October have been from Hermosillo and the agricultural areas east of Agiabampo. We think most Sonoran Cattle Egrets come from the south, particularly Sinaloa, where they have been abundant since at least the 1960s (Alden 1969). Many may be post-breeding wanderers. Some arrive from other areas, as indicated by recovery of a bird near Navojoa (at an unspecified date) that had been banded by A. Moorhouse as a nestling on 14 June 1972 at Powderly, Texas.

An odd record is that of nine Cattle Egrets standing on saguaros in the vicinity of Crater Elegante in the Sierra el Pinacate on 29 November 1987 (photographed, R. and E. Johnson).

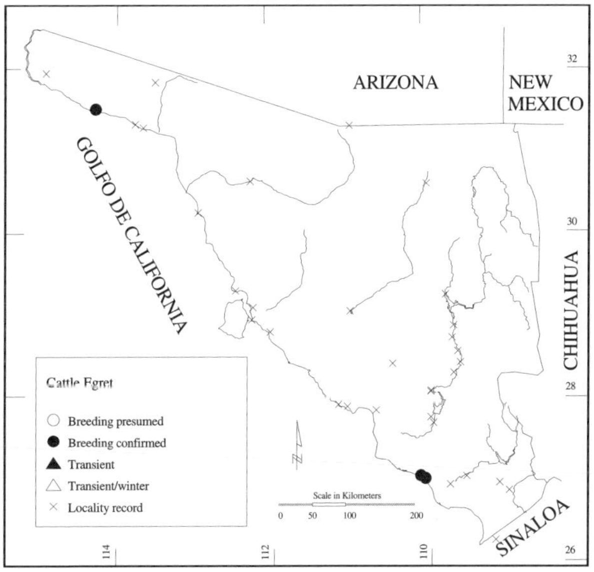

GREEN HERON
Garceta Verde
Butorides virescens

The small Green Heron is fairly common throughout the state along permanent streams (except in the higher mountains), in and near freshwater marshes (*ciénegas*), and in mangroves on the coast of El Golfo de California. It is not found in groups or flocks. It is a solitary bird; seldom is more than one flushed at a time. There are few records of nesting, but we think it is probably a permanent resident. During the winter, numbers are augmented by migrants from the United States.

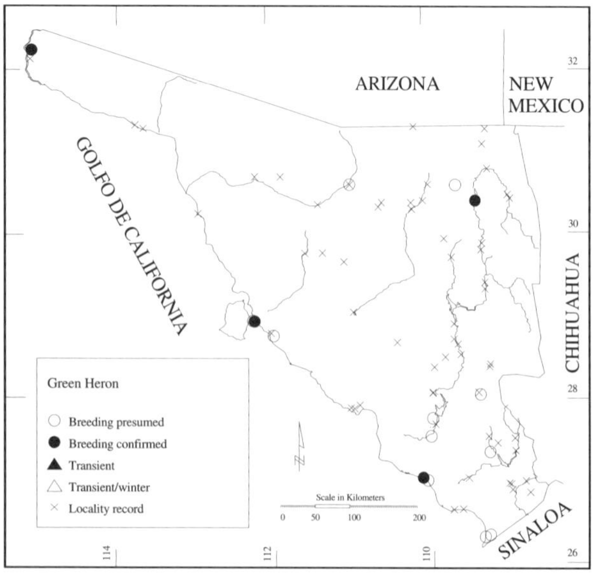

BLACK-CROWNED NIGHT-HERON

Pedrete Corona Negra

Nycticorax nycticorax

As its name implies, this squat, short-legged heron is seldom seen feeding during daylight. It roosts by day in a mangrove by the coast of El Golfo de California, or in a dense-foliaged tree or large bush along some inland waterway, or even at an isolated stock tank. Often it is not seen until it flies off with loud squawks that also are often heard by the camper or fisherman at night.

This is an uncommon bird in Sonora, even along the seacoast where it is regularly found. It is, however, widespread and may be seen from the Río San Bernardino in the extreme northeast west to the Río Colorado delta and southward in and near aquatic habitats, although north of Puerto Lobos it is rare along the coast. It disperses widely after the breeding period and during the winter, opportunely searching for places that might yield food. B. Brown recorded about 200 at the upper end of Presa Obregon on 11 November 1986, an unusually large aggregation.

G. Bancroft collected a clutch (WFVZ) of three eggs on 11 June 1930 from a nest in a mangrove 4 m above the water in Bahía Tóbari, where birds were also found nesting in May 1994 (Palacios and Mellink 1995). During the period 16 to 18 May 1975, about eight pairs had nests with eggs on rocky Isla Alcatraz at the entrance to Bahía Kino (M. Keasey, R. Perrill). We found a gathering of at least 70 in the Suaqui arm at the head of Presa Obregon in a large cormorant and heron rookery on 10 April 1986. At least two pairs were standing on nests, and others were holding sticks in their bills. We believe that these birds probably nested there.

H. Coss on 21 April 1984 photographed an immature Black-crowned Night-Heron far from permanent water as it fed on shield shrimp in a shallow flooded clay "pan" north of the Sierra el Pinacate.

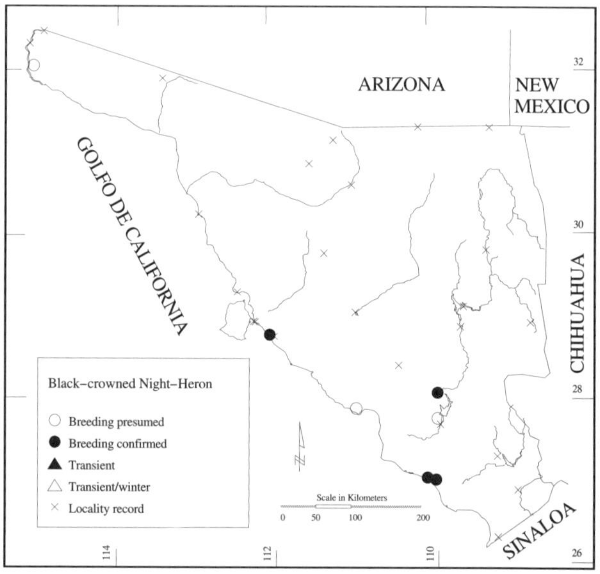

YELLOW-CROWNED NIGHT-HERON

Pedrete Corona Clara

Nyctanassa violacea

The distribution of the Yellow-crowned Night-Heron in Sonora is almost entirely coincident with the range of mangroves on the coast of El Golfo de California. Van Rossem (1945:40) called these birds "fairly common but apparently rather local resident[s]." We consider them uncommon, and ten is the maximum number we have ever seen (Bahía Kino on 28 April 1972), but R. Taylor estimated 50 at Bahía Kino on 7 June 1985. There are no records north of Puerto Lobos, the northern limit of mangroves. We have one inland record, a bird at the head of Presa Obregon on 21 September 1985.

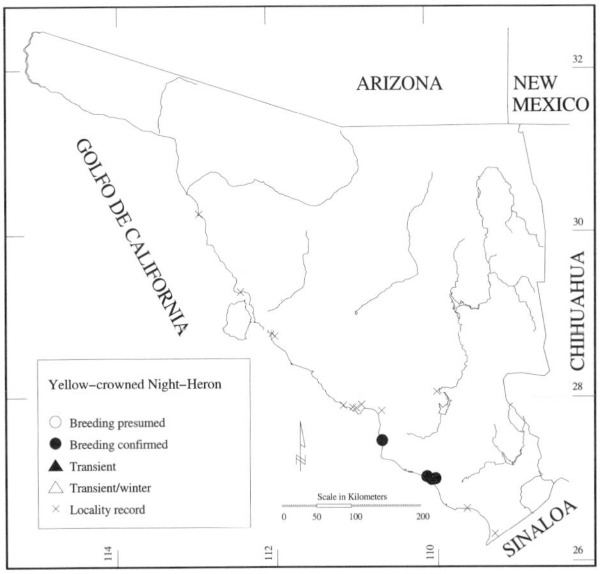

Nests, in mangroves and in mesquites close to the water, have been discovered only in colonies at Bahía Tóbari and Estero la Luna, in late April and May of 1930. Palacios and Mellink (1995) on 14 May 1994 counted at least 40 nests on Isla Huivulai and estimated that about 130 adults were present at the time on the island and the adjacent mainland shores.

FAMILY THRESKIORNITHIDAE

Ibises and Spoonbills

WHITE IBIS

Ibis Blanco

Eudocimus albus

The White Ibis is rare in Sonora except locally when nesting, with records only from coastal localities. Nesting records are from Laguna Guásimas 12–24 May 1930 and Isla Lobos 27–29 May 1930 (eggs, WFVZ). Some nests were built in cardón cactus; another was in a bush 20 cm from the ground. Palacios and Mellink (1995) visited Isla Huivulai heronries in Estero Tóbari on 14 May 1994, at which time they found about 75 nests and about 200 adult White Ibises. They also found another 20 adults and three immatures on the nearby coast. Otherwise, the maximum number noted has been 20 at Guaymas on 15 March 1986 (A. Moorhouse).

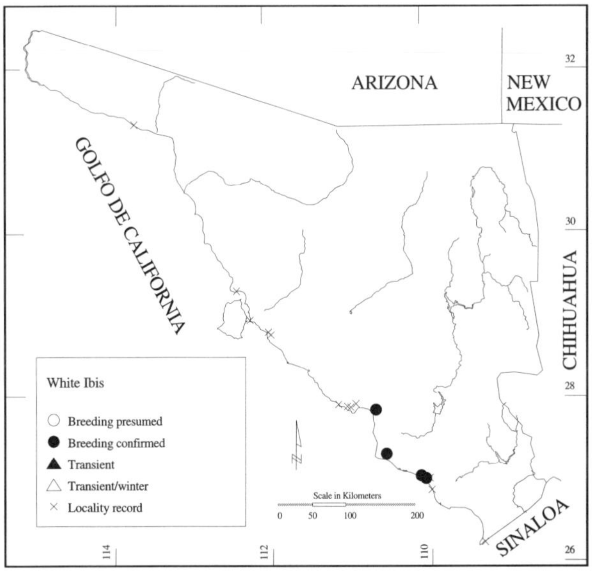

WHITE-FACED IBIS

Ibis Cara Blanca

Plegadis chihi

Despite its being a fairly common bird during migration in Arizona, especially in the lower Colorado River valley, the White-faced Ibis has not often been seen in Sonora in the last 40 years. Particularly surprising is its absence along the coast, where we have only four records, all of single birds in September and December. In the interior, most records are from major drainages in September. A flock of 200 was flying north above the Río Yaqui near Soyopa on 11 April 1990 (B. Brown). A flock of eight at Naco on the border on 10 June 1986 probably represented delayed migrants (A. Moorhouse). About 30 were observed between Navojoa and Cd. Obregon on 23 December 1990 (D. Stejskal). Van Rossem (1945) reported that the species was common at Santa Rosa Tesia, near Navojoa, in late June 1937 and that thousands were in the Río Yaqui valley west of Cd. Obregon on 24 November 1944. Apparently its numbers have declined greatly in recent years in Sonora. We know of no breeding records.

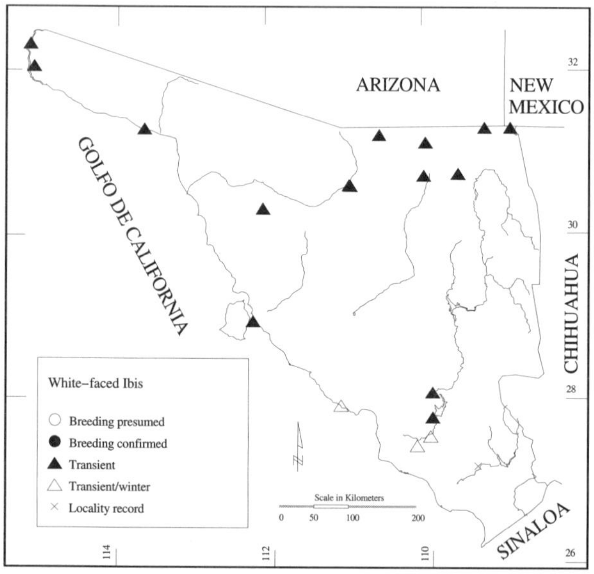

ROSEATE SPOONBILL

Espátula Rosada

Ajaia ajaja

This gorgeously tinted large wading bird is a southern visitant along the Sonora coast of El Golfo de California, as well as inland. Records extend from 3 March to 24 September. A large aggregation was 27 at Bahía Kino, 29 May–1 June 1987 (R. Taylor). Although van Rossem (1945) described this bird as fairly common, we consider it rare, perhaps indicating a general decline in its western Mexico population.

Fifteen pairs with nests were noted in a heronry on Isla Huivulai in Estero Tóbari on 14 May 1994 (Palacios and Mellink 1995). This is the only breeding record for Sonora.

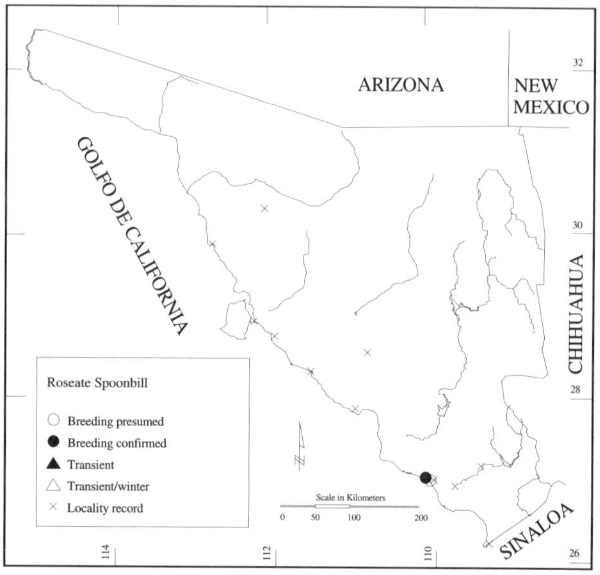

FAMILY CICONIIDAE

Storks

WOOD STORK

Cigüeña Americana

Mycteria americana

Van Rossem (1945:42) considered the Wood Stork "common in summer . . . from Guaymas southward," but we fear that its numbers have considerably decreased in recent years, just as they have along the lower Colorado River in Arizona (Monson and Phillips 1981). It is now rare in Sonora, and we cannot account for this change. We find no evidence of breeding in Sonora, but van Rossem (1945) stated that it likely nested at Laguna Guásimas south of Guaymas, where we have seen it only once (22 September 1985). Dates of record span from 17 May (1986) to 1 October (1984), and most involve only a single bird.

Three records are of special interest. About 250 were observed soaring over Pozo Crisanto, 19 km west of Carbo, 23 June 1942 (Neff 1947). On 17 May 1986 at Guajaray, on the river of the same name, a flock of six soared far overhead at a height estimated as 700 m (A. Moorhouse et al.). E. Johnson watched 11 on 1 October 1984 land on a hillside adjacent to Quitobaquito, Arizona, then take flight, disappearing to the south in Sonora.

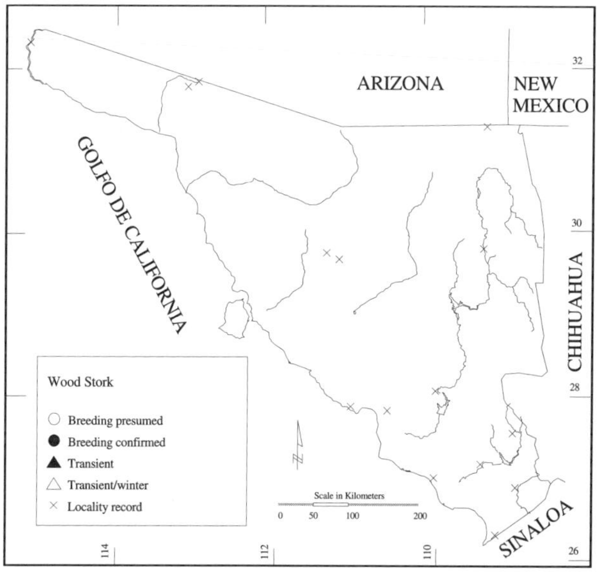

FAMILY ANATIDAE

Swans, Geese, and Ducks

FULVOUS WHISTLING-DUCK

Pijije Canelo

Dendrocygna bicolor

The Fulvous Whistling-Duck has been rare in Sonora historically. We have never seen it in Sonora, although we did find five on 11 July 1987 in semiflooded pastures just east of the pueblo of Bacorehuis in Sinaloa, west of Mexico Highway 15 and only about 2 km south of the Sonora-Sinaloa line. About 500 Black-bellied Whistling-Ducks were in the same pastures.

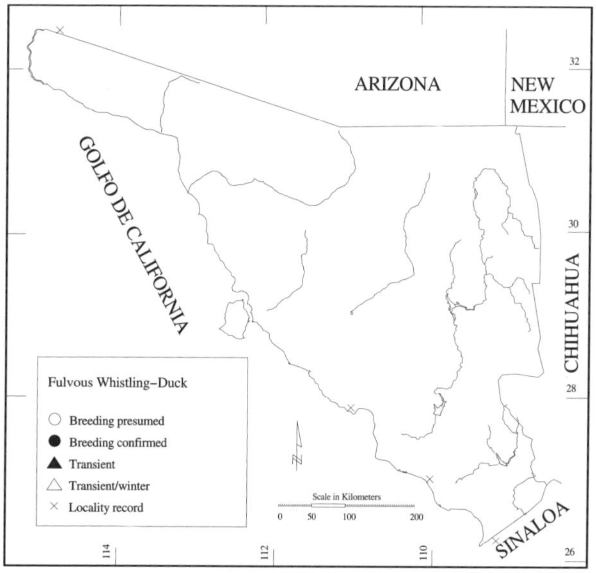

Only three records are presented by van Rossem (1945): one bird at Monument 205 on the Río Colorado on 19 March 1894 (E. Mearns), a flock of at least a hundred at Guaymas on 5 May 1930, and three at Bahía Yavaros on 25 November 1944 (both W. Sheffler).

Winter aerial surveys (USFWS 1947–1992) detected Fulvous Whistling-Ducks in the Yavaros-Agiabampo region in 1967 (150 birds), 1974 (850 birds), 1978 (130 birds), and 1981 (20 birds). The species becomes much more common south of the Topolobampo area in Sinaloa.

BLACK-BELLIED WHISTLING-DUCK

Pijije Ala Blanca

Dendrocygna autumnalis

In Sonora the Black-bellied Whistling-Duck is an uncommon summer resident; it arrives from more southern wintering grounds in early April and departs

Black-bellied Whistling-Duck

by late September. Some, seemingly looking for nest cavities, were seen in large cottonwoods along the Río Yaqui below Presa Obregon on 11 April 1986. We saw four newly hatched ducklings there on 10 July 1987, our only observation of very young birds. Their presence in numbers in spring and summer along the lower Ríos Yaqui and Mayo, as well as in the Río Fuerte drainage, suggests nesting in many suitable places in the southern part of Sonora.

Along the coast of El Golfo de California where trees are absent and there is developed farmland, we have seen them standing, incongruously enough, on the tops of hecho cacti. Russell startled a pair just as it was about to land on a whitewashed ledge near Guásimas. We think they nest on rock ledges in low hills that are near irrigated fields where they feed. Records from northern Sonora are not plentiful, although the species nests in the Santa Cruz and San Pedro valleys of adjacent Arizona.

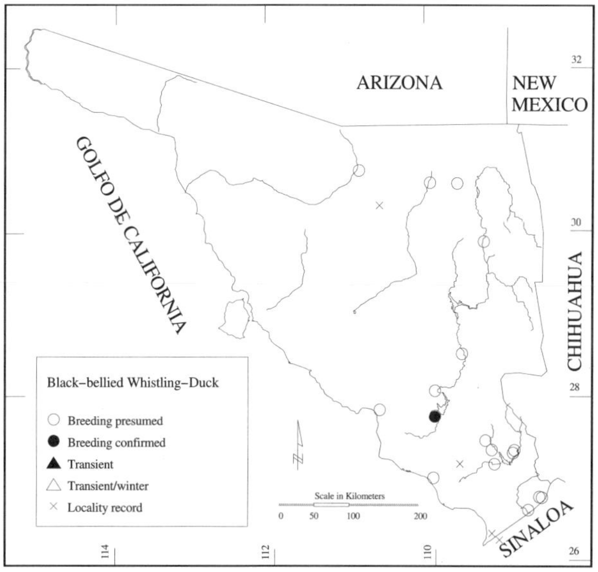

January aerial surveys (USFWS 1947–1992) since 1965 have usually located a few (but as many as 3500 in 1991) along the coast of El Golfo de California from Cd. Obregon south to the Sonora-Sinaloa border. Farther north, 80 were found scattered along 80 km of the coast south of Guaymas in 1985, probably in Bahía Guásimas.

A recent study of Black-bellied Whistling-Ducks wintering in Sinaloa discerned that they feed mostly on rice and corn, as well as barnyard grass seeds (Kramer and Euliss 1986). Because of the early maturation of cereal crops in southern Sonora, these may be the principal food of these ducks during the breeding season in that state. We believe the number of Black-bellied Whistling-Ducks may have grown with the expansion of agriculture following the construction of major dams and irrigation projects. Van Rossem (1945) considered its status uncertain.

TUNDRA SWAN
Cisne de Tundra

Cygnus columbianus

The Tundra Swan is accidental in Sonora. Two were seen near Cd. Obregon in 1958 (Saunders and Saunders 1981), a single immature flew across the border from the San Bernardino National Wildlife Refuge on the Arizona side to the Rancho San Bernardino on 5 December 1986 (B. Robertson), and two were seen at Rancho San Bernardino on 23 December 1991 (D. Fischer).

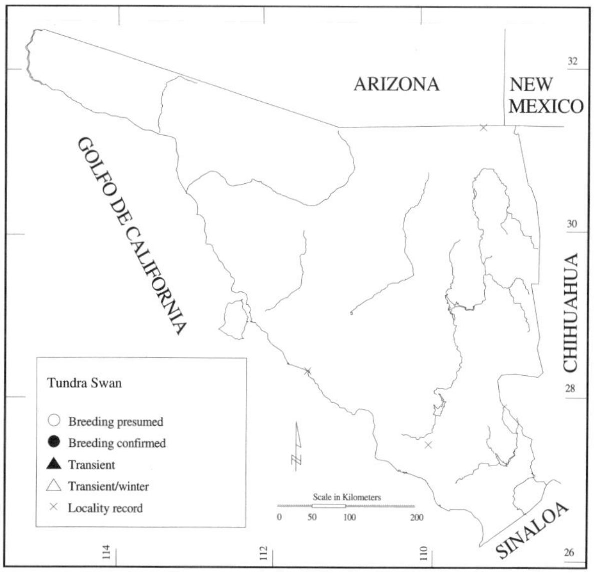

The remains of a swan were found at Estero Tastiota on 7 February 1982 by J. Edgar and were photographed. G. McCaskie (pers. comm.) inspected the photographs and suggested the carcass was that of a Tundra Swan on the basis of the culmen shape (concave rather than straight).

GREATER WHITE-FRONTED GOOSE
Ganso Careto-mayor
Anser albifrons

Saunders and Saunders (1981), discussing winter aerial waterfowl surveys in Mexico in 1947–1965, stated that in 1947 and years following, about 17 000 White-fronts were counted in the grain fields of the Cd. Obregon district. However, beginning about 1960 these large numbers fell off sharply. Since 1965 (USFWS 1947–1992), numbers counted in the Cd. Obregon–Agiabampo region ranged from zero to 1600 (in 1991). During the same period, winter numbers in Sinaloa and Nayarit averaged fewer than 2000. This decline agrees with an overall drop in the species' total New World population (e.g., D. Brown 1985; Garrett and Dunn 1981).

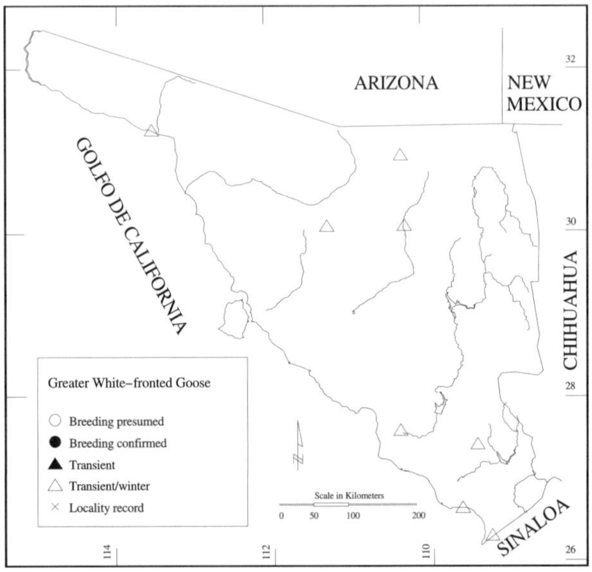

Van Rossem (1945:44) cited only a record by W. Sheffler, who found White-fronts in "great abundance" at Bahía Yavaros on 25 November 1944 and noted, "One flight . . . took two hours to pass." Fall arrivals in Sonora appear as early as mid-September, and spring returnees leave mainly in February and March. Once common during migration in the lower Colorado River valley of Arizona, today there are few or none. About 50 were seen 13 km east of Puerto Peñasco on 20 and 21 September 1990 (D. Smith, *fide* S. Levy).

From the interior of Sonora, we have three recent records: a bird standing by a roadside rainwater pool near Sinoquipe in the Río Sonora valley on 13 September 1986; two seen near Cananea on 31 January 1987 (A. Moorhouse); and a flock of 30 flying over El Seri near El Carrizo southwest of Benjamin Hill on 21 September 1990 (S. Dobrott).

SNOW GOOSE
Ganso Blanco
Chen caerulescens

The Río Colorado delta, including the Sonora side, was once a wilderness wintering ground for the Snow Goose. A. Leopold (1949) canoed and camped extensively there with his brother in the winter of 1921–1922 and saw numerous Snow Geese flying inland from the head of El Golfo de California to take gravel from a sandbar. Saunders and Saunders (1981) mentioned that about 5000 wintered there in 1947. The once numerous flocks now have vanished.

Saunders and Saunders (1981) indicated that Snow Geese averaged 1000 to 2000 birds in the area of Bahía Yavaros during the 1947–1964 aerial waterfowl surveys. Subsequent U.S. Fish and Wildlife Service (1965–1992) surveys found Snow Geese in 17 winters out of 23, with a high count of 470 in 1976. At Estero Tóbari in 1989, 300 were located, including a few of the blue phase; in 1991, 400 were found. A lone bird was noted at Puerto Peñasco on 16 December 1990 (D. Stejskal).

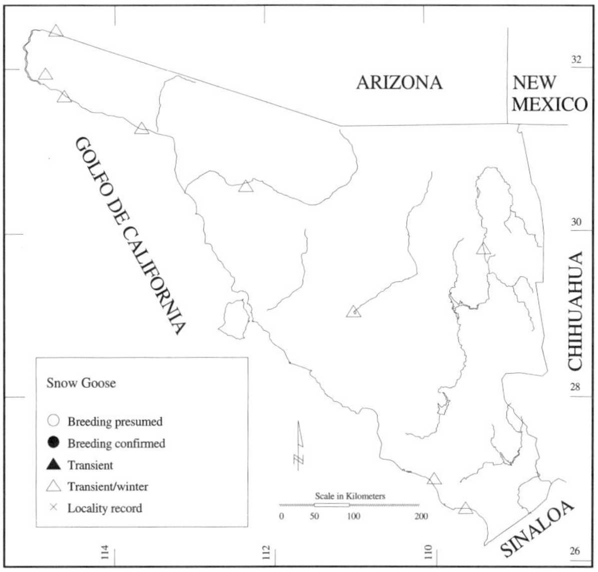

Recently, flocks have been seen on Presa Rodriguez at Hermosillo. A. Moorhouse noted about 500 there on 11 March 1986 and about 200 on 15 November 1986; and J. Levy saw 300 to 500 on 23 February 1987, at least three of which were of the blue phase. These flocks were likely migratory birds, moving from or to wintering grounds in central Mexico. Otherwise, inland, the Snow Goose has been recorded thrice: one on the Río Bavispe below Granados on 12 December 1963 (G. Brown *fide* J. Marshall), one about 13 km south of Granados on 12 December 1983 (B. Brown), and three at Caborca on 11 November 1986.

Although local hunters at Cd. Obregon have reported (Saunders and Saunders 1981) seeing several small white geese associated with flocks of Snow Geese, there is insufficient evidence to consider them Ross' Goose (*Chen rossii*), a species reported from Arizona and Chihuahua.

BRANT
Ganso de Collar
Branta bernicla

Brant have an interesting history in Sonora. Van Rossem (1945) did not include the species in his book on Sonoran birds. Price (1899) reported their presence on and near the Río Colorado delta without any particulars, but van Rossem (1945) considered his records unreliable. Definite records were lacking until January 1958, when aerial surveys initiated in 1947 found four Brant near Agiabampo (Saunders and Saunders 1981). By 1964–1965 their numbers had grown greatly, to about 25 000 in both Sonora and Sinaloa. No one seems to know what caused the Brant to begin wintering along the Mexican mainland coast, but human harassment in its former range is a likely factor (Smith and Jensen 1970).

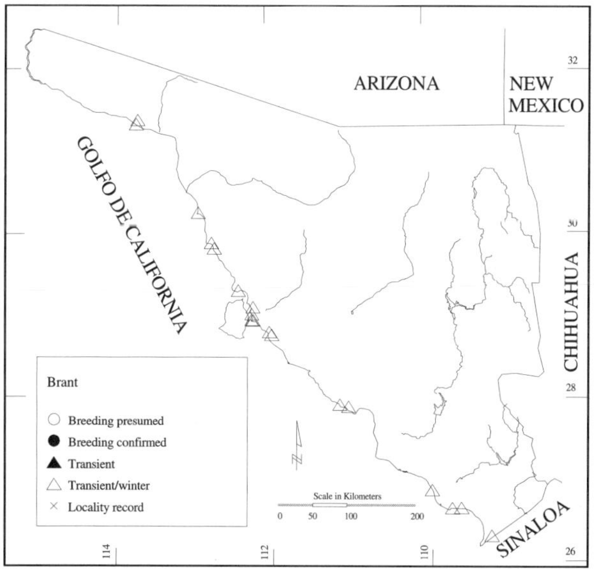

Since 1966, aerial surveys (USFWS 1947–1992) have attempted to count all Brant of the Pacific Coast in both the United States and Mexico. The results revealed that Brant rarely winter farther south than Bahía de Santa Maria on the northern coast of Sinaloa. Brant numbers in Sinaloa were greater than their numbers in Sonora in every year (except 1972) until 1977, when larger numbers appeared in Sonora and then dominated the totals except in 1981 and 1984. The survey data indicate considerable fluctuation from year to year. In Sonora, Brant have been most numerous, usually accounting for at least 75% of the survey total for the state, in the Canal el Infiernillo. The rest of the population has been in the general Yavaros-Agiabampo regions. We have seen Brant frequently in the Canal el Infiernillo from the shore at Punta Santa Rosa (as many as 2500 individuals) and at Bahía Kino.

The duration of stay in Sonora (and Sinaloa) is from early November until well into February, with some lingering into early March. Out-of-season dates range from 22 September to 8 June (Puerto Libertad, J. Levy) and include eight Brant photographed at Punta Cirio on 26 May 1984 (D. Brown).

The principal food plant used by Brant is eelgrass. The southward extent of Brant may be limited by the occurrence of this plant.

CANADA GOOSE
Ganso Canadiense
Branta canadensis

Until the U.S. Fish and Wildlife Service (1947–1992) undertook its January aerial waterfowl surveys in Mexico in 1947, Canada Geese were recorded in

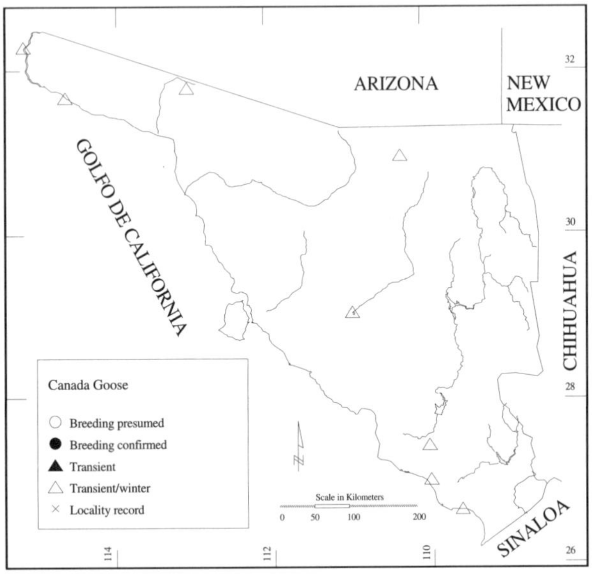

Sonora only by Mearns near Sonoita in January 1894 (van Rossem 1945) and subsequently by various observers in the Río Colorado delta. Saunders and Saunders (1981), in reporting the aerial surveys, indicated that Canada Geese were quite local, limited in Sonora to the Río Colorado delta and the Cd. Obregon district. Concerning the latter, they said that in January 1950 several hundred were located during ground studies. Inland, aside from Mearns's record, there are but two observations: one at Presa Rodriguez, Hermosillo, on 23 January 1987 (J. Levy); and seven near Cananea on 31 January 1987 (A. Moorhouse).

Winter aerial surveys from 1965 to the present yielded observations of only 45 Canada Geese: 20 at Bahía Yavaros in 1975 and 25 in the Río Colorado delta in 1976. There are no coastal records since 1976. We conclude that the Canada Goose is presently a casual wintering species in Sonora.

Of approximately 50 recoveries in Sonora of banded Canada Geese, only four are after 1979. Save for one banded in Arkansas, all recoveries were of birds from the Rocky Mountain states (and three from Alaska, as noted below). The southernmost recovery was from near Cd. Obregon.

Interestingly, three banded geese from the Aleutian Islands population in Alaska have been shot in Sonora. The Aleutian population, sometimes called the Aleutian Canada Goose (*B. c. leucopareia*, AOU 1957), is on the list of endangered and threatened birds (USFWS 1990). We are indebted to Paul F. Springer for information on the three birds, which were banded on Buldir Island in the western Aleutians on 22 July 1975. They were shot on 20 December 1975, about 28 km southwest of San Luis Río Colorado in the Río Colorado delta.

WOOD DUCK

Pato Arcoiris

Aix sponsa

The Wood Duck is rare in Sonora. One was shot from a flock by a hunter at a tank somewhere between Agua Prieta and Presa Angostura about February 1961 and was later prepared as a mounted specimen by S. Levy of Tucson, Arizona (J. Marshall). Two pairs were seen near Villa Hidalgo on the Río Bavispe about 25 km north of Huasabas on 6 February 1983. In all, 18 were observed on ten occasions on the Ríos Yaqui and Aros in 1986, 1987, 1989, and 1990 (B. Brown et al.). Extreme dates for all records are 14 December and 11 April. The status of the Wood Duck in Mexico was recently reported by Williams (1987).

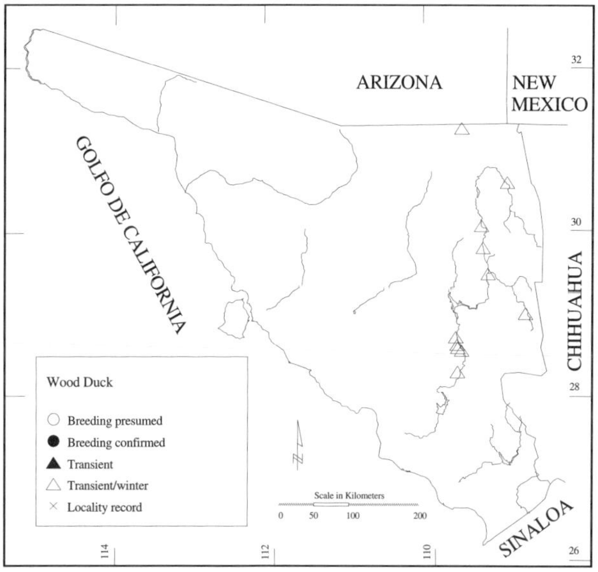

GREEN-WINGED TEAL

Cerceta Ala Verde

Anas crecca

Green-winged Teal are ubiquitous wintering ducks in Sonora wherever there is water, from the seacoast to as high as 1800 m, along even the smallest mountain streams. They begin arriving as early as mid-August and leave by mid-March, although some remain until late April. They are one of the more numerous wintering ducks along the coast of Sonora. Saunders and Saunders (1981) found them plentiful in coastal lagoons and marshes, with as many as 25 000 detected in the Cd. Obregon region during the 1950s. January aerial surveys (USFWS 1947–1992) since 1978 have provided counts of between 33 000 and 88 000 (but only 6730 in 1988) Green-winged Teal along the southern coast of Sonora (Yavaros, Agiabampo) and much smaller numbers farther north (Bahía Kino–Isla Tiburón). The average size of groups seen by us on inland waters is about 24.

Saunders and Saunders (1981) stated that gizzards of 42 Green-winged Teal from near Cd. Obregon contained 98.4% plant food, mainly the seeds of barnyard grass and cultivated rice.

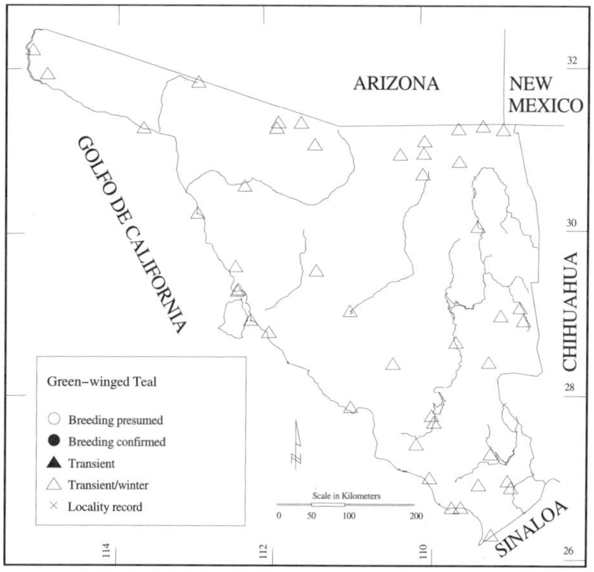

MALLARD
See text for Spanish names
Anas platyrhynchos

Two distinctive types (or subspecies) of the Mallard are found in Sonora. The two were known as separate species for many years, until it became apparent that they hybridize wherever they come in contact during the mating season. Such contact, however, is limited by ecological preferences of the two forms, and the likelihood that one may eliminate the other is remote (D. Brown 1985). The subspecies are *A. p. platyrhynchos*, the well-known common Mallard or "greenhead," and *A. p. diazi*, the Mexican Duck. The most obvious difference between them is that in the former the sexes are dissimilar in their breeding plumage, whereas in the latter they are virtually identical.

A. p. platyrhynchos. The Mallard proper is an uncommon winter visitant to Sonora and is recorded chiefly in the northern one-third of the state in open water,

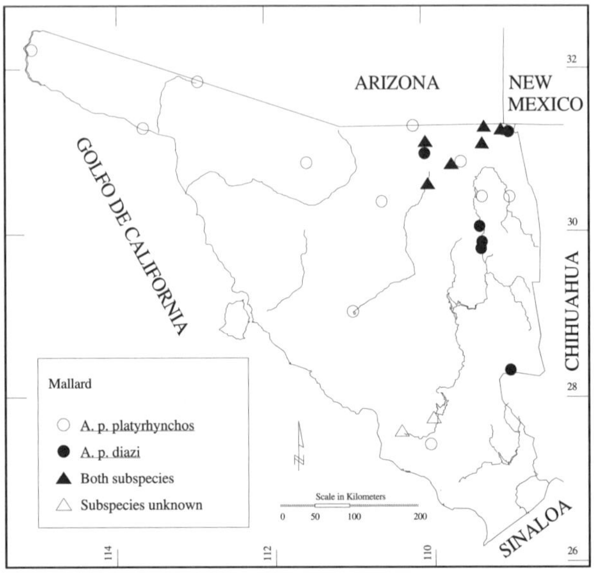

including the Río Colorado delta. Although apparently once wintering in some numbers in southern coastal Sonora, especially in the region of Cd. Obregon (Saunders and Saunders 1981), it is now rarely found there. This subspecies was noted along the Río Bavispe at Bavispe on 3–4 July 1954 by J. Marshall. The Spanish name is Pato de Collar.

A. p. diazi. The range is almost entirely the interior highlands of Mexico, where this subspecies is resident. We have found it regularly in small numbers in northeastern Sonora in the irrigated valleys of streams from the Río Magdalena eastward and southeastward to at least Granados and even Yécora. We found 14 on a reservoir in the Sierra los Ajos on 3 March 1984. Sixteen seen on 10 July 1987 on the Río Yaqui below Presa Obregon, all in "female Mallard" plumage, may have been Mexican Ducks, but there is the possibility that they were domestic birds. Although more or less resident wherever found, some individuals may wander during the winter. We have not found nests or flightless young in Sonora. Van Rossem (1945) did not mention the subspecies, hence we conclude that all records he quotes should be considered as *A. p. platyrhynchos*. The Spanish name is Pato Mexicano.

NORTHERN PINTAIL
Pato Golondrino
Anas acuta

The Northern Pintail is "the most abundant [wintering] duck in Mexico," according to Saunders and Saunders (1981:118). Prior to 1965, this was true in Sonora; since then the Green-winged Teal has been more numerous. Most are to be found in coastal waters adjacent to farmland. On the basis of January aerial waterfowl surveys (USFWS 1947–1992), from 1948 to 1956 the center of their winter abundance on the Pacific side of Mexico was the Cd. Obregon region, where they fed heavily on rice grains. Then changing

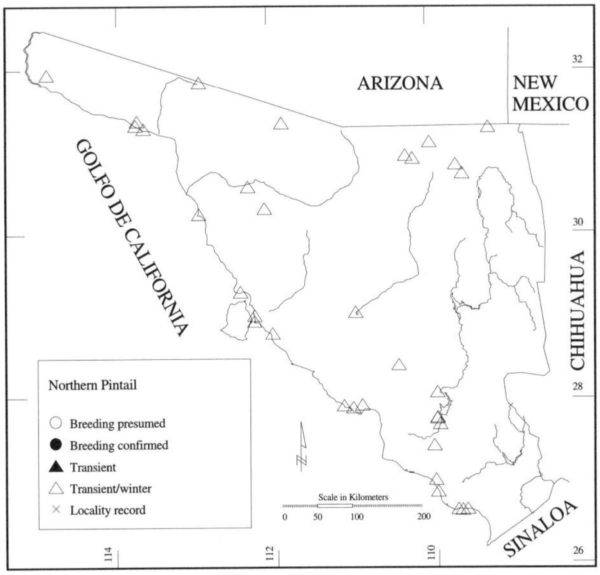

farming practices caused most pintails to move farther south, into Sinaloa.

January aerial waterfowl counts continued to show impressive numbers of pintails in agricultural areas of southern Sonora (reaching 160 000 in 1976), mainly in the Yavaros and Agiabampo areas. At the same time, smaller numbers were found in other coastal localities and in many interior localities as well.

Large flocks are seen in the spring migration, from early February to mid-March. In fall, southbound flocks are smaller, with migration extending over a longer period, from early September into November.

An occasional bird, perhaps disabled or diseased, is found in early summer to midsummer. There were two at Punta Chueca on 31 May 1971, two near Cananea on 13 June 1942 (Neff 1947), and one below Presa Obregon on 10 July 1987.

BLUE-WINGED TEAL
Cerceta Ala Azul
Anas discors

We have few observations of Blue-winged Teal, a casual migrant and winter visitant in Sonora. The greatest number noted on any date is 15 at Presa Agua Caliente on 4 January 1986. Dates range from 6 October at Puerto Peñasco (D. Stejskal) to 30 April at Puerto Peñasco (N. Boyajian), with most records from the coast of El Golfo de California. Van Rossem (1945) stated that Blue-winged Teal were fairly common on 24 November 1944 about 6 km southwest of Cd. Obregon, where they were seen by W. Sheffler.

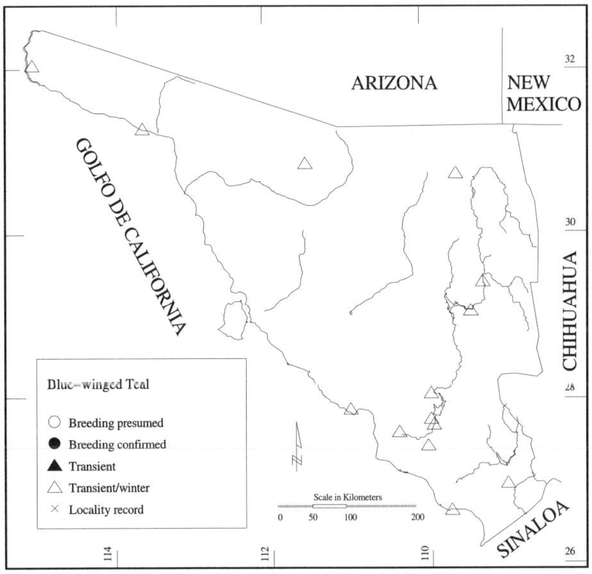

The U.S. Fish and Wildlife Service (1947–1992) reports on January aerial surveys along the Mexican coasts combined counts for Blue-winged and Cinnamon Teal because of the difficulty in distinguishing between them from the air, since they have similarly patterned wings. Our experience suggests that the Cinnamon Teal is a much more common species in Sonora.

CINNAMON TEAL

Cerceta Canela

Anas cyanoptera

The Cinnamon Teal is a fairly common migrant along the coast of Sonora, going south mainly in September and October and returning in February to April. It is an uncommon winter resident from Cd. Obregon south. Inland, the species is more scattered, and sightings generally are of six birds or less. There are no inland December records in the last 20 years, although November and January records are not infrequent.

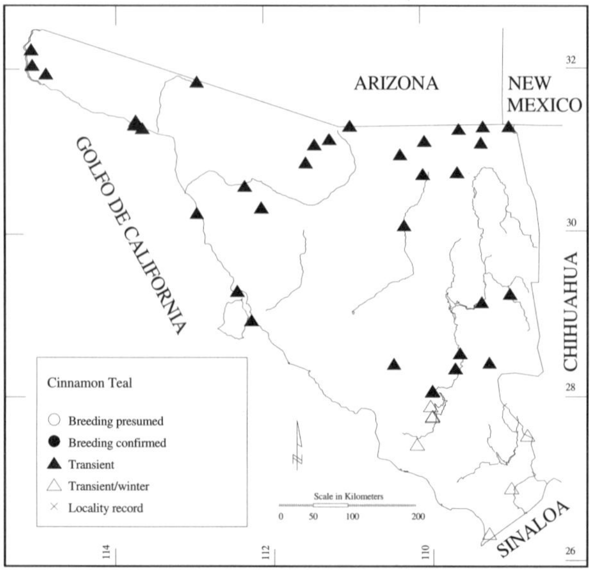

U.S. Fish and Wildlife Service (1947–1992) January aerial tallies of Blue-winged and Cinnamon Teal combined since 1965 in Sonora reveal much variation from year to year. Numbers range from only 100 in 1968 to 25 000 in 1985. The fluctuation is unexplained; we can guess it is related to food and water supplies, if not adequacy or unevenness of coverage by the surveys. The birds were located near and along the coast from the Río Yaqui delta south. Northward, very few are found in winter.

The species may nest in Sonora, but we have no evidence of it. Cinnamon Teal are known to nest in Mexico, including Chihuahua, Jalisco, and Baja California Norte. On 10 July 1987, we found 14 birds on the Río Yaqui below Presa Obregon and eight birds the next day at Agiabampo; these are extremely early dates for migrants. No others have been noted until 13 August (1892, on Rancho San Bernardino; van Rossem 1945), a more reasonable date for early fall migrants. In the spring, the latest date is 20 May 1986, near Rancho Quemado northeast of San Bernardo, when we saw two.

NORTHERN SHOVELER

Pato Cucharón-norteño

Anas clypeata

Saunders and Saunders (1981) considered this species to be the second most common waterfowl wintering on the Pacific coast of Mexico, although its principal wintering grounds are well to the south of Sonora. In Sonora it is fairly common in winter, chiefly along the shore of El Golfo de California and in the Río Colorado delta.

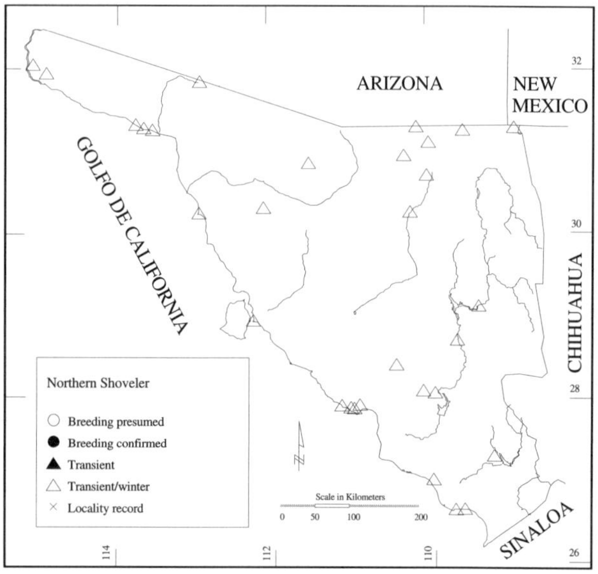

As with most waterfowl wintering in Sonora, by far the greatest number are south of the Río Yaqui delta. Since 1965, January aerial surveys (USFWS 1947–1992) disclosed an average of 15 000 shovelers present, with a maximum of 33 000 in 1969 and 1982 and 31 000 in 1991.

The Northern Shoveler is a common transient. It arrives in small numbers as early as 9 September (Bamuri, south of Pitiquito, 1973) and continues to pass through until early November. In the spring, it is on the move again in March and disappears from the state by late April. Some migrating birds fly over open salt water (e.g., near San Carlos on 17 April 1979).

As with almost any migratory duck in Sonora, one can expect to see one or two nonbreeding birds during the summer. An example is one seen at Naco on 10 June 1986 (A. Moorhouse).

GADWALL
Pato Friso
Anas strepera

The Gadwall is a fairly common transient and uncommon winter visitant, found mostly in the coastal plain, although quite a few records come from inland (including 100 birds at Hermosillo on 11 March 1986). Like many other migratory ducks, it winters in larger numbers south of Sonora. It also follows the pattern of other ducks by migrating into and through Sonora during September and October and out again in March and April. A pair observed in Sonora near Monument 67 on the Sonora–New Mexico border on 20 August 1951 (J. Marshall) may have represented birds that remained through the summer, or they may have been unusually early fall migrants.

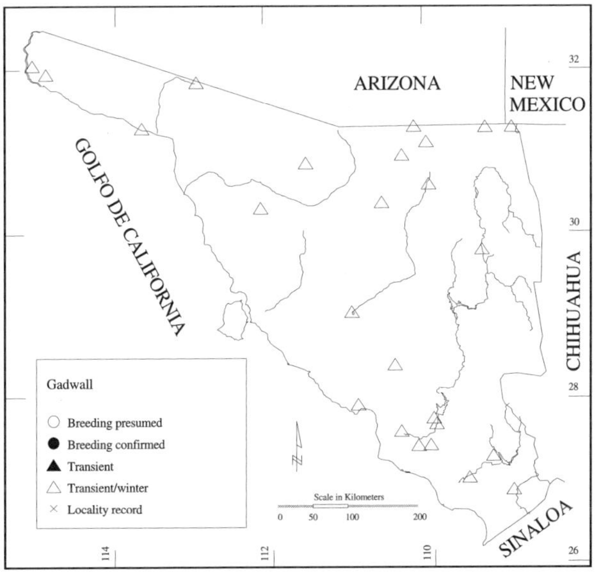

Annual January aerial surveys (USFWS 1947–1992) from 1965 through 1988 in Sonora did not find more than about 15 000 Gadwalls (in 1976); the average was 2500. Few of the birds were north of the Río Yaqui delta.

These vegetarian ducks are seldom found in salt or brackish wetlands except when loafing or sleeping. They are more frequently seen feeding in fresh water, even on relatively small streams.

[EURASIAN WIGEON
Pato Silbón
Anas penelope

A male was seen and carefully observed at Puerto Peñasco on 6 October 1979 by D. Stejskal. This is the only record for Sonora.]

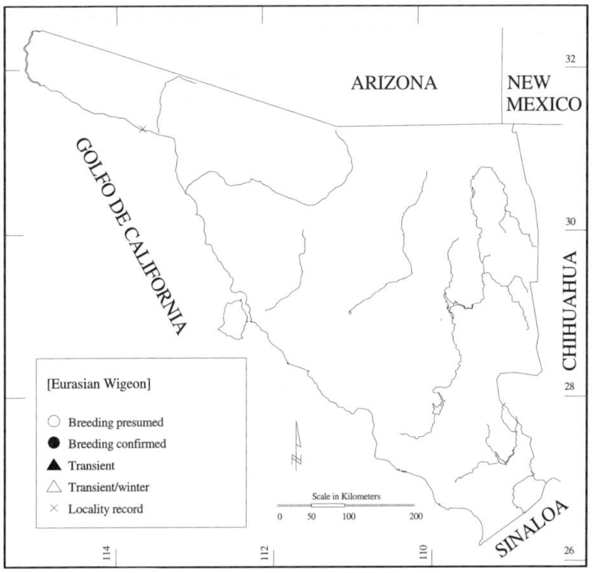

AMERICAN WIGEON
Pato Chalcuán
Anas americana

The American Wigeon is common along the coast of Sonora during migration and in winter. Inland it is much less frequent but, even so, well distributed. Two found on a small stock pond in foothills at the lower edge of oak woodlands (720 m) north of San Nicolás on 13 March 1984 were in rather unusual surroundings. Wintering numbers of the American Wigeon rival those of the Northern Pintail, according to January aerial surveys (USFWS 1947–1992). From 1965 to 1992, between 23 000 and 50 000 American Wigeon spent the winters in southern coastal Sonora, averaging 39 000 annually. In 1976 the estimate was 170 000. The wigeon is the third most numerous wintering duck in Sonora, following the Green-winged Teal and Northern Pintail. As indicated in previous accounts, the vast majority of ducks of the genus *Anas* wintering in Sonora feed in agricultural areas that are proximal to open waters and shorelines where they can roost and loaf.

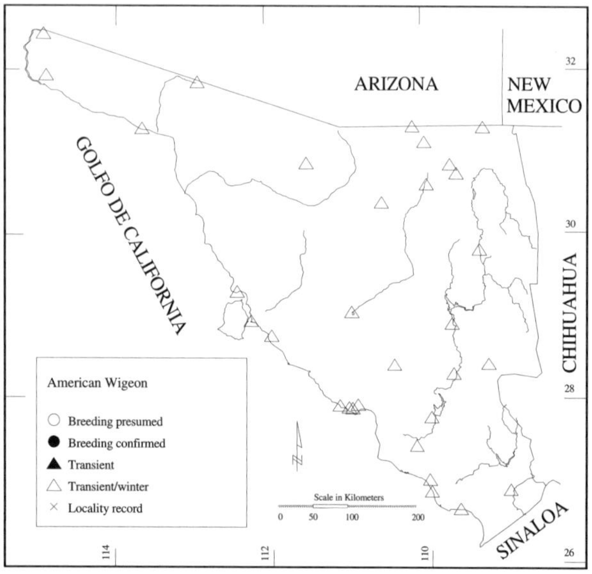

The earliest fall record is from the Río San Bernardino near the border on 2 August 1892 (van Rossem 1945). The latest spring sighting is of 50 on Presa Cuauhtémoc on the Río Altar on 3 April 1983.

CANVASBACK
Pato Coacoxtle
Aythya valisineria

The Canvasback is rare in Sonora. Van Rossem (1945) made no mention of it, aerial surveys (USFWS 1947–1992) located few birds, and we have few records. The highest counts were 40 at Puerto Peñasco on 14 September 1979 and 50 at a reservoir in El Valle de San Luis (1575 m) in the extreme northeast on 15 November 1983 (A. Moorhouse). The listing of 23 000 Canvasbacks in the Isla Tiburón area on the 1969 aerial survey (USFWS 1947–1992) was an error (J. Voelzer in litt.).

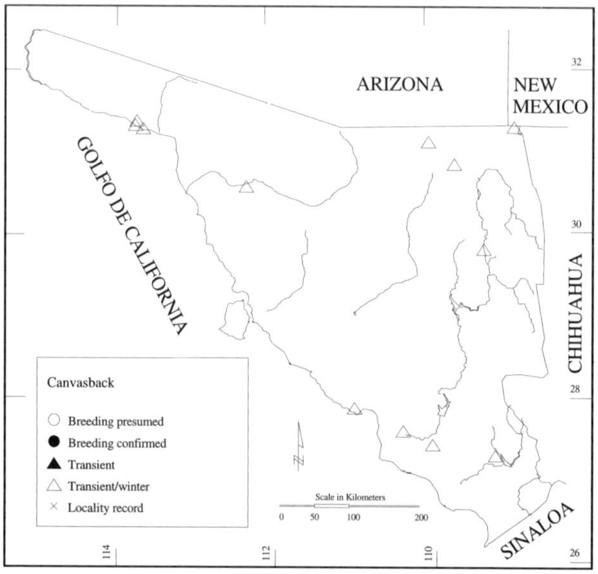

REDHEAD
Pato Cabeza Roja
Aythya americana

Redheads are common along the coast of El Golfo de California as migrants and wintering birds. According to the 1970 January aerial survey (USFWS 1947–1992), they were found "on the saline bays usually connected to the open ocean by 'bocas,' or passes." Considerable numbers were censused from 1965 to 1988 in the Canal el Infiernillo: as many as 30 000 in 1970 and 25 000 in 1975. Hundreds and occasionally thousands can be seen from the mainland shore of the strait (especially from

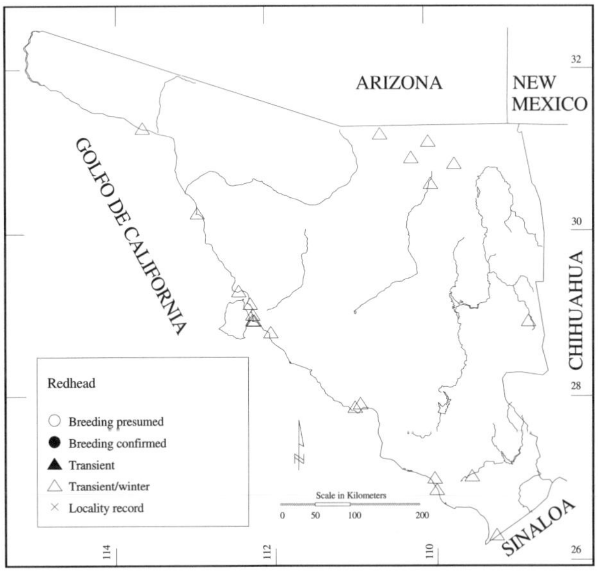

Puntas Santa Rosa and Chueca). Farther south, from the Río Yaqui delta to Agiabampo, counts ranged from a high of 20 000 in 1976 and 1983 to lows of 740 in 1978 and 300 in 1988. Low counts in Sonora corresponded to high counts in Sinaloa in some years.

The migration period extends from early October to late November, and from late February to late March. The few inland records likely do not reflect the true status of the Redhead in interior Sonora, which probably is that of a fairly common transient and winter visitant, as it is in southern Arizona.

RING-NECKED DUCK
Pato Pico Anillado
Aythya collaris

Although the Ring-necked Duck winters south as far as Panama, there are few records from Sonora, and we consider it a rare transient and winter resident. Records range from 3 November to 10 April; a bird on 10 June 1986 west of Naco is unusual. January aerial surveys (USFWS 1947–1992) along the coast since 1965 have seldom disclosed Ring-necked Ducks; the highest count was 80 at Agiabampo in 1981. Lesser Scaup are similar to Ring-necked Ducks, especially from the air; thus, the aerial surveys may have misidentified some birds.

This species prefers inland waters; it has not been seen in El Golfo de California. Typically, Ring-necked Ducks are found on ponds and lakes in small numbers; we have never found more than 27 (artificial pond at 1780 m in the Sierra los Ajos on 2 March 1984).

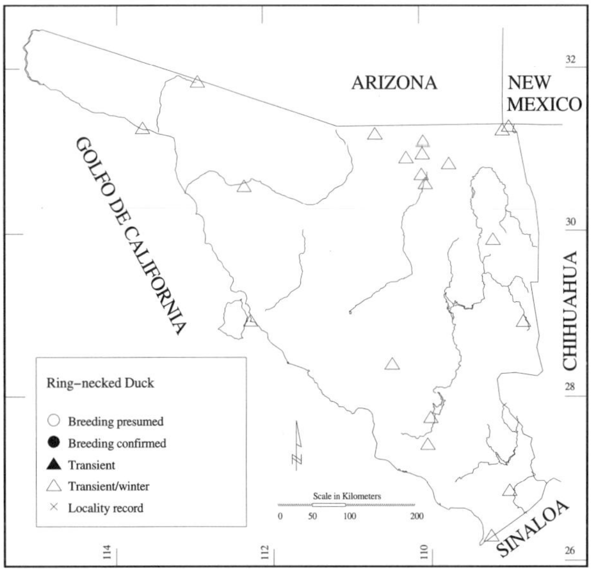

[GREATER SCAUP

Pato Boludo-mayor

Aythya marila

We have a single record for Sonora. One was listed in the 3 January 1981 Christmas Bird Count from Puerto Peñasco; this was a female seen by D. Stejskal.]

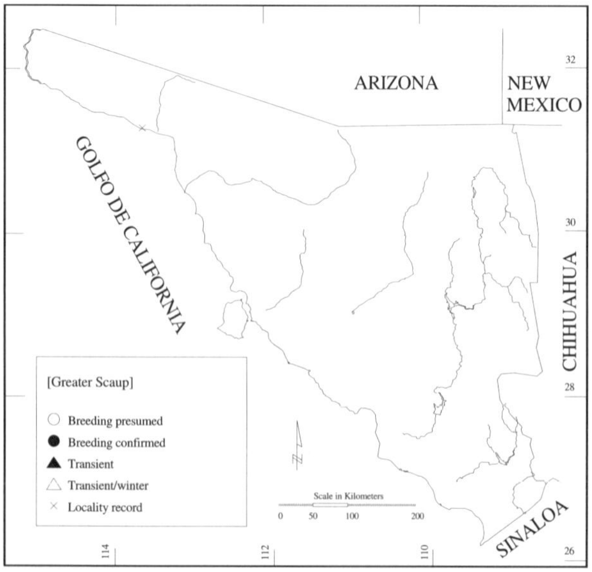

LESSER SCAUP

Pato Boludo-menor

Aythya affinis

The Lesser Scaup is a moderately common duck species, wintering in brackish and saltwater habitats along El Golfo de California. In 1965–1981, counts on aerial surveys (USFWS 1947–1992) in Sonora averaged 13 500 birds, compared to 20 000 in Sinaloa and southward. In 1982–1985, Sonoran totals increased to about 17 000 but had dropped to 6650 in 1991. These birds were found almost entirely from the Río Yaqui delta south, with the exception of 1983, when 1500 scaup were located southeast of Guaymas (probably at Bahía Guásimas). Migrating birds have been seen from early November to late April.

The species is uncommon inland in various freshwater habitats, although the largest number recorded is 405 at Villa Verde pond southwest of Naco on 11 March 1986.

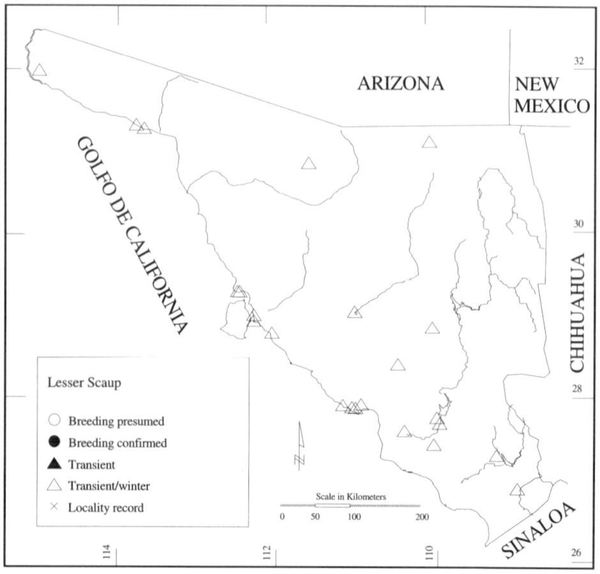

HARLEQUIN DUCK

Pato Arlequín

Histrionicus histrionicus

A drake Harlequin Duck was observed near Puerto Peñasco on 26–27 March 1977 by many, and it was photographed (Kaufman and Witzeman 1979). The bird stayed until at least 9 April. Two females were seen, also at Puerto Peñasco, on 12 January 1990 (T. Wurster). Subsequently, three females were noted there on 4

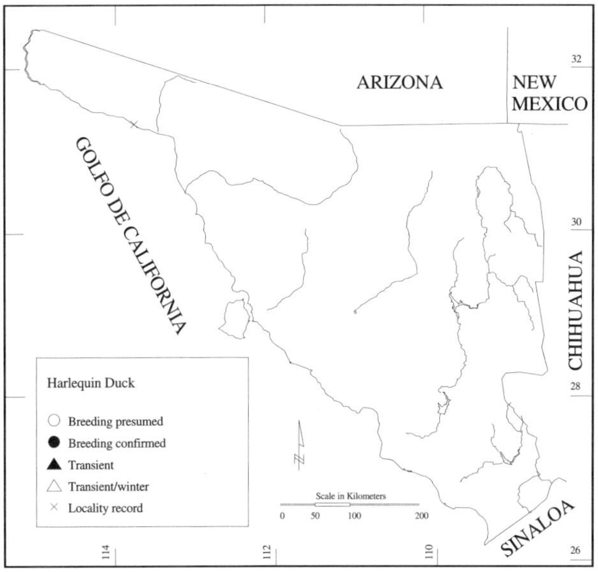

March 1990 (S. Burge et al.). The singularity of these records is attested to by the lack of records from either Baja California or Arizona.

OLDSQUAW
Pato Cola Larga
Clangula hyemalis

A male was seen in Bahía Sargento just offshore at El Dolar north of Isla Tiburón on 28 November 1970 (Russell and Lamm 1978). The observers were W. Clark,

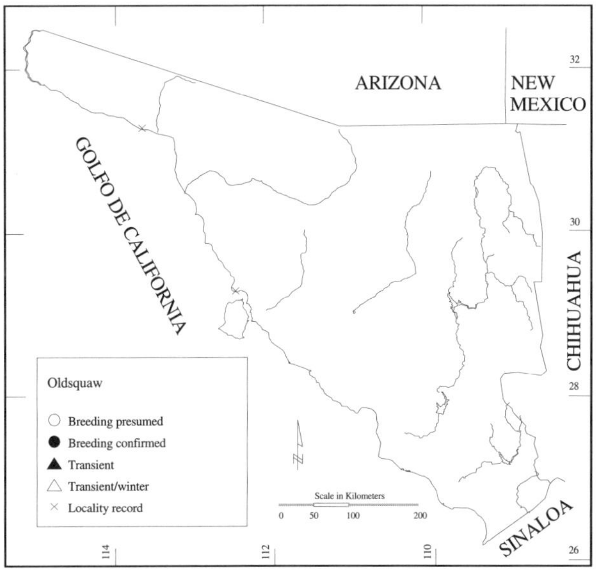

D. Lamm, and Russell. Another was reported by S. Ganley on 19 December 1993 at Puerto Peñasco on the annual Christmas Bird Count. The Oldsquaw has been recorded in southern and western Arizona at least 50 times since 1950.

[BLACK SCOTER
Negreta Negra
Melanitta nigra

On 2 January 1988, J. Bates carefully noted four Black Scoters in the bay at Puerto Lobos. This observation constitutes the only record from Sonora. The species is casual in Arizona and has not been recorded in El Golfo de California.]

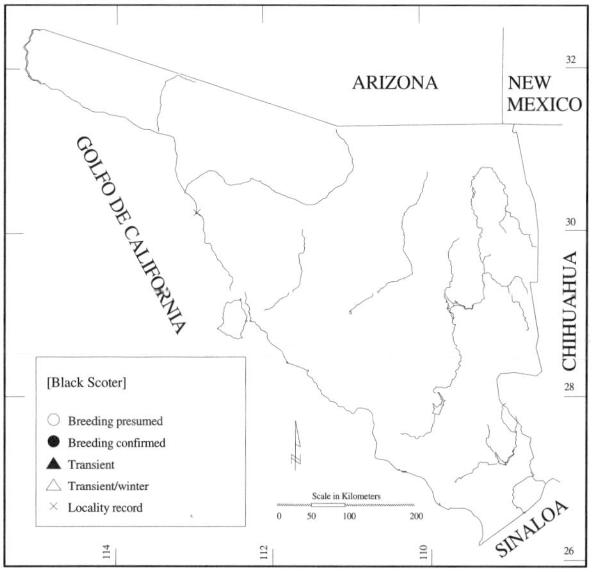

SURF SCOTER

Negreta Nuca Blanca

Melanitta perspicillata

The Surf Scoter is considered the most common sea duck of northwestern Mexico by Saunders and Saunders (1981). But it typically remains out of sight of the mainland shore, where we consider it an uncommon wintering species. Extreme dates for the few records are 6 October and 28 April.

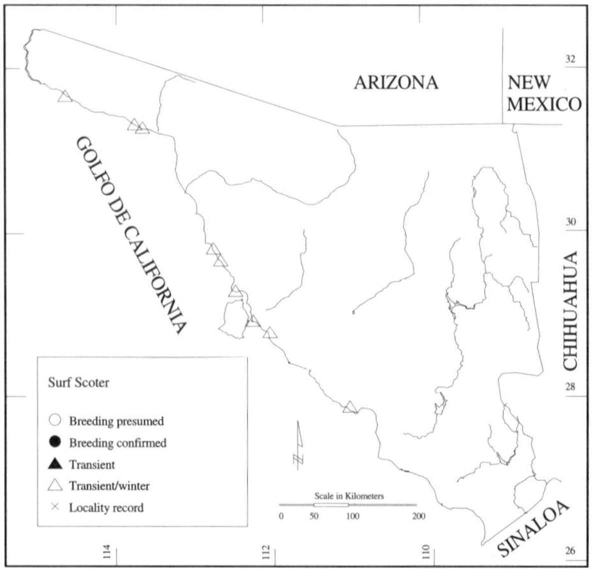

WHITE-WINGED SCOTER

Negreta Ala Blanca

Melanitta fusca

The White-winged Scoter is rare in Sonora, where it winters in El Golfo de California. We have six observations from the shore, each of fewer than five individuals, and in the period of 19 November through 1 April. Small numbers were seen during January aerial waterfowl counts (USFWS 1947–1992) in 1947–1964; the largest number counted at one time was 62, off Puerto Peñasco in 1947 (Saunders and Saunders 1981).

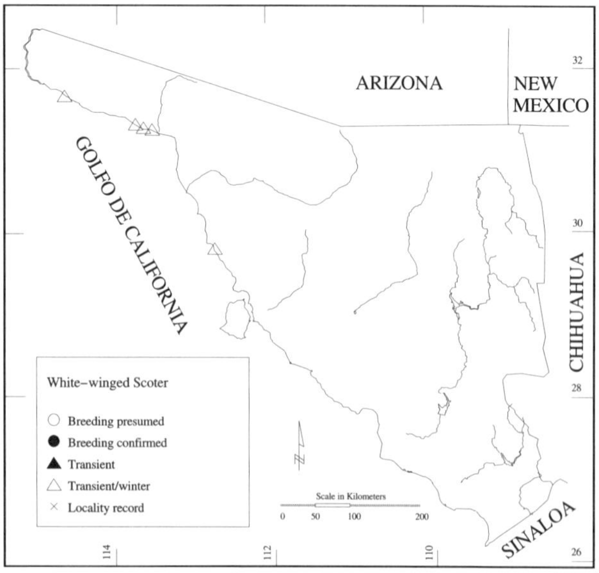

COMMON GOLDENEYE

Pato Chillón

Bucephala clangula

Only a few wintering Common Goldeneyes travel as far south as Sonora. All records come from in and along El Golfo de California and range from 9 November to 3 April. January aerial waterfowl surveys in 1947–1964 (USFWS 1947–1992) show that few were present in Sonoran waters, the maximum of 70 being reported in 1948 from the Río Colorado delta (state not indicated). Later aerial counts reached about 500 in Sonora in at least three years, although some years yielded no goldeneyes at all. The birds have been seen as far south as Agiabampo. Inland, one was observed on a small pond south of Alamos on 28 December 1983 (D. Stejskal).

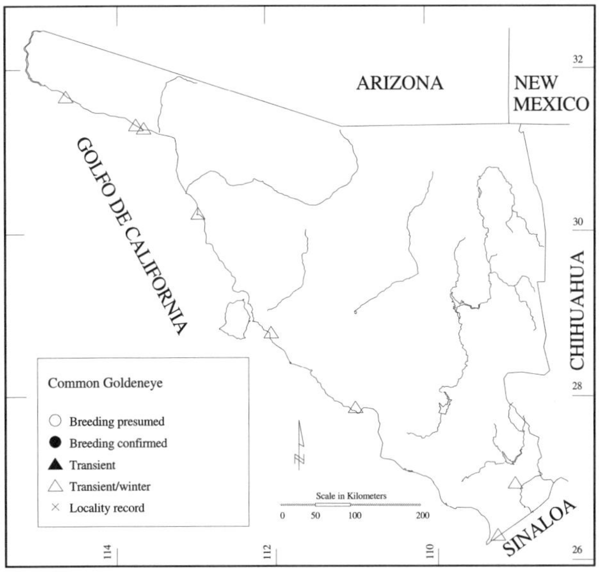

BUFFLEHEAD

Pato Monja

Bucephala albeola

Despite van Rossem's (1945) assertion that this small diving duck is rare in Sonora, we found it an uncommon wintering bird on both salt and fresh water throughout the state. Perhaps more individuals are wintering in Sonora than formerly. The earliest fall record is of seven on Presa Alatorre on 3 November, while the latest is of three near Naco on 18 April.

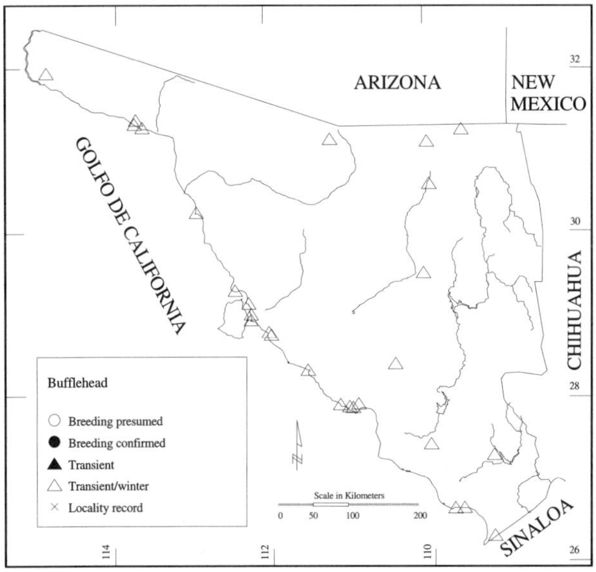

January aerial waterfowl surveys (USFWS 1947–1992) of the Mexican Pacific Coast in 1947–1964 seldom yielded more than a total of 1000 Buffleheads. However, in 1952 about 2900 were located in the area of Yavaros and Agiabampo, and in 1975–1977 from 3000 to 7300 were listed there and in the Isla Tiburón vicinity.

HOODED MERGANSER

Mergo Cresta Blanca

Lophodytes cucullatus

There are two records for this duck in Sonora: we saw a pair on the Río Mayo west of San Bernardo on 24 March 1977, and A. Moorhouse observed a female near Villa Hidalgo, on the Río Bavispe north of Huasabas on 6 February 1983.

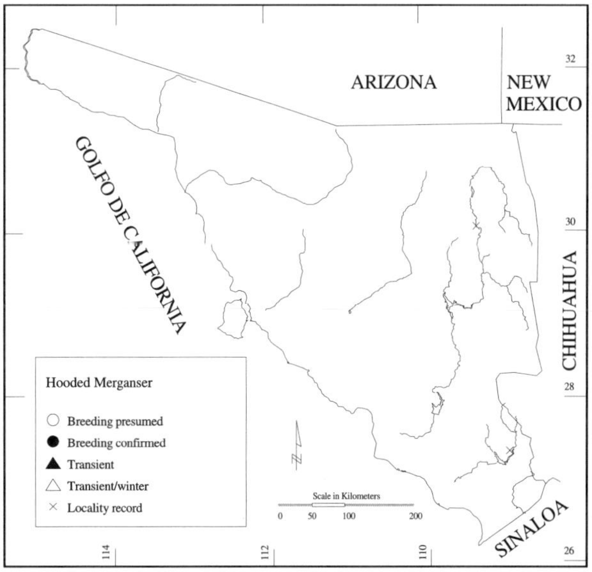

COMMON MERGANSER
Mergo Mayor
Mergus merganser

Van Rossem (1945) was dubious about the status of this merganser. Saunders and Saunders (1981) reflected this doubt. It is an uncommon winter visitant on mountain streams of the Sierra Madre as well as on the Río Aros and Río Bavispe. We are at a loss to understand Alden's (1969) assessment of the species as abundant at Bahía Kino, since we have not seen it there or anywhere else along the coast.

January aerial surveys (USFWS 1947–1992) of the coast did not distinguish between Common and Red-breasted Mergansers. The greatest number for both species combined in the period 1970–1985 was 13 000 in 1977. Some were found in most years near Isla Tiburón and Agiabampo. Red-breasted Mergansers are abundant along the coast, and we believe that virtually all mergansers noted on the aerial surveys were this species.

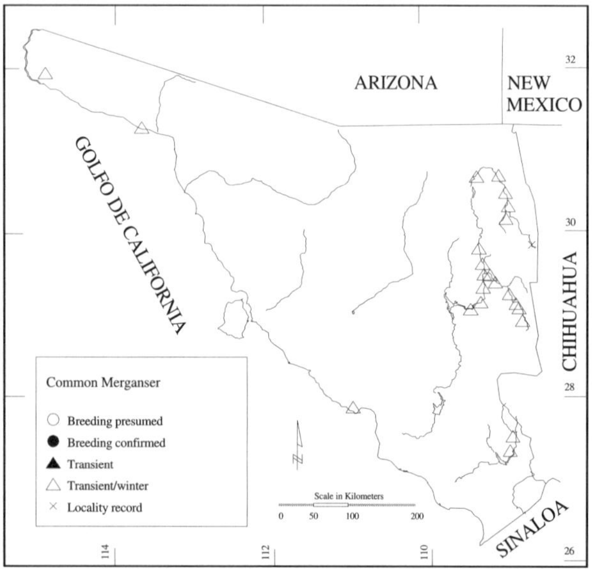

We found a few Common Mergansers along the Río Negro (upper Río Bavispe), a beautiful rushing stream at 1400 m in the Sierra Madre, northeast of Mesa Tres Ríos, on 21–22 March 1985 and conjectured whether they might breed there. The many records from the Ríos Aros, Bavispe, Mayo, and Yaqui are attributable to B. Brown and his colleagues, who made several surveys by canoe on these streams in 1981–1990, primarily in search of nesting Bald Eagles. Brown found Common Mergansers to be fairly common on these streams from January through March. The lowest inland record (160 m) is of five females on the Río Mayo west of San Bernardo on 24 March 1977, where they were in company with two Hooded Mergansers, other ducks, and cormorants. Nesting of this species in southern Chihuahua has recently been reported (B. Brown 1990).

RED-BREASTED MERGANSER
Mergo Copetón
Mergus serrator

Red-breasted Mergansers are common migrants and winter visitants on the coast of El Golfo de California, sometimes found in aggregations of 100 to 500 birds or more. They are chiefly seen inshore, swimming and diving as they pursue small fish at the edges of rocky beaches. Occasionally several individuals collectively chase their prey into shallow areas where they can more readily seize them.

They arrive in late October and normally remain until mid-April. Nonbreeders that for some reason spend the summer on the gulf may be seen outside these dates. Examples are three we saw at Puerto Libertad on 20 June 1978 and 35 in the vicinity of Punta Chueca on 8 September 1973. The greatest number reported, aside from aerial surveys (see account of Common Merganser), is 1511, on the 1983 Puerto Peñasco Christmas Bird Count.

Inland, records are almost lacking: one was seen near Sonoita on 7–8 November 1948 (Phillips and Amadon 1952), and three were observed on the Río Bavispe above Huasabas on 6 February 1983.

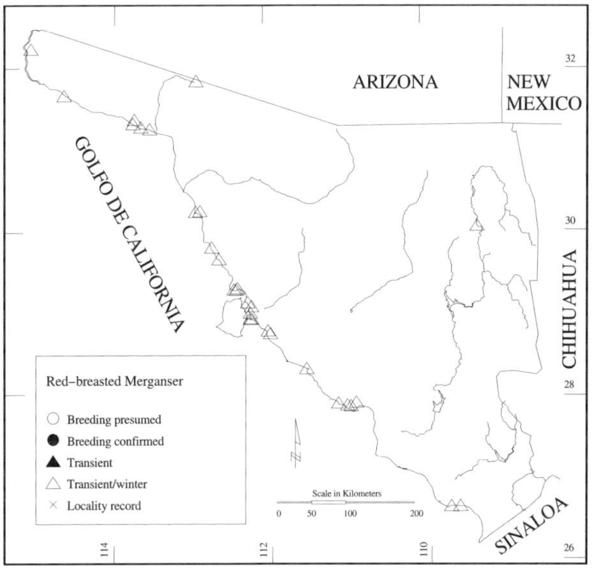

RUDDY DUCK

Pato Tepalcate

Oxyura jamaicensis

Van Rossem (1945) cited only three records of this species in Sonora, but we have found it a fairly common to common transient and winter visitant, chiefly on freshwater reservoirs. Dates of record range from 21 September to 15 May. Few were found on the January aerial coastal surveys (Saunders and Saunders 1981) in 1947–1964, although large numbers were found in Sinaloa. Later annual aerial surveys, 1965–1988 (USFWS 1947–1992), typically disclosed fewer than 1700 Ruddy Ducks in Sonora (that number was exceeded in four years). In 1983, 17 850 were counted. Almost every year larger numbers were found in Sinaloa and Nayarit, a combined maximum of 55 000 in 1985. Numbers fluctuated widely from year to year, even more so than those of other waterfowl species. The largest group observed inland was about 1000 near Naco on 18 April 1986 (A. Moorhouse), at which time the birds ostensibly were migrating.

The only nesting record of the Ruddy Duck in Sonora is of 12 downy young with adults at the Puerto Peñasco sewage ponds on 28 April 1994 (N. Boyajian). In the absence of summer records for Sonora, a flock of at least 110 at Presa Rodriguez, Hermosillo, on 29 June 1959 (A. Phillips) is of exceptional interest.

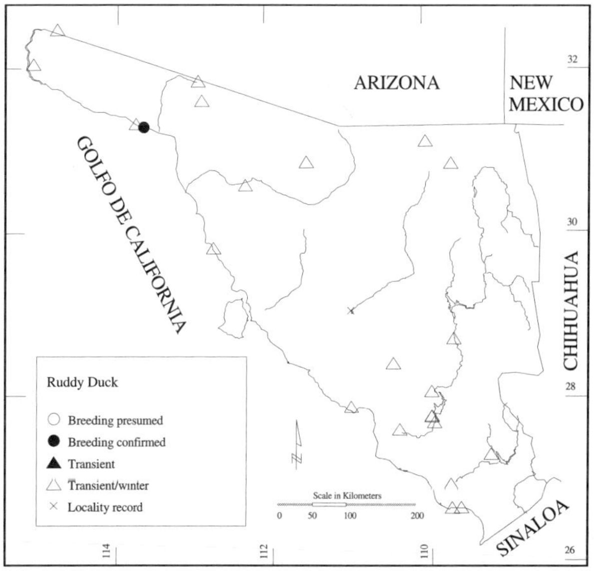

FAMILY CATHARTIDAE

American Vultures

BLACK VULTURE

Zopilote Común

Coragyps atratus

El Zopilote, the Black Vulture, is a common and familiar sight in Sonoran skies, as it alternately flaps its wings and glides to gain elevation so it can circle with others in a "kettle," usually near villages and cities where it feeds at garbage dumps or abattoirs. Or it is seen clustered in numbers about a dead cow or burro or, if evening is approaching, roosting in groups in the tops of large trees along permanent streams.

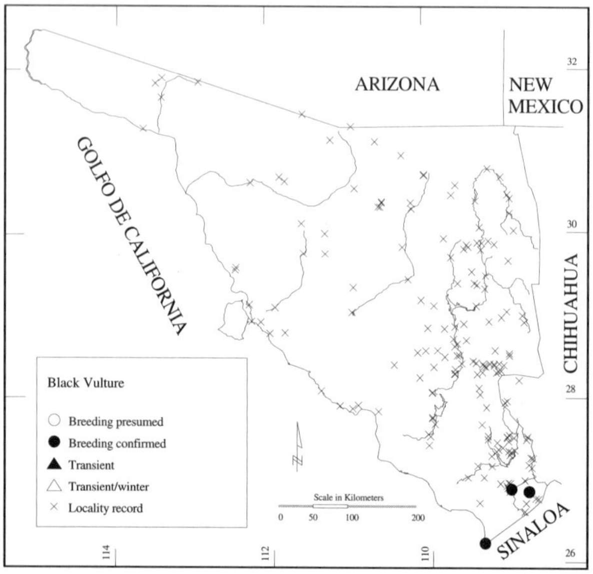

Lowlands under 1500 m harbor the most Black Vultures. At higher elevations they are widely distributed but in smaller numbers. They are resident where found and are typically associated with pueblos and farmland. Along the coast and elsewhere, they are numerous in the vicinity of poultry farms, where the tops of nearby giant cacti are whitened by their excrement. They are scarce along the coast from Punta Chueca northward. Five nests (WFVZ) are known; they contained eggs or small young in late April and the first half of May. Three of the nests were from an island in Estero Agiabampo.

Van Rossem (1945) remarked that the Black Vulture must have extended its range during this century, as early explorers did not encounter it until they arrived at tropical localities. The species did not extend into Arizona until the 1920s.

A. Phillips has informed us, "Van Rossem's notes mention that the Black Vultures have good night vision; they flew to inspect him at night near Empalme in February 1946. Similarly at the roost [20 km north of Bavispe] these birds flew and maneuvered very well in a dark cliff when we shined a light at them." This vulture uses its eyes to find food, not olfaction, as does the Turkey Vulture.

TURKEY VULTURE

Zopilote Aura

Cathartes aura

Except in the far northeast in the winter, one can expect to see a Turkey Vulture or two in the sky any time of year, drifting along close to the hillsides or sailing effortlessly in a thermal current. Now and then groups of up to 20 or 30 are seen, especially during spring and fall when migrants are in transit. On occasion one or more will lift off the highway at the approach of a car, leaving a road-killed jackrabbit or other animal. No matter how hot and dry the countryside may be, Turkey Vultures course over it in their search for food. Turkey Vultures also gather with Black Vultures at garbage dumps, poultry farms, and other places where carcasses have been discarded. Migratory flights occur in late February, March, and April, and again in September and October. These semiannual movements are conspicuous in northern Sonora, but we have seen few presumed migrants in southern Sonora. We have made trip counts from Nogales to Navojoa from 1970 to the present, and we do not find evidence that numbers have changed. Turkey Vultures are abundant residents and migrants.

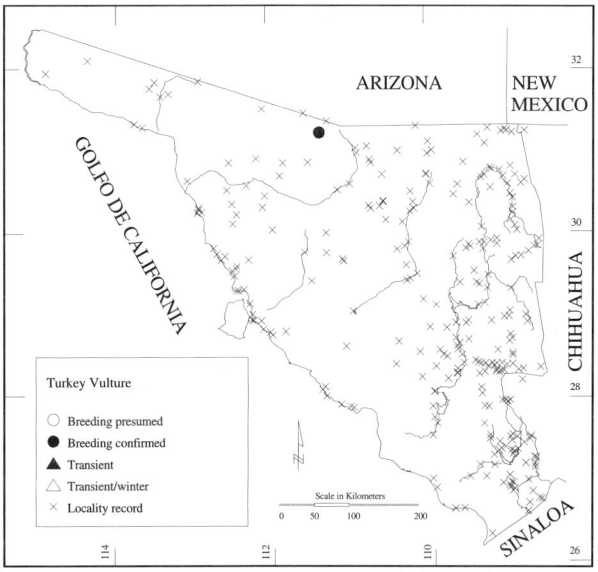

Only one nest of this vulture has been found in Sonora, near Saric in July (van Rossem 1931), an indication of the secretiveness of its breeding behavior and of the rugged nature of the terrain in which it nests. It is likely that Turkey Vultures breed in or near most localities where seen.

FAMILY ACCIPITRIDAE

Kites, Eagles, Hawks, and Allies

OSPREY

Gavilán Pescador

Pandion haliaetus

The Osprey is a fairly common permanent resident along the entire coast of Sonora. It builds its nests on cardóns and saguaros, on utility line posts, and on rocky stacks along the shore. On the basis of aerial and ground surveys made in 1977, Henny and Anderson (1979) estimated there were 124 active Osprey nests in coastal Sonora, none of them south of Guaymas. They found the densest concentrations of Ospreys in the region of El Golfo de California just inland from Punta Sargento, where the birds nest exclusively on cardóns. Nesting is primarily in the period of February through May, although we have observed courtship and copulation in late December. Our data indicate no increase in the number of Ospreys in winter on the coast, indicating the absence of birds from the north.

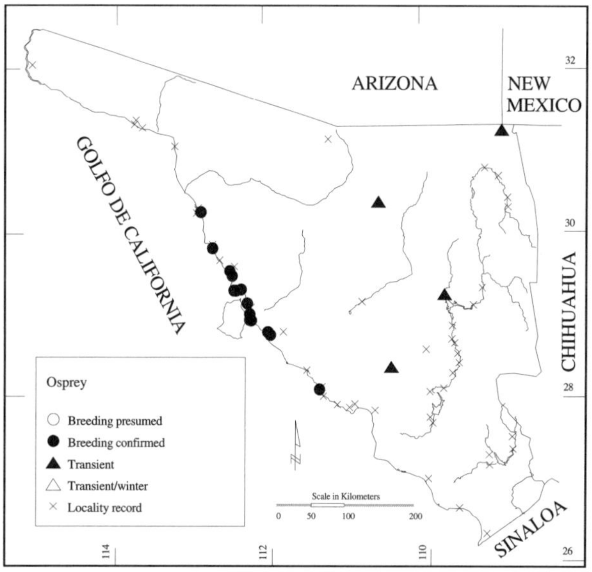

Many Ospreys from North America spend the winter on the Ríos Bavispe and Yaqui and on the upper Río Mayo in Sonora; between 30 December 1985 and 11 January 1986, during a canoe trip down the Río Yaqui from El Novillo to the head of Presa Obregon, B. Brown and colleagues recorded a total of 55 Ospreys. We saw 12 at Presa Agua Caliente on 4 January 1956. Most birds found in the interior, especially in April–May and September–October, are probably migrants. Although there are several records during the summer along the Río Yaqui from Tonichi south, and one at Presa Mocuzari, nests have not been located in the interior.

WHITE-TAILED KITE
Milano Cola Blanca
Elanus leucurus

Only recently has the White-tailed Kite extended its range into Sonora. This is consistent with an expansion into the southwestern United States in the 1970s and into Central America about 1960 (Gatz et al. 1988). The history of this kite indicates that it entered Sonora from the south. The first one was found and photographed near Alamos on 23 March 1979 (D. Stejskal).

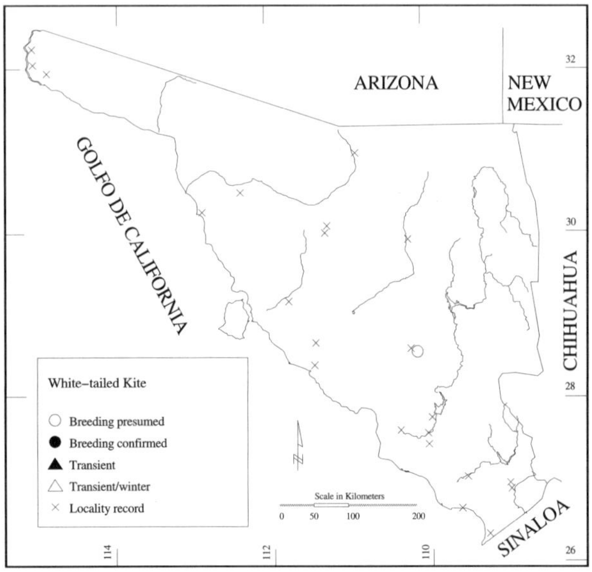

Subsequent records are from all sections of the state, except the far northeast, at elevations below 1030 m. The birds favor farmlands and open grasslands, including those established after brush-clearing operations. Although most records are of single birds, four were between Banámichi and Huepac in the Río Sonora valley on 13 September 1986, three were in grassland west of Tecoripa on 11 March 1986 (A. Moorhouse, D. Danforth), and four were in agricultural land southwest of Hermosillo on 3 January 1993. We found them in the Río Colorado delta, one each near Riito and El Doctor, respectively, on 9 January 1985, as well as one at Ciénega de Santa Clara on 9 April 1994. In all, about 40 birds had been seen in Sonora to mid-1994, none of them in May or June.

Although we think this species has nested in Sonora (a number of nests have been found in adjacent Arizona beginning in 1982), the only evidence is of a pair copulating on top of a tree morningglory 10 km west of Tecoripa on 27 February 1986 (P. Martin).

BALD EAGLE
Águila Cabeza Blanca
Haliaeetus leucocephalus

The Bald Eagle was unknown in Sonora until 2 April 1976, when G. Gregg observed an immature bird in Arroyo Cajón Bonito. Subsequently A. Rea saw two adults about 8 km above Onabas on the Río Yaqui in May 1978. On 2–3 January 1982, B. Brown and P. Warren discovered five adults and two immatures along the Río Yaqui between Soyopa and Onavas (Brown and Warren 1985). Brown subsequently led several eagle surveys of the Río Yaqui from El Novillo to Presa Obregon; of the same river from its formation at the confluence of the Ríos Aros and Bavispe to Presa Novillo; of the Río Aros from the village of Natora downstream 105 km to the Río Yaqui; and of the Río Bavispe from Huachinera to the head of Presa Angostura and from Granados downstream to the Río Yaqui. He even canoed the Río Mayo in February 1989 from its source in Chihuahua to San Bernardo, but he found no eagles along that stream.

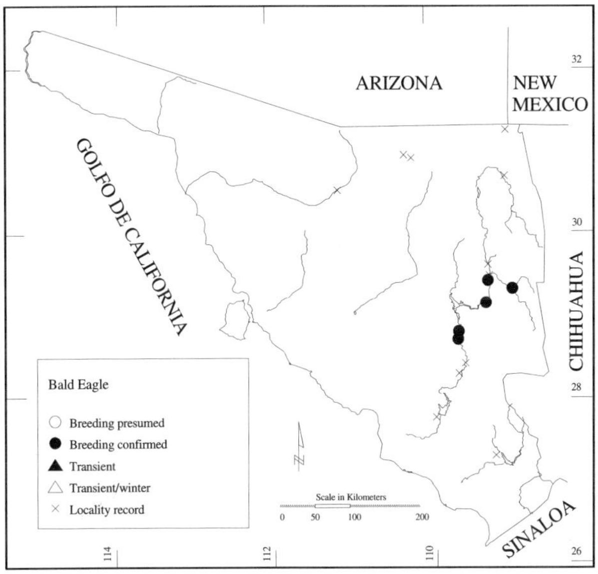

That the Bald Eagles found in 1982, and those seen in following years, were not entirely winter visitants but included permanent residents was ascertained by the discovery and study of active nests in 1986–1990 (Brown, Warren, and Anderson 1987; Brown and Arenas 1988; B. Brown 1988; B. Brown et al. 1990). Five active nests were found: one on a cliff along the lower Río Aros; one on a cliff along the Río Yaqui not far below the confluence of the Ríos Aros and Bavispe; one on a cliff on the Río Yaqui not far above the head of Presa Novillo; and two on the Río Yaqui near Soyopa below El Novillo. Of the last two nests, one was in an hecho and the other in a mesquite that was subsequently burned in agricultural operations. In 1990, a survey that did not include the Río Aros revealed that only the nest near the confluence of the Ríos Aros and Bavispe and the nest on the Río Yaqui not far above the head of Presa Novillo were active.

Aside from the nesting birds in the Río Yaqui drainage, it is possible that some winter birds observed along the Río Yaqui may have been from localities in the United States. Birds were seen in winter at some sites well away from known nests, including an adult on the Cedros arm of Presa Mocuzari on 7 January 1986, a pair of adults on the Río Bavispe just north of Tres Ríos on 9 November 1987 (B. Brown and Arenas 1988), and an adult near Colonia Oaxaca on the great northern loop of the Río Bavispe on 13 February 1988 (B. Brown and Arenas). We are not aware of the basis for the statement (L. Brown and Amadon 1968) that wintering Bald Eagles have been found along the coast of El Golfo de California in Sonora.

NORTHERN HARRIER

Gavilán Rastrero

Circus cyaneus

The Northern Harrier is a fairly common transient and winter visitant in Sonora; a few of those passing through may winter as far south as northern South America. Generally, harriers reach Sonora by early October and are seen until the following April. Extreme records are 21 August and 20 April.

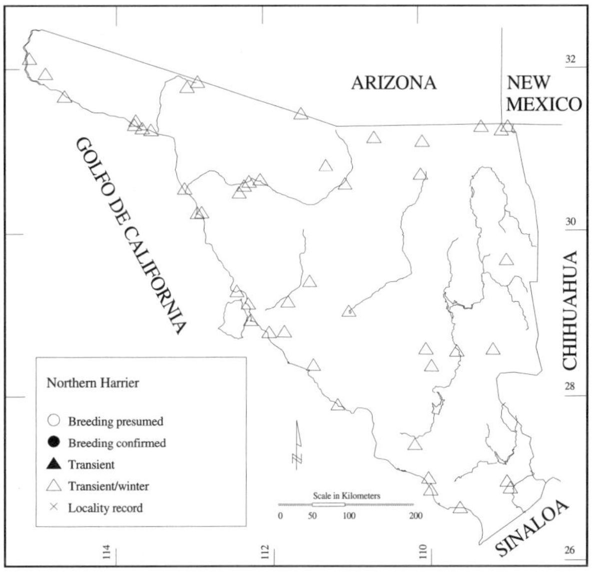

Northern Harriers, also known as Marsh Hawks, fly low in open country, with wings outstretched and tilted upward, effectively using even a slight breeze to help keep aloft. They prefer grassland, mixed shrub-grassland, irrigated farmlands, salt marshes, and estuaries. They range from sea level on the coast to 1600 m in the grasslands of the northeast.

SHARP-SHINNED HAWK
Gavilán Pecho Rufo

Accipiter striatus

This small fast-flying hawk is a fairly common winter visitant and transient throughout Sonora, found most often in tall, dense vegetation. The earliest fall bird was on the coast on 17 September, and the latest spring one was on the upper end of Presa Novillo on 12 April 1988 (R. Mesta).

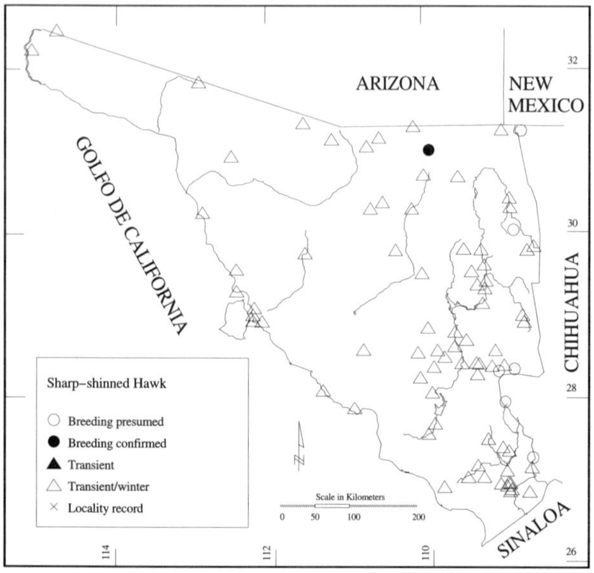

J. Marshall found evidence of breeding on the west fork of the Río Claro (1830 m) in the Sierra los Ajos, where he collected a juvenile male and an adult male on 22 July 1953. He had noted the species earlier in the same locality on 28 May–1 June. The stomach of the adult contained the remains of a House Wren and a juvenile Bewick's Wren. Although Marshall's is the only definite record of the nesting of this species in Sonora, the map indicates other probable breeding areas, based upon the presence of birds in summer at elevations from 1500 to 2000 m.

Field identification of Sharp-shinned and Cooper's Hawks is in some cases difficult, and we have rejected some records.

COOPER'S HAWK
Gavilán de Cooper

Accipiter cooperii

The Cooper's Hawk is a common transient and winter resident everywhere in Sonora, found most often in wooded areas and riparian habitats and less often in open desertscrub. The fall migration is mainly in October; the spring migration mainly in March and early April. This hawk is more common along the coast in fall than in winter and spring.

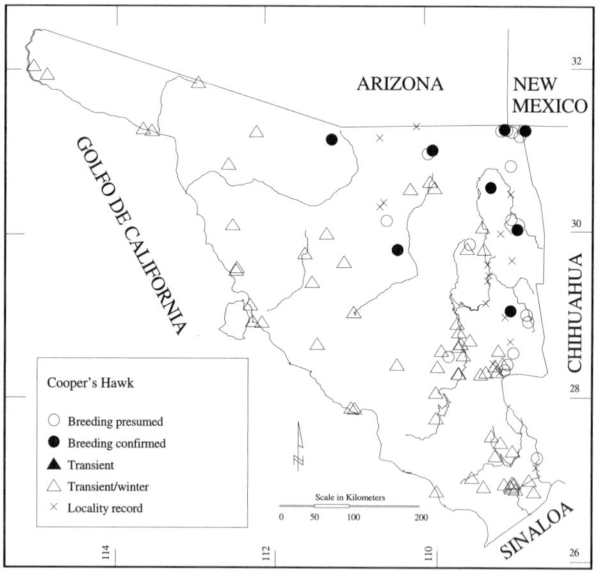

As a nesting bird, it is fairly common and presumably resident in the mountains of the northeast, less so to the south. It also nests along main drainageways where there are sizable trees, which may include large mesquites and chinos as well as cottonwoods and sycamores.

A Cooper's Hawk banded near Ferguson, Nevada, on 28 September 1981 as a hatching year male was found dead at El Novillo on 23 March 1982 (S. Hoffman, pers. comm.).

NORTHERN GOSHAWK
Gavilán Azor
Accipiter gentilis

The Northern Goshawk is found only in the pine-oak woodland and higher: it is a mountain bird in Sonora. We found it to be rare and saw it only from March through July, although it is undoubtedly resident. Observed elevations ranged from 1200 m (in oak woodland at the south end of the Sierra la Madera, 21 July) to 2100 m (Mesa Rodriguez, 9 June). Van Rossem (1945) cited but one record, an adult collected at Yécora on 13 April 1888. Goshawks from north of Mexico would be expected to migrate through Sonora and would be indistinguishable in the field from the breeding population. Russell watched a nest containing fresh green oak and pine twigs and attended by an adult on Mesa de Enmedio on 13 March 1989, but the contents could not be ascertained.

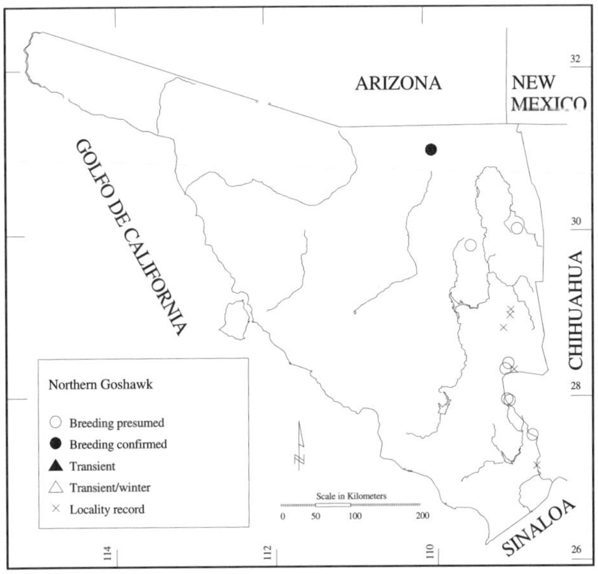

Goshawks feed on Band-tailed Pigeons, among other prey. Marshall (1957) collected a begging juvenile goshawk on 22 July 1953 in the Sierra los Ajos that had its stomach and esophagus bulging with the remains of a Band-tailed Pigeon. At Tabacote in the mountains east of Sahuaripa we watched a goshawk make a spectacular chase of a Band-tailed Pigeon through our camp on 16 March 1984.

CRANE HAWK
Gavilán Zancón
Geranospiza caerulescens

A slender red-eyed raptor with long reddish legs and a small rounded head, the bizarre Crane Hawk is a rare resident of a few localities (80 to 1000 m in elevation)

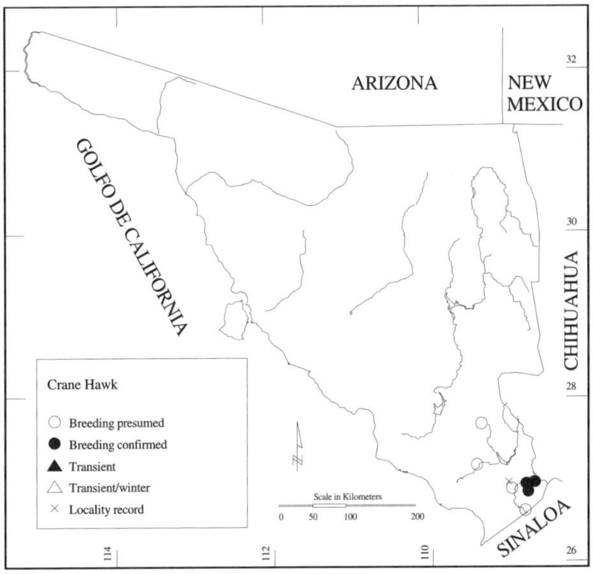

in tropical deciduous forest not far from permanent streams. When found in a tree, it flies reluctantly, and when restless it may pick up one foot, then the other, side-stepping down the branch on which it stands and ultimately crouching before jumping to another branch. The northernmost record is of one seen by J. Bates on a hillside above the Río Cedros about 7 km south of the village of Cedros (4 January 1986).

Five nests (WFVZ) found in the Güirocoba area in May and early June contained one or two eggs or young. All of the nests were situated about 10 m above the ground in Montezuma baldcypresses.

COMMON BLACK-HAWK
Aguililla-negra Menor
Buteogallus anthracinus

Common Black-Hawks are seldom seen away from streams with permanent water supporting cottonwoods, sycamores, or Montezuma baldcypresses. They habitually perch well-concealed in dense foliage and are difficult to observe; once on the wing they are excellent fliers. Their food consists of amphibians, reptiles

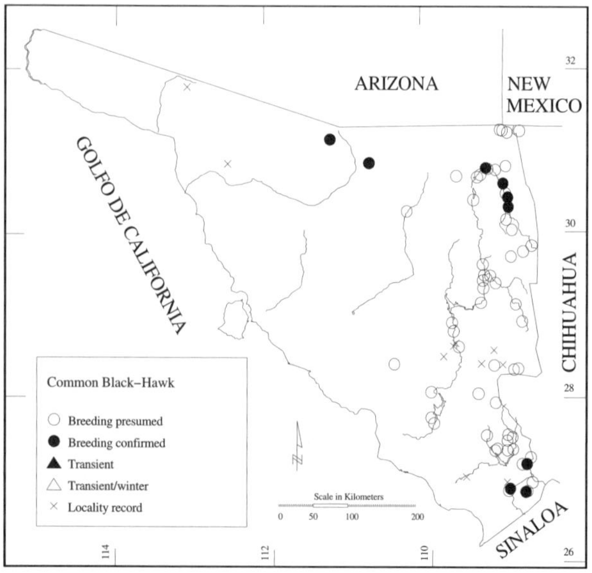

that frequent aquatic habitats, fish, and occasionally birds. This black-hawk is fairly common to common at elevations to 1600 m, but its range is being threatened by the destruction of large native trees to make way for ever-increasing expanses of agricultural land. It breeds from mid-April through July, placing the nest in the crown of a tall tree.

The Common Black-Hawk migrates, being found regularly in winter only as far north as the head of Presa Obregon on the Río Yaqui, and Quiriego and San Bernardo in the Río Mayo watershed. Numbers are no greater in this area in winter than they are in summer. Migratory Sonoran birds may leave the state entirely. Farther north, where it is a summer resident only, this hawk arrives in late February or early March, and most birds leave before the end of September.

GREAT BLACK-HAWK
Aguililla-negra Mayor
Buteogallus urubitinga

This hawk has been found only five times in Sonora, always along streams with large trees. Two specimens were taken at Alamos on 8 March 1888 (van Rossem 1945), and since then it has been recorded on the Río Cuchujaqui southeast of Alamos (28 November 1975,

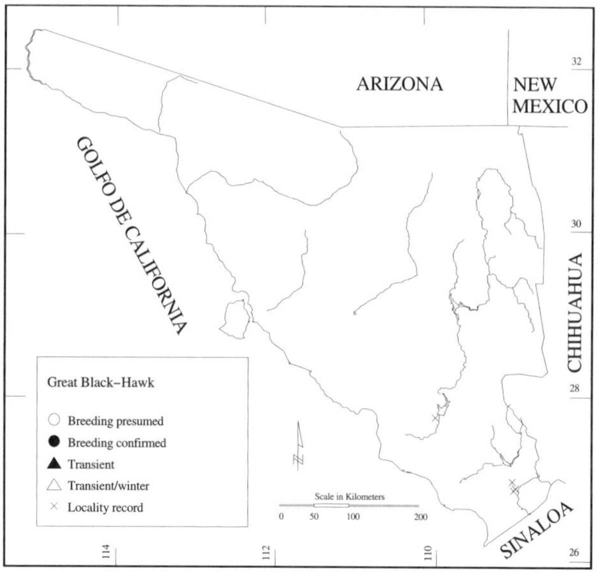

G. Gregg, D. Danforth et al.; about 23 November 1979, D. Stejskal; and on a 29 December 1978 Christmas Bird Count). Another was found along the Río Yaqui below Presa Obregon on 11 April 1986 (Russell). There is no evidence of breeding. The widely distributed Common Black-Hawk is often misidentified as a Great Black-Hawk, thus reports of the latter species must be carefully documented.

HARRIS' HAWK
Aguililla Rojinegra
Parabuteo unicinctus

This dark-chocolate hawk with vivid white in the tail is a fairly common resident below 1000 m almost throughout desertscrub and tropical thornscrub

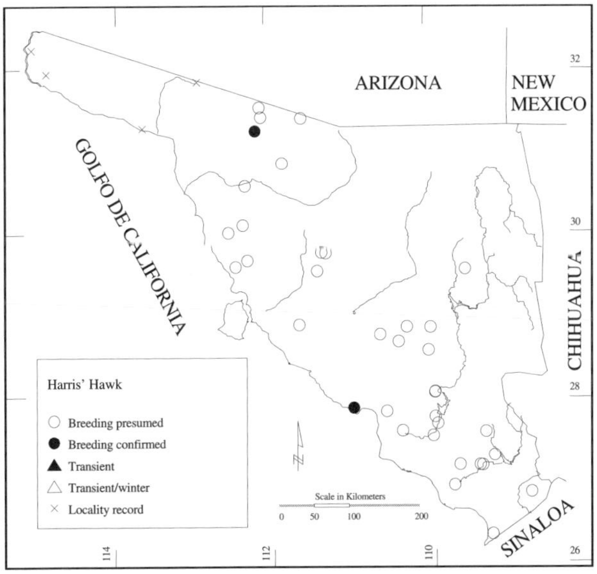

communities. It was common in the Río Colorado delta before large dams were built; except for one we saw near El Doctor on 9 April 1994, there are no recent records from the delta. The usual nest site is a large cactus—saguaro, cardón, or hecho—or desert tree.

Harris' Hawks are often found in social groups of three to six or more, which act as hunting units. During the nesting period, pairs may be augmented by an extra bird which shares in the rearing of the young.

SOLITARY EAGLE
Águila Solitaria
Harpyhaliaetus solitarius

The Solitary Eagle is a rare resident of extreme southeastern Sonora and is perhaps no longer present. Harrison and Kiff (1977) described the finding of two nests: one in 1947, the other in 1958. The first was in a

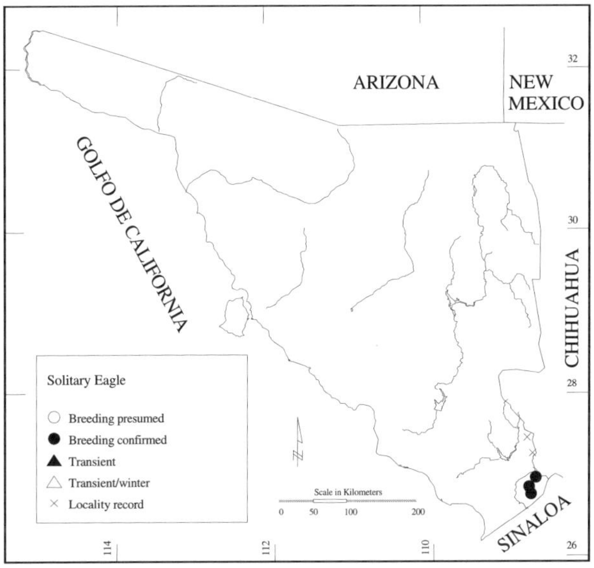

ponderosa pine about 8 km east of the village of El Mirasol; a female parent was collected and the single eaglet photographed on 7–8 June. The second nest was in a huge fig tree in the bottom of a canyon about 15 km southeast of Güirocoba at an elevation of 1000 m; the male parent and the single egg (WFVZ) were collected on 9 May. A third nest was found near Güirocoba on 10 May 1949. It contained a nestling and one egg; the adult male was collected (LSUMZ). In addition to these records, an adult male was taken at Los Algodones on 1 March 1948 (Harrison and Kiff 1977).

Because the Solitary Eagle is such a rare bird anywhere in Mexico, and because southern Sonora is at the extreme northern tip of its range, any Sonoran record should be thoroughly documented. A very large raptor, possibly a Solitary Eagle, was watched by several of our colleagues (Bates, Moorhouse, Suter, Weatherly) over Barranca las Colas (1400 m) near Sahuarivo on 17 July 1987. Other than this observation, there have been no reports since 1958.

Gray Hawk

GRAY HAWK
Aguililla Gris
Buteo nitidus

This confiding, small, agile buteo is common in much of Sonora to an elevation of 1300 m. Its frequently uttered and characteristic shrill cry often attracts attention. It eats mostly small reptiles and large insects.

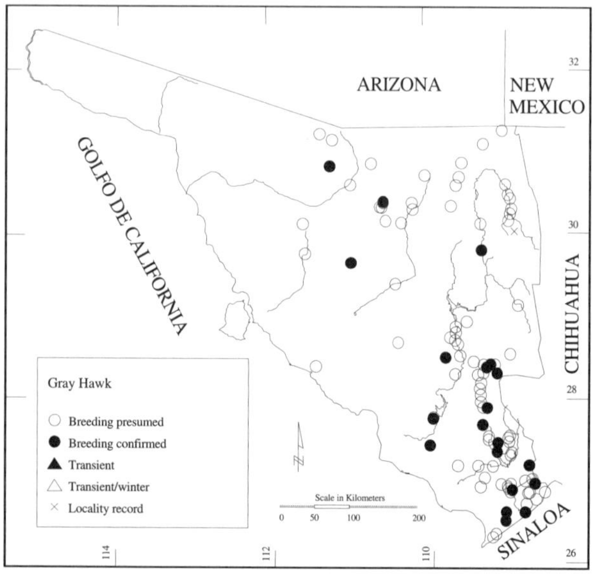

Nests in northern Sonora are typically in trees along permanent streams, usually in large cottonwoods but occasionally in large ashes and walnuts. From the Trincheras area southward we have found the principal nest tree to be the chino, which generally grows along watercourses that are dry almost year-round. In the southern and more eastern parts of the tropical thornscrub, usually along streambeds that contain at least a little permanent water, nest sites include other large trees, particularly Montezuma baldcypress. Gray Hawks may nest in oaks and in oak–mesquite savanna, as at Rancho Jacalón (650 m) about 20 km northeast of Rosario. At Rancho Jacalón not only were two nests in

Chihuahuan oaks less than 0.8 km apart, both holding young on 11 June 1984, but one nest was adjacent to the occupied ranch house. The two or three eggs are normally laid in May and young fledge in July (coinciding with the onset of the summer rainy season), but earlier nests have been observed.

Our records indicate that Gray Hawks migrate southward in September and return in early March. One may occasionally be found in winter in the valley of the Río Yaqui as far north as Tonichi. There are also several winter records from the Alamos region.

The AOU (1997) has placed the Gray Hawk in the genus *Asturina;* the name of the species becomes *A. nitida.*

SHORT-TAILED HAWK

Aguililla Cola Corta

Buteo brachyurus

The Short-tailed Hawk appears to have only recently extended its range northward into Sonora. Monson observed the first one soaring over the outskirts of Alamos on 21 March 1982. The next birds were found in 1986, when we noted two on 16 May over the Arroyo Banderita northeast of Tepahui, one on 19 May at El Terreito northeast of San Bernardo, one on 11 July near Quiriego, one on 12 July near Sejaqui, and one on 15 July near Presa Mocuzari. In 1987 and 1988 one or two birds were found at three additional locations in May, July, and August. One bird was near Nacozari in May 1995.

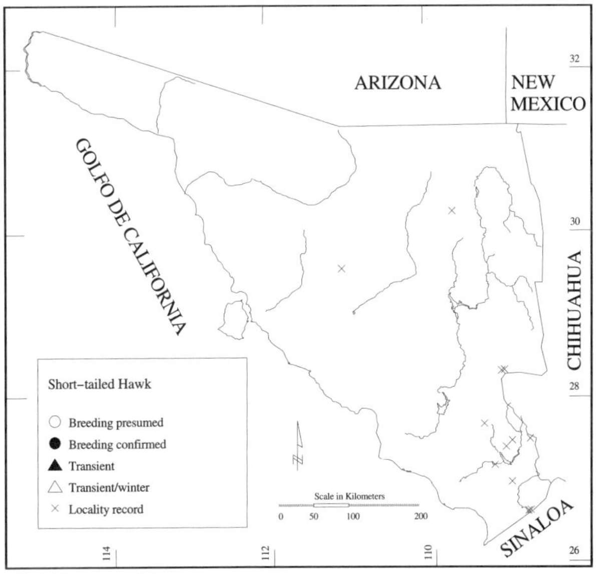

The eleven localities reflect an altitudinal range from 125 to 2000 m and habitats from tropical deciduous forest to pine-oak in the mountains. Although we have seen two individuals at once, there is no evidence of breeding. Nine of the 13 individuals have been dark phase birds.

SWAINSON'S HAWK

Aguililla de Swainson

Buteo swainsoni

The Swainson's Hawk is an uncommon and local summer resident of the more grassy plains and wide valleys of north-central Sonora. Its current limited number is contrary to earlier observations by van Rossem (1945) and Neff (1947), who described it as common everywhere. Our summary of its presumed breeding range is based on records from mid-May to early August, as we have only two records of nests. One was found on 5 August 1983 in a blue palo verde near El Arpa, about 37 km west-northwest of Carbo (S. and J. Levy). The nest held two or three young about two weeks old. Neff (1947) found a pair at a nest in the same general area on 20 June 1942. This appears to be unusually late nesting; perhaps this species in Sonora times its nesting to match the summer rainy season.

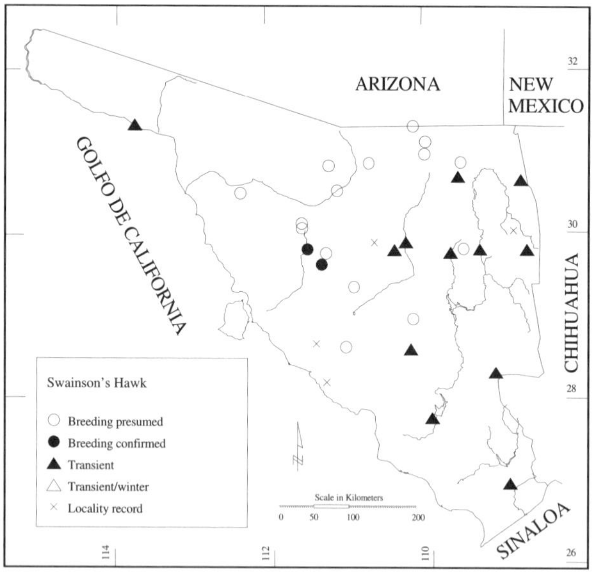

Individuals from breeding areas to the north of Mexico are transients through Sonora, but they pass through in ones and twos and are never conspicuous. The earliest records are of one near Alamos on 20 March 1982 (D. Stejskal) and one high over the Sierra Madre village of Mesa Tres Ríos on 21 March 1985. The spring migration through the state seems to last until about the middle of April. Fall migration is probably under way about 20 August. The latest acceptable record is of a bird seen near Huepac in the Río Sonora valley on 14 September 1986.

Russell saw an adult bird in agricultural lands 40 km west of Hermosillo on 19 June 1974 and another at San Miguel about 45 km northwest of Guaymas on 11 June 1984. We doubt that the species breeds this far south in Sonora.

WHITE-TAILED HAWK

Aguililla Cola Blanca

Buteo albicaudatus

The status of the White-tailed Hawk in Sonora is puzzling. An adult male was collected near Alamos on 21 October 1934 and three adults were seen between Empalme and Potam on 12 May 1937 by van Rossem (1945). Neff (1947) saw at least four daily at Agua Caliente about 70 km east-northeast of Guaymas in the last week of May 1942, and one was collected there on 6 May 1945 (CMNH). Alden (1969:26), writing of travel from Hermosillo to Guaymas, stated, "White-tailed Hawks vie with the Red-tails, Harris' Hawks, and Swainson's Hawks for the tops of cacti and telephone poles." Short (1974) saw several over fields near Alamos on the road to Navojoa on 29 July and 1 August 1971.

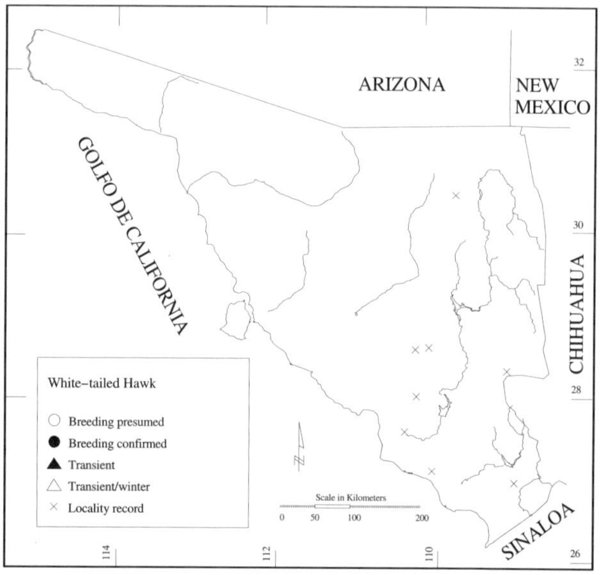

More recently, an adult was seen flying over oak woodland about 20 km north of Nacozari on 29 April 1978 (Clark 1984). A single bird was photographed near San José de Pimas on 18 August 1984, and another was seen there on 10 June 1986 (both by A. Moorhouse).

One was observed on Mesa de Enmedio over conifers on 3 April 1986 (S. Levy), and another was at Alamos on 5 April 1987 (J. Coons, R. Bowers). The authors have yet to see one in the state.

We conclude that White-tailed Hawks are not as common now as they were 20 to 30 years ago, but we do not know enough about them to speculate on causes of the decline. Seasonally, records extend from 3 April to 21 October, as given above. No nesting records exist.

ZONE-TAILED HAWK
Aguililla Aura
Buteo albonotatus

The Zone-tailed Hawk is fairly common at elevations up to 2100 m in the pine-oak woodland and along permanent streams with tall deciduous trees (as exemplified by Arroyo Cajón Bonito in the northeast). It is uncommon throughout most of the lowlands, except in the extreme northwest, where it is absent. Soaring birds may be seen over the desertscrub as well as over the higher mountains, canyons, and foothills.

Usual sites for nests are tall cottonwoods, sycamores, and pines. Nests in the lowland desert have been found in palo verdes, palms, and cottonwoods.

Zone-tailed Hawks are migratory and are largely absent from Sonora in winter; outside dates are 29 September and 12 March. There are few winter records. Decidedly extralimital records are one immature at Puerto Libertad on 25 November 1977 (J. Barlow) and one at San Ignacio near Magdalena on 27 January 1980 (Terrill 1981).

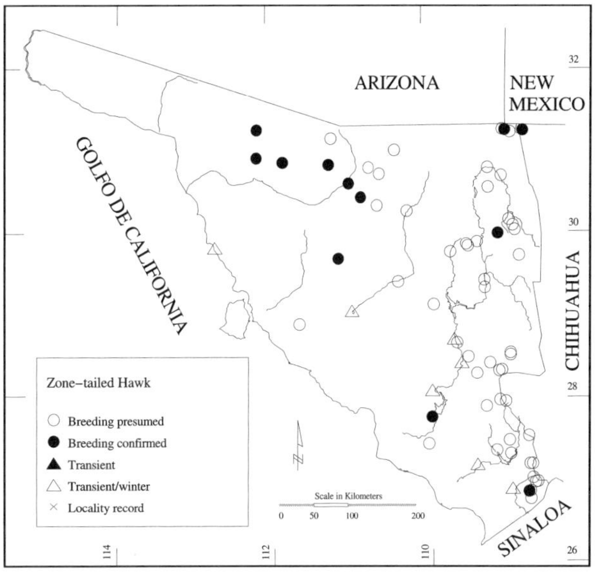

RED-TAILED HAWK
Aguililla Cola Roja
Buteo jamaicensis

The Red-tailed Hawk is the most common and conspicuous raptor in most of North America, including much of Sonora, where it ranges from the mountaintops of the east to the central and northern shores of El Golfo de California. It builds its nest in a variety of trees, including large cacti, and on rock ledges. Successful nests on the ground in the lee of large sand dunes south of the Sierra el Rosario, the northwesternmost (albeit small) range of mountains in Sonora, were photographed in March 1976 and again in mid-March 1992 (Cowgill 1976, 1992). The nesting season runs from about mid-March to late May. We have no evidence that this hawk nests in the coastal plain south of Bahía Kino; indeed, the species is rare even in the southern Sierra Madrean foothills, except as transient and wintering birds.

Migrants are at peak numbers in March and October. In the winter, birds visiting from the north augment the numbers of residents considerably. At that season, one may see many hawks perched on utility poles and electric power transmission towers; almost all are Red-tailed Hawks.

The Red-tailed Hawks in Sonora, as in much of western North America, vary considerably in the color and extent of their dark markings. They are sometimes accused of killing poultry, but they rarely do so. Their usual prey consists of small mammals such as cottontails and ground squirrels.

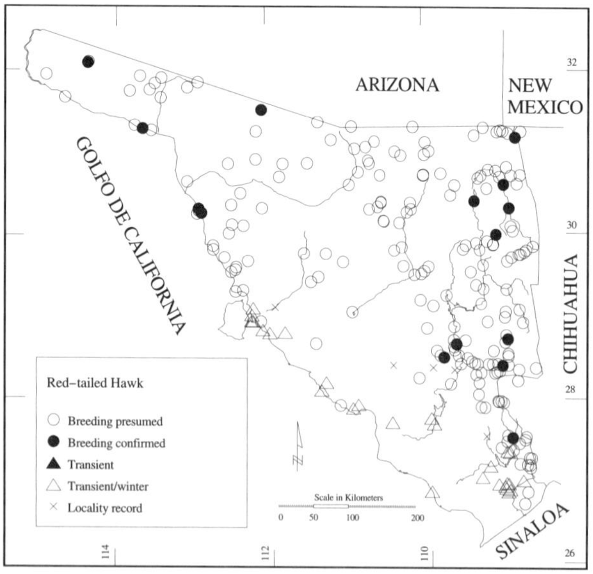

FERRUGINOUS HAWK

Aguililla Real

Buteo regalis

A large, pale buteo that favors open country, especially farmland, the Ferruginous Hawk is found infrequently (about 30 records in all) in northern Sonora as a winter visitant. The first record for the state came from the sandy arid desert of the extreme northwest, along Mexico Highway 2 about 60 km east-southeast of San Luis Río Colorado on 25 March 1952 (A. Halloran). Most birds have been seen at Puerto Peñasco (where the Ferruginous Hawk is now regularly seen) and the agricultural area southwest of Hermosillo (Distrito de Riego de Hermosillo). Seasonal extreme dates are 6 October and 29 March.

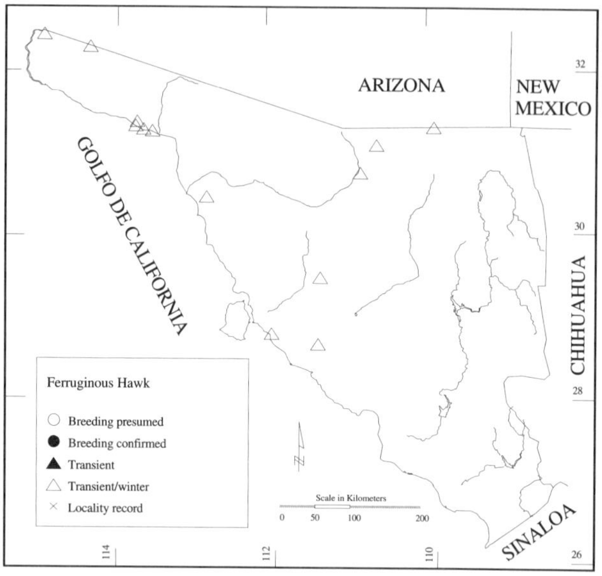

ROUGH-LEGGED HAWK

Aguililla Ártica

Buteo lagopus

The Rough-legged Hawk is a rare winter visitant, found so far only in the extreme northern portion of eastern Sonora. It was first found on 3 December 1953 when one was seen near Fronteras (A. Phillips). Three records have come from near Cananea, from 1973 through 1986, and one was from about 21 km south of Nogales, where two adults were noted on 6 January 1973 (P. Alden). Dates range from 3 December to 17 February.

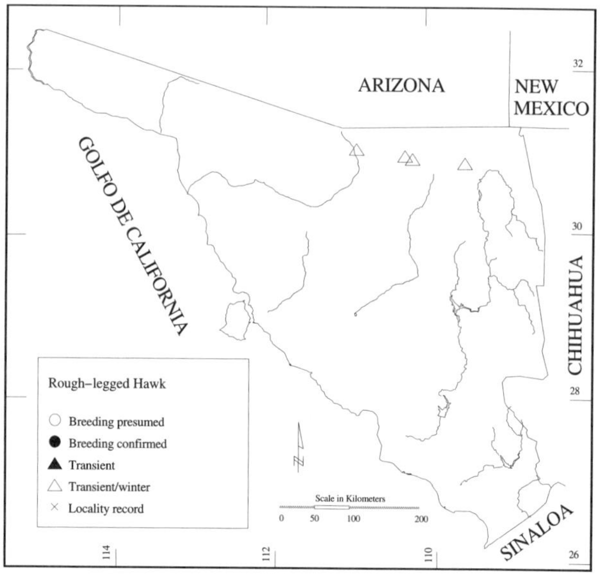

GOLDEN EAGLE
Águila Real
Aquila chrysaetos

Surprisingly few Golden Eagles are seen in Sonora, where we consider the species a rare permanent resident of mountainous country. Numbers may have declined during this century due to poisoning and trapping by ranchers who hoped to protect their calves. In recent years, none has been discovered in the Sierra Madre south of the Sierra San Luis. In adjacent southern Arizona, where the Golden Eagle is protected, numbers are such that it is considered uncommon rather than rare.

A cliff nest with one large young was located in Arroyo Cajón Bonito on 3 June 1975 (R. Glinski, Monson); the same nest was also in use on 21 May 1982 (R. Russell).

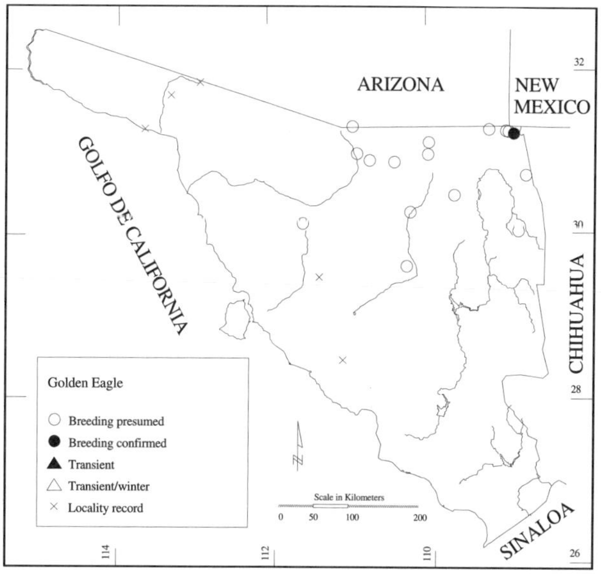

FAMILY FALCONIDAE
Caracaras and Falcons

CRESTED CARACARA
Caracara Quebrantahuesos
Caracara plancus

Caracaras are fairly common to common residents throughout much of Sonora below 1000 m. As van Rossem (1945) noted, they and Black Vultures are often found in numbers around human habitations, where they consume refuse. We have never observed a caracara pursue and capture any bird. Caracaras are more scavengers than predators; prey that is captured is apt to be of a sluggish kind.

Considerable gatherings of these birds may sometimes be seen. As many as nine were in one locality a few kilometers west of Presa Obregon on 21 September 1985; ten were following a road-working machine near Quiriego on 6 January 1986; ten were seen at Piojillo at the head of Presa Obregon on 10 April 1986; 12 were found along the Río Cuchujaqui

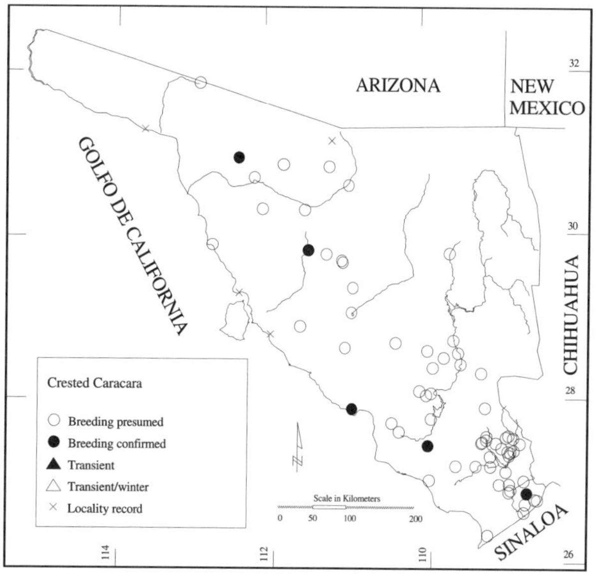

Crested Caracara

near Alamos on 13 March 1982 (D. Stejskal); and 20 were near Tepahui on the Río Cedros arm of Presa Mocuzari on 7 January 1986. A. Phillips writes us that on one occasion he counted 50 caracaras in as many kilometers while driving from Navojoa to the Sinaloa line.

Records indicate that caracaras lay their eggs in late April or early May. We found a nest in a saguaro at Félix Gómez that held two half-grown young on 5 August 1986. In our experience most nests are to be found in saguaros, hechos, or pitahaya cacti.

In Sonora, caracaras are invariably known as "Quelele," after the bird's call, and are never called "el Aguila Mexicana." The Crested Caracara has been erroneously referred to as the Mexican national emblem in the literature, even as recently as 1988 (Palmer 1988). Actually, the bird on the flag and the shield of Mexico is an uncrested eagle of no one species; a decree of 14 April 1823 states that the bird of the flag and shield is "el Aguila mexicana." The national bird so depicted is an abstract eagle; the framers of the decree had no knowledge of eagles to speak of, and it is almost certain they had no particular kind of eagle in mind.

LAUGHING FALCON
Halcón Guaco
Herpetotheres cachinnans

As far as we know, the Laughing Falcon is no longer a Sonoran species. Van Rossem (1945) wrote that he saw one near Güirocoba on 2 June 1937; that three pairs were nesting in the same locality on 9 and 11 May 1941 (two adults, one immature, one downy young, and two eggs obtained by W. Sheffler); and that an immature was also at Güirocoba on 16 November 1944. Van Rossem (field notes) saw the species in February 1948 in "Texas Canyon." Subsequently Sheffler collected a female and an egg ready to hatch (WFVZ) at a nest of leaves in a niche of a sheer cliff on 8 May 1949 near Güirocoba. These records constitute all that is known of the history of the Laughing Falcon in Sonora. We do not know the cause of its disappearance. The Güirocoba area has been greatly cleared in recent years, but Laughing Falcons may be found in similar disturbed habitats to the south. A firsthand account of the collection in 1941 of the two adults and downy young, the latter from a cliff nest, appears in a delightful book (Hilton 1947).

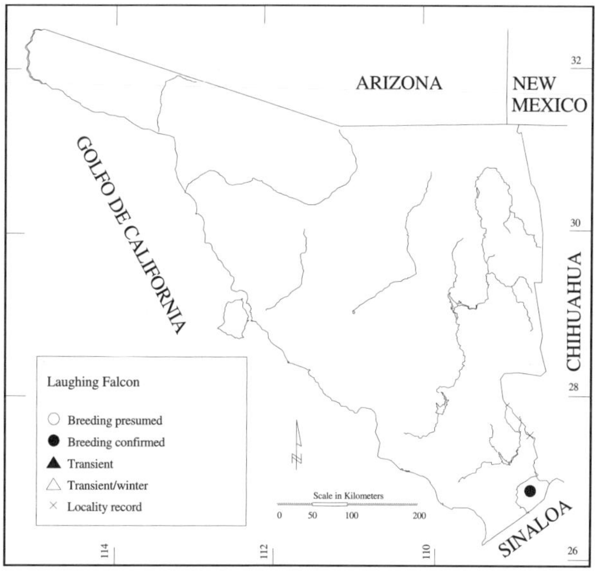

AMERICAN KESTREL
Cernícalo Americano
Falco sparverius

The American Kestrel is a common transient and winter resident and a fairly common breeder. As a winter visitant, it is more common in open desert and agricultural areas than it is in woodland or forests, being found only occasionally in the eastern mountains above 1500 m. As a breeding bird, it is most common in desertscrub, where it nests in old woodpecker holes in saguaros, cardóns, and hechos. Along major water-courses it nests primarily in holes in large cottonwoods. It does breed in pine-oak woodland, but only sparingly. Nesting is in April and May. We do not know if the breeding birds are also residents.

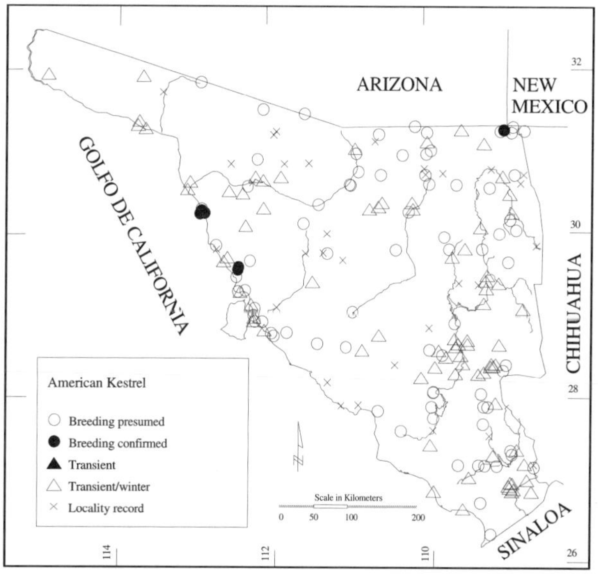

Fall migration is under way in late September. Roadside counts of American Kestrels between Nogales and Hermosillo increase from about 10 birds in summer to 40 and 50 in fall and winter. The spring migration is inconspicuous. An unusual concentration for early summer was a flock of seven in an alfalfa field a few kilometers north of Caleta Venecia on 12 June 1984. Twenty presumed migrants were concentrated in an irrigated field near Caborca on 5 April 1987. Typically kestrels are not in groups.

MERLIN
Halcón Esmerejón
Falco columbarius

This circumpolar small falcon, which breeds in latitudes far north of Mexico but winters south to northern Peru, is a rare transient and winter visitant in Sonora. Observations have been from scattered points the breadth of the state, except in the mountains of the east. Most birds were observed in winter. Extreme dates are 27 April (1957) near Navojoa (J. Marshall) and 3 May (1974) at Punta Chueca (D. Lamm), and 11 October (1973) at Punta Chueca and 16 October (1934) at Camoa near Navojoa (van Rossem 1945).

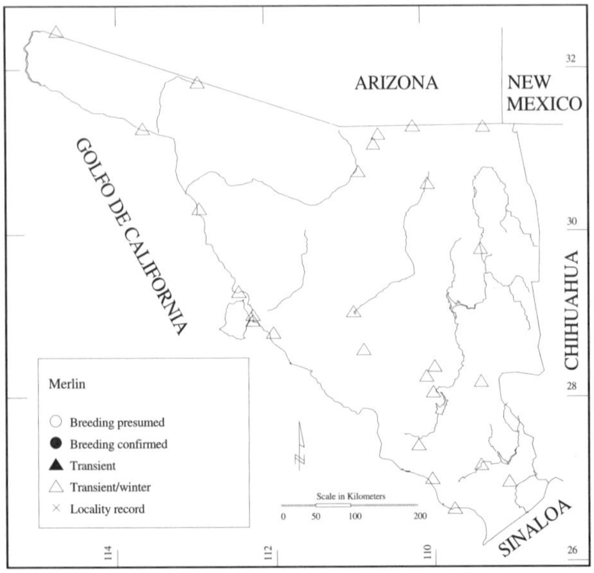

[APLOMADO FALCON
Halcón Fajado
Falco femoralis

There are no documented records for this falcon from Sonora. Alden (1969) claimed it was seen at Guaymas in the summer of 1966, but he gave no supporting details. A report by Gallucci (1981) of an immature seen about 45 km northeast of Imuris on 2 August 1979 is unsubstantiated. The Aplomado Falcon formerly nested in southern Arizona, but there are no acceptable records for the state since 1940.]

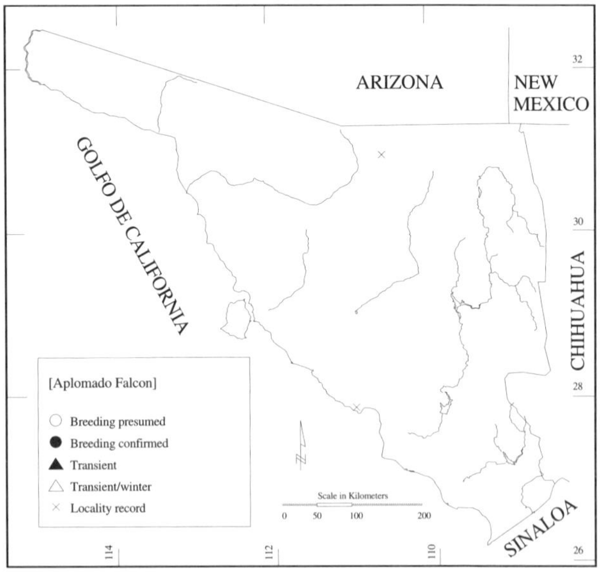

BAT FALCON
Halcón Enano
Falco rufigularis

This small falcon is rare on the Pacific slope of Mexico, and it may be extirpated in Sonora. All Sonoran records are from the period of 1937–1949, when ten were collected (LSUMZ, MVZ, WFVZ) near Güirocoba (May, June, November) and one near Cd. Obregon (November). Two clutches, each of two eggs (WFVZ) almost ready to hatch, were taken on 10 May 1941 and 9 May 1949. In each nest, the eggs were laid on bare gravel in a crevice 15 m above the ground in a cliff face. Another nest contained a seven-day old chick on 4 June 1945.

In the 1940s the Güirocoba area was a farming-ranching community, less populated than today. A small number of Bat Falcons apparently lived very close to the houses and fields. It is possible that the disappearance of the species was hastened through the collecting of adults and eggs in such a small population.

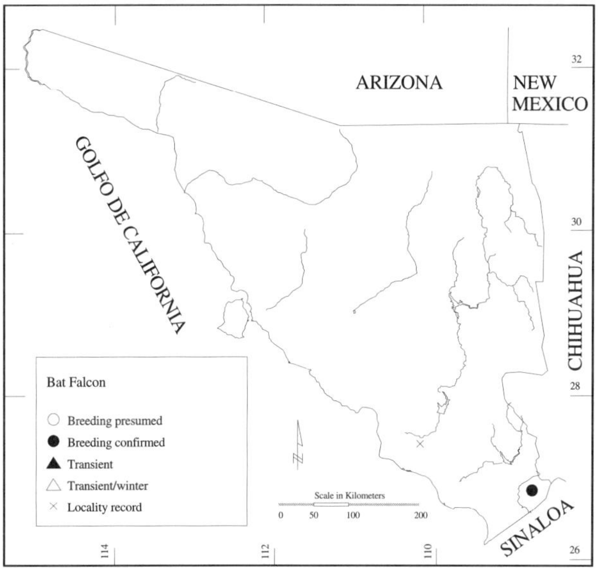

PRAIRIE FALCON
Halcón Mexicano

Falco mexicanus

The AOU Check-list (1983) does not include Sonora in the Prairie Falcon's breeding range, and van Rossem (1945) implies that it is only a winter visitant. We have three May records that may be of late migrants or possibly locally breeding birds: one at El Desemboque at the mouth of the Río San Ignacio on 29 May 1971; one in Arroyo Cajón Bonito on 21 May 1982; one east of Morelos in El Pulpito on 2 May 1977.

It is an uncommon to rare winter visitant from the latitude of Hermosillo north, being seen most regularly along the coast of El Golfo de California from Bahía Kino north, including the Sierra el Pinacate. Outside dates for winter visitants are 10 September 1986 northeast of Cucurpe and 6 April 1990 southwest of Hermosillo.

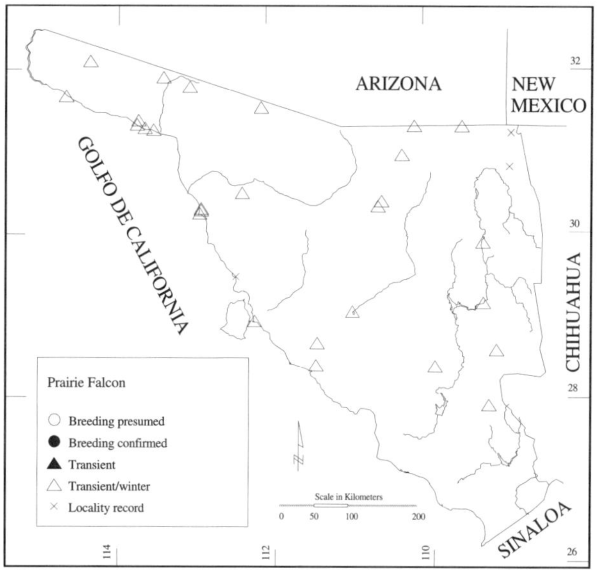

PEREGRINE FALCON
Halcón Peregrino

Falco peregrinus

Van Rossem (1945:63) described the Peregrine Falcon as a "fairly common resident of suitable Gulf islands and probably of the mainland coast." He cited only four inland records, presumably of transients and winter visitants. Although we have not found a nest, B. Brown and colleagues located one on a ledge along the Río Bavispe about 11 km below the mouth of Arroyo Bacadehuachi on 26 March 1987. Two adults were at the nest, which contained young. We have a number of records (including pairs) from suitable habitat from February through August. We think this species is an uncommon resident along the coast, in isolated desert ranges, and in mountainous eastern Sonora.

Finding Peregrine Falcons in the nesting season in eastern Sonora was not entirely unanticipated, as at least seven cliffs with pairs were identified on the east side of the Sierra Madre Occidental about 1975 (Harris 1977). There is every indication that the great barranca country in Sonora, Chihuahua, Sinaloa, and Durango is an important stronghold of the Peregrine Falcon.

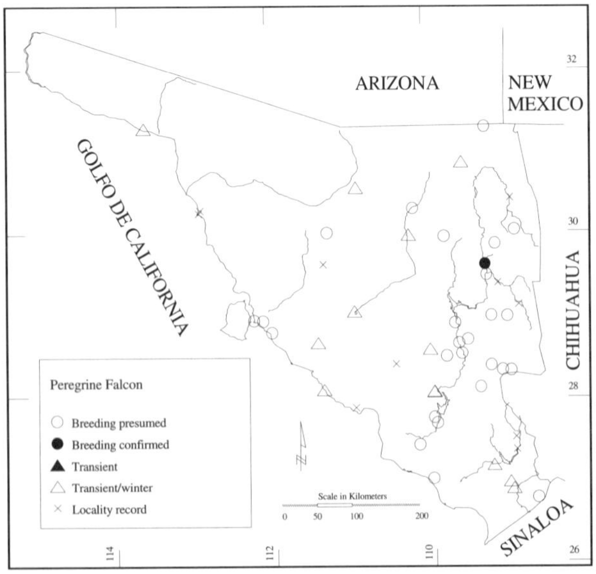

FAMILY CRACIDAE

Curassows and Guans

RUFOUS-BELLIED CHACHALACA

Chachalaca Vientre Castaño

Ortalis wagleri

The Rufous-bellied Chachalaca is an uncommon resident of densely wooded hillsides in tropical deciduous forest. Until 1986, the species had not been found north of the Alamos region. On 5–6 January 1986 we observed a few on the Río Cedros a few kilometers above Quiriego. Then on 14–15 May 1988 we found chachalacas another 85 km farther north, in the east-west running Arroyo el Reparo. We have no reason to believe that this species is extending its range. It is more likely that no one has been in these areas to report them.

Chachalacas are noted for the loudness of their calls. Their ear-splitting cackles break out suddenly and may continue for several minutes before abruptly ceasing. The sound carries for some distance, and the birds seem much closer than they actually are. Although the loud cackling is produced mainly in the courting season, the birds may call at any time of year. Typically, calling chachalacas perch in a tree 5 m or more aboveground well up on a hillside or on a ridge, often silhouetted against the sky in the dry season. At such a time, the forehead crest and the naked red throat are quite visible.

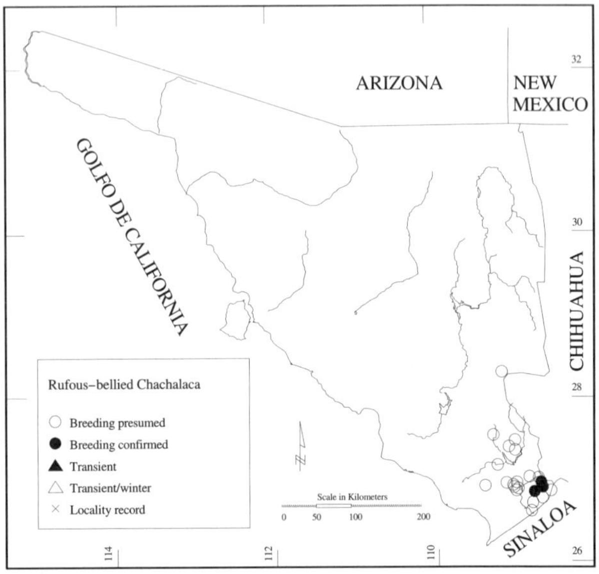

Nineteen clutches (WFVZ), each of three eggs, were all taken in the Güirocoba area from 8 to 15 June. The dates indicate that the chicks hatch in time to take advantage of improved food conditions brought on by the summer rains.

Chachalacas often feed on ripe figs. At El Reparo, we saw two feeding on fruits of *Bumelia persimilis* in May. The pair dropped to the ground and ran up the adjacent hillside. The bumelias are scattered along the arroyo (here more a barranca) bottom, and their fruit is popular with a variety of birds—flycatchers, thrushes, tanagers, grosbeaks, and orioles included. Now and then one will see a domesticated chachalaca in a farm or ranch yard. In such cases, the chachalaca was caught as a small chick, or perhaps the chachalaca egg was brought home to be hatched under a chicken.

FAMILY PHASIANIDAE

Partridges, Grouse, Turkeys, and Quail

[RING-NECKED PHEASANT

Faisán de Collar

Phasianus colchicus

On 9 April 1994, G. Monson, R. Russell, and S. Russell saw a male Ring-necked Pheasant in a wheat field southwest of San Luis Río Colorado. We do not know if the species has become established, but there are several undocumented reports of releases of pheasants in this area.]

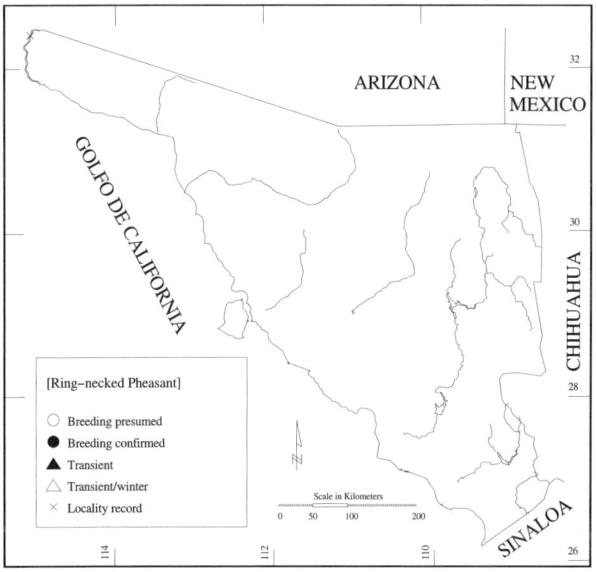

WILD TURKEY

Guajolote Norteño

Meleagris gallopavo

Although there has been overhunting, overgrazing, and heavy logging of the pine-oak habitat preferred by Wild Turkeys, they are locally common in much of this habitat in eastern Sonora, from the Sierra San Luis south to at least the Sierra Oscura. Their gobbling in the spring mornings beginning in early April was a familiar sound at our camps in the pines. Van Rossem (1945) mentioned their presence in mountains in extreme southeastern Sonora. We know of no recent observations of the Wild Turkey in the Sierra los Ajos or the Sierra la Madera.

An indication of their abundance in the Sierra San Luis region was the observation of a flock of 32 (eight adults, 24 young) on 23 August 1985 a few km northeast of Rancho Nuevo in Arroyo Cajón Bonito (A. Moorhouse, S. Schemnitz).

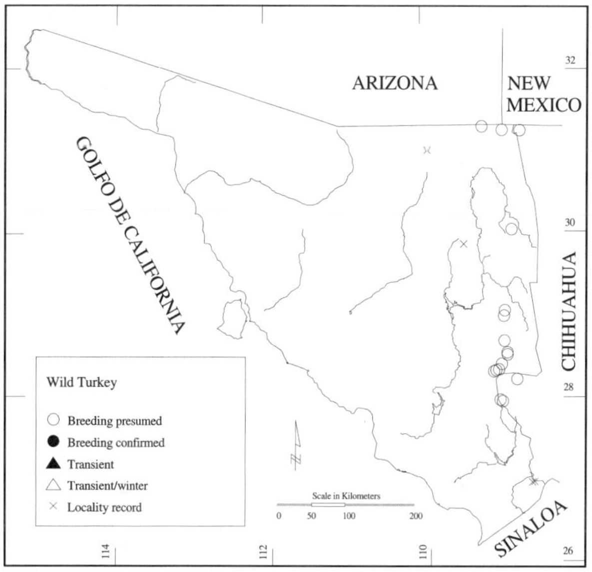

Wild Turkeys in Sonora are the so-called Mexican or Gould's Turkey, *M. g. mexicana,* whose rectrices and upper tail coverts are tipped with white rather than cinnamon or buff.

MONTEZUMA QUAIL
Codorniz Moctezuma

Cyrtonyx montezumae

The Montezuma Quail is a common resident from oak-grassland savannas up into the highest forests of Sonora. Without the aid of a good hunting dog it is difficult to find Montezuma Quail, for they stick to cover and do not fly until they are nearly stepped upon, when they rocket away with low clucking whistles and are speedily out of sight behind the nearest cover. They nest well after the onset of the summer rainy season, in late July and August. We have found newly hatched young in mid-September.

They range primarily between 1200 and 2000 m. There seems to be some altitudinal migration downward for short distances in the winter months. Their food is principally bulbs (mostly *Oxalis*) and the larvae and pupae of insects, the latter being scratched up from oak leaf litter; mast also forms a part of their diet (D. Brown 1989; S. Levy, pers. comm.). They might well be called "oak quail," for they are almost invariably associated with oaks.

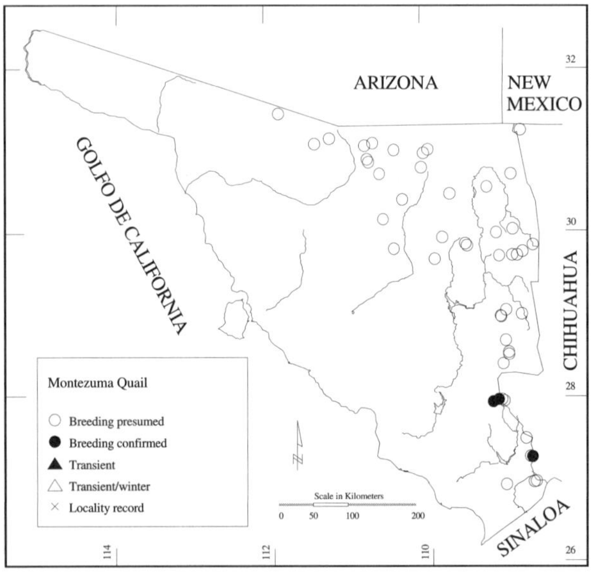

W. J. Schaldach Jr. collected (WFVZ) two males on 2 January 1961 at a locality identified on the specimen label as "18 miles west of Sasabe." This location (820 m), the most northwestern point on the map, is in mesquite grassland far from oaks and is an unlikely area for Montezuma Quail. It is possible the birds were taken closer to Sasabe, or even in the Sierra el Humo about 30 km south, where there are oaks.

NORTHERN BOBWHITE
Codorniz Cotuí

Colinus virginianus

The Northern Bobwhite has a disjunct subspecies that exists in the wild state only in Sonora. This is the Masked Bobwhite, the name commonly used for the subspecies *C. v. ridgwayi* (Spanish, Codorniz Mascarata).

The best life history summaries for the Masked Bobwhite are those of Tomlinson (1972a, 1972b). From them we learn that the birds once ranged from the Santa Cruz and Altar valleys in Arizona southward to within 50 km of Hermosillo; in parts of the valleys of the Ríos Sonora and Moctezuma; and in the drainages of the Ríos Matape, Tecoripa, and Agua Caliente.

By the 1960s, the known range of the Masked Bobwhite had been reduced to small areas between Santa Ana de Magdalena and Hermosillo, a single ranch near Mazatán, and possibly the Valle de Agua Caliente. It was in this decade that Tomlinson carried out his life history studies. In the 1970s, bobwhites could be found only in two restricted areas: south-southwest of Benjamin Hill and to the west-northwest of El Oasis.

Northern (Masked) Bobwhite

They have survived in these two areas into the early 1990s (S. Dobrott [pers. comm.] and W. Kuvlesky Jr. [1994], both of the U.S. Fish and Wildlife Service; Garza-Salazar et al. 1992). As many as 200 Masked Bobwhites were counted in the winter of 1991–1992 on Rancho el Carrizo south-southwest of Benjamin Hill. In the winter of 1993–1994, it was estimated that almost 1500 birds occupied the 1600 ha of the ranch judged to be suitable Masked Bobwhite habitat (Kuvlesky 1994).

Survival data for other ranches, such as Rancho Sausa adjacent to Rancho el Carrizo, and Ranchos el Arpa and Grande west-northwest of El Oasis, are not available, but bobwhites were still present on Sausa in 1992, on El Arpa in 1991, and on Grande in 1988 (S. Dobrott, pers. comm.).

Gustavo Camou of Rancho el Carrizo is keenly interested in the survival of the Masked Bobwhite and has made himself an authority on the birds. He gathers information on their habits and population and has instituted a combination of management practices (including discing, seeding of food and cover plants, and limiting grazing by cattle) designed to benefit the birds (Dobrott, pers. comm.; Kuvlesky 1994).

In the Rancho el Carrizo and Rancho el Arpa localities, the semi-open Sonoran desertscrub consists of ironwood, palo verde, mesquite, hackberry, and *Acacia angustissima,* with some grass, mostly false grama and Rothrock grama. The false grama resists overgrazing because of its low growth form. Unfortunately, in this general area much native vegetation has been bulldozed away, and the clearings seeded to the exotic buffel grass. This grass establishes itself densely, and the resulting monotonous plant cover is not acceptable to bobwhites.

Stocking of birds on ranches east of Hermosillo in 1980–1982 (Brown and Ellis 1984) was unsuccessful. No birds were located when these ranches were

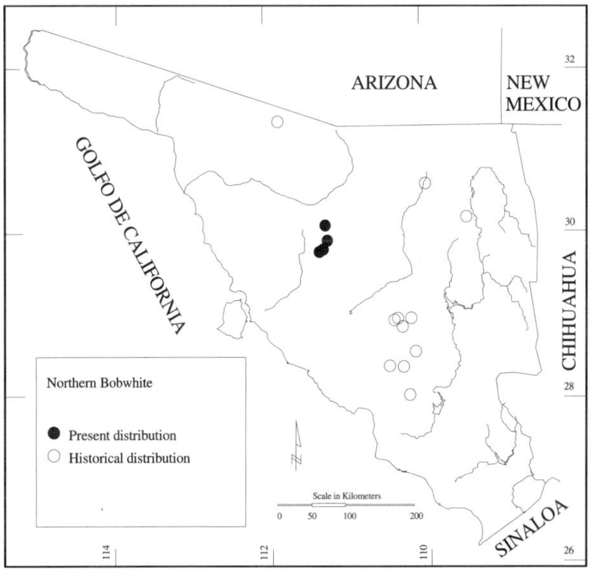

surveyed by Dobrott and colleagues in 1990.

The disappearance of Masked Bobwhites in Sonora can be attributed to persistent overgrazing. The original excellent cover of grasses was replaced by eroded topsoil, gullies, and invading woody plants. An indication of the bobwhites' former abundance is the trapping of 130 near San Marcial in 1938 by J. S. Ligon.

The peak of calling by male Masked Bobwhites is not reached until mid-August, well after the commencement of summer rains. The peak of hatching, according to Tomlinson (1972b), is the middle third of September. Tomlinson (1972b:305) commented that "when most birds have concluded their nesting season, and migratory birds are in the midst of their southern travels, masked bobwhites are just beginning to raise their broods. Although the short nesting season is a liability in some respects, the peak of hatch is ideally timed to exploit the invertebrate foods, seeds, and green feeds which are then abundant."

The Masked Bobwhite was the object of unsuccessful efforts over the years to restock the species in Arizona with birds trapped in Sonora. Not until the passage of the first endangered species legislation in the United States in the 1960s did funds and personnel became available to study the Masked Bobwhite on its much reduced range in Sonora and to develop techniques for successful restocking in Arizona (Brown and Ellis 1984). Following nearly 15 years of research, the large Buenos Aires National Wildlife Refuge was established in 1985 in southern Arizona as an adequate place for Masked Bobwhite reintroduction and propagation, using descendants of birds captured in Sonora. Only time will tell if a self-sustaining population will result.

SCALED QUAIL

Codorniz Escamosa

Callipepla squamata

The Scaled Quail's range in Sonora is confined to a narrow strip along the northern border. The birds are present in the more open semidesert grassland where good stands of grasses persist. Such stands are discontinuous, and the quail's distribution is correspondingly patchy. The quail's existence in Sonora is seriously threatened by overgrazing, which unfortunately tends to be increasing as a mode of land management.

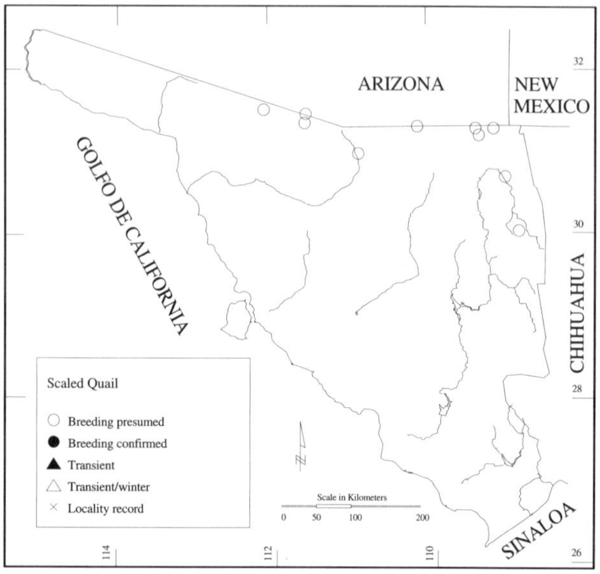

No nests of the Scaled Quail have been found in Sonora. Farther north, the breeding season extends from April through September. In most years it is believed that the principal nesting coincides with the summer rainy season, in July and August; but in years with ample winter rain resulting in good spring food conditions, nesting probably takes place frequently in April and May, as is true of Gambel's Quail.

ELEGANT QUAIL
Codorniz Cresta Dorada
Callipepla douglasii

This quail with a distractingly loud call is a resident of tropical thornscrub and tropical deciduous forest, where it is common to almost abundant. We have found it as far north as Rancho Charaibabi (at 1160 m), southeast of Cucurpe, where in September 1986 it was present in mesquite-oak-grassland. It is present at other localities where oaks occur, as at the south end of the Sierra Oscura (1200 m), the Sierra el Tiznado east of Mazocahui (1150 m), and at 1500 m near Sahuarivo in the southeast. It is absent from flatlands along and inland from El Golfo de California. Although van Rossem (1945) mentioned more than one record at Guaymas, we have not found it there.

The Elegant Quail, so named for its pale tawny topknot, does not frequent open spaces as much as the Gambel's Quail; it is an occupant of dense, heavy brush, particularly along fencerows and field edges. It is not as easily seen as the Gambel's Quail, although cocks do show themselves by perching upon bush tops while calling. It flushes reluctantly, as does the Montezuma Quail, when it does not choose to run ahead of the intruder. We have little information regarding the breeding period, except that we have found a brood of half-grown young on 22 June near Mazocahui and a clutch of 10 eggs in the same region on 11 September.

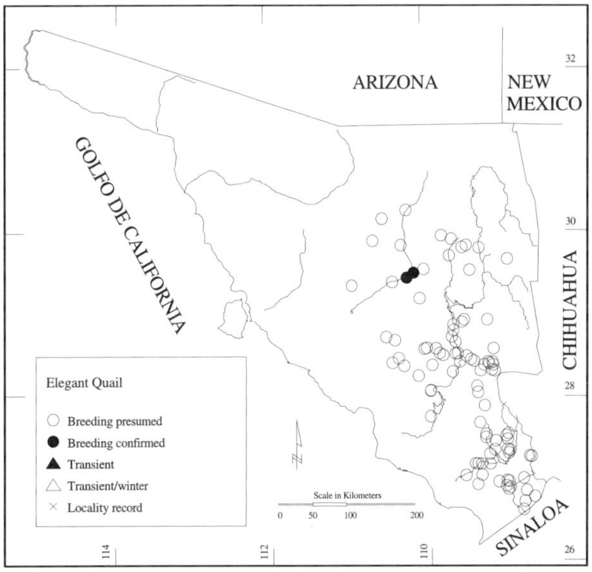

GAMBEL'S QUAIL
Codorniz Chiquiri
Callipepla gambelii

Gambel's Quail is one of the most common resident birds of Sonora. Often it is abundant, particularly in scrubland near water and in agricultural districts. Almost always there is mesquite present. It is found all along the coast of El Golfo de California, often in the absence of fresh water. We find it absent only in the extreme southeast at elevations above 200 m, as at Nuri, Alamos, and Güirocoba, and in mountains from the lower part of oak woodlands upward. Frequently it

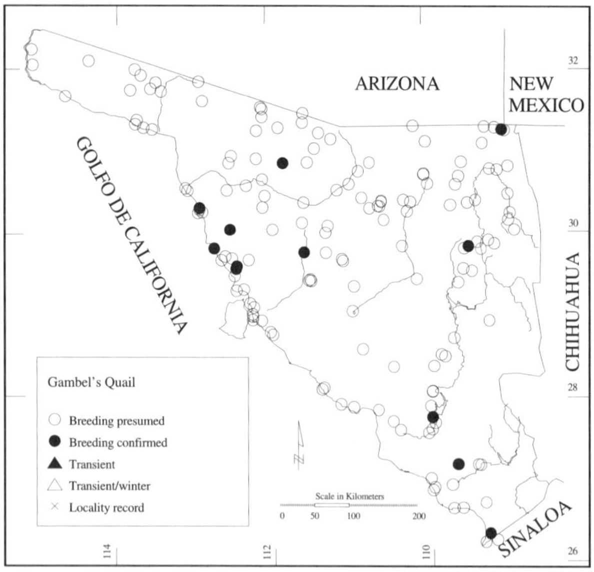

exists in the same localities as the Elegant Quail, including along the Río Sonora and the Río Yaqui. We have seen no interactions between the two species.

The spacing of rainfall and the resulting vegetation govern nesting. If winter rains have been generous, Gambel's Quail nest freely from March through the dry season and summer. If winter rains are meager, few pairs will nest before the summer rainy season. The querulous loud calling of the male is heard beginning in February and lasts well into summer. It is during the summer rainy period, from July through mid-September, that most nesting activity takes place. We have found tiny young tended by their parents as early in the year as 9 April and as late as 20 September. In contrast, Elegant Quail, though common, are much more secretive and inhabit denser cover; their young are almost never viewed.

FAMILY RALLIDAE

Rails, Gallinules, and Coots

CLAPPER RAIL
Rascón Picudo

Rallus longirostris

As pointed out by Todd (1986), the Clapper Rail's range on the coast of El Golfo de California in Sonora is identical with that of the black mangrove, almost without exception. Thus, in Sonora, it is found from Punta Sargento south to the Sonora-Sinaloa border. In such habitat it is common to abundant. For instance, Russell estimated 100 present in a limited area of Punta Santa Rosa on 31 May 1971, when there was much calling. In winter, early spring, and late fall, there are few observations, because outside of the breeding season Clapper Rails are very secretive, remain in the interior of the mangrove swamps, and seldom vocalize.

In 1930, G. Bancroft collected 28 sets of eggs between Miramar (Guaymas) and Agiabampo in the

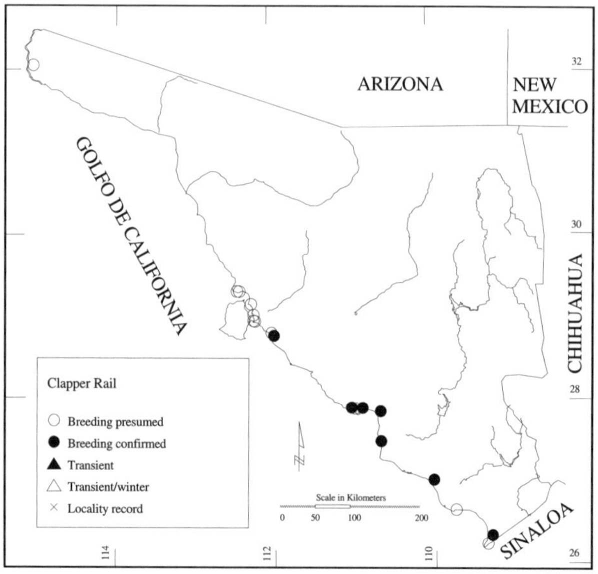

period 25 May through 14 June. His notes indicated that the nests of twigs were typically placed just above high-water levels in a mangrove. Nests were lined with a few leaves and held two to seven eggs, with 64% of the nests having five or six eggs. Farther north, at Bahía Kino, R. Taylor saw downy young about 1 June in 1987.

The studies of Todd and colleagues strongly suggest that the majority of Clapper Rails breeding in the lower Colorado River valley and around the Salton Sea in the United States depart for the winter and presumably join those already present in Sonora (and possibly Sinaloa). Migrations are thought to take place in September–October and March–April.

We will never know if Clapper Rails formerly inhabited the sloughs and swamps of the Río Colorado delta in Sonora, for suitable habitats were largely wiped out by the construction of dams that diverted virtually all of the Colorado's waters before ornithological studies were made in the area. The Ciénega de Santa Clara presently supports a large population (Eddleman 1989).

VIRGINIA RAIL

Rascón Limícola

Rallus limicola

This rail is a rare transient and winter visitant in salt- and freshwater marshes throughout Sonora; most records are from the coast. Outside migration dates are 3 September (1979, Puerto Peñasco; D. Stejskal) and 1 May (1988, several, both at a marsh near Imuris and at Laguna Tular at Guaymas; R. Taylor).

A bird collected at Santa Cruz on the Río Santa Cruz east of Nogales in June 1855 (van Rossem 1945) may have been breeding.

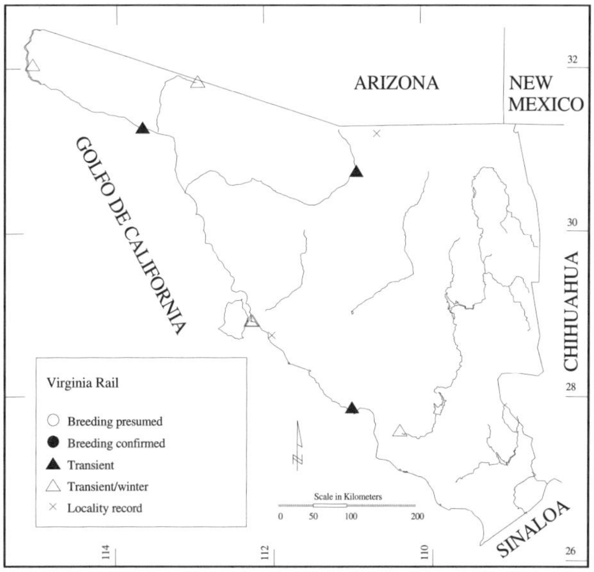

SORA

Polluela Sora

Porzana carolina

This small rail is an uncommon winter resident of marshes, particularly freshwater cattail marshes and tidal salicornia marshes. It is also present in cattails that often rim reservoirs or the backwaters of rivers. It is more often heard than seen, its easily stimulated whinnying call betraying its nearness.

The Sora has been recorded in Sonora from 10 September (at the Saracachi ciénega) to 11 April (heard along the Río Yaqui below Presa Obregon).

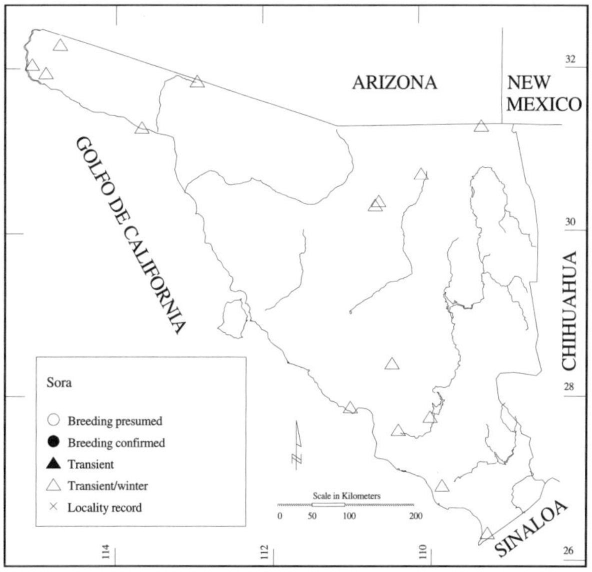

PURPLE GALLINULE
Gallineta Morada
Porphyrula martinica

J. Wright collected (OC) a male near Güirocoba on 23 July 1933. Although the species is common in places in central and southern Sinaloa, there are no other records from Sonora.

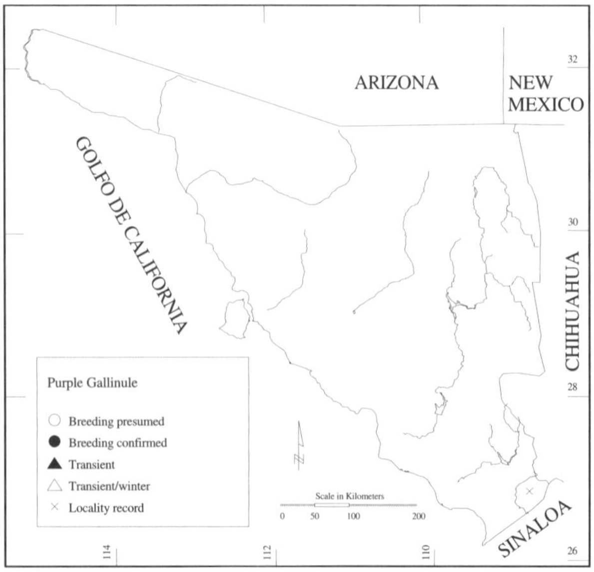

COMMON MOORHEN
Gallineta Frente Roja
Gallinula chloropus

The Common Moorhen is almost cosmopolitan in temperate and tropical regions of the world. In Sonora, it is an uncommon permanent resident of cattail-dominated fresh water associated with irrigation drains, ponds, reservoirs, rivers, and streams. We have no records for saltwater habitats. The highest elevation at which it has been found in Sonora was 1140 m at Mututicache in the upper valley of the Río Sonora. We have seen juveniles in the cattails along the Río Yaqui below Presa Obregon in July and September, and Eddleman (1989) observed chicks in the Ciénega de Santa Clara in April. Palacios and Mellink (1995) found eight chicks in three broods in agricultural drains near Estero Tóbari on 14 May 1994.

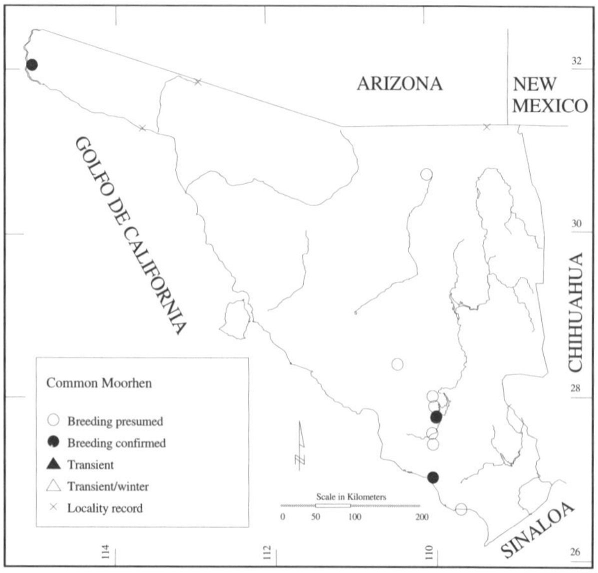

AMERICAN COOT
Gallareta Americana
Fulica americana

Except in summer, American Coots are fairly common freshwater birds in Sonora, found everywhere there is open water, excluding high mountain streams. They are seen in small groups, or pairs, or as individuals, steadily searching for food they find on and under the surface of the water, often by diving. In contrast to Common Moorhens, they do not require places to hide, although they feed in and out of cattail stands. Coots from farther north winter in Sonora, which makes them most common in that season.

Ideal nesting sites for American Coots are cattail-fringed waters that may be either stagnant or flowing. We have found young birds in July along the Río Yaqui and its associated backwaters below Presa Obregon.

There are a number of localities where we are confident the birds will be found breeding, including most reservoirs from Presa Mocuzari in the south to Presa Cuauhtémoc in the northwest. A. Moorhouse found an adult on a nest and another adult with three downy young at a reed-choked pond at Mututicache on the Río Sonora on 18 April.

American Coots are infrequently found on salt water. There are few records on or along El Golfo de California; those few that were found were on ponds back of the beach, exemplified by Laguna Tular at Guaymas. This lagoon, once an excellent place to find birds, is rapidly becoming unsuitable for them because of dumping and heavy pollution.

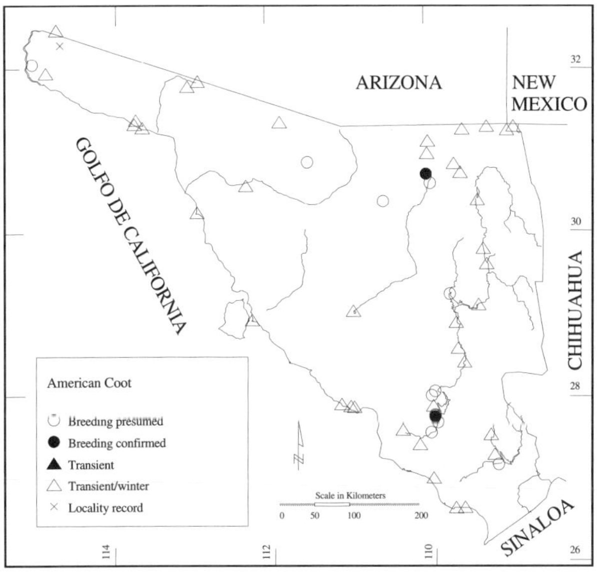

FAMILY GRUIDAE

Cranes

SANDHILL CRANE

Grulla Gris

Grus canadensis

One can expect to find Sandhill Cranes in winter in Sonora in the extensive irrigated farmlands of the lower Río Sonora basin that lie southwest of Hermosillo. Here about 200 were observed, about 60 km from Hermosillo, on 19 January 1974 (R. Craig); we are told they have been observed regularly there since at least that year. A flock of about 50 was flying over fields in the same area on 6 April 1990, a late date. Otherwise, the only recent records are of 30 counted on the coast southwest of Cd. Obregon on 23 January 1971 (Knoder et al. 1980), two on the Río San Pedro near Cananea on 13 October 1971 (S. Speich, E. L. Smith), and one flying over Puerto Peñasco on 22 November 1980 (J. Dunning, S. Goldwasser). Van Rossem (1945) stated that Sandhill Cranes were abundant in the Río Colorado delta around the turn of the century, and Leopold (1949) mentioned seeing them there. Audubon (1906) wrote of seeing a crane or cranes at Sonoita on 14 September 1849, which would be an exceptionally early fall date. Van Rossem (1945:75) further recorded, "According to residents, cranes formerly wintered in abundance in the Mayo River valley but of late years have markedly decreased in numbers."

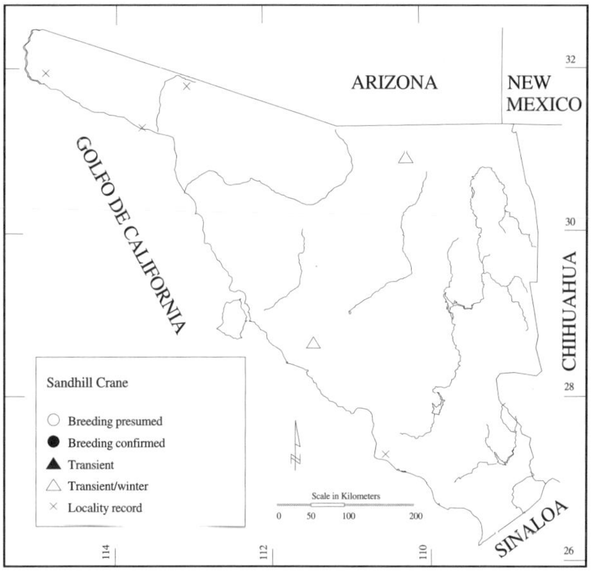

FAMILY CHARADRIIDAE

Plovers and Lapwings

BLACK-BELLIED PLOVER
Chorlo Gris
Pluvialis squatarola

The Black-bellied Plover is a common transient and winter visitant on the beaches and tidal flats of El Golfo de California. One or more always seem to be in sight in such places, and 10 to 50 or more are not unusual. The largest number counted in one area is 506, listed on a Puerto Peñasco Christmas Bird Count on 3 January 1981.

Principal migration periods are late August to mid-October, and April and May. A few birds do not migrate but linger through the summer; for example, 15 were noted at Punta Sargento, north of Isla Tiburón, on 16 July 1975.

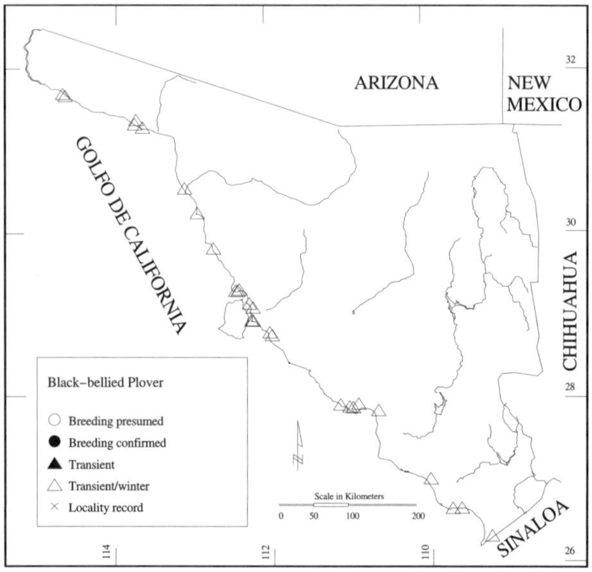

SNOWY PLOVER
Chorlo Nevado
Charadrius alexandrinus

This small plover is quite common in fall, winter, and spring along the coast of El Golfo de California, on beaches and tidal flats and along open shores of lagoons. Sometimes considerable numbers are found, such as 300 at Punta Sargento on 24 December 1974, and 243 on the Puerto Peñasco Christmas Bird Count on 19 December 1981. Migrants have been noted as early as 17 August and as late as 17 April.

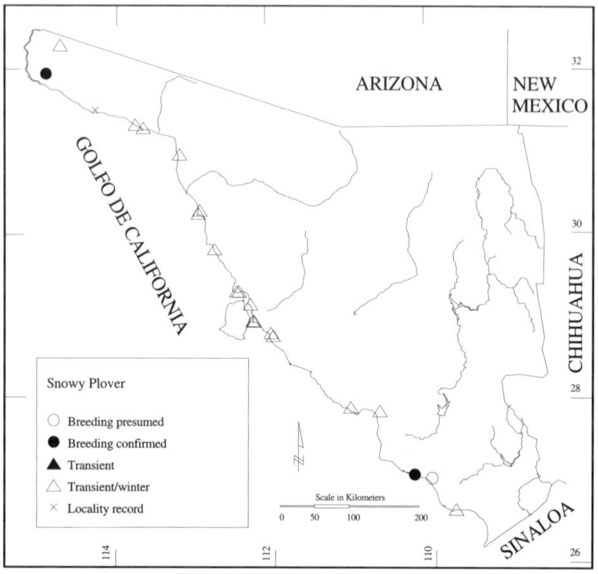

Eddlemann (1989) found nesting Snowy Plovers in the Colorado River delta west of El Doctor. Mellink and Palacios (1993) report seeing an unattended juvenile feeding on the salt flat at La Salina in June 1991. The possibility exists that the bird had already dispersed from breeding grounds elsewhere. On 14 May 1994, Palacios and Mellink (1995) found a pair calling and performing a distraction display on a salt flat at San José on Estero Tóbari; there was an empty shell-lined nest. We saw two at Punta Santa Rosa on 15 July 1975, and D. Stejskal saw two at Puerto Peñasco on 28 July 1974; these were probably nonbreeding birds.

Wilson's Plover

WILSON'S PLOVER
Chorlo Pico Grueso
Charadrius wilsonia

Wilson's Plover is a common resident of the coast of El Golfo de California north to Puerto Peñasco. Its preferred habitat is tidal flats with salicornia, salt grass, sea blite, and so forth, but it will also be seen along the edges of tidal lagoons and on beaches. The largest number we observed is 150, at Punta Sargento on 24 December 1974. Nests have been found as early as 4 April (1987, Puerto Lobos) and as late as 20 June (1974, Punta Chueca). On the latter date, one bird incubated three eggs (the typical clutch size) while other pairs attended tiny chicks or full-sized juveniles. Nests are merely bare depressions in the sand, often protected by vegetation such as salicornia. In one nest, the eggs rested on a lining of small dead leaves. Gallucci (1981) reported seeing downy young at Puerto Peñasco on 27 July 1979.

Inland appearances must be rare. One was observed on 9 September 1973 at a pond at Bamuri, about 75 km inland and 40 km south of Pitiquito, following two days of strong southerly wind.

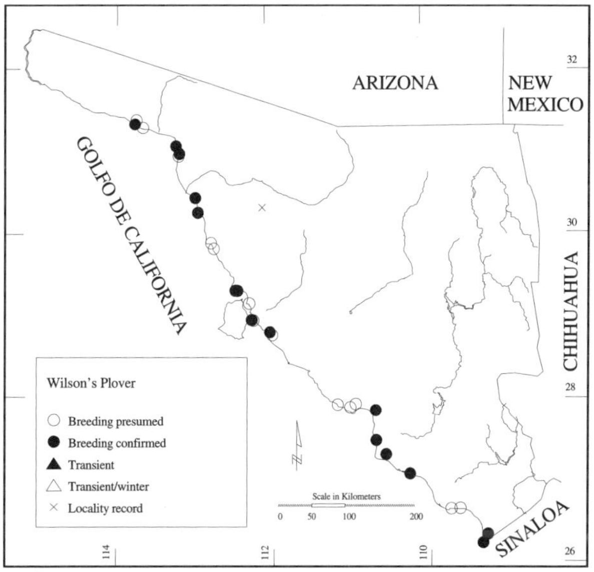

SEMIPALMATED PLOVER

Chorlo Semipalmeado

Charadrius semipalmatus

We see this attractive small plover almost exclusively along the coast of El Golfo de California, where it prefers mudflats to sandy or rocky beaches. It is a fairly common winter resident and transient. Outside dates recorded for migration are 3 September and 1 May. We have no June or August records. A half-dozen July records (including two birds at Agiabampo on 11 July 1987) may be early migrants or summering birds.

The Semipalmated Plover is perhaps more common inland than the one record indicates (one seen at Presa Agua Caliente on 4 January 1986; J. Siegel).

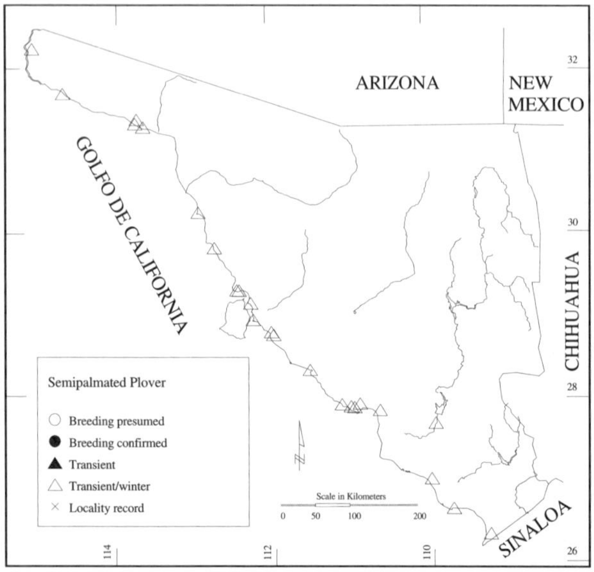

PIPING PLOVER

Chorlo Chiflador

Charadrius melodus

The Piping Plover has been reported seven times in the Puerto Peñasco vicinity; these are the only records known to us from western Mexico, except for three or four near San Blas, Nayarit, in 1980–1982 (Howell 1993). First found in Sonora in 1965 (Russell and Lamm 1978), one to five individuals have been seen on seven dates between 14 November and 17 April, but none has been seen since 1982. Diagnostic photographs were taken of one on 3 February 1977 (R. Witzeman) at Bahía la Choya.

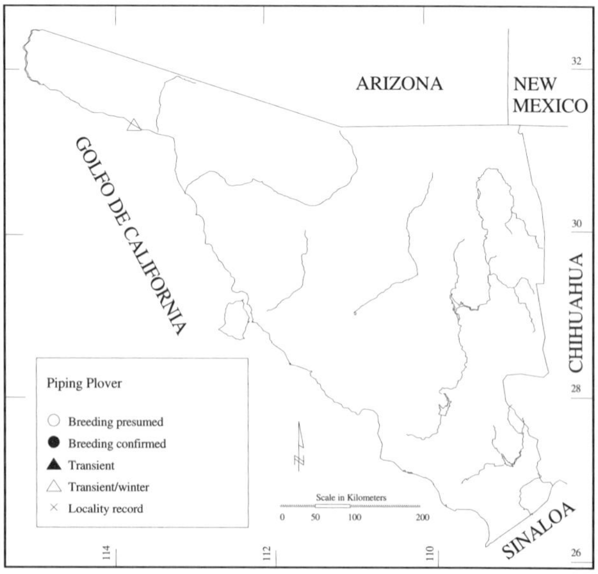

KILLDEER

Chorlo Tildío

Charadrius vociferus

No shorebird is more omnipresent in wet places in Sonora than the Killdeer, if we except the beaches and salty backwaters of El Golfo de California, where it is infrequent. In appropriate freshwater habitat, the Killdeer is a fairly common resident. Its numbers are greatest in the winter months, augmented by birds from the north. Migrations are difficult to detect but are thought to be primarily in August and September and

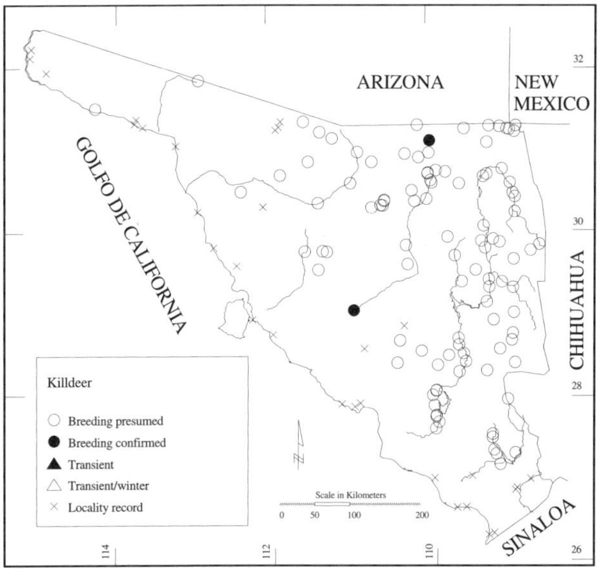

in March. Although Killdeer are typically conspicuous while nesting, we know of only two nesting records: eggs found at Presa Rodriguez, Hermosillo, on 29 June 1959 (A. Phillips), and young birds noted at the Villa Verde pond south of Naco on 29 May 1988 (A. Moorhouse). The highest elevation at which we have found it is 2100 m in the Sierra Oscura, northeast of Rosario, on 10 June 1984. Along the coast of El Golfo de California it is limited to shallow-water or seepage areas inshore.

"Tildío" is, of course, a rendition of the bird's call, as is "Killdeer."

MOUNTAIN PLOVER

Chorlo Llanero

Charadrius montanus

The Mountain Plover is a rare winter visitant in Sonora, known only from seven observations, two of them in the last century and cited by van Rossem (1945; specimens from Hermosillo and also from 16 km southeast of Sonoita). Russell and Lamm (1978) found about 45 in three groups in extremely barren terrain near Volcan Colorado in the Sierra el Pinacate on 22 November 1973, and 20 to 25 were seen east of the Pinacates in December 1980 (K. Kaufman, S. Terrill, et al.). We flushed a flock of 41 a few kilometers northeast of the mouth of the Río Concepción on 7 January 1985 from a field as it was being plowed. About six were seen near Puerto Peñasco on 10 February 1989 (J. Coons). K. Kaufman saw three at Puerto Peñasco near the airport on 26 February 1990, the latest seasonal record, and has suggested that Mountain Plovers may be regular in winter in the very dry country between Sonoita and Puerto Peñasco.

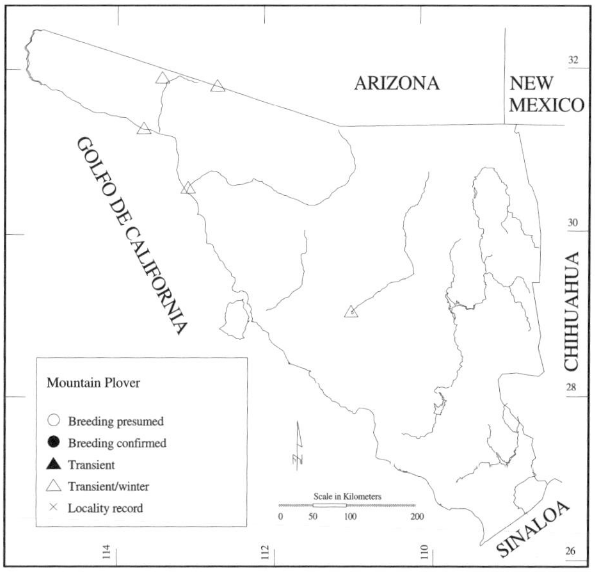

FAMILY HAEMATOPODIDAE

Oystercatchers

AMERICAN OYSTERCATCHER

Ostrero Americano

Haematopus palliatus

The American Oystercatcher, strikingly red-billed and striped-winged, is a fairly common resident along the entire length of the coast of El Golfo de California in Sonora. It feeds both on sandy beaches and on rocky shorelines. Ordinarily seen in singles or pairs, sometimes as many as 25 to 30 are grouped in the nonbreeding season. In 1930, G. Bancroft collected eggs (now in WFVZ) from 17 nests between 17 April and 26 May between Isla Alcatraz and Estero Lobos. Typically the nests were shallow depressions on sandy or pebbly beaches and were lined with bits of shell. Most nests contained three eggs. Incubation had just begun in the April nests, but late May nests included clutches at various stages of incubation. Palacios and Mellink (1995) found several pairs nesting at Estero San José and at various points on Estero Tóbari on 14 May 1994. Most nests contained only eggs, but one pair had three chicks.

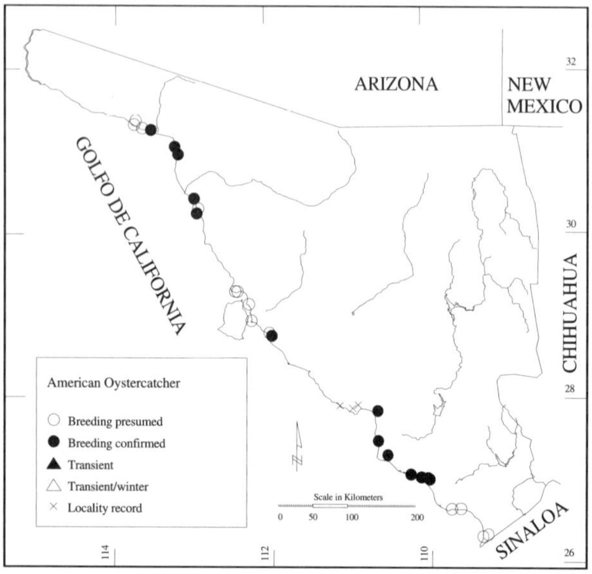

[BLACK OYSTERCATCHER
Ostrero Negro
Haematopus bachmani

We have three undocumented sight records, all from Puerto Peñasco. J. Marshall saw an unspecified number with American Oystercatchers on 19 February 1955. S. Demaree reported three seen in March 1970 and one seen in May 1970.

Black Oystercatchers are resident along the Pacific coast from the Aleutian Islands south to central

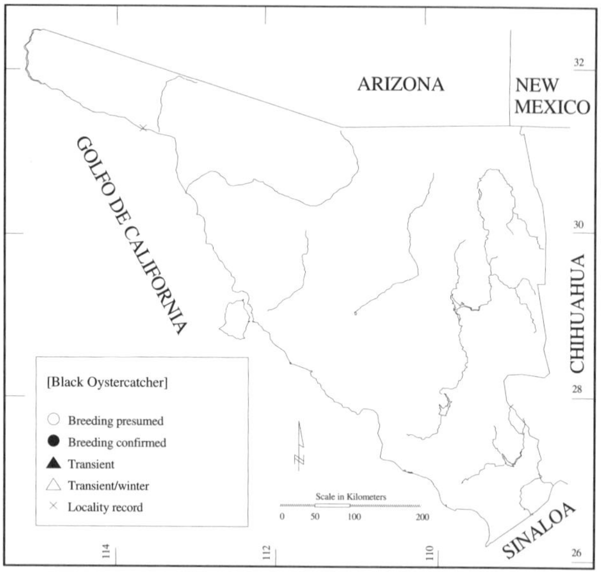

Baja California. American Oystercatchers breed from El Golfo de California and Massachusetts southward on the Pacific and Atlantic coasts to southern South America. The two species hybridize in Baja California and may represent strikingly different populations of a single species.]

FAMILY RECURVIROSTRIDAE
Stilts and Avocets

BLACK-NECKED STILT
Candelero Americano
Himantopus mexicanus

Black-necked Stilts are fairly common transients and winter residents around lagoons and estuaries of the coast of El Golfo de California. They are found inland throughout the state at ponds and reservoirs, as well as along streams. Groups of 10 to 20 and even 50 birds are not unusual, and Russell once found 240 at Bamuri, about 40 km south of Pitiquito (9 September 1973). Outside dates of appearance are 27 August and 18 May. Stilts have nested in the Colorado River delta west of El Doctor (Eddleman 1989), and we found a nest containing eggs on the edge of Ciénega de Santa Clara on 9 April 1994. There is no evidence that ten observed at Puerto Peñasco on 7 June 1975 (D. Stejskal) were breeding. A single bird was seen there on 25 July 1979 (Gallucci 1981).

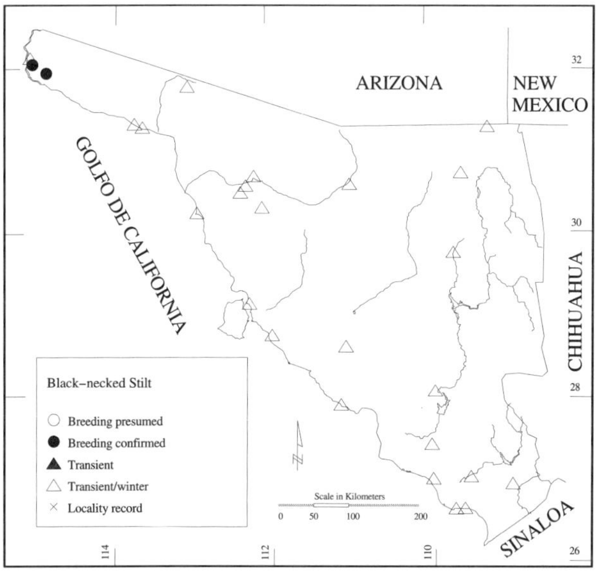

AMERICAN AVOCET

Avoceta Americana

Recurvirostra americana

This bird is a common winter resident and transient, chiefly along the coast of El Golfo de California; inland, it is almost entirely a migrant. It is at home in any shallow lagoon or estuary but, like the Black-necked Stilt, is not found on sandy or pebbly beaches. The greatest number we recorded in one place was 800 near Golfo de Santa Clara on 20 March 1969. Arriving migrants were seen as early as 16 August, while most departing birds have left by mid-April.

Apparently, nonbreeding American Avocets remain into early summer in sizable groups in some years, as suggested by the following: 22 on 31 May 1971 at Punta Chueca; eight on 20 June 1978 at Punta Sargento; and four on 23 June 1975 at Puerto Peñasco.

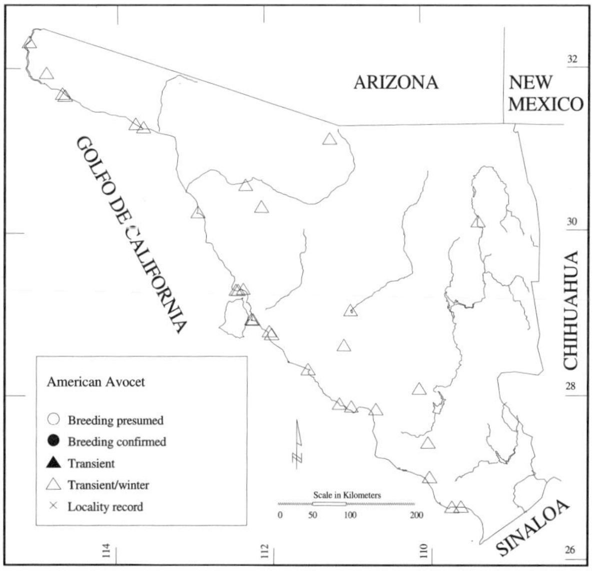

FAMILY JACANIDAE

Jacanas

[NORTHERN JACANA

Jacana Norteña

Jacana spinosa

There is a skeleton of the Northern Jacana in the Los Angeles County Museum labeled "Sonora." Apparently the bird was obtained in 1961 by W. Rook and K. Wolfe, who gave it to D. Bleitz. It was ultimately donated to the museum. In view of the uncertainties of the origin of the specimen, we consider the species hypothetical in Sonora. Jacanas do occur in southern Sinaloa and would be expected occasionally in Sonora.]

FAMILY SCOLOPACIDAE

Sandpipers, Phalaropes, and Allies

GREATER YELLOWLEGS

Patamarilla Mayor

Tringa melanoleuca

We have found the Greater Yellowlegs to be an uncommon transient and winter visitant in Sonora. While the bulk of our records are from the coast of El Golfo de California, the birds are fairly regular inland at reservoirs and streams. Because a few Greater Yellowlegs, in common with some other shorebirds, stay the entire

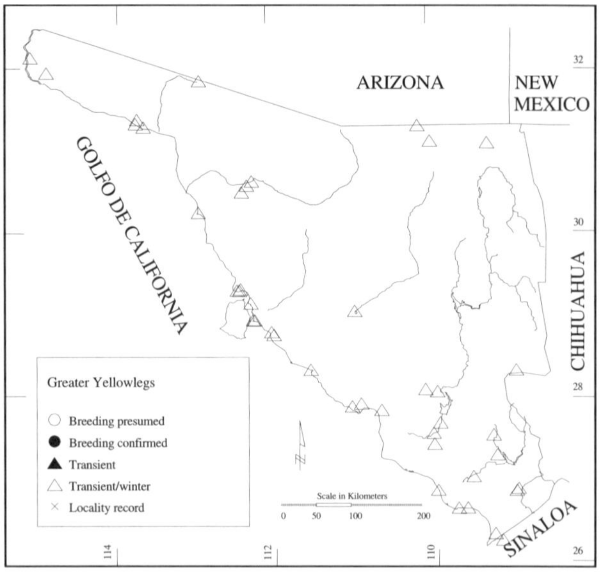

summer far south of their breeding area, it is difficult to determine when the first migrants arrive and the last depart. The migration period appears to begin in the fall by mid-July and end in the spring in late April and early May. The greatest number we found in one locality was 20. An unusual record, because of the elevation, is of one along the river at Yécora (1520 m) on 10 May 1983.

LESSER YELLOWLEGS

Patamarilla Menor

Tringa flavipes

This shorebird is a rare fall transient and winter visitant in Sonora. It has arrived in fall as early as 3 July 1991 (west of Naco; A. Moorhouse). Most records are of fewer than six individuals, although Russell saw 30 at

Bamuri south of Pitiquito on 9 September 1973. There is only one spring record, a single bird at Punta Sargento on 12 April 1970. In winter this species has been found north to Puerto Peñasco.

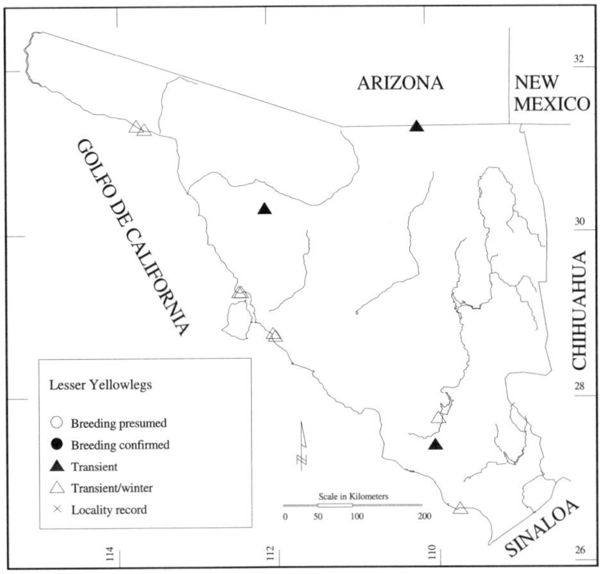

SOLITARY SANDPIPER
Playero Solitario
Tringa solitaria

The Solitary Sandpiper is an uncommon fall transient in Sonora, preferring small bodies of freshwater having shaded or grassy banks. These sandpipers have been seen occasionally, usually solitarily or in twos but never in flocks. Most have been seen in the last half of September. Outside migration dates in the fall are about 6 August at Granados (A. Phillips) to 3 November. We are aware of only five spring observations, in the period of 19 March to 7 May. The single winter record is one collected near Baviácora on 29 December 1952 (Cornell University; R. Dickerman).

The only coastal record is of two found at Punta Chueca on 23 September 1977 in a muddy mangrove-lined channel. We are unable to account for Alden's (1969) assertion that the Solitary Sandpiper is common in migration at Bahía Kino and rare in migration at Guaymas.

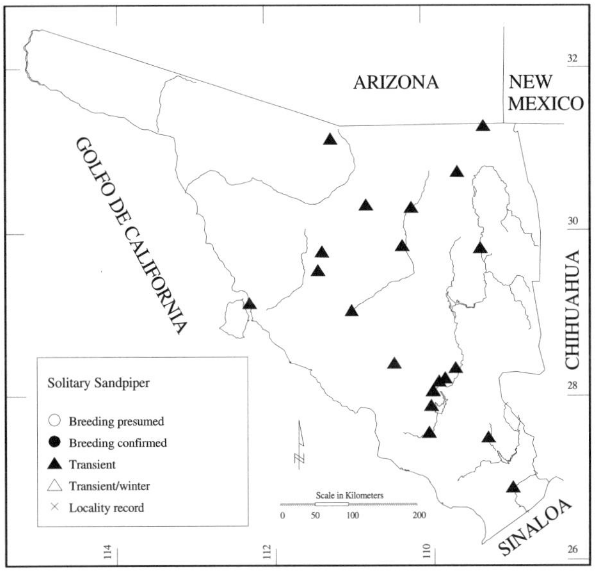

WILLET
Playero Pihuiuí
Catoptrophorus semipalmatus

At least one or two Willets characterize nearly every beach along El Golfo de California in Sonora from mid-July to the following May. They often stand on one leg as the incoming tide approaches, flying or hopping away (still on one leg) as the water becomes belly deep. Even during June and early July, numbers of nonbreeding birds remain in Sonora—as many as 36 at Puerto Peñasco on 23 June. Willets are common transients and winter visitants, sometimes gathering in large numbers (e.g., about 650 at Punta Sargento on 27 May 1975 and about 700 at Bahía Kino on 29 December 1978).

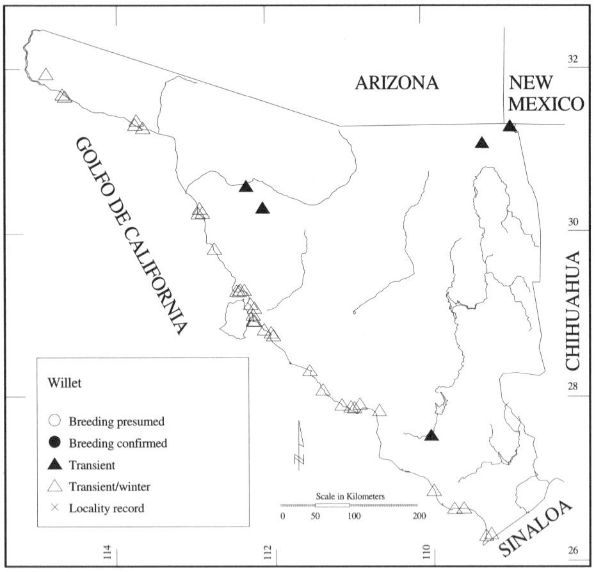

Inland we have only five records, although on the basis of their frequency in southern Arizona we would expect more. Four observations were in the period of 20 August–15 September, and one was on 30 April.

WANDERING TATTLER
Playero Vagabundo
Heteroscelus incanus

This interesting shorebird, which nests in arctic mountains and winters on ocean shores, is an uncommon wintering bird on the coast of El Golfo de California. It frequents rocky shorelines and has not been seen on the predominantly sandy beaches from Guaymas southeastward. Migratory flights have not been detected; winter numbers are about the same as during spring and fall. Curiously, we know of no records between 25 October and 18 December, and there are only two winter records from Puerto Peñasco. We have never seen more than 12 at one time. The earliest fall record is one at Puerto Peñasco on 3 September 1979 (D. Stejskal); the latest spring record is from the same place on 23 April 1981 (R. and J. Witzeman). One seen at Puerto Peñasco on 28 July 1974 (D. Stejskal) and an observation of two, one in breeding plumage, at San Carlos on 1 August 1979 (Gallucci 1981) are exceptional, although the species is known to occasionally summer within the winter range.

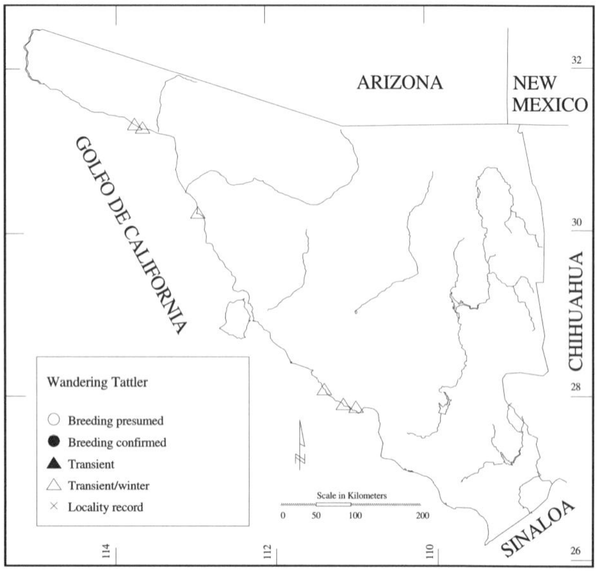

SPOTTED SANDPIPER
Playero Alzacolita
Actitis macularia

The familiar Spotted Sandpiper is a common transient and winter resident in both fresh- and saltwater habitats of Sonora. It is generally observed in ones or twos at water edges, being most common on streams and mangrove channels. We have no June records, but fall migrants appear by mid-July (earliest date, 10 July). In the spring, most birds have departed by the first week of May (latest date, 29 May).

The largest number seen in one area on one date is 18, for which we have three dates: 28 November 1970, Punta Sargento; 13 March 1982, Río Cuchujaqui near Alamos; and 4 January 1986, Presa Agua Caliente south of Presa Obregon.

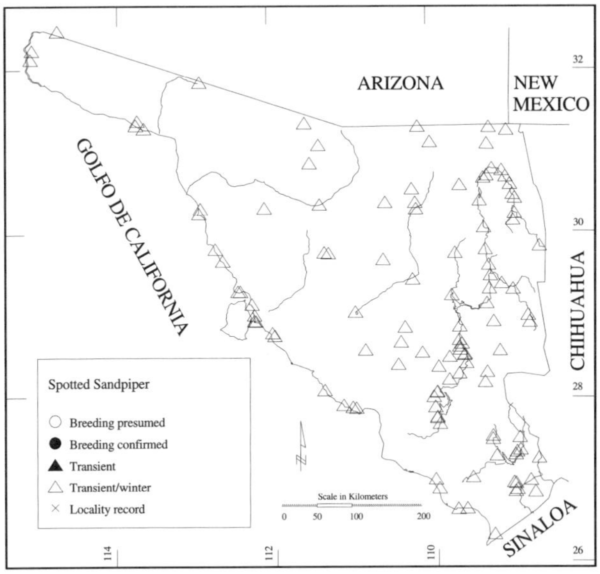

[UPLAND SANDPIPER

Zarapito Ganga

Bartramia longicauda

One was seen in grasslands about 16 km south of Benjamin Hill on 31 July and 8–9 August 1969 by R. Tomlinson (Russell and Lamm 1978). The observer was quite familiar with Upland Sandpipers, having seen them often in the United States, and was able to carefully study this bird.]

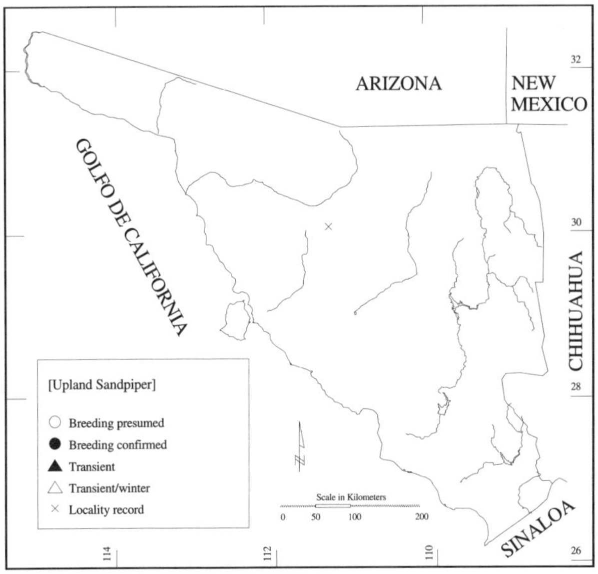

WHIMBREL

Zarapito Trinador

Numenius phaeopus

A common migrant and winter resident on the beaches of El Golfo de California, the Whimbrel has also been found three times in June. The southward migration takes place from July through September, while the return migration is predominantly in late March and April, but 20 were observed at Estero Tóbari on 14 May 1994 by Palacios and Mellink (1995). Most daily counts are of fewer than 25 birds, but on 14 April 1973 more than 1500 were assembled on a mudflat in mangroves near Punta Sargento.

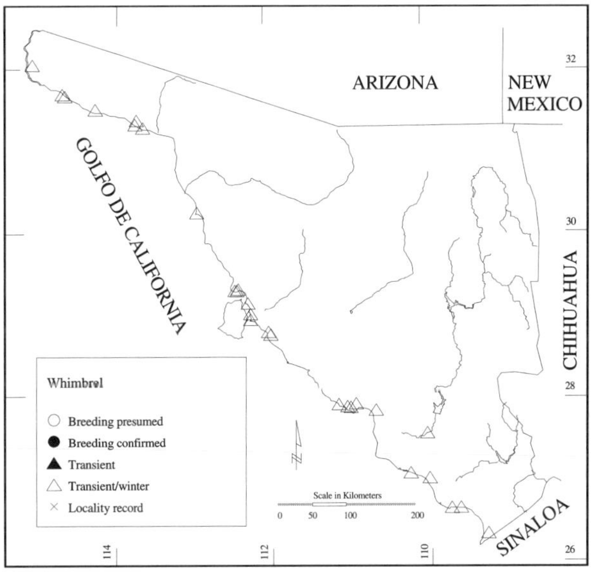

LONG-BILLED CURLEW

Zarapito Pico Largo

Numenius americanus

Most Sonoran reports of Long-billed Curlews come from the beaches and tidal flats of El Golfo de California, where it is a common transient and winter visitant.

Fall migration starts in mid-July, but most individuals pass through in August and September. Spring migration begins in March and is largely over in early May, although 100 were still present at Punta Sargento on 27 May 1975. As we have noted in the case of other shorebirds, a few linger through the summer. Concentrations of 300 or more at any time are unusual but have been noted at various localities in March, July, October, and December.

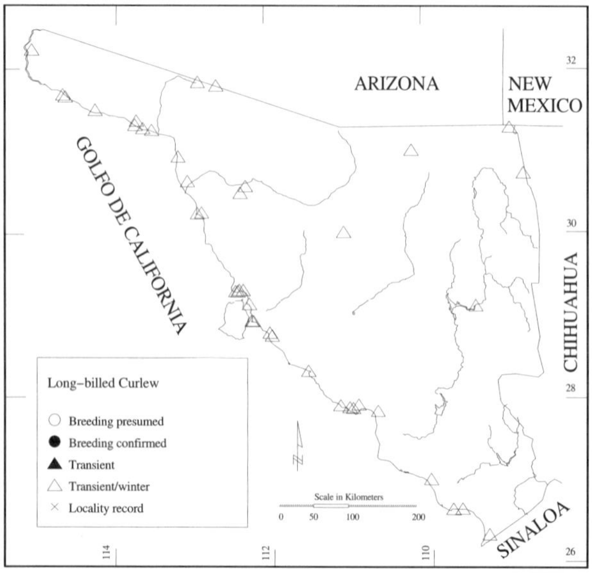

There are few inland records, but one of the most interesting is of 40 to 50 found in a Black-tailed Prairie Dog (*Cynomys ludovicianus*) town about 13 km west-southwest of Pancho Villa, Chihuahua (in the grasslands of extreme northeastern Sonora), on 12 November 1984. "The curlews probed the short vegetation, using their bills like pincers to pick up bugs. They moved like waves with the front edge moving like a wave over the rest" (Joel Brown *fide* R. Sidner). About 100 were noted in the same locality on 8 November 1985 (J. Turner).

Long-billed Curlews would no doubt be found to be regular migrants and winter visitants throughout much of inland Sonora if sustained locality observations were made; that is their status in Arizona away from the Río Colorado.

MARBLED GODWIT
Picopando Canelo

Limosa fedoa

Large numbers of Marbled Godwits spend the winters and pass as transients along beaches and tidal flats along El Golfo de California. W. Sheffler (van Rossem 1945) found thousands at Puerto Peñasco on 21–23 January 1934. Monson estimated that 10 000 birds were massed near Bahía Kino on 17 November 1972. Most observations are of fewer than 25 birds, however. The fall migration is well under way by mid-July, with the spring migration not over until late May. In addition to being common to abundant migrants and winterers, Marbled Godwits are frequently seen in summer. Although inland records do not exist, some birds must migrate across mainland Sonora.

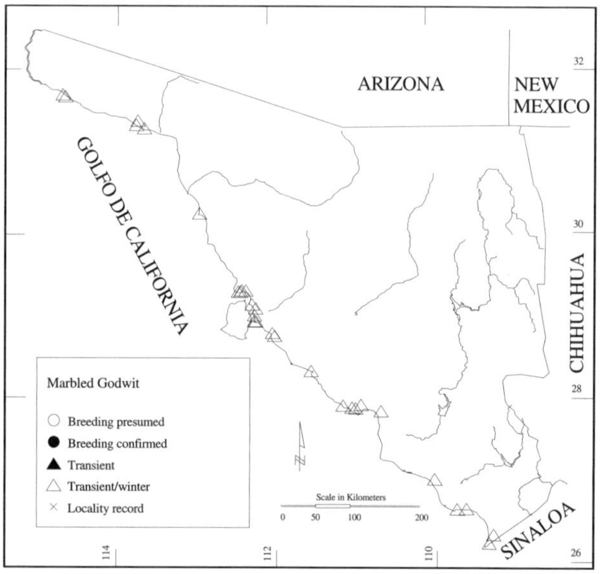

RUDDY TURNSTONE
Vuelvepiedras Rojizo
Arenaria interpres

The Ruddy Turnstone frequents beaches and tidal flats along the coast of El Golfo de California, where it is common during migration and in winter. It searches for food by using its bill to flick seaweed and small pebbles aside; it eats the tiny invertebrates that are exposed. Thus it is most often found on rocky shores or where beaches are littered with seaweed. The principal migratory movements take place in September and early October, and again in late March to mid-April. Occasional birds arrive as early as 24 July. Birds seen in June (four records) were probably summering. The maximum number we have seen in spring is 90 at Puerto Lobos on 29 March. In December as many as 145 have been counted at Puerto Peñasco.

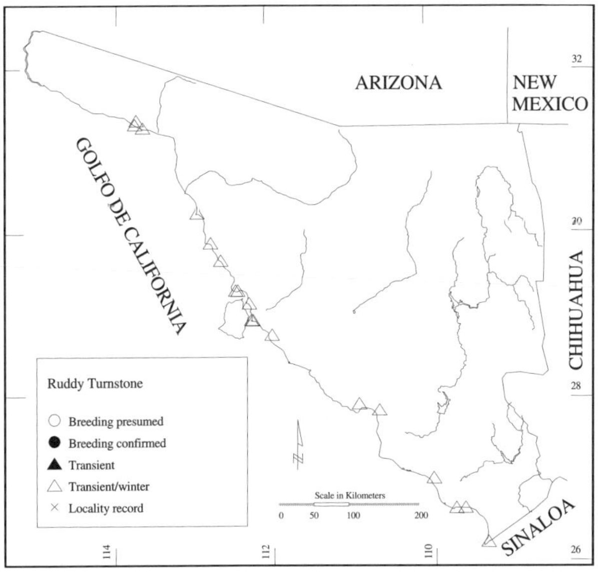

BLACK TURNSTONE
Vuelvepiedras Negro
Arenaria melanocephala

The Black Turnstone is fairly common but is restricted to the rocky shores and reefs of El Golfo de California, occasionally as far south as Huatabampito. Migration

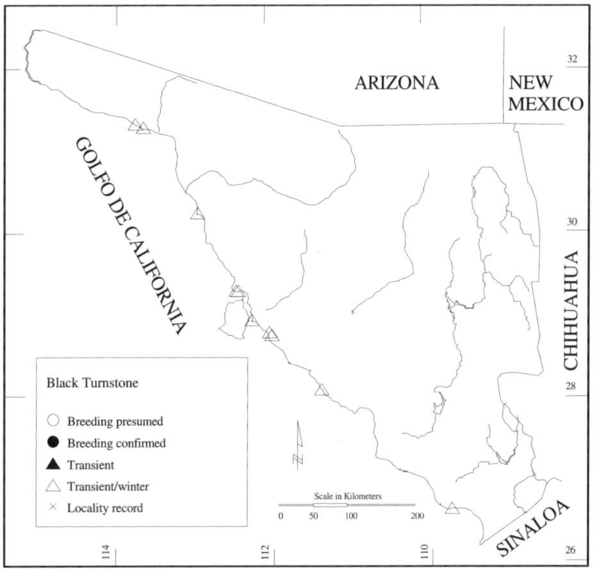

periods of this transient and winter visitant are chiefly in September and late March–early April. Extreme dates are 3 September and 19 May. There are three observations in July, from Puerto Peñasco and Bahía Kino. The largest number seen together at one time that we have recorded is 80 at Caleta Venecia on 23 March and at Puerto Lobos on 29 March.

SURFBIRD
Playero Roquero
Aphriza virgata

The grayish Surfbird, flashing white in the wings and tail when it takes wing, gathers in small groups on rocks and reefs on the coast of El Golfo de California, where it may be accompanied by Wandering Tattlers and Black Turnstones. As with most shorebirds in Sonora, it is a transient and winter visitant, fairly common in spring migration but less numerous in fall and winter. Although it winters to the tip of South America, it has not been found in Sonora south of Guaymas. Rocky shorelines are scarce south of Guaymas. Extreme dates

Surfbird

for the Surfbird are 3 September and 21 April. The largest number counted is 340 seen at Bahía Kino on 3 April 1982 (S. Howell); this was perhaps a group readying to depart on a long nonstop flight across the western United States en route to breeding grounds in arctic mountains. There is one report from summer: five birds seen by D. Stejskal at Puerto Peñasco on 28 July 1974.

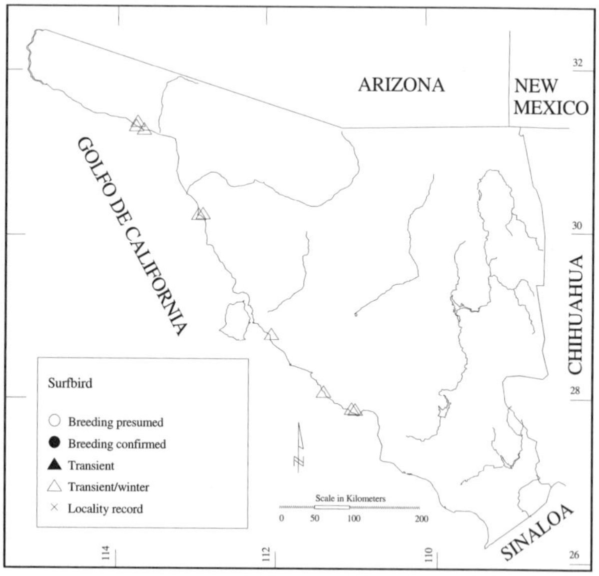

RED KNOT
Playero Canuto
Calidris canutus

Van Rossem (1945) reported one instance of the Red Knot in Sonora. At Estero Tóbari south of Cd. Obregon from 26 April to 1 May 1930, he found it to be the most abundant shorebird, swarming over the tidal flats in flocks of hundreds. We now know it is a fairly common transient and winter resident (Russell and Lamm 1978), found on beaches and tidal flats all

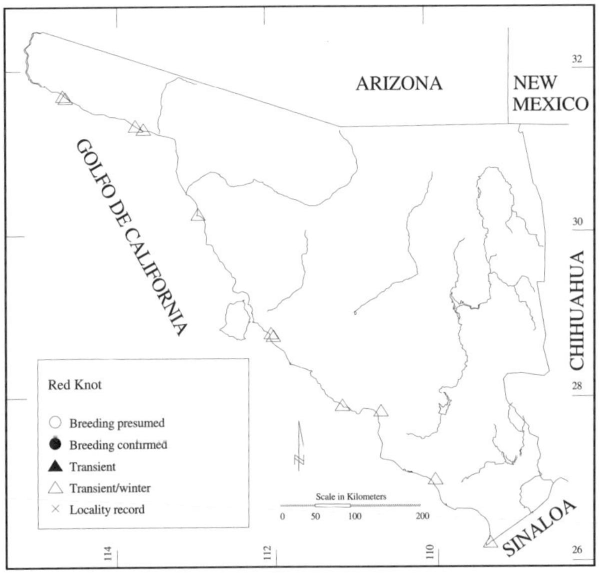

along the coast of El Golfo de California. Sometimes it is quite numerous. One hundred forty-four were listed on the Puerto Peñasco Christmas Bird Count on 19 December 1981, and more than 100 were seen at the same place on 13 April 1973 (S. Levy). At Golfo de Santa Clara, at the head of El Golfo de California, more than 375 were feeding with about 450 Sanderlings on 9 January 1985. The mixed flock would run up the beach in front of the houses, hastily feed on something in the sand, then run back to the water's edge. This act they repeated again and again, sometimes breaking into short flights (Monson). The earliest fall record is of one at San Carlos on 31 July (Gallucci 1981), yet there are no observations in August. The latest spring record is for 1 May.

SANDERLING
Playero Blanco
Calidris alba

The Sanderling is the quintessential sandpiper, following the ebbing surf, then scampering just barely ahead of the returning water, on most of the ocean beaches of the world. On sandy shores of El Golfo de California it is an ubiquitous and common transient and winter visitant. Usually scattered in small groups, it occasionally is seen in larger gatherings, such as 450 observed with Red Knots at Golfo de Santa Clara on 9 January 1985, and 615 at Guásimas southeast of Empalme on 22 September 1985. Another large group of at least 300 was at Puerto Lobos on 2 January 1988. Migration is under way in mid-August and continues through September; northward movement is in late March and April. There are no June records and only six in May and July, including 56 birds counted at Estero Tóbari on 14 May 1994 by Palacios and Mellink (1995).

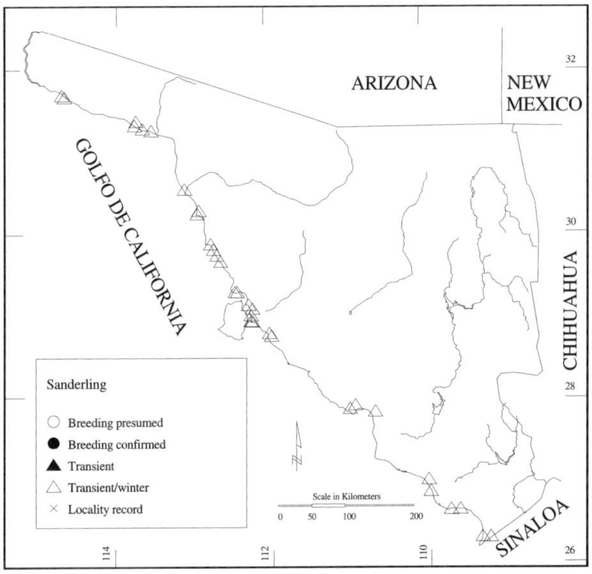

[SEMIPALMATED SANDPIPER
Playero Semipalmeado
Calidris pusilla

Two observations (D. Stejskal) at Puerto Peñasco (3 September 1979 and 12 October 1975) constitute the only reports from Sonora. In view of specimens taken in adjoining southern Arizona, the Semipalmated Sandpiper must be a migrant in Sonora.]

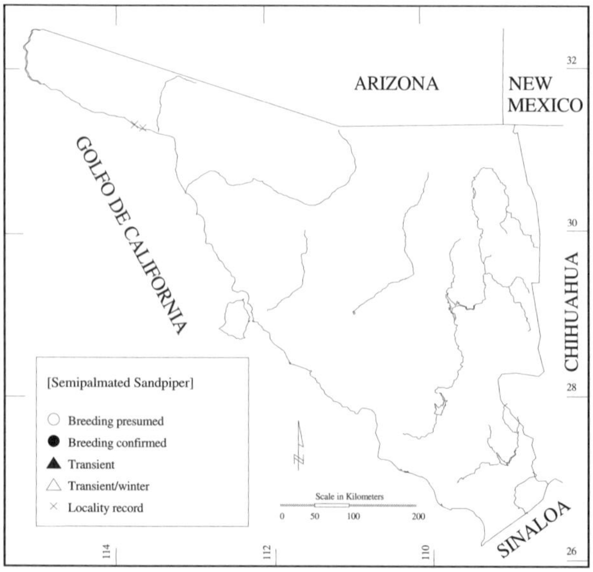

WESTERN SANDPIPER
Playero Occidental
Calidris mauri

The Western Sandpiper is a common, sometimes abundant, transient and winter resident of the coast of El Golfo de California, but it is somewhat less common in the interior, where it is rare in winter. It favors tidal flats along the coast, and the muddy shorelines of reservoirs and ponds inland. Often the Western Sandpiper associates with the slightly smaller Least Sandpiper, which it typically outnumbers except at inland locations. Fall migrants have appeared by 15 July, and some spring transients are not gone until mid-May. There are no June records. More than 1000 were observed on at least four occasions, in December and January, with the maximum of 7700 in the Yavaros area on 10 December. Sandpipers rarely perch, but we once saw six Western Sandpipers on dead mangrove branches along an inundated channel at Punta Sargento.

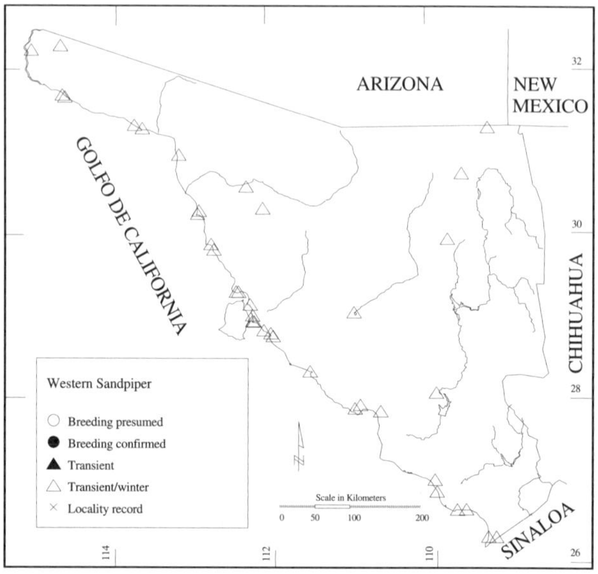

LEAST SANDPIPER
Playero Chichicuilote
Calidris minutilla

This is the most widespread and smallest sandpiper found in Sonora. In migration and in winter, it is common and dwells on tidal flats and sometimes beaches along the coast of El Golfo de California, as well as inland along the muddy or grassy fringes of ponds and streams or by shallow-water areas of receding reservoirs. Only during the months of May and June is it entirely absent. This sandpiper appears on southward migration before mid-July, but it does not become numerous for another six weeks. In spring, its numbers conspicuously decline by mid-April; our latest record is of 35 at Bahía Kino on 28 April. Aggregations seldom exceed 50 individuals.

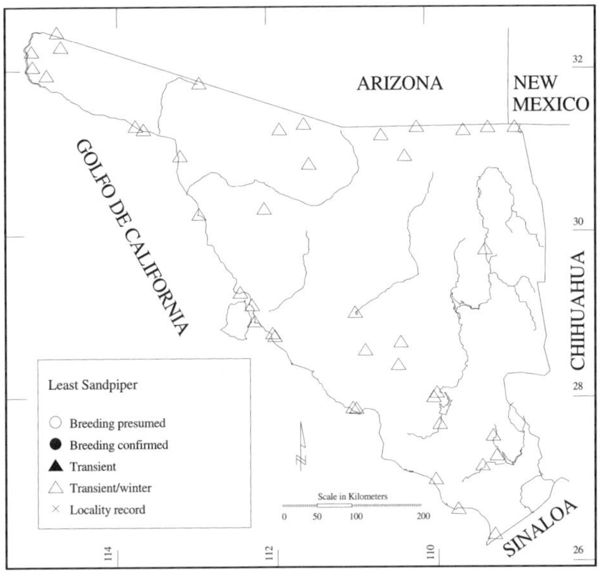

[BAIRD'S SANDPIPER
Playero de Baird
Calidris bairdii

Baird's Sandpiper went unrecorded by van Rossem (1945). Nevertheless, it must be an occasional fall migrant at least in the interior of Sonora, because it is fairly common then in southern Arizona. We have only two locality records. Four birds were seen east of Sonoita on 25 August 1967, and three were at the same place five days later (Groschupf, Brown, and Johnson 1988). Eight were noted at a pond northeast of Cananea on 10 October 1987 (D. Stejskal).]

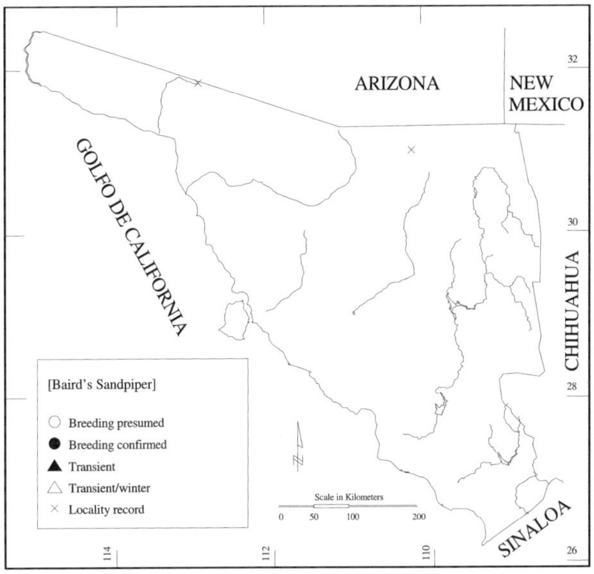

[PECTORAL SANDPIPER
Playero Pectoral
Calidris melanotos

Although the Pectoral Sandpiper is a regular fall transient in southern Arizona, including the lower Colorado River valley, its presence in Sonora has not yet been adequately documented. Alden (1969) listed it from Sonora. D. Stejskal observed two at Puerto Peñasco on 3 September 1979, and Gallucci (1981) reported one there on 24–25 July 1979. Stejskal also found two at a pond northeast of Cananea on 10 October 1987.]

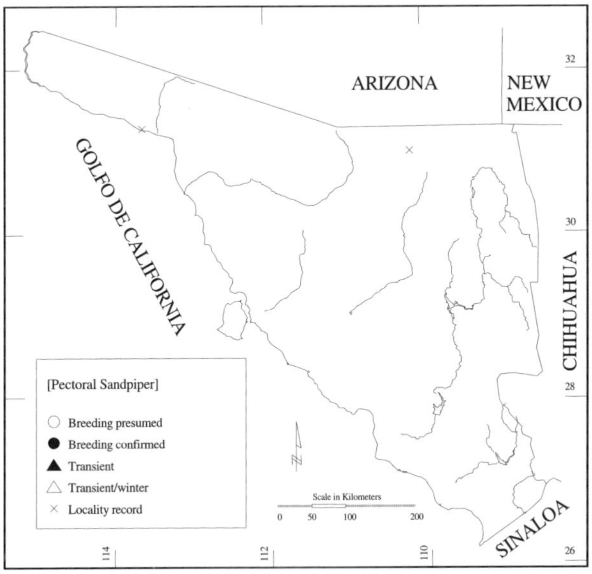

DUNLIN
Playero Dorso Rojo
Calidris alpina

The Dunlin is a fairly common migrant and winter resident along the coast of El Golfo de California, although van Rossem (1945) noted it only once, at Estero Tóbari south of Cd. Obregon, where he found it abundant from 26 April to 1 May 1930. It is a bird of tidal flats and sandy beaches. Migration periods extend from mid-March to late April and again from late September through October. Usually Dunlins are in groups of 10 to 60 birds; single birds are rarely observed. Large gatherings are sometimes found: 530 at Bahía Kino on 29 December 1978, and about 300 at Puerto Peñasco during Christmas Bird Counts in 1980 and 1981. We know of no records between April and September.

The only inland record, and one not very far inland, is of at least 20 at a roadside pond 55 km east of Empalme on 16 November 1972.

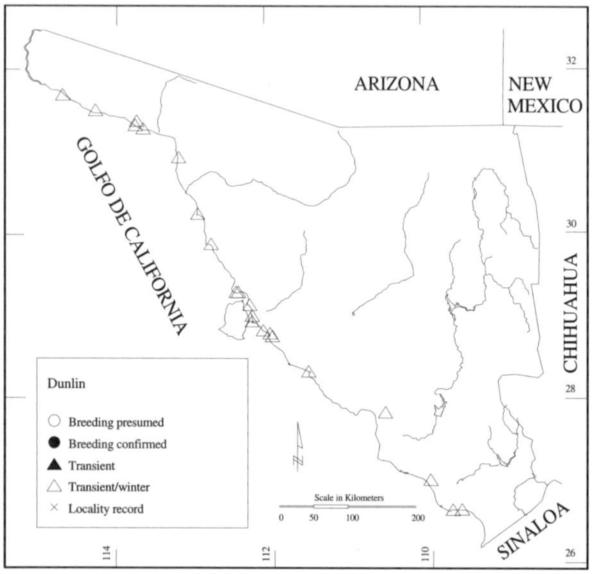

[STILT SANDPIPER
Playero Zancón
Calidris himantopus

We have only a single record: one Stilt Sandpiper carefully studied by Russell and Lamm (1978) as it foraged in a flooded irrigated field at Bamuri, south of Pitiquito, on 9 September 1973. With time there will undoubtedly be more records, because the Stilt Sandpiper is seen annually in both southern California and Arizona as a spring and fall transient, as well as in winter at the Salton Sea in California.]

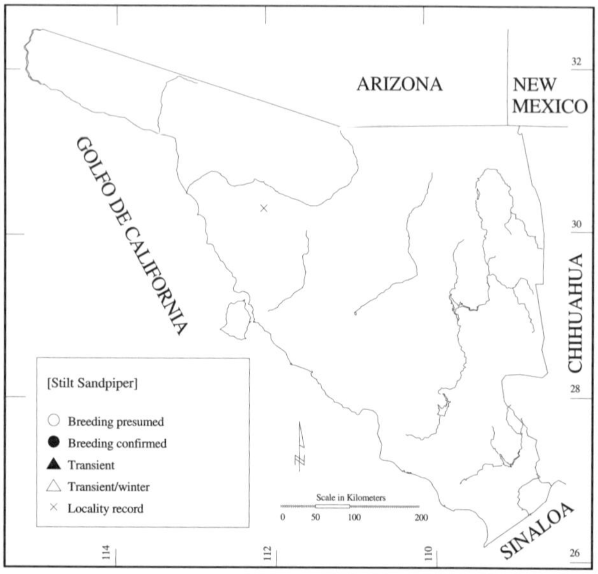

SHORT-BILLED DOWITCHER
Costurero Pico Corto
Limnodromus griseus

The status of this dowitcher in Sonora is uncertain. Few observers can competently name the two species of dowitcher, and many records are not positive as to the species. The calls are distinctive and provide certain recognition. We know of verified reports from 18

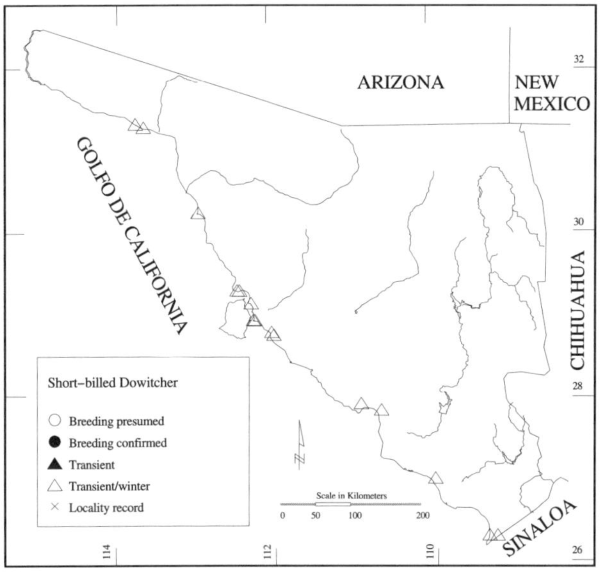

August through the winter to 14 April. We think Short-billed Dowitchers are uncommon in Sonora. Van Rossem (1945) reported four specimens, all still in winter plumage: two taken near Agiabampo on 16 May 1937, one at Agiabampo on 7 June 1930, and one at Estero Tóbari on 27 June 1930. There are specimens (UA) from Puerto Peñasco in March and from Bahía Kino in July.

LONG-BILLED DOWITCHER

Costurero Pico Largo

Limnodromus scolopaceus

On the basis of calling, carefully identified individuals, we have established that the Long-billed Dowitcher is a fairly common shorebird of the coast of El Golfo de California, especially on estuarine tidal flats, and inland wherever there is suitable shallow muddy water.

Statewide, it is a transient and winter visitant, even in the far northeast as at Rancho San Bernardino, where eight were found on 29 January 1987. Spring migration is in March and April; fall migration from late July through September. Definite May–June records are lacking. The only unusually large aggregation was at Bahía Kino on 8 April 1976, when at least 530 were present. A flock of 600 dowitchers at Bahía Kino on 9 November 1984 and more than 5000 near Golfo de Santa Clara on 20 March 1969 and reported by Russell as "dowitcher sp." were probably Long-billed Dowitchers. Small flocks of from five to as many as 30 individuals are usual.

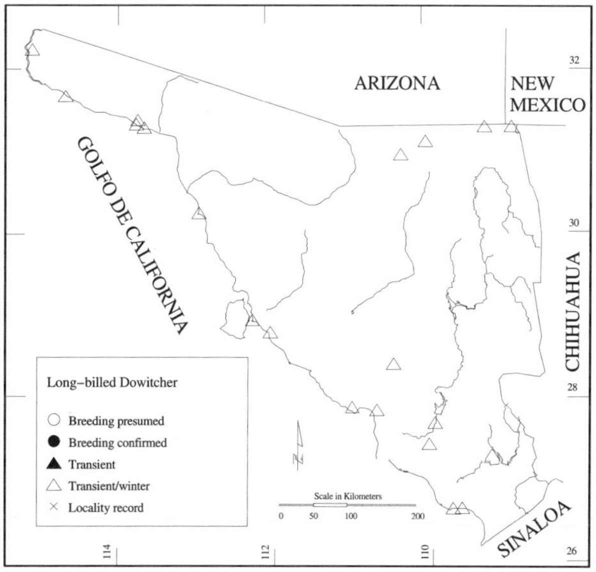

COMMON SNIPE

Agachona Común

Gallinago gallinago

The Common Snipe is an uncommon winter bird of freshwater habitats where shoreline growth offers concealment. These habitats include ponds and reservoirs with weedy or brushy edges, streams and river backwaters with grassy banks, and swamps and marshes. It is found also in thickets along streams; seven were jumped from dense alders along the stream (1780 m) in Cañon Evans of the Sierra los Ajos on 3 March 1984. We flushed one from a dry grassy hillside just outside of Alamos on 30 December 1977; we assumed that it hid there after feeding along a nearby stream, where human and livestock traffic was heavy. Extreme fall arrival and spring departure dates of 19 August and 6 May, published by van Rossem (1945), remain valid. We have found it in saltwater environments seven times, including twice in the undergrowth of mangrove swamps.

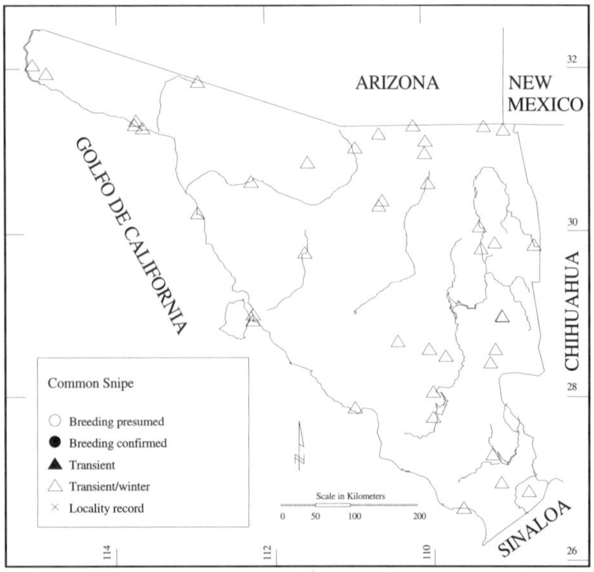

WILSON'S PHALAROPE

Falaropo Pico Largo

Phalaropus tricolor

Despite the paucity of records for Sonora, Wilson's Phalarope is a common transient in adjacent Arizona. Fewer than a dozen records are available for Sonora. In spring it has been observed only in late March; in fall, it has been seen from 31 July to 3 October (van Rossem 1945). Shallow freshwater ponds and reservoir edges are its preferred habitat. Rather surprising is van Rossem's (1945:89) report that he saw it "migrating abundantly off Tiburón Island, April 19, 1925." Two taken at Bahía Kino on 23 June 1963 (UK) were probably southbound migrants. There is one winter sighting: a single bird at Puerto Peñasco on a Christmas Bird Count on 1 January 1980.

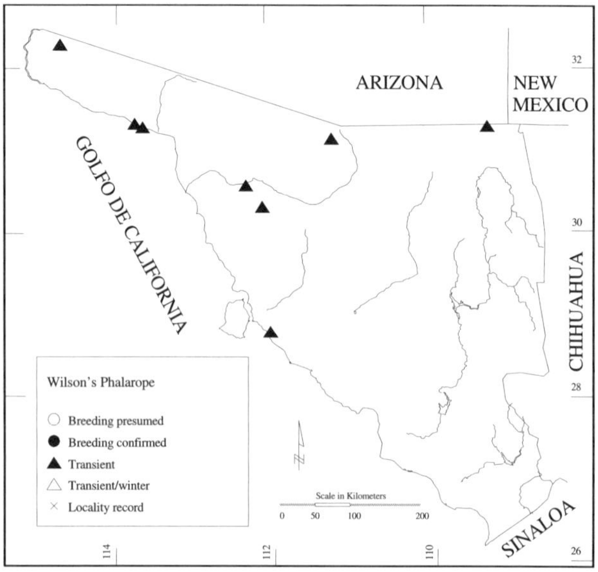

RED-NECKED PHALAROPE

Falaropo Cuello Rojo

Phalaropus lobatus

The Red-necked Phalarope appears to be an abundant transient over the waters of El Golfo de California but is rare (ten records) on shore and inland. There are seven fall reports, from 3 September to 6 October. In spring, two were seen in a roadside freshwater pool

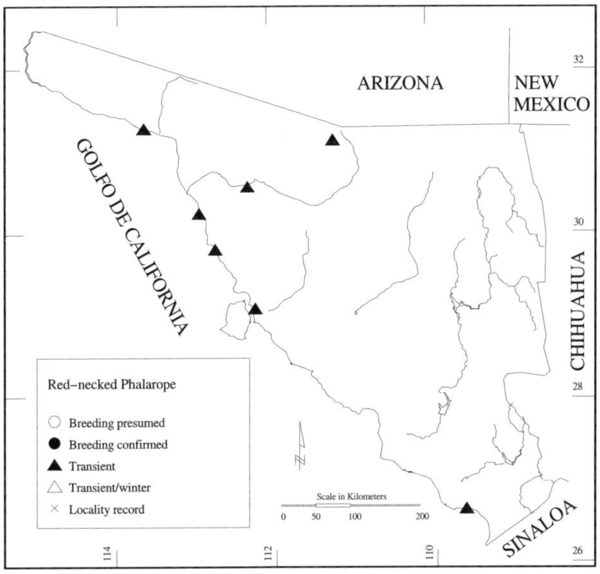

about 8 km north of Punta Chueca on 4 March 1983 (R. Taylor). The only other spring record is on 5 April 1986 at Puerto Lobos. We have seen one in winter, a bird wading in a tiny brackish roadside pond near Yavaros on 10 December 1983.

Van Rossem (1945) observed this phalarope offshore in the gulf in April 1925, migrating northward in immense numbers. On a late March boat trip off Guaymas, Alden (1969) saw thousands. In the fall, thousands more were seen 5 to 6 km off Bahía Kino in early November (R. Taylor).

RED PHALAROPE

Falaropo Pico Grueso

Phalaropus fulicaria

The Red Phalarope is known from only three records in Sonora. D. Stejskal saw 30 at Puerto Peñasco on 6 October 1979. One was seen at Puerto Lobos on 28 February 1986 (J. Bates, R. Bowers); another found at Puerto Peñasco on 12 January 1990 (T. Wurster) may have been wintering.

The Red Phalarope is a pelagic species in migration and winter. It may be present in El Golfo de California more frequently than observations indicate. Van Rossem (1945:88) described "small flocks noted at frequent intervals" in the gulf on 26 December 1931.

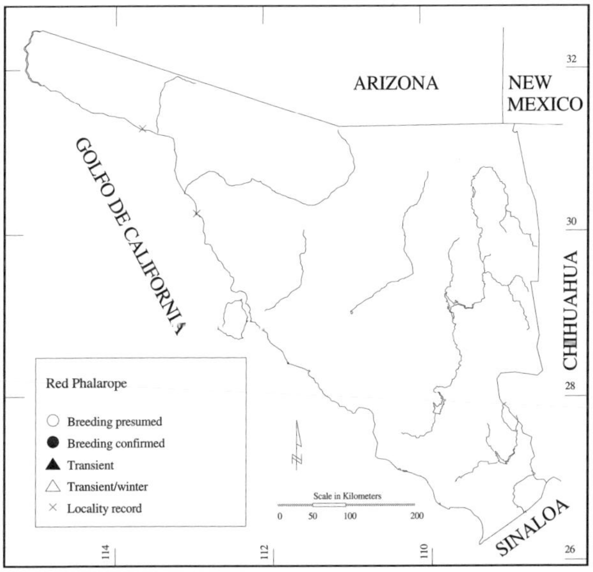

FAMILY LARIDAE

Skuas, Gulls, Terns, and Skimmers

[POMARINE JAEGER

Salteador Pomarino

Stercorarius pomarinus

Van Rossem (1945) reported no jaegers of any species. Two Pomarine Jaegers were observed by K. Kaufman and S. Terrill from a boat while participating in the Puerto Peñasco Christmas Bird Count on 3 January 1981. A. McGill watched two adults harassing an Elegant Tern at Bahía Kino on 11 April 1976.

The field identification of immature jaegers is difficult. We have not accepted several reports of birds observed.]

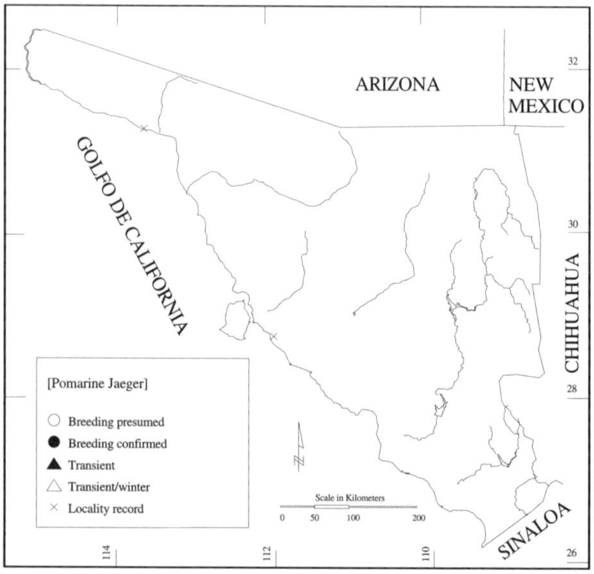

[PARASITIC JAEGER

Salteador Parásito

Stercorarius parasiticus

Our records for this jaeger are limited to two winter observations at Puerto Peñasco: one light-phase adult on 29 December 1974 (D. Stejskal) and one of unknown age on a Christmas Bird Count on 3 January 1981 (seen from a boat by K. Kaufman and S. Terrill). Neither record is documented by a specimen or photograph.]

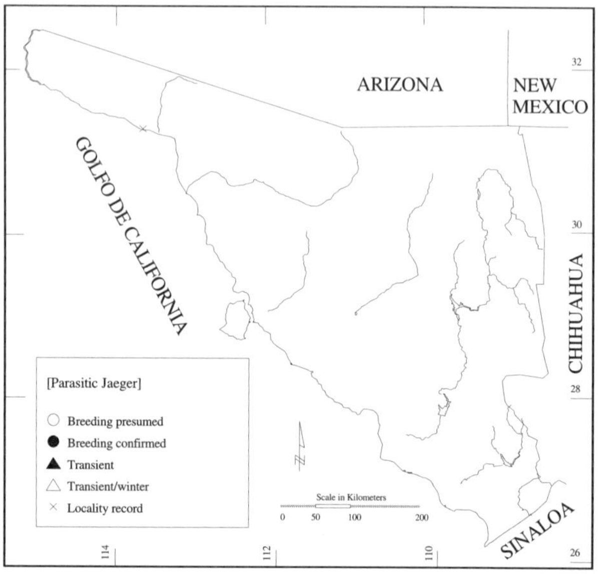

LAUGHING GULL

Gaviota Reidora

Larus atricilla

The Laughing Gull's status in Sonora is quite different from that of other gulls. It is rare in late fall and winter (Puerto Peñasco on 23 October 1964, 16 December 1990, 10 January 1994, and 12 January 1990). It is uncommon from late winter through summer and has not been found nesting, contrary to the AOU Check-list (1983). The earliest spring record is of an adult seen at

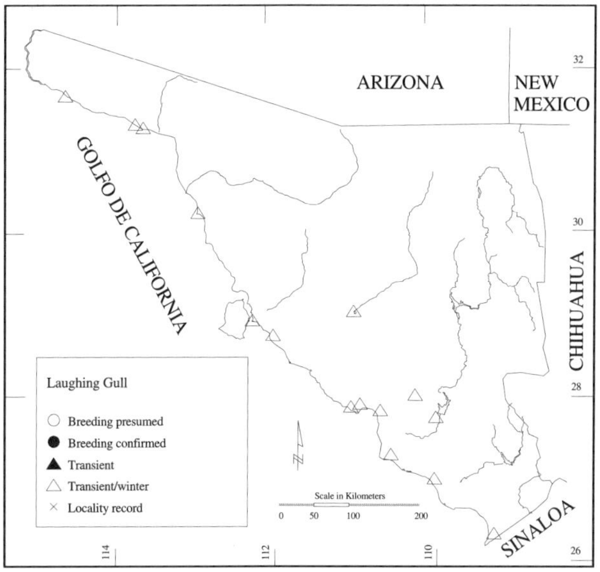

Puerto Peñasco on 20–21 February 1993 (L. Dombrowski). K. Kaufman noted at least three in basic plumage at Puerto Peñasco on 25–27 February 1990; one bird was starting to develop a black head. The latest summer record is of one at Bahía Kino on 24 September 1977, but there are no coastal records between 22 May and 11 July. Inland records come from the Río Yaqui below Presa Obregon, where we saw two on 11 April 1986 and about 140 on 10 July 1987; from Presa Agua Caliente, where S. Levy saw about 20 on 19 June 1988; and from Presa Rodriguez at Hermosillo on 5 April 1990.

The inclusion of Sonora in the breeding range apparently is based on the statement by van Rossem and Hachisuka (1937:333) that "specimens collected at Tóbari Bay [26 April to 1 May 1930] were in breeding condition, and the action of many pairs observed strongly indicated they were either breeding or about to do so." Although it is possible these birds did nest, breeding was never confirmed. Palacios and Mellink (1995) visited Bahía Tóbari on 14 May 1994 and found only two Laughing Gulls (at San José) and no suggestion of nesting. The absence of this species in late May, June, and early July would seem to preclude nesting, at least in recent years.

FRANKLIN'S GULL
Gaviota de Franklin

Larus pipixcan

Franklin's Gull is a rare spring migrant in Sonora, with only four definite records. At Puerto Peñasco, K. Kaufman and S. Terrill observed two adults on 4 April 1980. The same observers also saw an adult at Guaymas on 6 April 1980. We saw seven at Golfo de Santa Clara on 9 April 1994. The single inland observation is of an adult at a pond southwest of Naco on 29 May 1988 (A. Moorhouse).

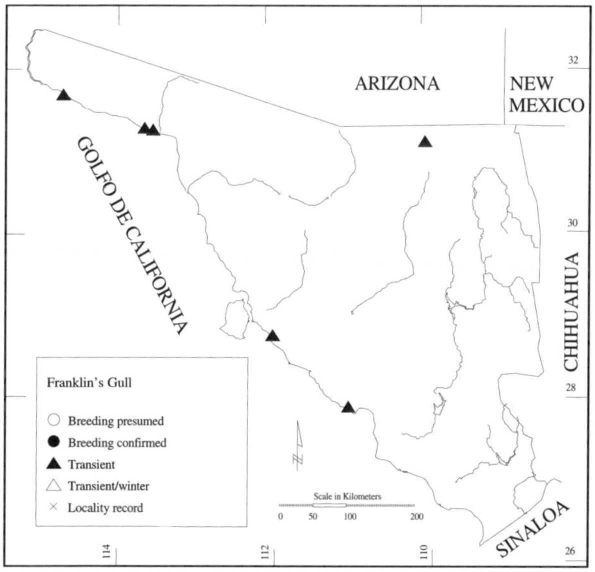

BONAPARTE'S GULL
Gaviota de Bonaparte
Larus philadelphia

This gull is a common and sometimes locally abundant transient and a fairly common winter resident, mainly along the coast of El Golfo de California from Puerto Lobos southward. It is uncommon at Puerto Peñasco. Groups of more than 50 birds are unusual, but at least 4000 were estimated at San Carlos on 28 March 1982, resting at one point on the beach and flying inland to feed at the lagoons. More than 3000 were reported from Puerto Lobos on four dates ranging from 27 February to 7 April, the largest number exceeding 5500 (most in breeding plumage) on 7 April 1984. Migration takes place in September and October and again in March and early April; extreme dates are 8 September (Punta Sargento) and 4 May (Guaymas; van Rossem 1945).

Inland records include one on the Agua Caliente arm of Presa Obregon on 21 September 1985, one at Presa Alatorre on 3 November 1985, 51 on the Río Cedros inlet of Presa Mocuzari on 7 January 1986, and two at Hermosillo on 15 November 1986 (J. Dunning).

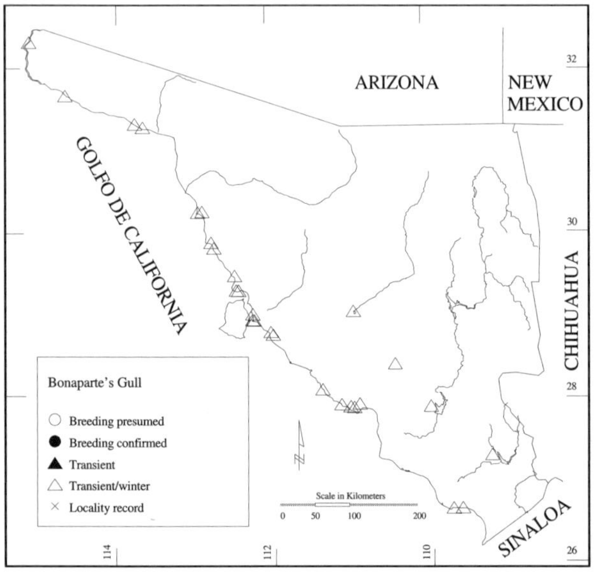

On 9 March 1976, six were hovering over a dense feeding flock of 45 Red-breasted Mergansers, now and then diving into the water among the ducks, at Guaymas.

HEERMANN'S GULL
Gaviota Ploma
Larus heermanni

Heermann's Gull is endemic to the Pacific coast of Mexico and its islands, as well as El Golfo de California. All year this gull is abundant on the gulf coast, with numbers falling off somewhat between mid-March and late June, when many adult birds are breeding on the islands. Hundreds are sometimes gathered at one

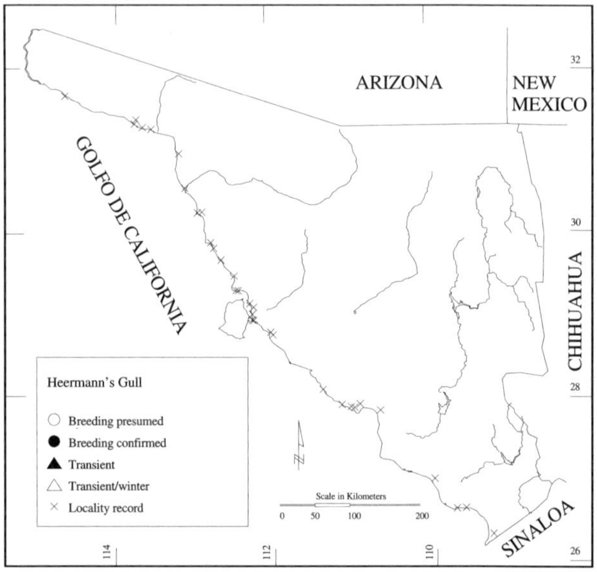

Heermann's Gull

locality: for example, we saw more than 1000 at Punta Cirio on 26 December 1974 and over 2000 at Puerto Lobos on 7 April 1984. On 7–8 November 1987, between 2000 and 3000 were on Isla Alcatraz at the entrance to Bahía Kino (R. Taylor). They often associate closely with Brown Pelicans and adroitly steal fish from them.

RING-BILLED GULL
Gaviota Pico Anillado

Larus delawarensis

This gull is common as a migrant and winter visitant along El Golfo de California. It is particularly numerous near coastal cities and towns; more than 1000 birds have been counted at a time at Puerto Peñasco, Bahía Kino, and San Carlos, with the largest number, 2500, being reported at Puerto Lobos on 13 December 1986 (J. Bates). From mid-April to early October this gull is largely absent, but an occasional bird can be seen in summer. Forty-nine were found at Puerto Peñasco on 23 June 1975 (D. Stejskal).

The Ring-billed Gull is the only common gull of interior Sonora, to be found chiefly on large reservoirs in late fall and winter. Among high counts were more than 350 on the Río Cedros inlet of Presa Mocuzari on 7 January 1986 and at least 60 at Presa Alatorre on 3 November 1985. It is regularly found on Presa Rodriguez at Hermosillo, where it is present occasionally even in summer.

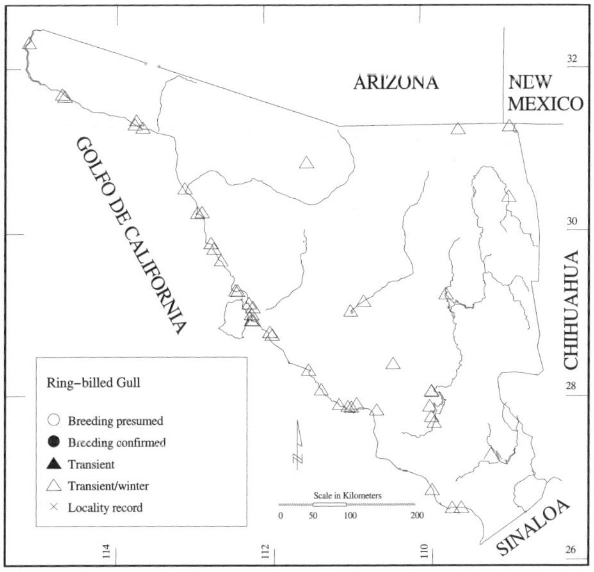

CALIFORNIA GULL
Gaviota Californiana

Larus californicus

The California Gull is an uncommon fall migrant and a fairly common winter visitant and spring migrant on the central and northern coast of Sonora. Most observations have been of fewer than 50 birds, but this has been the most abundant gull at Puerto Peñasco on four of eight Christmas Bird Counts; total numbers there have ranged from 500 to 6300 birds. Ring-billed and Heermann's Gulls have been more numerous on the other counts.

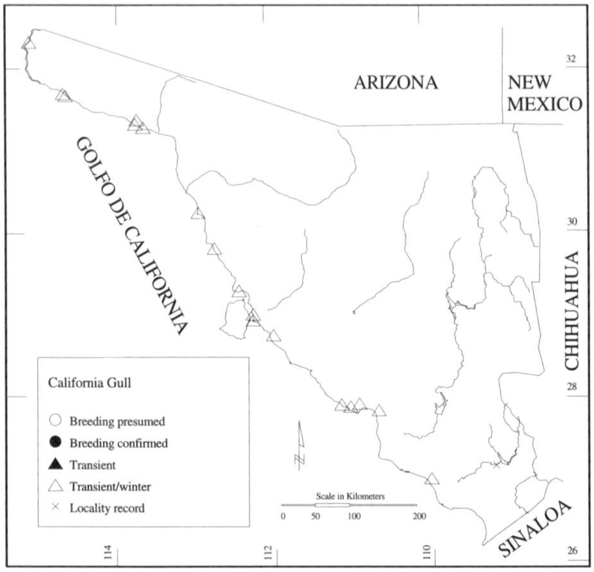

Migration periods are mainly mid-March to early May, and again in October. There are no recent records south of the Guaymas area, but van Rossem (1945:91) stated that these gulls were "migrating northward in large flocks at San Esteban Island, April 17 to 19, 1930, and at Tóbari Bay, April 26 to May 1, 1930; common in Guaymas Harbor as late as May 14, 1930, although in this last instance most individuals seen were immature." There are two summer records: an immature at Punta Chueca on 16 July 1975 and a single bird of unknown age at Puerto Peñasco on 23–24 July 1979 (Gallucci 1981). The only inland Sonoran record is a bird reported at Presa Mocuzari on 25 November 1979 (D. Stejskal).

HERRING GULL
Gaviota Plateada

Larus argentatus

The Herring Gull is a fairly common transient and winter visitant from November through early April on the coast of El Golfo de California, mostly north of Bahía Kino. Occasionally numbers may be especially large from Puerto Lobos north; more than 1000 have been listed on Puerto Peñasco Christmas Bird Counts. There are two May records: we saw one on 22 May 1986 at Guásimas, and D. Stejskal found six at Puerto Peñasco on 19 May 1974. The solitary inland record is of 23 birds at Presa Cuauhtémoc near Tabutama on 3 April 1983.

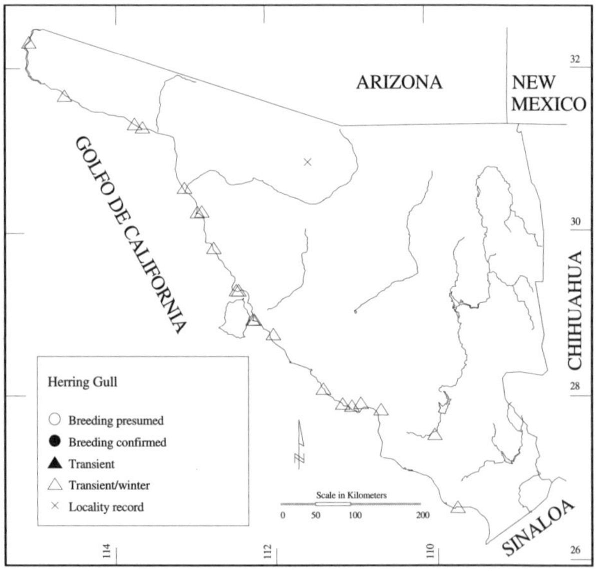

THAYER'S GULL
Gaviota de Thayer
Larus thayeri

The only reports are from Puerto Peñasco, where (with one exception) one to three individuals have been observed on eight dates in December, January, February, and March. K. Kaufman and D. Stejskal counted 15 on 12 February 1982. The species was unknown from Sonora prior to 1981. Thayer's Gull may have long been present in El Golfo de California in winter, but for many years (until 1973) the species was considered a race of the Herring Gull and few people attempted to identify it in the field. Even now, the species is probably overlooked and individuals are mistakenly identified as other large gull species.

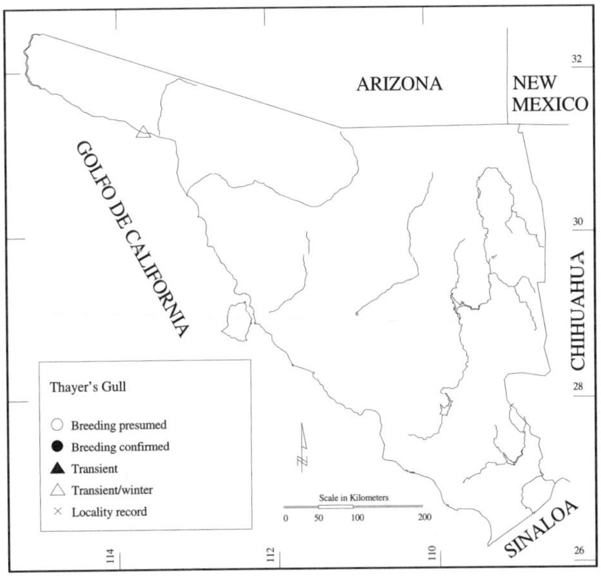

YELLOW-FOOTED GULL
Gaviota Pata Amarilla
Larus livens

Prior to 1983 the Yellow-footed Gull was included in the AOU Check-list as a subspecies of the Western Gull. It is common year-round on the coast of El Golfo de California in Sonora but breeds only on islands in the gulf, including Isla Alcatraz in Bahía Kino. Among the high counts are more than 900 on a Christmas Bird Count at Puerto Peñasco on 19 December 1981 and nearly 1000 at Puerto Lobos on 7 April 1984 and 18 February 1989. Observation of fewer than 100 individuals is typical.

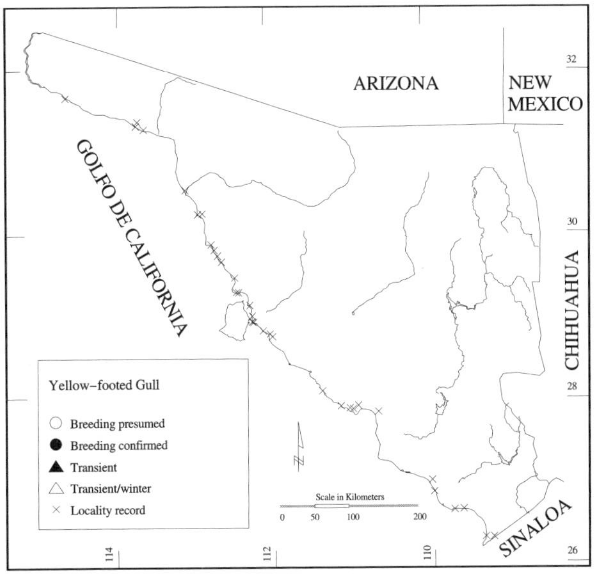

WESTERN GULL
Gaviota Occidental
Larus occidentalis

The first report of a Western Gull in Sonora was of an adult observed at Bahía San Pedro north of Guaymas on 25 December 1931 (van Rossem 1945). Since then, there have been reports of one to four individuals, ranging seasonally from 4 September (1986) at Puerto Lobos (J. Bates) to 3 May (1988) 10 km south of Bahía Kino (R. Taylor). The southernmost observation is of one at Bahía Guásimas on 9 January 1986. Other months represented are November, February, March, and April. Western Gulls, especially immatures, may be regular in small numbers at Puerto Peñasco but are undoubtedly overlooked among the large numbers of Yellow-footed Gulls. K. Kaufman, on about six visits to Puerto Peñasco in recent years, has found one to four (mostly immature) Western Gulls on each occasion associating with Yellow-footed Gulls around the fish docks.

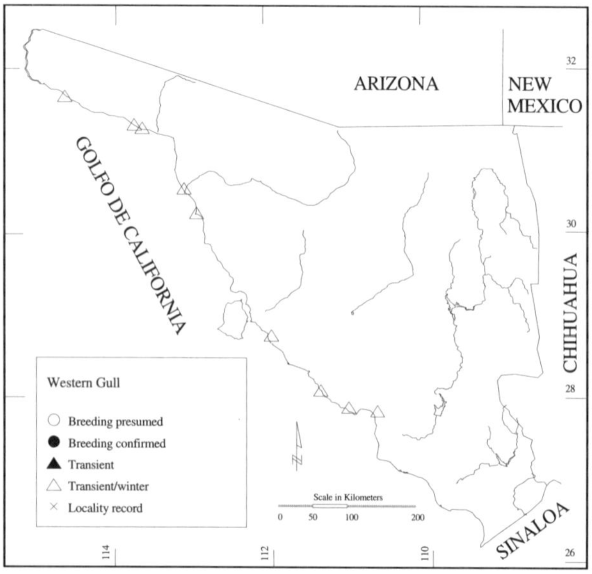

GLAUCOUS-WINGED GULL
Gaviota Ala Glauca
Larus glaucescens

The Glaucous-winged Gull is rare in Sonora, as indicated by only 14 reports, from 19 August to 28 April. Most observations have been of one to three birds in December, January, and February at Puerto Peñasco, but K. Kaufman and D. Stejskal found seven there on 12 February 1982.

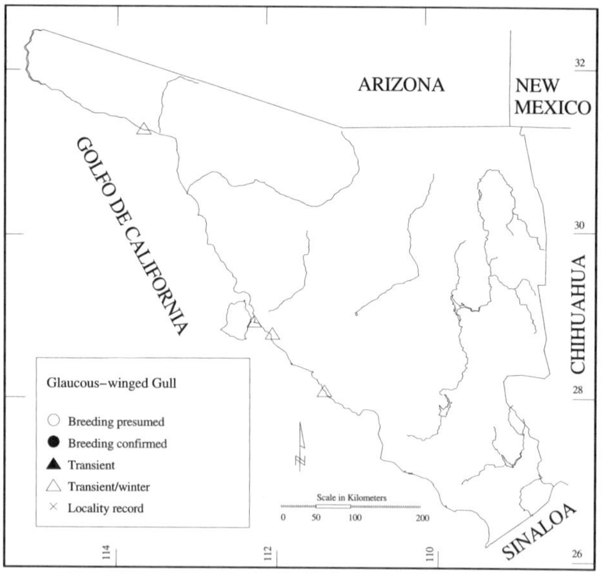

GLAUCOUS GULL
Gaviota Blanca
Larus hyperboreus

We know of four records from Sonora, all from Puerto Peñasco. A first-year Glaucous Gull was identified on 21 or 22 March 1978 (D. Stejskal, P. Burch). A third-year bird was photographed on 10 March 1979 (B. Whitney, photograph UA). A first-winter individual was seen on 11 February 1989 (J. Coons). On 25 February 1990, a bird in first-winter plumage was observed on the fish docks (K. Kaufman).

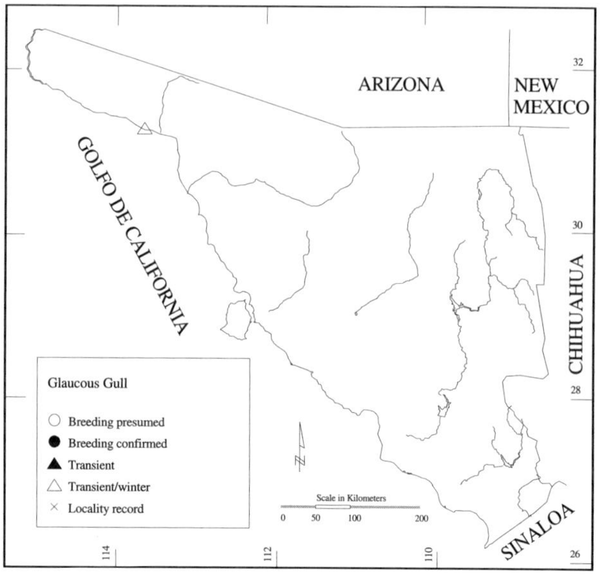

BLACK-LEGGED KITTIWAKE
Gaviota Pata Negra
Rissa tridactyla

A bird in first-winter plumage was seen in the Guaymas harbor on 26 January 1982 by S. Goldwasser, who provided good details of the observation. Another first-winter individual (photograph UA) was noted at Puerto Peñasco on 25–27 February 1990 by K. Kaufman.

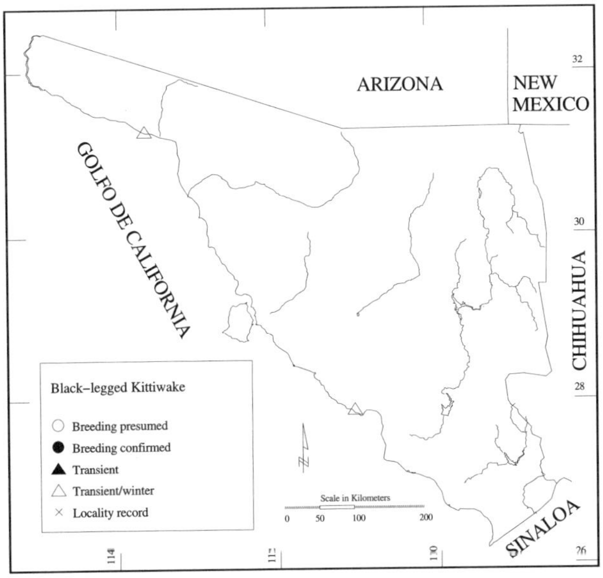

[SABINE'S GULL
Gaviota Cola Hendida
Xema sabini

We know of only two records (Russell and Lamm 1978), both near Puerto Peñasco: three observed on 5–6 October 1969 (R. Witzeman) and an immature bird found on 13 October 1969 (R. Cunningham).]

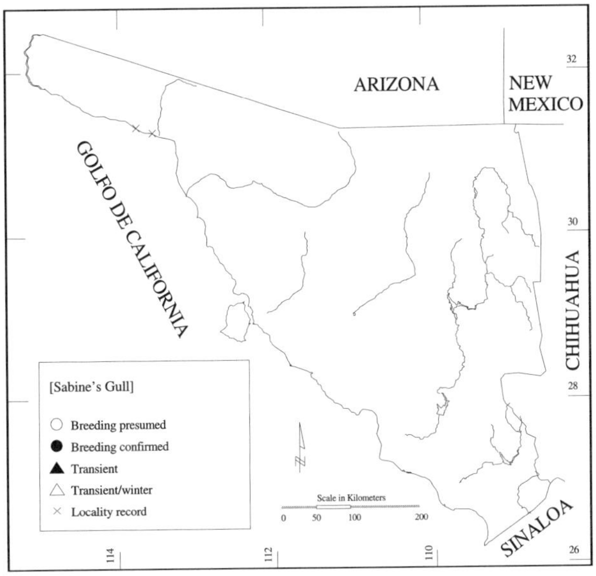

GULL-BILLED TERN
Charrán Pico Grueso
Sterna nilotica

Although the 1983 AOU Check-list states that the Gull-billed Tern breeds on the coast of Sonora, such has not been documented. The conclusion is based on van Rossem's observation (van Rossem and Hachisuka 1937:333), "Gull-billed Terns were noted as not uncommon at Tóbari Bay from April 26 to May 1 1930. They were usually in pairs and it was the distinct impression they were preparing to breed on one or more sand islands at the entrance to the bay." Aside from van Rossem's observations, we are aware of ten records, including dates from 20 February to 5 October (except May and June). No more than seven individuals were noted on any occasion, and there are no records from extreme southern Sonora except those of van Rossem. We consider the species to be a rare summer visitant.

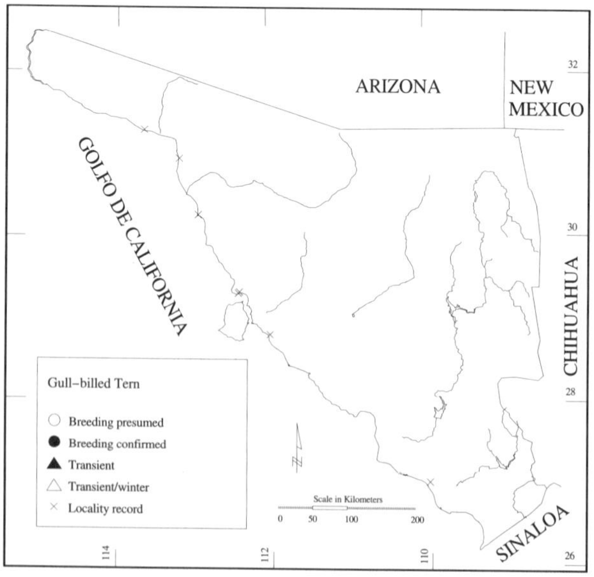

CASPIAN TERN
Charrán Caspia
Sterna caspia

Van Rossem (1945) cited only two records of the Caspian Tern for Sonora. It has apparently increased markedly in numbers in recent years, and we now consider it uncommon in May through July and common in other months. It has not been found nesting, although nests have been reported in adjacent Sinaloa. Principal migration periods are April and September. Only 15% of all reports have been of 15 or more individuals at any one place. We observed 300, the largest concentration, at Bahía Kino on 24 September 1977.

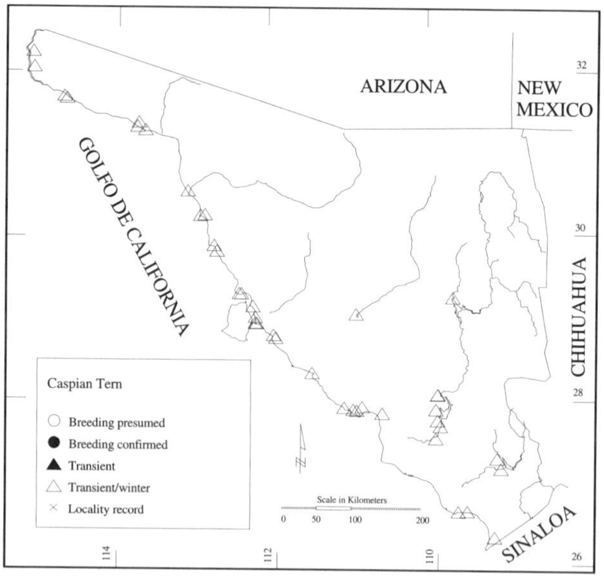

ROYAL TERN
Charrán Real
Sterna maxima

This large tern is uncommon throughout the year on rocky and sandy beaches of El Golfo de California. It is less frequently seen in summer, when presumably most individuals have moved to insular breeding grounds in the gulf. Van Rossem (1945) reported pairs that may have nested in Estero Tóbari in 1930 and Estero Agiabampo in 1937, all in May. We have found no positive evidence of nesting and believe that if it ever nested in Sonora it no longer does so. It seems never to be found in large numbers; the largest groups have been seen at Punta Chueca in December (80 birds) and September (75 birds), but fewer than six are usually noted at a locality.

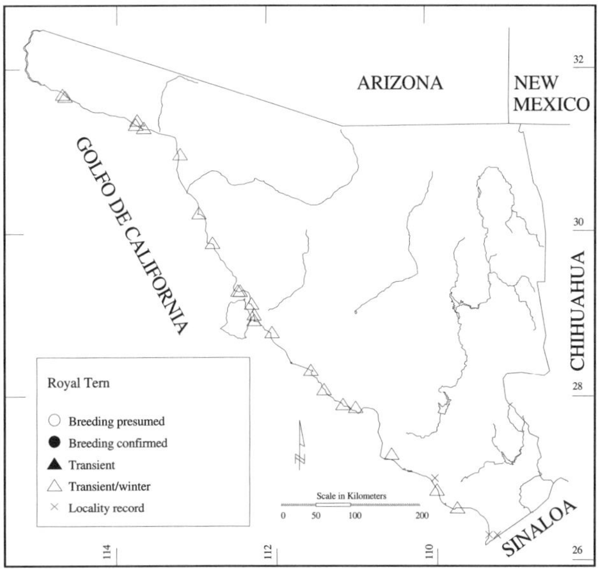

ELEGANT TERN
Charrán Elegante
Sterna elegans

This tern is chiefly a bird of the Pacific coast of Baja California and El Golfo de California, with a well-known breeding colony on Isla Raza, where thousands nest. Nonbreeding individuals are fairly common along the coast of Sonora from spring through fall. In winter, Elegant Terns are uncommon north to Canal el Infiernillo and rare farther north. The largest gatherings have been seen prebreeding—including 3800 individuals, some courting and copulating, at Puerto Lobos on 7 April. Some of the 700 at Bahía Kino on 8 April were engaged in courtship activity, with males flying in from the sea with fish to present to the females. The assemblage was noisy; the most common call resembled a single note of the Gila Woodpecker (Monson). Smaller groups have been observed in autumn (a maximum of 350 at Puerto Lobos on 5 October 1986; J. Bates).

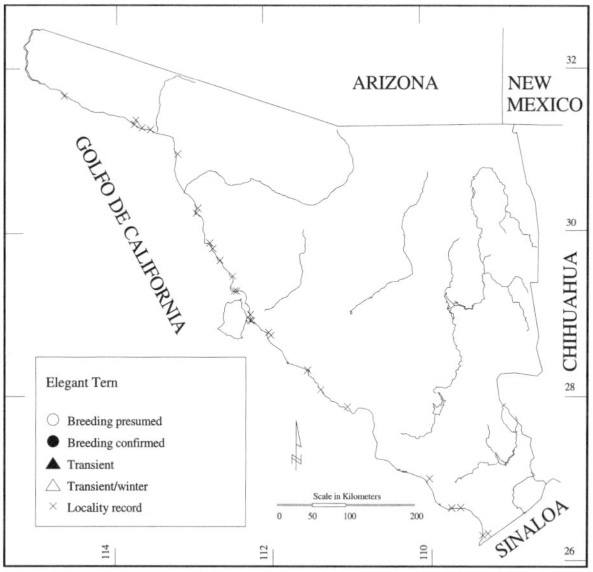

[SANDWICH TERN
Charrán de Sandwich
Sterna sandvicensis

On 4 April 1987, H. Hobart studied a Sandwich Tern as it rested on a sandy beach at Puerto Lobos, carefully noting the yellow-tipped black bill through his telescope. Hobart had previously done research on a nesting colony that included this species and had no reservations concerning the identification. This tern is accidental on the Pacific coast north of Oaxaca.]

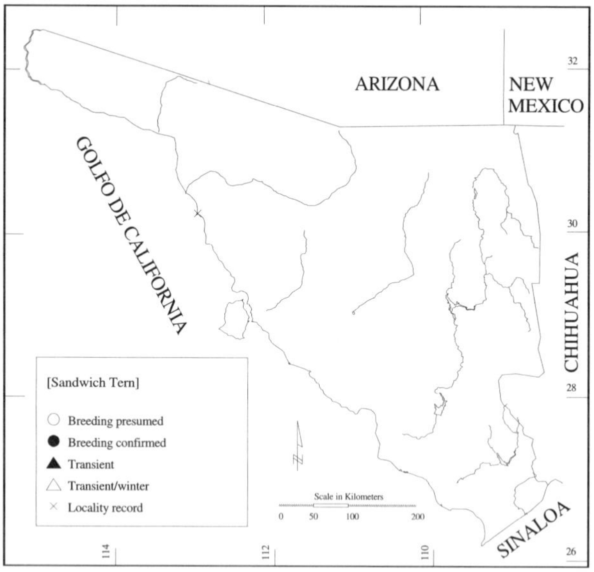

COMMON TERN
Charrán Común
Sterna hirundo

The Common Tern is a rare transient and casual winter visitant in Sonora. More than half of the 19 records of which we are aware are from the months of March–April and September–October. The species has not been reported in May and December. On 13 October 1969, R. Cunningham counted 120 Common Terns flying past Puerto Peñasco in 4.5 hours. Two reports from June (one a specimen, UK) and one in July suggest that the species occasionally summers in Sonora.

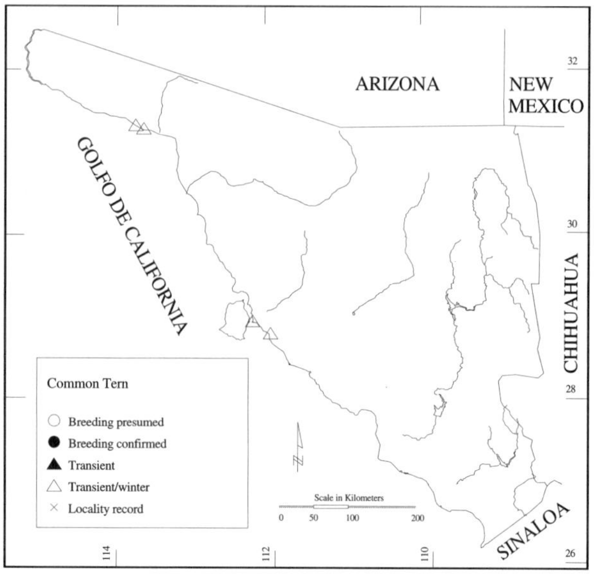

FORSTER'S TERN
Charrán de Forster
Sterna forsteri

Forster's Tern is a common transient and winter resident of the coast of El Golfo de California in Sonora. The southbound migration occurs in September and early October, and birds return northward in March and the first half of April. Many may be present in summer. Maximum numbers found in one place on one date are 530 observed at Puerto Peñasco on 3 September 1979 (D. Stejskal) and 500 at Puerto Lobos on 2 October 1986 (J. Bates). Although there are no inland records, the frequency with which this species is observed on migration throughout Arizona would make it probable that individuals also pass through interior Sonora.

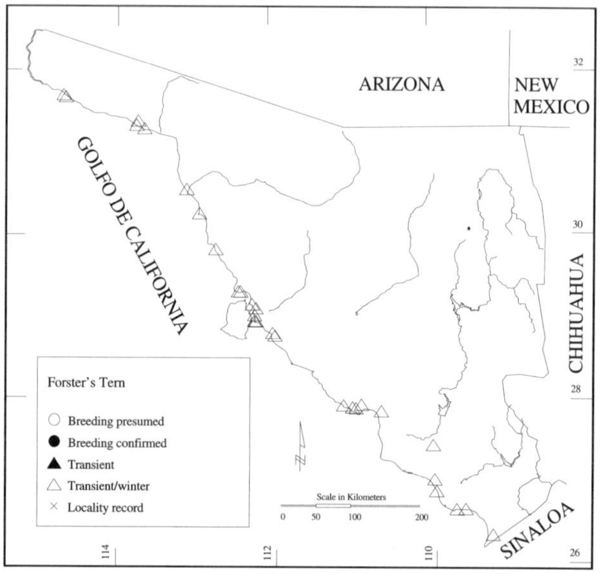

LEAST TERN
Charrán Mínimo
Sterna antillarum

This small tern is an uncommon summer resident of the coast of Sonora, with scattered nesting colonies. The earliest arrivals were one bird at Puerto Peñasco on 26 March 1983 and two at San Carlos on 28 March 1982; most arrive in early April. Breeding birds depart in July, but some transients have been seen until 2 October (1986; J. Bates). D. Stejskal saw 48 migrating at Puerto Peñasco on 3 September 1979.

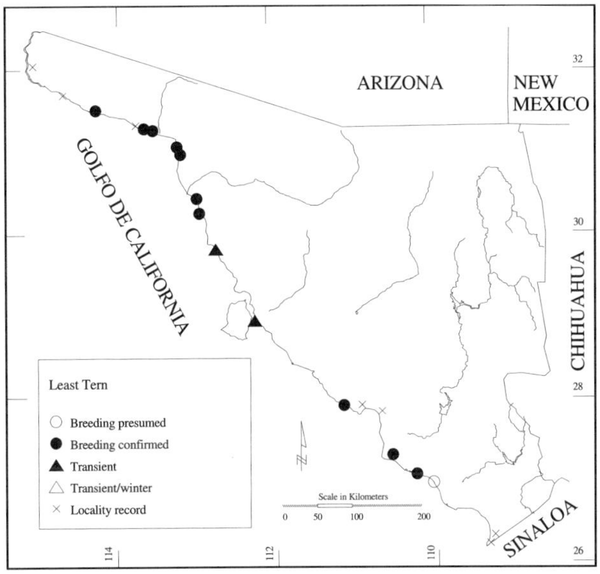

Approximately 25 pairs nested in a colony at San Carlos in 1979. Eggs were laid on the ground on a barren expanse at the north end of the lagoon area just to the east of the San Carlos development. Wilson's Plovers were nesting in the same locality. A small colony at Puerto Peñasco was reported by Russell and Lamm (1978). The birds were nesting in June and absent by 13 July. Three pairs nested about a kilometer north of the Puerto Lobos lighthouse in June 1986, followed by four nesting pairs in July (J. and R. Hanson). The nests were situated in sand dunes just above the storm tide line, and the prevalence of off-road vehicle tracks suggested that some nests had likely been destroyed. Three nesting colonies were found in the Puerto Peñasco and Bahía de San Jorge areas in 1991 and 1992 (Mellink and Palacios 1993). On 14 May 1994, Palacios and Mellink (1995) found 15 pairs nesting in each of two colonies near San José. The number of breeding colonies and individuals may be declining, and their status should be carefully ascertained and monitored. Several sets of eggs (WFVZ) were taken by G. Bancroft in May 1930 at "San Juan Lagoon," a locality which we cannot find on maps.

Van Rossem (1945) stated that Least Tern specimens were taken, but later lost, at Puerto Peñasco on 21–23 January 1930. There are no other winter records.

BLACK TERN
Charrán Negro
Chlidonias niger

Black Terns are rare spring coastal migrants and uncommon fall migrants on the coast and in the interior. Each of the seven spring records is of fewer than five birds (with one exception) and within the period of 28 February through 7 June. Palacios and Mellink (1995) noted 37 birds at Bahía Tóbari on 14 May 1994. D. Stejskal noted 77 individuals at Puerto Peñasco on 23 June 1975 and 69 there on 20 July 1975. These were presumably very early fall migrants. Most records are of migrants in August through October; the latest fall bird was collected near Caborca (Phillips and Amadon 1952) on 31 October 1948.

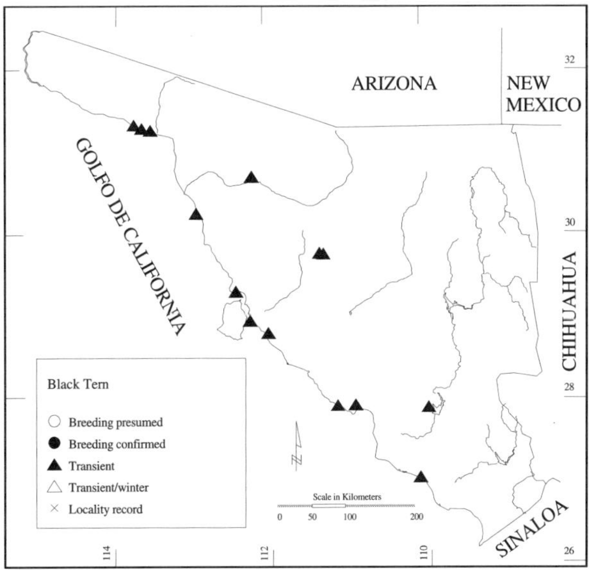

BLACK SKIMMER
Rayador Americano
Rynchops niger

The Black Skimmer, a rare transient and winter visitant along all of the coast of El Golfo de California, was not detected north of Guaymas until 24 April 1974, when one was noted at Puerto Peñasco by J. and R. Witzeman (Russell and Lamm 1978). Subsequently it has been seen at Puerto Peñasco and Puerto Lobos eight times. The largest number found in Sonora to date is 53 near Golfo de Santa Clara on 9 April 1994. The only record for mid-May through August is that of a single bird at Puerto Peñasco on 27 July 1979 (Gallucci 1981).

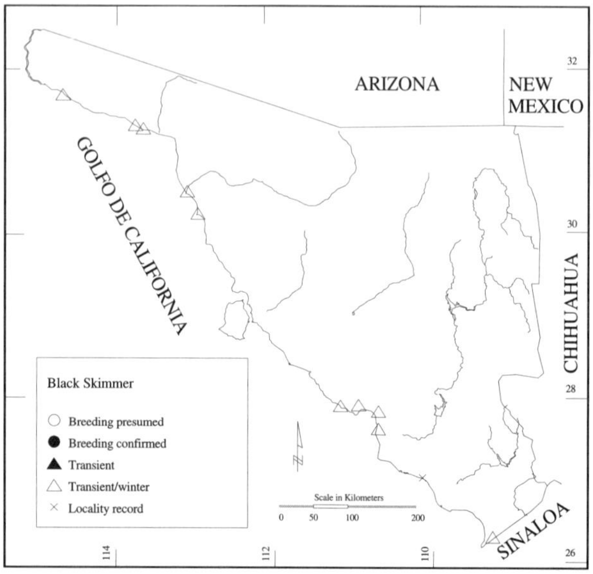

The assertion in the 1983 AOU Check-list that it breeds along the coast of Sonora is probably based on van Rossem (1945:95): "Six specimens in the Dickey collection, taken at Algodones Lagoon, May 1, 1930, and Guásimas Lagoon, May 12, 1930, were just finishing the prenuptial molt and were apparently nearly ready to breed on the above dates." Nests have not been found in Sonora, but Palacios and Mellink (1995:102) found 50 adults in the presence of "nest-like depressions in the sand" at Bahía Tóbari on 14 May 1994.

FAMILY ALCIDAE
Auks, Murres, and Puffins

CRAVERI'S MURRELET
Mérgulo de Craveri
Synthliboramphus craveri

Craveri's Murrelets breed primarily on islands in El Golfo de California. They were found nesting on Isla Alcatraz, 1.5 km from shore at Bahía Kino (DeWeese and Anderson 1976), where an egg was collected (MVZ) on 21 May 1975.

We have never seen Craveri's Murrelets from shore. A specimen (LACM) was taken from the bay at San Carlos on 25 March 1949 by J. Garth, and another (UK) was collected near Bahía Kino on 27 June 1963 by R. Johnston.

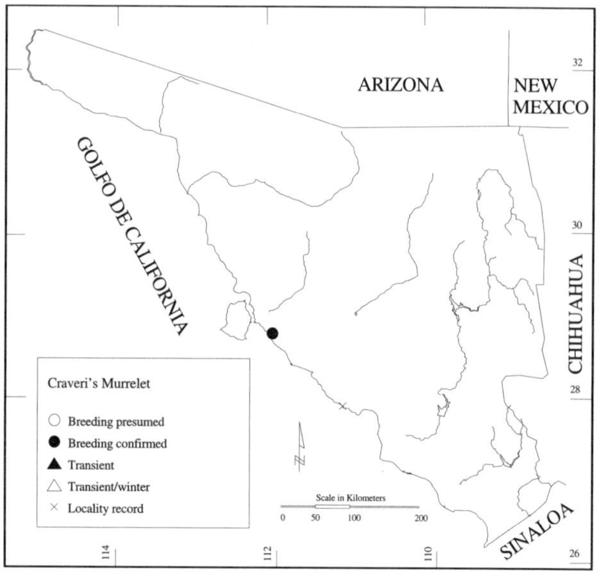

FAMILY COLUMBIDAE

Pigeons and Doves

ROCK DOVE
Paloma Doméstica
Columba livia

Scarcely a village or city in Sonora is without permanently residing Rock Doves. Monson has never seen these birds as numerous anywhere as on 9 January 1985 in farmland between San Luis Río Colorado and Riíto, when he saw more than 5000 while driving about 25 km. The birds presumably fed upon grain in the vicinity of feedlots and in harvested fields.

In America, where they are not native, Rock Doves display an endless variety of plumage. Because they do not nest anywhere in Sonora except in association with man and his edifices, few people have included them in their accounts of birds. Van Rossem (1945) chose not to cite them. We do not know when they were first brought to Sonora, and our map does not portray their extensive current distribution.

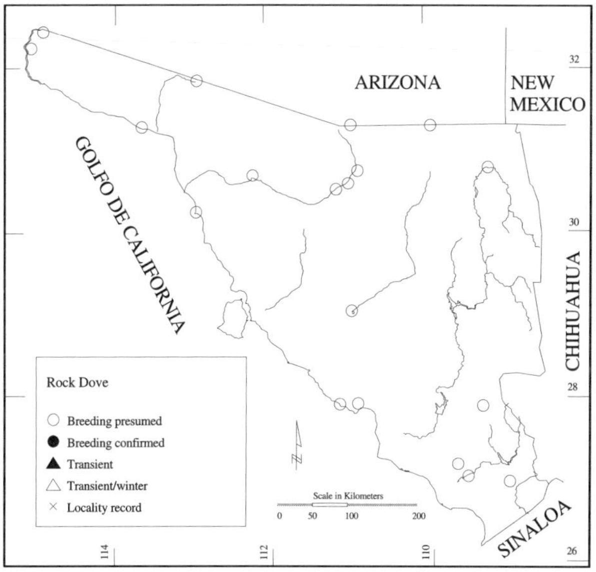

RED-BILLED PIGEON
Paloma Morada
Columba flavirostris

Red-billed Pigeons are fairly common residents of the tropical deciduous forest, where they are generally associated with large fruiting trees, especially along streams and around clearings. We have seen nest building as early as 11 April and fledged young to mid-July. Only one egg is laid. These pigeons seldom call in the nonbreeding season, and they can stay hidden in dense trees.

About a dozen birds observed on 20 June 1988 in irrigated farmland about 70 km southwest of Hermosillo (S. Levy) may have been nonbreeding individuals, as this is distant from any supposed nesting locality.

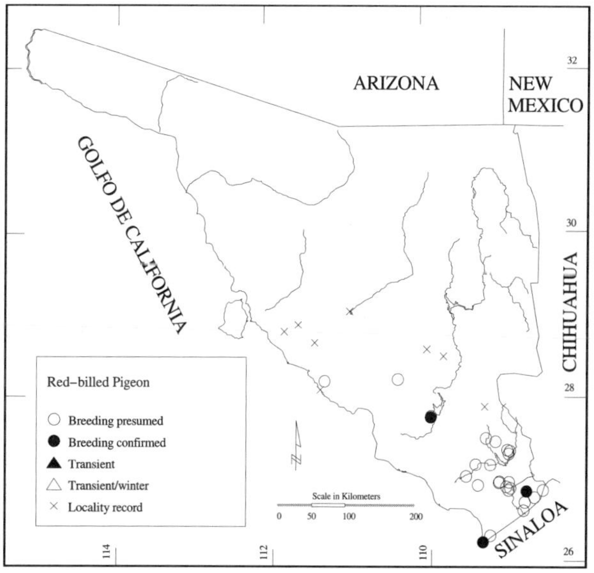

Commencing in September, many birds move northward coastwise and inland as far as Hermosillo. About 200 were found from 22 to 24 September 1981 in fields south-southwest of Hermosillo, flying in at evening with large numbers of White-winged Doves to roost in citrus groves; they were said to feed on grapes being raised for the brandy industry (S. Levy). Quite a few are killed during the hunting season (October to early March) in farmlands southwest of Hermosillo, where birds examined had fed on sorghum and grapes, and local hunters claimed that the pigeons were especially fond of garbanzo beans (Foust 1987). That this northward movement is not entirely confined to the period following the start of large-scale irrigation in the Hermosillo region is indicated by a record from Rancho Costa Rica on 15 December 1932 (van Rossem 1945) and one collected at Hermosillo in early February 1946 by van Rossem.

BAND-TAILED PIGEON
Paloma de Collar
Columba fasciata

These pigeons of the mountains are found throughout the year in oak and pine-oak woodland in Sonora, although they are less often recorded in winter. They are notorious wanderers, influenced in their movements by the relative abundance of mast and berries. Even in the nesting period, which may be irregular and can last from March to September but usually peaks in August (J. Marshall), they may fly some distance to feeding grounds. Regardless of their proclivity to wander, Marshall (1957), in his interesting account of Band-tailed Pigeon habits in the northeastern mountains of Sonora, noted birds nesting at the same localities in succeeding years.

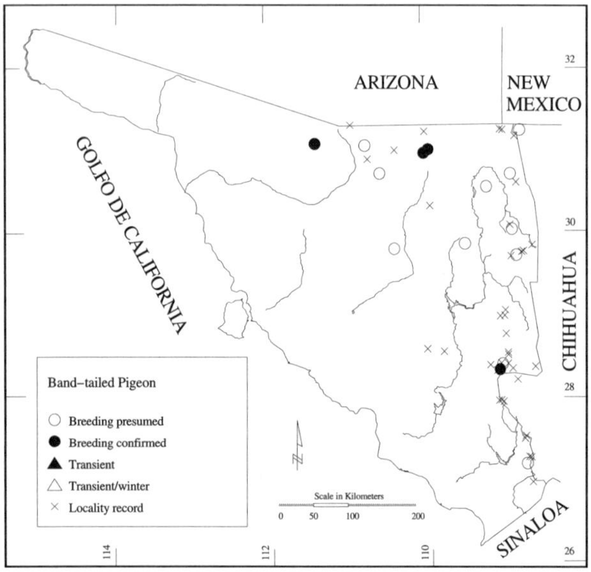

The largest number we have seen at one time was on 30 April 1977 from 1640 to 1810, when several thousand flew from east to west in oak woodland about 8 km southeast of El Trigo, north of Yécora. J. Levy saw about 250 gathered in a dense flock eating gravel or salt on an earthen cliff in the Sierra Cananea at 0730 on or about 10 October 1984. Mast is likely the predominant food. Mexican elderberry fruits lure the pigeons, especially in early summer, to somewhat lower than usual elevations, such as Arroyo Cajón Bonito in the northeast and the Río Magdalena south of Nogales. We have not seen them at elevations below 760 m (San Nicolás, west of Yécora), but van Rossem (1945) cited an early April record for Tecoripa (400 m).

WHITE-WINGED DOVE
Paloma Ala Blanca
Zenaida asiatica

Everywhere in Sonora from oak woodland (1600 m) down to the seashore, the White-winged Dove is a common to abundant breeding bird. It is uncommon in winter north of latitude 30° N, but occasional small flocks may be present. It is common in southern Sonora in winter. At that season many can be found in agricultural areas, feeding upon unharvested grains. We do not know if the breeding birds of the south migrate or are resident.

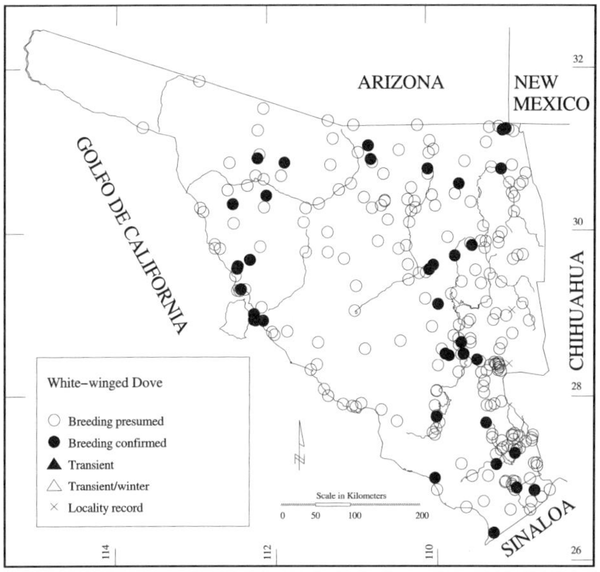

The nesting season is long, extending from March to late August. The flimsy nest of twigs may be placed in almost any tree, shrub, or cactus. On the coast this dove frequently nests in mangroves. Clutches are almost invariably of two eggs. In areas of dense riparian vegetation within a few kilometers of good foraging in agricultural fields, White-winged Doves may nest so close to one another that they may be called "colonial." In such situations, the distinctive song of this species comes from seemingly everywhere in an overwhelming chorus. Several thousand pairs nested in huge mesquite thickets (now cut down) between Pitiquito and Caborca in July 1942 (Neff 1947). Nesting often coincides with the ripening of cactus fruits, particularly those of saguaro, organ pipe, and cardón, which are important items in the bird's diet.

MOURNING DOVE
Paloma Huilota
Zenaida macroura

The Mourning Dove is a common bird throughout much of Sonora and may be found in all major plant communities, including the high-elevation coniferous forests. It does not remain in the pines through the winter, and it is uncommon in tropical thornscrub and deciduous forest in summer, as well as in the most arid parts of the desertscrub in northwestern Sonora. It is outnumbered by White-winged Doves throughout much of their shared breeding range. We do not know if breeding birds are resident year-round, but numbers everywhere increase in September with a corresponding decrease in March. Banded Mourning Doves taken by hunters in Sonora had their origins mainly in Arizona, California, Nevada, and Utah, but some were from Washington, Montana, and Nebraska.

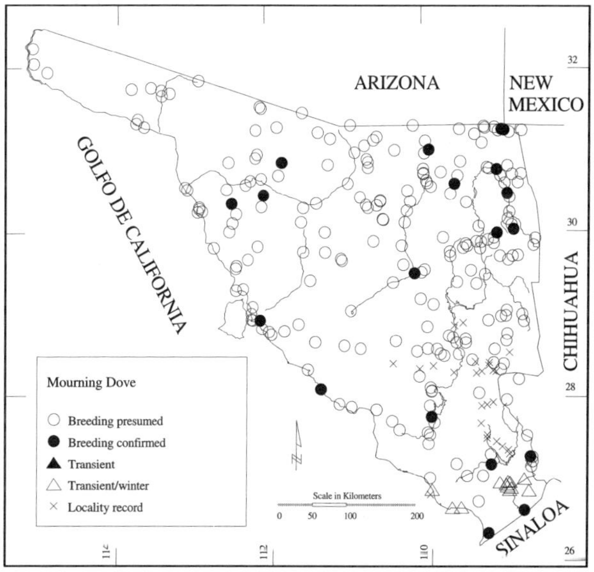

Like others of its family, the Mourning Dove has a prolonged nesting season, from about mid-March to mid-September. It lays its two eggs in a frail stick nest placed in cholla cacti, if available, or on a branch in the lower 2 or 3 m of almost any tree or shrub. Interestingly, it is absent in summer, and nests have not been found in much of lowland southeastern Sonora, including Güirocoba and the Alamos area, where much fieldwork has been conducted.

Occasionally, gatherings of several hundred birds occur in and around agricultural areas, especially in winter. At other seasons, Short (1974) described Mourning Doves as abundant along the highway between Cd. Obregon and Navojoa in late July and early August. We found doves to be abundant along the Tecoripa highway east-southeast of Hermosillo on 20 September 1985 for the first 40 km out of the city; six weeks later there were none along the same stretch of road. Usually such gatherings as these are attributable to blowing weed seeds that collect on the road edge, but grain spilled or blown from trucks may also be the attraction.

INCA DOVE
Tórtola Cola Larga
Columbina inca

The Inca Dove is associated with cities, villages, ranch buildings, corrals, and farmhouses, where it is a common resident. It also lives in dense thornscrub. Its low monotonous double call, interpreted almost invariably in English as "no hope," is heard throughout the year but less frequently from November through January.

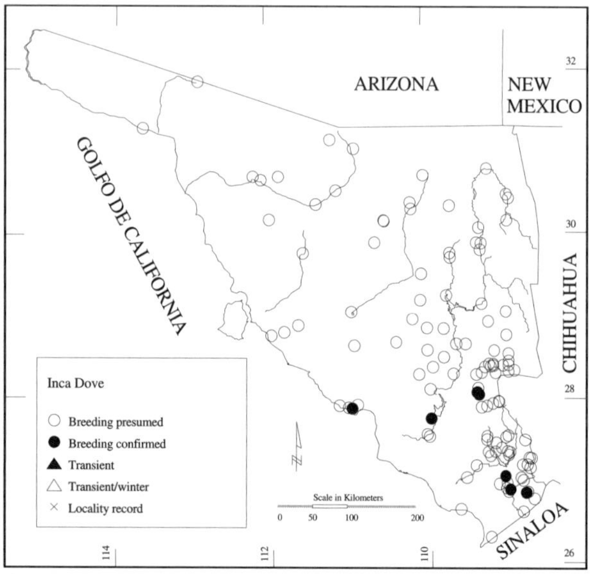

Throughout much of Sonora the Inca Dove is a familiar resident. It lives from sea level at the coast to over 1800 m in the sierras. We have few nesting records but surmise it may nest in every month. This is a ground-feeding dove, invariably seen pecking at almost invisible seeds in open areas.

COMMON GROUND-DOVE
Tórtola Coquita
Columbina passerina

The tropical thornscrub and tropical deciduous forest are well populated by the Common Ground-Dove. In some places, one has the impression that these ground-doves are as numberless as grasshoppers, as they fly up before you when you make your way through the brush. Along the coast of El Golfo de California they are uncommon, with only a few records north of Bahía Kino. We found single birds at Black Mountain north of Puerto Peñasco on 12 March 1978 and 6 October 1979 and at Punta Cirio on 25 November 1977. D. Stejskal observed one at Puerto Peñasco on 16 December 1989. It is possible that the birds were fall- or winter-dispersing young of more northerly populations.

North and east of the southern coast and the tropical plant communities, Common Ground-Doves become relatively scarce, occurring chiefly in agricultural areas and along stream bottoms. We have no evidence that any Sonoran birds are migratory. At Lo de Campa, where we have visited regularly throughout the year, this species is sedentary.

Common Ground-Doves occupy brushy habitats up into pine forests at 1500 m, although at the higher elevations they are local and limited to ranches and small settlements. Like others of its family, the ground-dove has a protracted nesting season; we have found nests with eggs from early April to September. The nest may be placed in small trees or shrubs from 1 m above

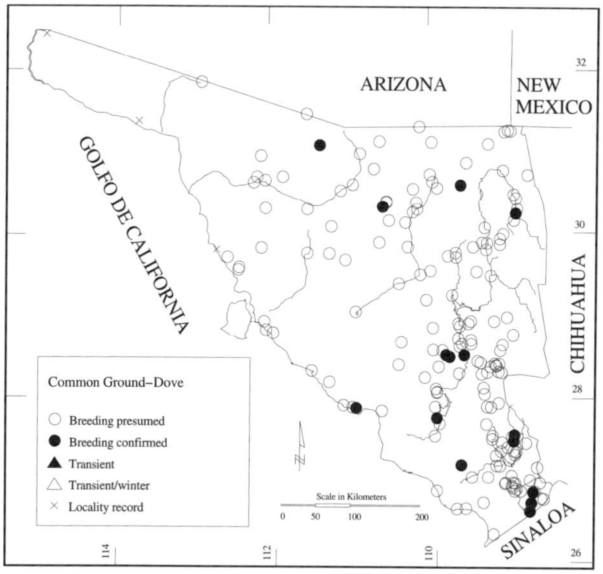

the ground to as high as 6 m. We have even found nests in prickly pear and the formidable cholla cactus. The low unobtrusive cooing of this dove is heard throughout the nesting season.

RUDDY GROUND-DOVE

Tórtola Rojiza

Columbina talpacoti

A few Ruddy Ground-Doves, until then recorded no farther north than southern Sinaloa, appeared in Sonora in the early 1980s. The first Sonoran record was a bird on 10 November 1982 on the Río Cuchujaqui east of Alamos (R. Taylor). This sighting was followed by other singles at Alamos on 7 April 1983 (Taylor) and on the Río Cuchujaqui on 23 November 1984 (eight members of the Maricopa Audubon Society of Phoenix, Arizona). Two were seen west of Alamos on 29 December 1984 (D. Stejskal, N. Krug), while six, two of them copulating, were recorded at Alamos from 3 to 7 April 1987 (Stejskal, R. Bowers, J. Coons). One was also noted during the Alamos Christmas Bird Count period, on or about 30 December 1987. Subsequently birds of this species have been seen regularly. It appears that the Ruddy Ground-Dove is established in the Alamos area.

Coincident with the appearance of this species in Sonora, it was found in southern Arizona and California. There are now many records from those two states, yet there are only two records from Sonora north of Alamos: one in March 1993 and one in March 1994. We surmise that northward-dispersing birds have been overlooked in Sonora.

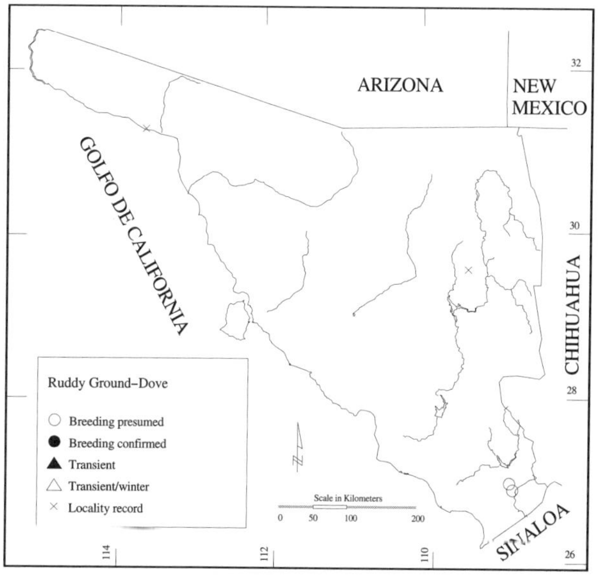

WHITE-TIPPED DOVE

Paloma Arroyera

Leptotila verreauxi

A visitor in the hills of southern Sonora for the first time is likely to be puzzled by a sepulchral, hollow, low three-syllable call heard repeatedly from thick thornscrub. After some careful stalking, one may find a White-tipped Dove a moment before it explosively bursts into flight. Once heard, the call is not forgotten. This dove vocalizes year-round but more repeatedly in the breeding season from March to September. The distinctive vocalization facilitates detecting this bird, which stays on the ground or very close to it in the thickest brush and on the steepest slopes, never flying for more than 30 m before disappearing, a real dodger of shrubs and tree trunks, as Leopold (1959) wrote.

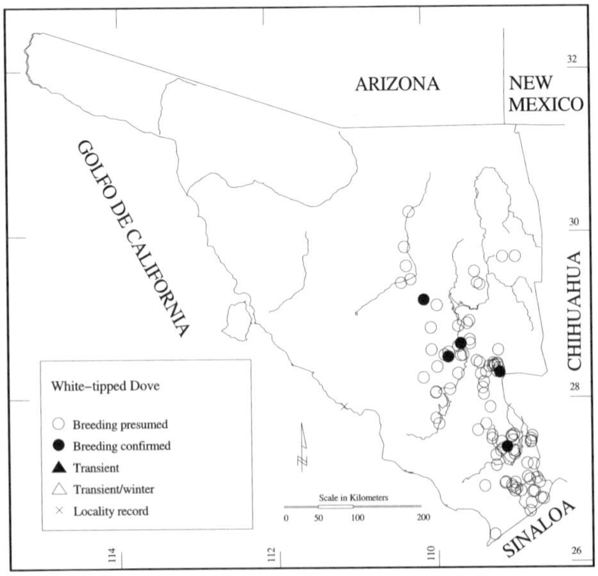

The White-tipped Dove is a resident of tropical thornscrub and tropical deciduous forest, where it prefers hills, arroyos, and barrancas. It frequents the coast of El Golfo de California sparingly, but it occurs at Agiabampo, and four were observed in a canyon just north of San Carlos on 2 May 1988 (R. Taylor). The greatest elevation at which we have found it is 1630 m, at El Terreito on the edge of the pines northeast of San Bernardo. Although this dove is typically solitary, it does aggregate at isolated water sources. We saw 20 waiting to drink at a shrinking pool in a streambed on a hot day in late September north of Tonichi. One to three birds would fly to the water's edge and drink, to be shortly replaced there by others.

FAMILY PSITTACIDAE

Lories, Parakeets, Macaws, and Parrots

GREEN PARAKEET
Perico Mexicano
Aratinga holochlora

The Green Parakeet has been found once in Sonora, away from its breeding range, which includes southwestern Chihuahua and northeastern Sinaloa. Van Rossem (1945) cited the sole record, that of eight specimens from Hacienda de San Rafael at about 460 m in the Sierra Madre foothills northeast of Alamos. They were taken between 5 and 7 May 1888 by M. A. Frazar, collecting for William Brewster. These eight birds may have been wanderers, or they may have been among the last of a now extirpated northern breeding population.

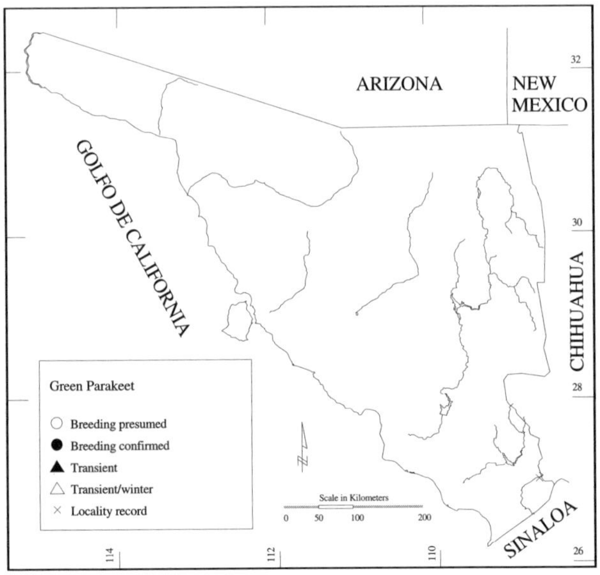

Military Macaw

MILITARY MACAW
Guacamaya Verde

Ara militaris

With resplendent plumage, a long tail, a massive bill, great size, and strident, ear-splitting calls, the Military Macaw is a spectacular bird. It reaches its northernmost range in Sonora as an uncommon summer resident. In the more rugged foothills and barrancas of the Sierra Madre as close as 240 km to the Arizona border, one may see or hear a pair every few days. Usually they are heard first and then spotted as they fly at ridgetop level.

With two exceptions, all records are in the period of 12 March through 23 October. Two birds were heard

(D. Stejskal) on the 1982 Alamos Christmas Bird Count and reported during the "count period" of the 1987 version of the same count. A group of about 40 birds (P. Jenkins) in the headwaters of an arroyo northeast of Alamos on 23 October 1992 may have been about to migrate.

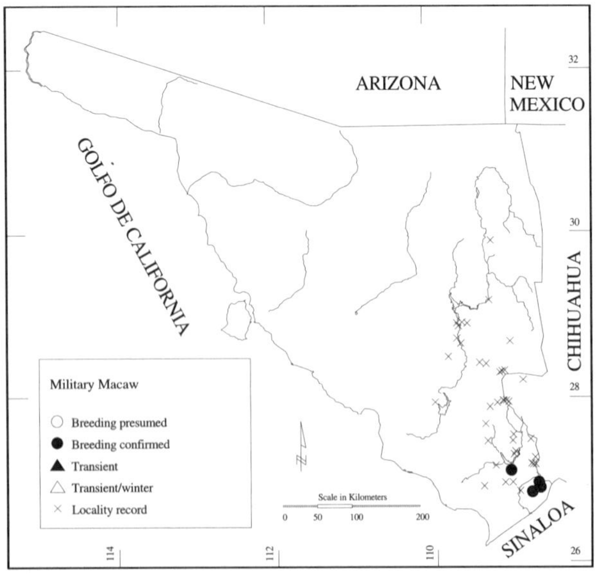

The Military Macaw favors steep cliff faces. On 29 July 1988, four were watched as they flew around a cliff face near the high point along the highway then being built from San Nicolás to Yécora. During their maneuvers near the cliff they chased and were chased by a dark-phase Short-tailed Hawk (J. Holloway, J. Kingery). Three eggs were taken from a cavity in a large sycamore on a Río Mayo tributary about 20 km northeast of Alamos on 6 June 1967 (L. Huber). Macaws may have formerly been more common. In 1947, seven clutches (WFVZ) of one to three eggs were collected in the Güirocoba region from 5 to 11 June. Three of the nests were 12 to 18 m above the ground in very large fig trees.

THICK-BILLED PARROT
Cotorra-serrana Occidental
Rhynchopsitta pachyrhyncha

We have spent much time in the pine country of Sonora, from the United States border south to the Chihuahua boundary northeast of Alamos, without seeing Thick-billed Parrots. There seems little doubt that the birds are now extremely scarce or indeed extinct as a breeding species in Sonora. Nor do the limited number still in existence in the Sierra Madre Occidental wander to Sonora. The last reported were a flock of 90 seen at an elevation of 480 m near El Taymuco in the barranca country northeast of Alamos on 8 January 1981 (Russell).

Marshall's (1957) experience in the mountains of the northeast in the early and mid-1950s was different. In 1951 and 1952, he noted 18 pairs and flocks of 8, 50, and 60; the localities included the Sierra Huachinera, the Sierra Nacori Chico, and even the Sierra la Madera. Numbers varied from year to year: for example, they were numerous in the Huachineras in 1954 but none was seen there in 1953 or 1955. He found no evidence of nesting.

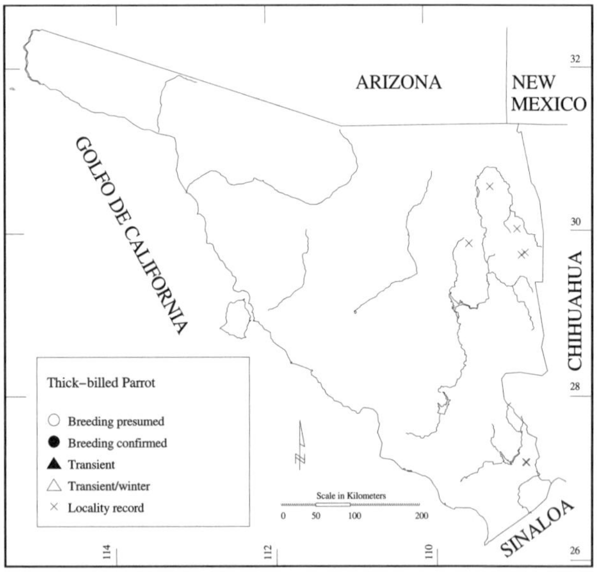

Lanning and Shiflet (1983) summarized the status of Thick-billed Parrots in Chihuahua and Durango. The nesting density was no doubt far below that of earlier years. Human settlement and thorough logging of pines had taken their toll. Lanning and Shiflet's review stated, in part,

> During a four-month search for nesting Thick-billed Parrots . . . in 1979, we found 55 active nests at elevations of 2,300 to 3,070 m in northwestern Mexico's Sierra Madre Occidental [in Chihuahua and Durango]. All nests were in cavities in live trees or standing dead trees (snags). Pine (*Pinus* spp.) snags contained over one-half (58%) of these nests. These parrots laid two to four eggs . . . between mid-June and late July and young flew from their nests between early September and late October. Nesting density was apparently related to availability of suitable cavities and was variable, with some nests as close as 2 m apart in the same tree. The nesting season corresponded with the maturing of pine seeds, the parrots' principal food. Commercial logging of live pines for lumber, and of pine snags for pulpwood, is eliminating large numbers of proven and potential nest sites. Forest management practices should be modified to leave some suitable trees throughout the forest as potential nest sites for Thick-billed Parrots.

Thick-billed Parrots appear to be doomed to early extinction unless immediate efforts are made to protect them.

MEXICAN PARROTLET
Perico Catarina

Forpus cyanopygius

This tiny bright green-and-blue short-tailed parrot buzzes along in tight flocks like so many flying jewels, chattering as they go. Alden (1969) wrote of huge flocks in Alamos. Since then its numbers have diminished noticeably throughout the Sonoran portion of its range and it is now uncommon, but small flocks may still be found in far southern Sonora in valleys and foothills arroyos. With the exception of a few individuals that reside in Alamos, there are no records in the past 60 years from October through March. The scarcity of Mexican Parrotlets in winter leads us to believe that this extreme northern population may be migratory, with some individuals remaining where conditions are favorable.

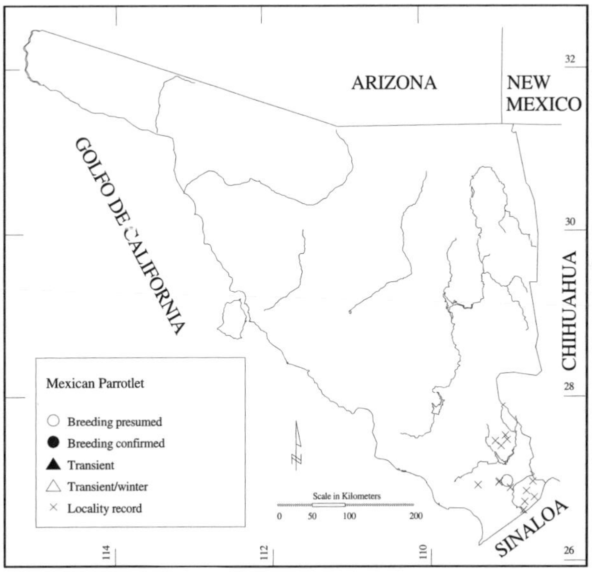

Little is known of their feeding and nesting habits in Sonora. We have seen them feeding on small-fruited figs. We have never found them nesting. No nests are known from Sonora, but gonads of birds collected in

April (SWC) were moderately enlarged, suggesting the possibility that the birds would soon breed. A female collected in early June (UK) had enlarged follicles and oviduct, indicating that it could have been nesting. The recent decline in numbers and the lack of information on behavior suggest that the Mexican Parrotlet is in need of protection and study in Sonora.

WHITE-FRONTED PARROT
Loro Frente Blanca
Amazona albifrons

This is a fairly common resident parrot in lowland southern Sonora, not often found above 600 m. It ranges throughout the tropical deciduous forest and riparian corridors in thornscrub. It nests in cavities in trees and in large cacti (saguaros, hechos, and organ pipe). It begins to breed in April, and young fledge in June. On 11 April 1986, T. Huels noted three nests in a large grove of tall cottonwoods below Presa Obregon; a bird peered from each cavity (10, 16, and 16 m above the ground) as its mate perched nearby. Several clutches of eggs (WFVZ) were taken near Güirocoba in May. Two specimens (UK) taken on 9 June 1958 from the Río Cuchujaqui southeast of Alamos are of nestlings that were almost ready to fledge.

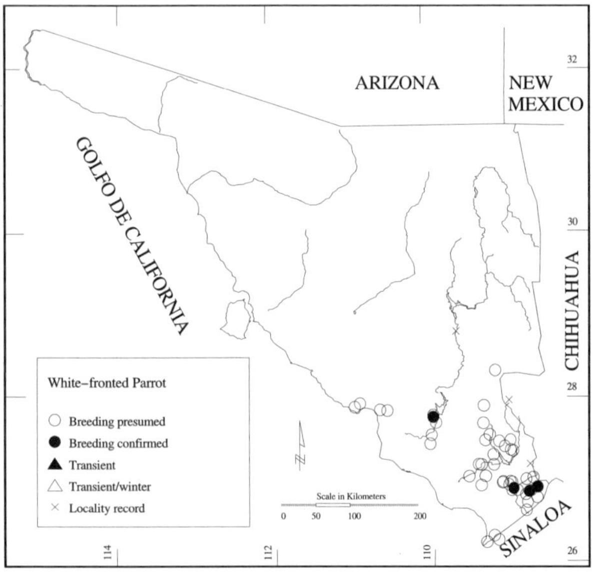

Although chiefly resident where found, some records indicate that White-fronted Parrots may wander. B. Brown observed at least six on the Río Yaqui about 12 km above Soyopa on 31 December 1981. Van Rossem (1945) noted their presence in pines at 1500 m at Rancho Santa Barbara about 50 km northeast of Alamos in early June 1937. Farther north, on the Sonora-Chihuahua boundary in the Sierra Oscura, we found two in pines at 2020 m near a sawmill community called El Serruchito on 15 March 1982. A remarkable total of 308 was reported on the 1978 Alamos Christmas Bird Count.

The diet of the White-fronted Parrot includes ripe fruits of various plants, including hecho and saguaro. Two were eating saguaro flower buds 25 km east of Empalme on 14 March 1986 (A. Moorhouse, D. Danforth), and a half-dozen or more were feeding on willow catkins at Presa Agua Caliente on 3 January 1986. These observations merely hint at the large variety of buds, flowers, and fruits consumed by these parrots.

LILAC-CROWNED PARROT
Loro Corona Lila
Amazona finschi

The large Lilac-crowned Parrot has a more restricted range in Sonora than does the smaller White-fronted Parrot. It has been found nesting in tropical deciduous forest (above 250 m), upward into pine forest (1600 m). We think the species is probably resident in southeastern Sonora, although individuals tend to wander in the nonbreeding season. Numbers have declined since about 1970, and it is presently rare. It is now more often seen in mountains than in lowlands. We noted several pairs in the mountains north of Los Algodones in 1986. One pair was nesting (21 May) in a knothole 8 m above the ground in a pine 16 m tall that had been

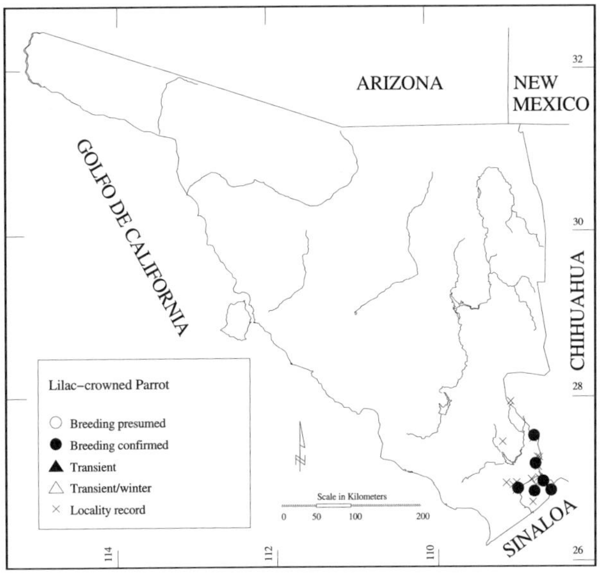

killed and split by lightning. We saw few in the Sahuarivo area in 1987. The natural history of this parrot is poorly known in Sonora, and its behavior should be studied and its numbers monitored to determine its status in order to plan for its continued existence.

FAMILY CUCULIDAE

Cuckoos, Roadrunners, and Anis

YELLOW-BILLED CUCKOO

Cuclillo Pico Amarillo

Coccyzus americanus

Yellow-billed Cuckoos arrive in Sonora in the first week of June (earliest date, 2 June, at Bacoachi); some are migrants en route to the western United States, but many stay to breed after the summer rains begin. Through July and into early August they are everywhere in the interior plains and foothills of Sonora, revealed by their curious, characteristic calls. After their young fledge they are much quieter. Some breeding activity extends into September. A pair gave a very impressive distraction display to lure the human intruder away from a bob-tailed young on 12 September 1986 in Arroyo Santo Domingo. The earliest nest reported was north of Ocuca on 18 June 1942 (Neff 1947) in an extensive mesquite bosque dominated by trees 20 m tall (now cut). The latest date observed was 1 October 1984 southwest of Benjamin Hill.

The species is a common summer resident in tropical deciduous forest, tropical thornscrub, and locally in desertscrub. Its favored habitat is willow-cottonwood woodland and large leguminous trees or shrubs along desert washes. The nesting-season density of this cuckoo in Sonora is truly imposing when compared to its frequency in Arizona or other parts of the United States; we have noted as many as 40 calling individuals in a morning at Lo de Campa.

The Yellow-billed Cuckoo is found in the pine-oak woodland of northeastern Sonora only as a belated migrant or wanderer (Marshall 1957).

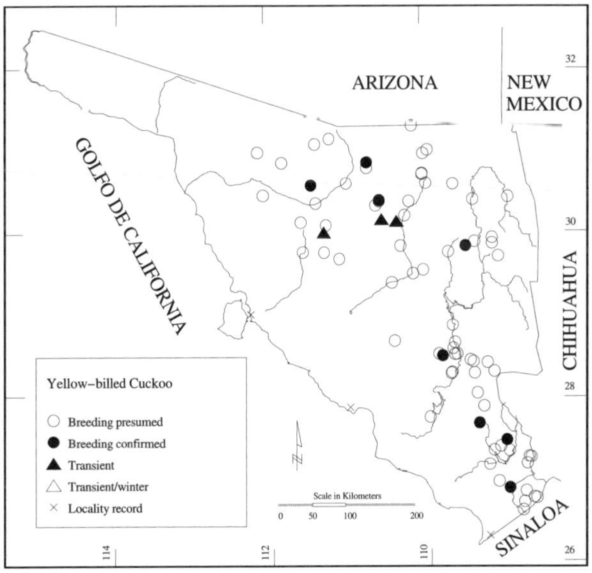

MANGROVE CUCKOO
Cuclillo Manglero
Coccyzus minor

The Mangrove Cuckoo is an uncommon and local summer breeding resident in patchy tropical deciduous forest below 700 m. The species was first reported from Sonora by Alden (1969) near Alamos in June 1965. On 27 August 1965, M. Buchanan collected (UCLA) a female about 10 km west of Alamos. It contained a large egg that would have been laid soon. Subsequently, the Mangrove Cuckoo was found twice within 24 km of Alamos in July 1971 (Short 1974). On 24 June 1986, S. Levy saw one east of Tonichi. On 28 July 1988, we found one in an arroyo 18 km east of Tecoripa. R. Russell watched a bird for several minutes as it carried a large insect larva on a luxuriantly green hillside in tropical thornscrub. In mid-July 1987 we observed a total of 24 Mangrove Cuckoos in the hills between Alamos and the Sonora-Sinaloa border.

In Sonora the appellation "Mangrove" is a misnomer, for this cuckoo has never been found in mangroves there.

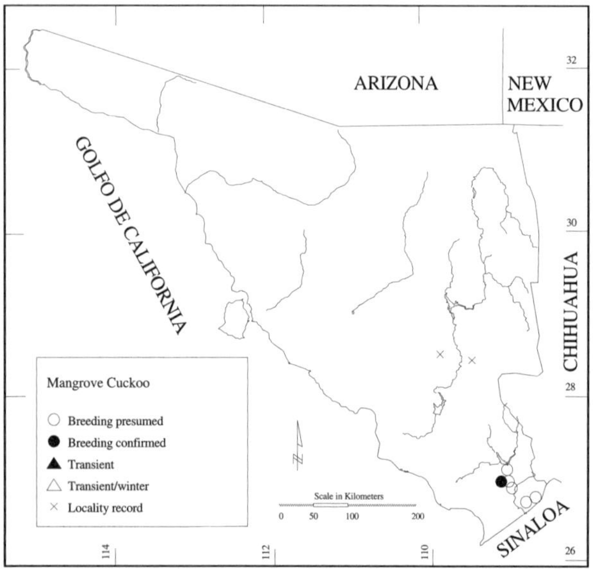

SQUIRREL CUCKOO
Cuclillo Canela
Piaya cayana

The Squirrel Cuckoo is an uncommon and local resident of dense thickets in tropical deciduous forest. Although it is an impressively elaborate bird with bright cinnamon rufous upper parts and an outlandishly long graduated tail, it is difficult to see. It calls infrequently and moves deliberately along branches. Birds in breeding condition have been collected (LSUMZ) in the Güirocoba region from early May to early June.

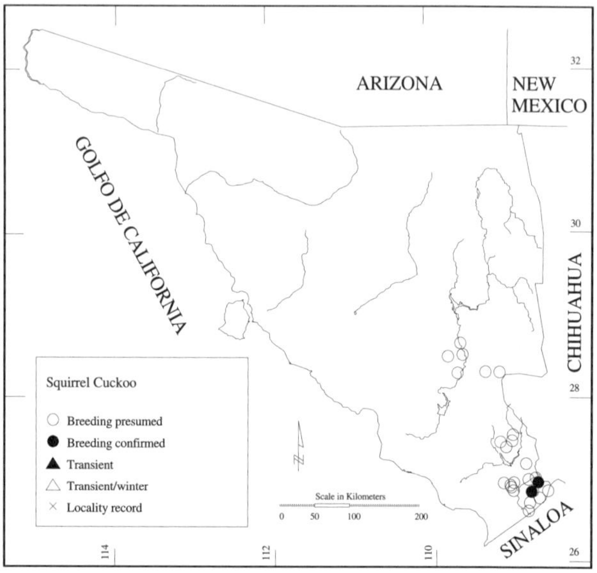

Squirrel Cuckoo

LESSER ROADRUNNER

Correcaminos Tropical

Geococcyx velox

This poorly known species is a rare resident of open grassy patches in tropical deciduous forest and is often found on steep rocky slopes above arroyos at elevations of 280 to 1160 m.

Two sets of eggs (WFVZ) were taken in June in the Güirocoba area. We found a pair near El Zapote on 14 July at a nest in the crotch of an hecho 5 m above the ground. The birds carried sticks, called, displayed, and copulated.

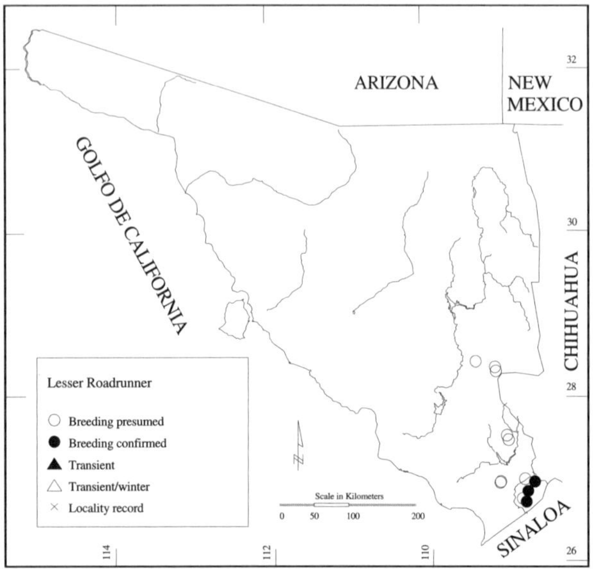

GREATER ROADRUNNER

Correcaminos Norteño

Geococcyx californianus

Much has been written about this ground cuckoo of odd appearance and habits. It is a well-known, widespread, and common resident of Sonora from seashore to mountains, but it is most common in well-developed desertscrub and tropical thornscrub. It is less common in high desertscrub of the northeast, pine-oak woodland, and rocky open areas of pine forest to 2100 m.

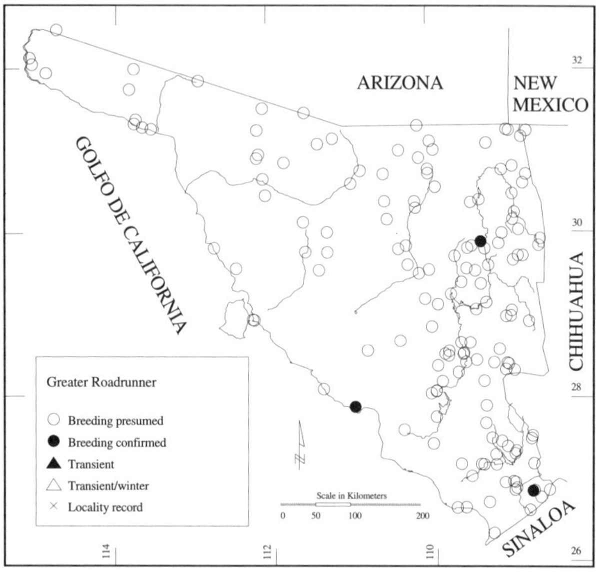

Breeding takes place from May into July, the coarse stick nests being built within 2 to 3 m of the ground in a dense shrub or tree. One clutch (WFVZ) was taken from a nest near Güirocoba that was placed in an old Crested Caracara nest 2 m up in an hecho cactus. Another clutch (WFVZ) taken near Guaymas was from a nest padded with cow dung.

In Sonora, the Greater Roadrunner is invariably known as "Churea"—a representation of the call. Marshall (1957:74) gives a delightful account of the Greater Roadrunner as he observed it in the pine-oak woodland of northeastern Sonora.

GROOVE-BILLED ANI

Garrapatero Pijuy

Crotophaga sulcirostris

Although this strange all-black cuckoo is large, conspicuous, and usually in groups, little is known about it in Sonora. It is an uncommon and local summer visitant in brushy areas, usually near streams and often in the company of cattle. It returns in early June and remains until late September. A few individuals winter in Sonora, especially in and near Alamos. We have found it no higher than 650 m (northeast of Rosario). Nesting begins in late June and continues into August. We have seen bob-tailed juveniles as late as 29 September (near Tonichi). Far northern observations may be of wandering birds; Arizona, where the species does not breed, has a number of instances of single anis and even small groups of them.

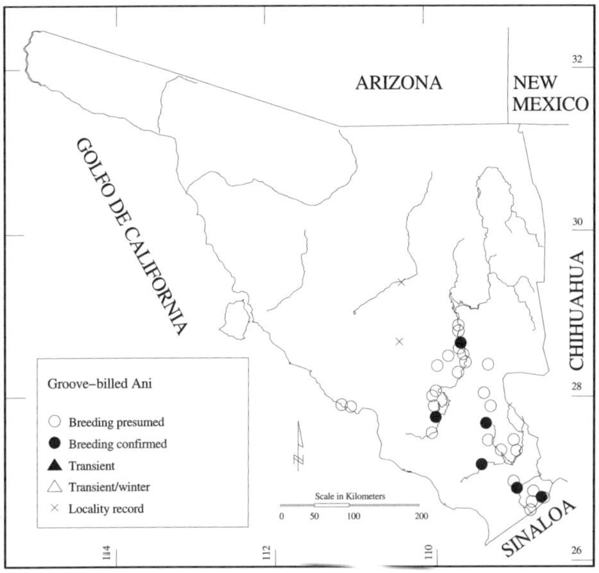

FAMILY TYTONIDAE

Barn Owls

BARN OWL

Lechuza de Campanario

Tyto alba

The Barn Owl is widely distributed over Sonora at elevations below 1100 m and is probably resident. It is often located by its calls, as it is seldom seen because of its habit of hiding in heavily leafed trees (including palm-tops), unoccupied buildings, well- and mine-shafts, and similar cover during the daytime. For nesting sites, it often chooses holes in ledges and stream banks and corners in dark interiors of caves and man-made structures. We know of no nesting sites in Sonora, but it probably nests in every locality where found. The highest locality where we have found it is at 1910 m in pines 12 km southwest of Mesa Tres Ríos, on 20 March 1985.

On Isla Alcatraz at the entrance to Bahía Kino, R. Perrill and M. Keasey on or about 17 May 1975 saw gulls attack a Barn Owl so violently that some feathers were knocked off; the owl disappeared over the slope of the island.

A hand-reared nestling from Chandler, Arizona, was banded and released as a hatching-year bird by K. A. Ingram at Glendale, Arizona, on 20 July 1979. It was shot and killed at Huatabampito in extreme southern Sonora on 5 May 1981. This incident serves to illustrate that although the Barn Owl is generally regarded as a permanent resident where found, young birds may disperse widely.

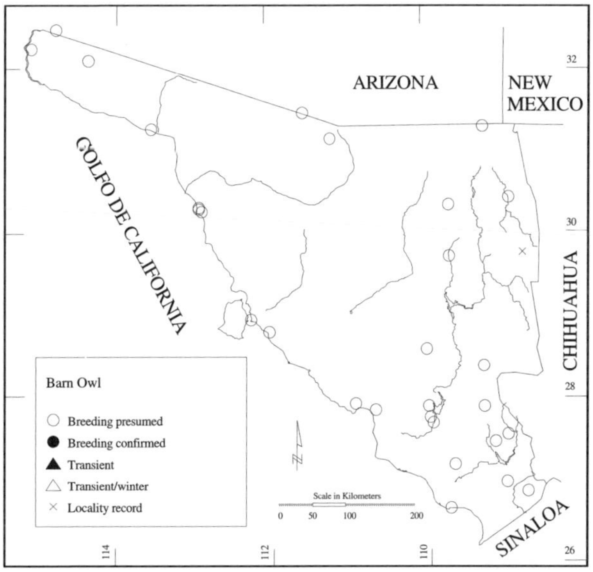

FAMILY STRIGIDAE

Typical Owls

FLAMMULATED OWL
Tecolote Ojo Oscuro
Otus flammeolus

This small, dark-eyed screech-owl was not recorded in Sonora until J. Marshall found it in August 1951 as he initiated his fieldwork in the mountains of the northeast (Marshall 1957). He regarded it as a bird of the pine forest but stated that it is also found in pine-oak woodland. All records are between 1600 and 2000 m. Closer study will perhaps reveal that it is more widespread than Marshall's records indicate. We have never encountered it.

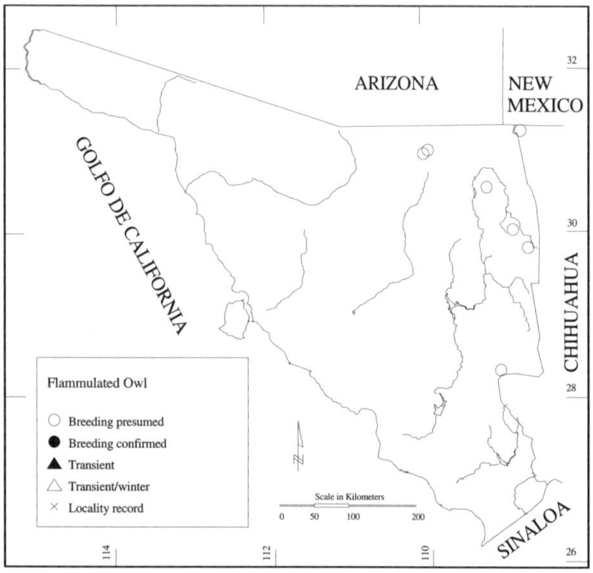

An arthropod eater, the Flammulated Owl is migratory and presumably winters no farther north than Sinaloa (AOU 1983). J. Marshall found it in Sonora from 6 April to 3 September. Arizona birds migrate in late March–April and again in September–October.

WESTERN SCREECH-OWL
Tecolote Occidental
Otus kennicottii

The Western Screech-Owl is a common and widespread resident of Sonora below 1300 m. It favors dense trees (including mesquite and ironwood as well as willows and cottonwoods) along streams and the dry washes that cross the desertscrub and the tropical thornscrub communities. It is scattered throughout the tropical deciduous forest. It is rare in the more sparsely vegetated extreme northwestern desert. It was once abundant in the Río Colorado delta but now is much less so because of the disappearance of willows and cottonwoods due to extensive agricultural development.

Marshall (1967) noted its absence in sycamores, oaks, and pine-oak woodlands at middle elevations in central eastern Sonora and in southeastern Sonora at the Chihuahua boundary. This agrees with our observations. Marshall (1957:76) mentioned the Western Screech-Owl as "rare and irregular from year to year in pine-oak" up to 1800 m, where it is replaced by the Whiskered Screech-Owl.

The Western Screech-Owl nests in woodpecker holes and other cavities in any tree or large cactus in May and June. It feeds chiefly on large insects, spiders, scorpions, small nocturnal reptiles, and so forth, most of which it captures on the open ground. Not much is known of its winter food, which probably includes

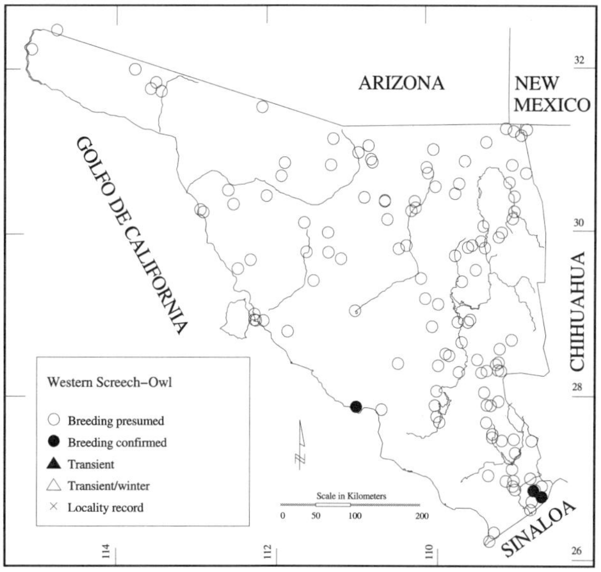

fewer arthropods and more small rodents and birds. Several of these owls were regularly present in the mangroves at Punta Chueca from early September to mid-March. We never found them there at other times; there are no potential nest cavities. Perhaps they were birds from the local desertscrub community that had moved a short distance to enhance their foraging opportunities.

This owl's voice is best known for the "bouncing ball" call, but it has other vocalizations in its repertoire. It may call in any month of the year. We owe much of what we know about all phases of the life history, distribution, and taxonomy of this bird to the remarkable work of J. Marshall, summarized in his 1967 monograph. This work includes the results of his studies of the races of Western Screech-Owl found in Sonora.

WHISKERED SCREECH-OWL
Tecolote Rítmico

Otus trichopsis

This small owl is found entirely in the pine-oak woodland and is invariably associated with oaks. There it is a common resident. Marshall (1967:22) described its domain as "dense groves of oaks within pine-oak woodland." We have found it literally abundant in such places and have heard as many as six at a time. Its altitudinal range is from 1270 to 2160 m.

The Whiskered Screech-Owl's nesting season is during April and May. Two adults were perched with two fledglings in an oak sapling in the Sierra Oscura on 8 June 1984 (T. Huels et al.). The calls are characteristic and easily told from those of other night-birds. Marshall (1957:76) wrote, "This owl hunts in foliage and branches of dense groves on mountain slopes and in canyons." He went on to state that its prey is principally medium to large insects such as caterpillars and beetles.

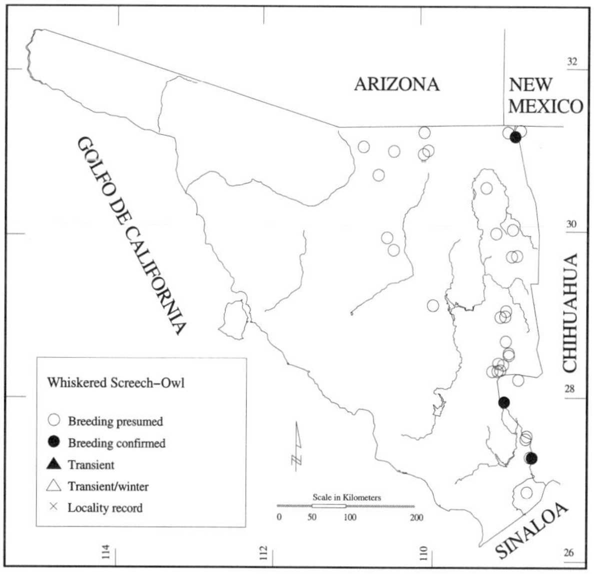

VERMICULATED SCREECH-OWL
Tecolote Vermiculado
Otus guatemalae

This uncelebrated screech-owl, of the same size and same general coloration as the Western and Whiskered Screech-Owls, is known from Sonora by 15 specimens (LSUMZ, WFVZ) taken in the vicinity of Güirocoba in March, April, and May between 1930 and 1958. It has not been seen subsequently, and most of our knowledge of the bird comes from Marshall (1967), who was the last to see it in Sonora.

Some specimens were taken from steep slopes of tropical deciduous forest above deep arroyos supporting Montezuma bald-cypresses and figs. W. Sheffler collected clutches of three and four eggs on 17 and 12 May 1949 respectively, but the nests were not described. A female collected by Marshall on 11 May 1958 had a brood patch.

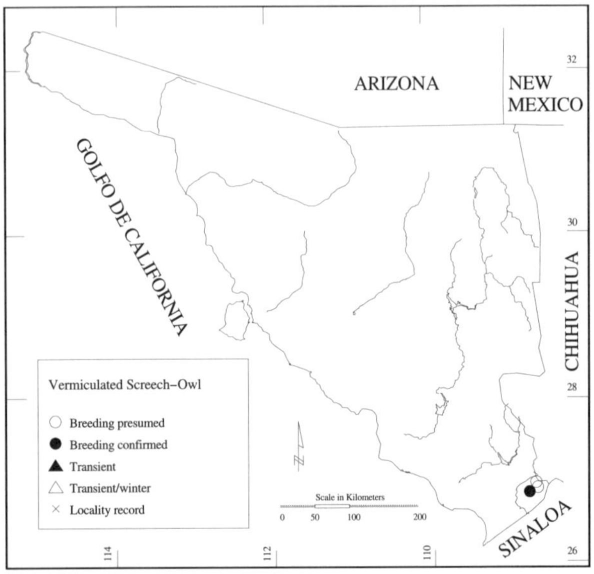

GREAT HORNED OWL
Búho Cornudo
Bubo virginianus

Almost anywhere in Sonora one may be within earshot of the solemn and resonant yet hushed hooting of the "tiger of the night." Except in the highest mountains (above about 1600 m), where pines stand dominant, the Great Horned Owl is a common resident, from the mangroves of the coast to well up into the mountains and mesas of the east. Although recorded from elevations up to 2300 m, it is rare at high elevations and perhaps only wanders there.

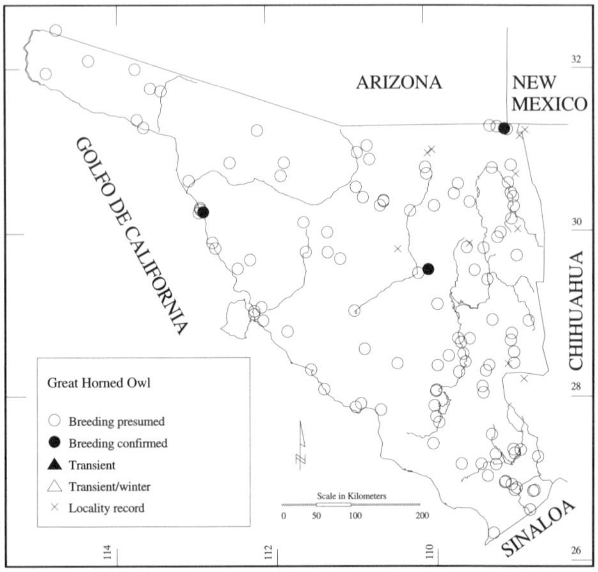

It nests in the spring, late February through May, utilizing old hawk and raven nests in trees, large shrubs, and cacti, but it prefers ledges on cliffs or high banks. Near Puerto Lobos, a ledge site about 15 m above a sandy arroyo was in use in 1973 and 1985 (and at least some of the intervening years). Along with the Red-tailed Hawk, the Great Horned Owl is the most widespread of the large raptors.

NORTHERN PYGMY-OWL

Tecolote Serrano

Glaucidium gnoma

Van Rossem (1945) did not list this small diurnal owl, but it is common throughout the pine-oak woodland, including lower elevations where it intrudes on the tropical thornscrub or tropical deciduous forest. It is uncommon in the mountains of the extreme southeast. Oaks of any species are essential plants in its breeding habitat, although juniper woodland is also used. The Northern Pygmy-Owl ranges upward to 2380 m, but most live between 1000 and 2100 m. Although generally this pygmy-owl is a permanent resident where found, some individuals or populations must disperse to considerably lower elevations. This would account for birds noted, chiefly in March, at such locations as Lo de Campa (490 m), the Río Carrizo north of Tonichi (260 m), or a deciduous riparian site (820 m) near Imuris (Terrill 1981).

The Northern Pygmy-Owl captures birds, mammals, and reptiles, as well as insects, especially

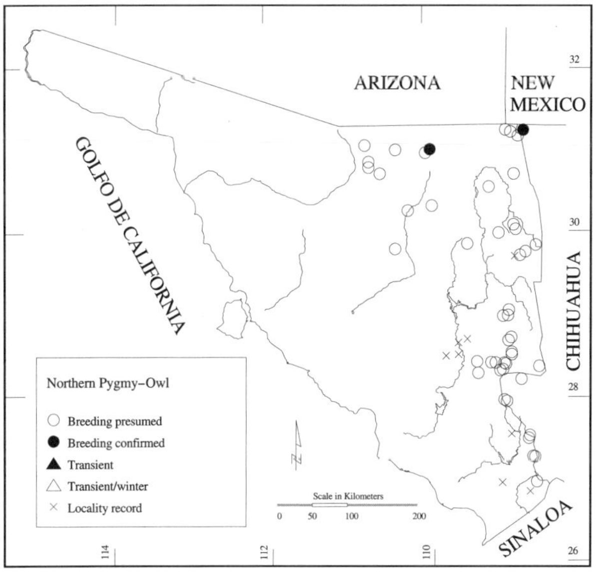

grasshoppers and crickets (Bent 1938). Marshall (1957) found that it has a large territory, about 260 ha per pair. Its call is a repetitive somewhat musical hooting or, as some authors describe it, a cooing, lasting as long as an hour or more. The call may be heard at all times of the year but particularly in the spring mating season. At that time the birds are extraordinarily active and vociferous. Calling birds are frequently mobbed by innumerable small birds.

LEAST PYGMY-OWL

Tecolote Colimense

Glaucidium minutissimum

This small rare owl, first found in Sonora in 1945 by W. Sheffler, is restricted to elevations of 400 to 1000 m mostly in the extreme southeast. In all, 24 specimens (LACM, LSUMZ, WFVZ) were taken in the Güirocoba region (1945–1959). The only recent documented record is of a bird photographed about 8 km north of Taymuco in the barrancas northeast of Alamos on 6 April 1987 (R. Bowers et al.). S. Howell and S. Webb heard two calling in foothills east of Hermosillo on 22 June 1991 (Howell and Robbins 1995). They did not record the birds, but their studies on *Glaucidium* vocalizations suggest familiarity with the species. Its habitat is tropical thornscrub where oaks extend down

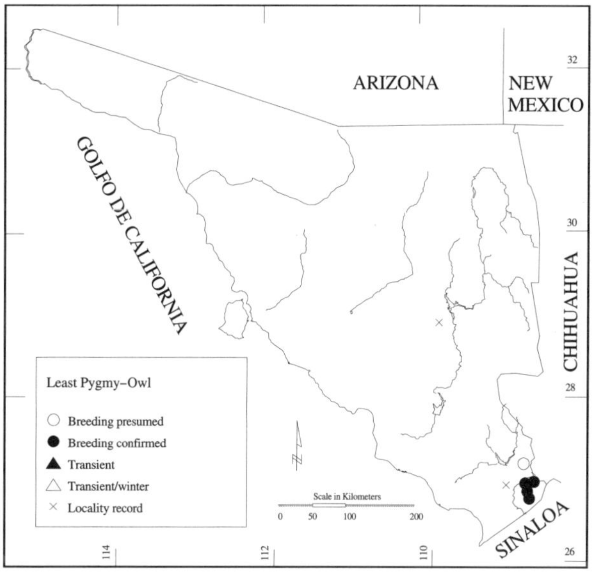

Least Pygmy-Owl

from higher elevations on steep slopes. The call is a series of whistles with a frequency of about 1400 Hz. Like other pygmy-owls, the Least is abroad during daylight hours, especially at sunrise and sunset. Much remains to be learned of its habits in Sonora.

Except for occasional reports without adequate details on Alamos Christmas Bird Counts, all records have been from March to mid-June. Four clutches (WFVZ) of three and four eggs were taken between 18 May and 3 June. Egg laying begins at least as early as 6 May, when a female was taken with a shelled egg in the oviduct. Two of the nests were in oaks in cavities excavated by woodpeckers 6 and 10 m above the ground.

Glaucidium minutissimum has recently been split (AOU 1997). The species in Sonora becomes *G. palmarum*, the Colima Pygmy-Owl.

FERRUGINOUS PYGMY-OWL
Tecolote Bajeño
Glaucidium brasilianum

The uncommon Ferruginous Pygmy-Owl is resident throughout most of the desertscrub, tropical thornscrub, and tropical deciduous forest, being most common in the latter association. There are only three records from above 1000 m (below Sahuarivo, northeast of Santa Ana de Yécora, and on the south end of the Sierra la Madera), each at the lowest edge of oaks. We are aware of only two records from latitudes north of Hermosillo between mid-September and mid-April. Though presumed to be resident in Arizona by Phillips, Marshall, and Monson (1964), most northern birds may be migratory. We have documented no decline in numbers in Sonora in recent years, but populations in the United States have diminished significantly (Hunter 1988).

The Ferruginous Pygmy-Owl typically discloses its presence, especially in the morning and evening, by persistent and distinctive high-pitched rapid call notes. Nesting is in woodpecker holes or natural cavities. In the Güirocoba area, 11 clutches (WFVZ) of two to five eggs (most had three or four) were taken between 10 May and 17 June. Some nests were in oak and Montezuma baldcypress, although large cacti are often used. Little is known of the food habits of this pygmy-owl in Sonora. It is frequently attacked by small passerines, and once we saw one harassed by a Ladder-backed Woodpecker.

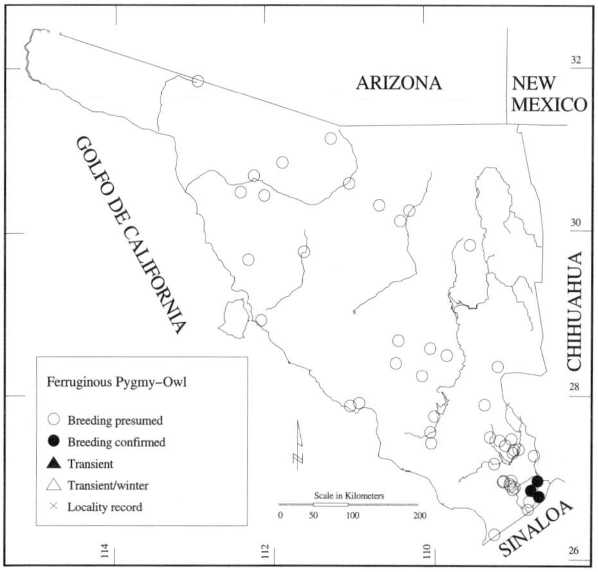

ELF OWL
Tecolote Enano
Micrathene whitneyi

The smallest owl of the world (mean weight 41 g; Dunning 1993) is a common summer resident in most parts of Sonora. It is absent only from grassland and desertscrub in the extreme northeast and from the coniferous stands of the highest mountains. Marshall (1957) presumed it was breeding in pine-oak woodland in the Sierra los Ajos and the Sierra Nacori. Two were found in oak woodland just west of Tarachi at 1180 m on 30 April 1977. Below the oaks it is invariably common, and it is probably the most numerous owl in Sonora.

Elf Owls arrive in the spring from their winter home farther south in mid-March, and they leave Sonora again by mid-October. However, some may remain through the winter. One was netted on 7 January 1983 at an elevation of 780 m a few kilometers west of Tepoca. One or two were noted at Lo de Campa 8 and 9 December 1983, three were found near Alamos 17 November 1986 (D.Stejskal), and three were banded near Piojillo on the Suaqui arm of Presa Obregon on 1 November 1985. All may have been wintering birds. None has been reported on Alamos Christmas Bird Counts. An Elf Owl called from a cardón at Puerto Lobos on 18 February 1989, an exceptionally early date.

The typical clutch of three eggs is laid in April or May in a cavity. Woodpecker nest-holes in large cacti seem to be preferred sites, although small cavities in trees such as cottonwoods and sycamores are acceptable.

The call notes of Elf Owls are readily heard in spring and summer, especially at dusk and dawn. Most commonly heard are a single-note sharp *kew* and a rapid series of the same abbreviated call.

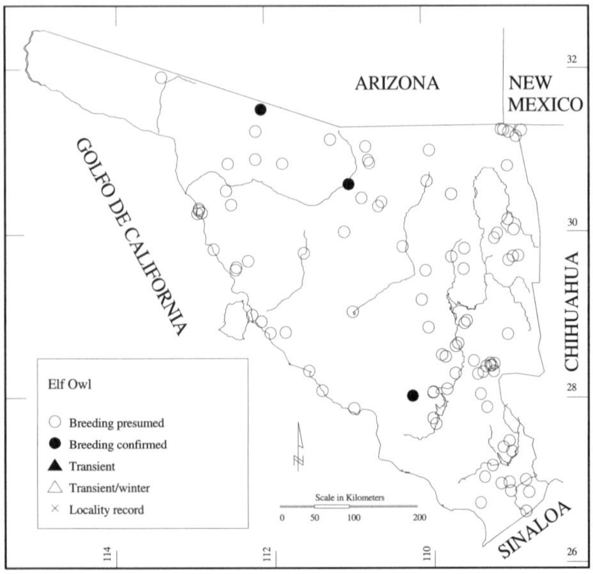

BURROWING OWL

Tecolote Llanero

Speotyto cunicularia

Burrowing Owls are uncommon and locally distributed in Sonora. Presumably they nest and are resident in many localities, although we have not found eggs or young. Their numbers are possibly increased in winter by birds coming from farther north. They are sometimes seen along the raised banks of irrigation ditches, along the edges of farm fields, and in hard-packed coastal dunes. As such, the species is recorded mainly from the Río Colorado delta and along the coast inland no farther than Hermosillo. Along irrigation ditches it may nest in holes or "pipes" formed by overflowing or leaking water; elsewhere it utilizes rodent burrows.

In the extreme northeast, in the drainage of Arroyo las Carretas, about 12 km west of the Chihuahuan town of Pancho Villa but on the Sonoran side of the state line, we saw at least four Burrowing Owls in classic habitat, a large Black-tailed Prairie Dog (*Cynomys ludovicianus*) town, on 5 September 1983. West of Naco, it probably uses burrows of large kangaroo rats (*Dipodomys* sp.).

One image of a Burrowing Owl stands out in our recollections of this bird. While driving through the coastal plain village of Juan de la Barrera just north of the Sonora-Sinaloa border about sunset on 10 July 1987, we noted a Burrowing Owl perched on a fence post at the edge of the road. Clouds of dust were being raised by passing traffic, children were shouting and playing close by, and barking dogs were everywhere, but the doughty little owl nevertheless stood gamely on his post and continued to do so as we drove by. The village and its surroundings were barren in the extreme, but irrigated farmland was not far away.

The AOU (1997) places the Burrowing Owl in the genus *Athene*.

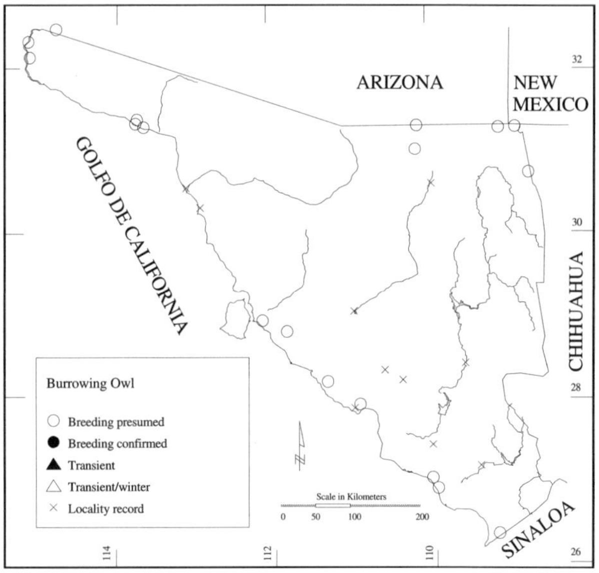

MOTTLED OWL

Búho Café

Ciccaba virgata

Mottled Owl

In Sonora, the tropical Mottled Owl has been found most often in riparian areas with large trees in tropical thornscrub and tropical deciduous forest below 1000 m. We know little of its density, but the paucity of records suggests it is rare. With the exception of a single bird flushed from an hecho a few kilometers west of San Bernardo on 8 January 1986 by J. Bates, all records fall between 14 March and 20 August. These dates suggest that the species may be migratory, but if not calling it would be difficult to detect; we think it may be resident. At times Mottled Owls will call frequently during the night, making their presence obvious.

Three two-egg clutches (WFVZ) were taken from nests 3 to 8 m high in oaks near Güirocoba in mid-May and mid-June.

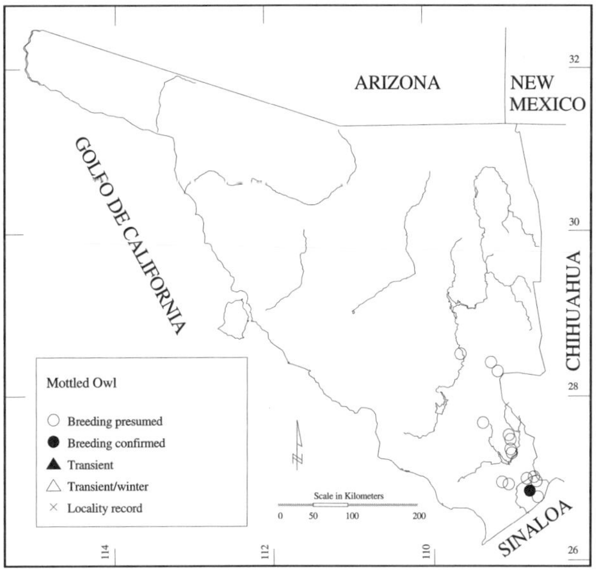

SPOTTED OWL

Búho Manchado

Strix occidentalis

The Spotted Owl is rare in Sonora, confined to coniferous forest and pine-oak woodland of steep canyon sides of the higher mountains. We presume it is resident in the nine mountain ranges from which it is known. The only reference to this species made by van Rossem (1945) was of a specimen (MCZ) taken by J. C. Cahoon in the Sierra la Madera on 11 June 1887. J. Marshall did not find the species there in the 1950s, although he did locate Spotted Owls in the Sierras San

Luis, el Pinito, los Ajos, Azul, Huachinera, and Aconchi. In our fieldwork in the 1980s, we found Spotted Owls in only two additional localities (Sierra Oscura and north of Algodones). The species occupies an altitudinal distribution from 1500 to 2100 m. Nests have not been found in Sonora.

Marshall (1957) collected 30 pellets under a roost in the Sierra Madre (Río Gavilan drainage barely in Chihuahua). The pellets contained such varied items as large scarabs, a bat, wood rats and mice, a young rabbit, and a large frog or toad.

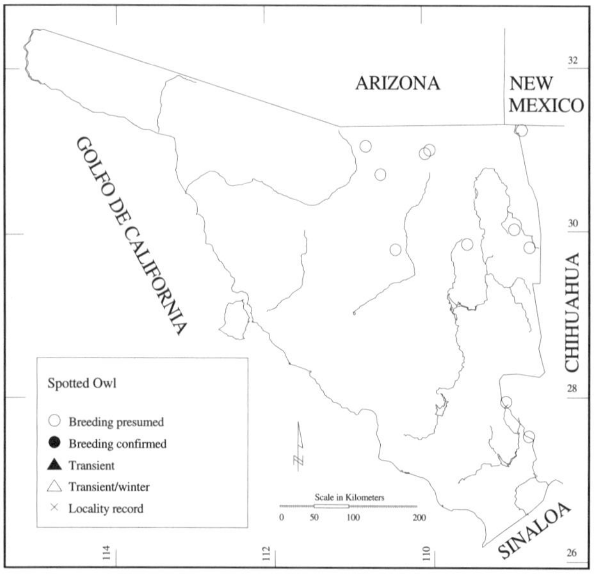

LONG-EARED OWL
Búho Cara Café
Asio otus

On the basis of only four records, the Long-eared Owl is considered a casual winter visitant in Sonora. W. Bulmer found one in mesquites at Puerto Peñasco on 23 October 1964. Another was seen along the Río Magdalena between Magdalena and Imuris on either 16 or 27 January 1980 (Terrill 1981). Several were reported at San Carlos from 8 to 15 February 1987 by K. Jones.

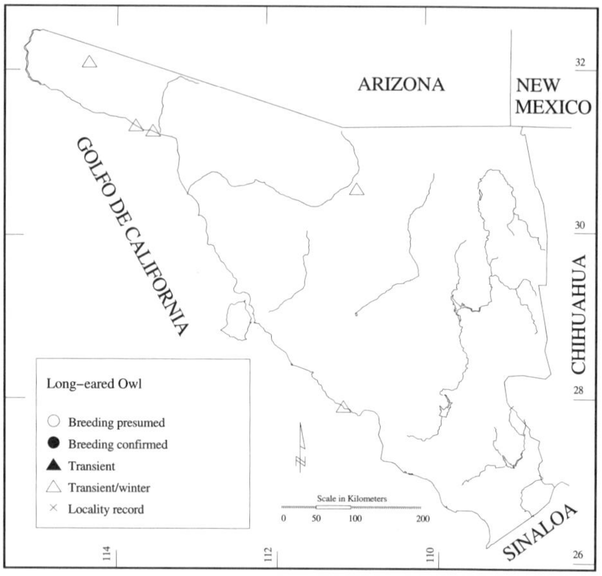

Russell found one roosting in an ironwood northwest of the Sierra el Rosario on 27 February 1993. The species nests occasionally in southern Arizona not far from the border.

SHORT-EARED OWL
Búho Cuerno Corto
Asio flammeus

The Short-eared Owl is a rare winter visitant in northern Sonora. There are seven records in all: from the Río Colorado delta (1898), the Río Santa Cruz east of Nogales (1892), and Puerto Peñasco (five dates, including five on a 3 January 1981 Christmas Bird Count). The earliest fall date is 6 October 1979 at Puerto Peñasco (Russell); there are no spring records. Short-eared Owls may roost on the ground in open country, including marshy areas and even open creosote bush desert.

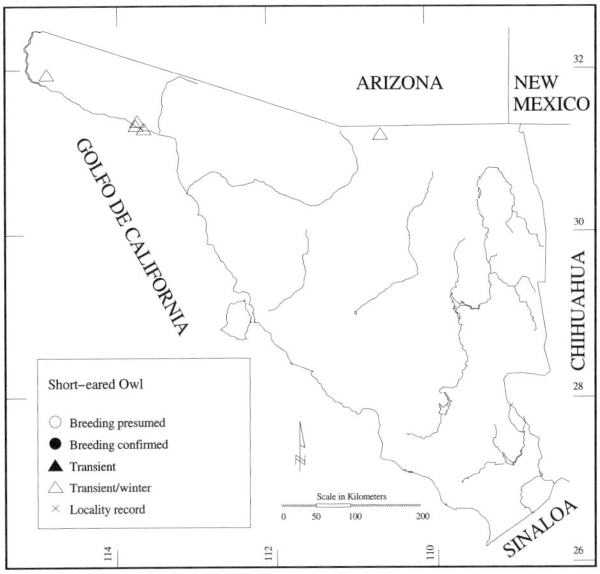

NORTHERN SAW-WHET OWL

Tecolote Afilador

Aegolius acadicus

The only record of a Northern Saw-whet Owl in Sonora is of a freshly killed bird picked up in a road near Puerto Peñasco on 12 November 1977 by J. Short (UA). Marshall (1957) did not find this owl in the mountains of northeastern Sonora; it is resident in the Chiricahua Mountains just across the border in Arizona.

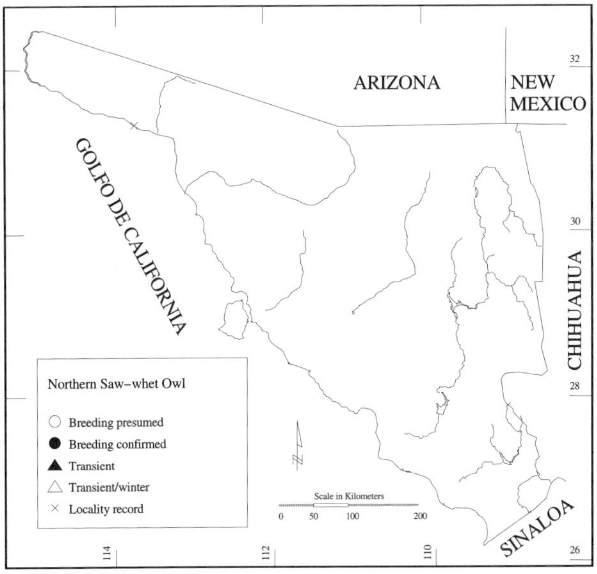

FAMILY CAPRIMULGIDAE

Goatsuckers

LESSER NIGHTHAWK

Chotacabras Menor

Chordeiles acutipennis

The Lesser Nighthawk is a common summer resident from sea level to about 1150 m throughout the desertscrub and open portions of the tropical thornscrub and tropical deciduous forest. It is less common in the north and east and in the barren desert from the Sierra el Pinacate region west to the Río Colorado delta, where it is again common. Small numbers apparently winter in the extreme south; it is regularly reported on Alamos Christmas Bird Counts. One was seen at San Carlos on 23 December 1982 (C. Green).

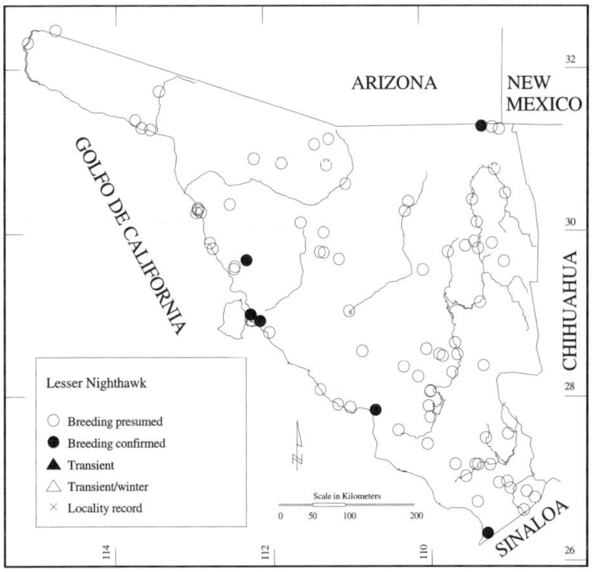

Spring arrival is in early April (occasionally late March), and the birds depart in late September. About 20 were migrating with a considerably larger number of Common Nighthawks on Rancho el Carrizo southwest of Benjamin Hill on 30 September and 1 October 1984. Other migrants include at least 50 flying in the early morning along the shore of the Agua Caliente arm of Presa Obregon on 22 September 1985, 160 over alfalfa fields below Presa Obregon in the evening on the same date, and about 300 flying over the Río Yaqui below the Tonichi bridge on 18 September 1985 after a late afternoon thunderstorm.

At times, nonmigratory concentrations of Lesser Nighthawks are seen. Twenty-five were observed foraging in late afternoon near Tonichi on 23 July 1985, and 48 were counted at Puerto Peñasco on 7 June 1975 (D. Stejskal).

The nesting season is from May into July. Eggs or young (WFVZ) were found almost within a stone's throw of the shore along El Golfo de California at Guásimas (22 May), at Punta Chueca north of Bahía Kino (30 May), and on a hummock in the marsh near Agiabampo (8 June).

COMMON NIGHTHAWK
Chotacabras Zumbón

Chordeiles minor

This goatsucker is a fairly common summer resident of open grassland above 1200 m in eastern Sonora. It migrates in late April and May, and again in mid-September to early October, when birds perhaps nesting as far away as northern Canada are passing through to wintering grounds in South America. Some records indicative of migration are two perched on fence posts and four flying around at 0900 west of Tecoripa on 17 May 1988 and about 85 in the evening and about 50 in the morning on Rancho el Carrizo southwest of Benjamin Hill on 30 September and 1 October 1984. These last birds were going south at a considerable height, along with smaller numbers of Lesser Nighthawks that were flying relatively near the ground.

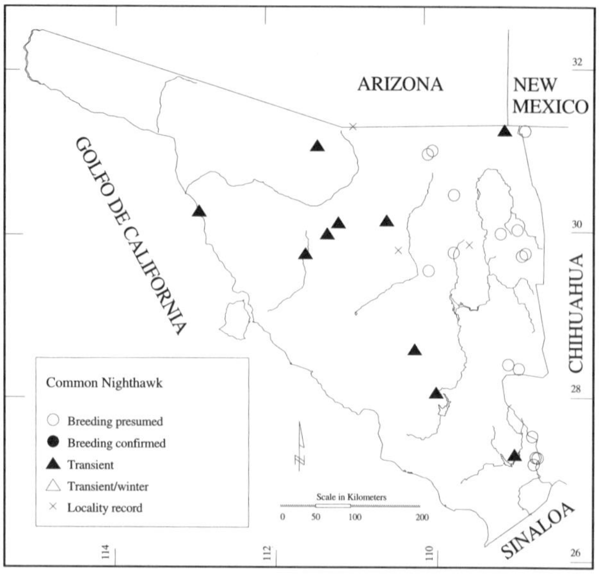

Nesting presumably commences in early June and continues into July, but no nests have been found. J. Marshall noted that birds were booming from 0330 to dawn and through the evening to 2200 on 11 June 1960 at a grassland location about 26 km by road southwest of Moctezuma, in an area where Lesser Nighthawks were also present. He watched one drinking at a cow trough while on the wing at 0800 in the Sierra los Ajos on 24 July 1953.

PAURAQUE
Chotacabras Pauraque

Nyctidromus albicollis

This bird was not discovered in Sonora until 26 May 1980, when it was considered common along the Río Yaqui at Mexico Highway 15 north of Cd. Obregon (Terrill and Terrill 1986). Subsequently, we found at least one about 1.6 km below Presa Obregon on 10 and 11 April 1986, and we recorded calls from a dense mesquite thicket and an adjacent alfalfa field. On a subsequent visit on 7 July 1986, J. Bates found a nest containing two eggs on the ground in the thicket. At the same place on 9 and 10 July 1987, R. Bowers photographed a young bird that could fly. Three called vigorously for about five minutes at dusk on the Río Cuchujaqui south of Alamos on 19 October 1988 (R. Taylor). We think the Pauraque will be found at other places, especially in the Río Mayo drainage.

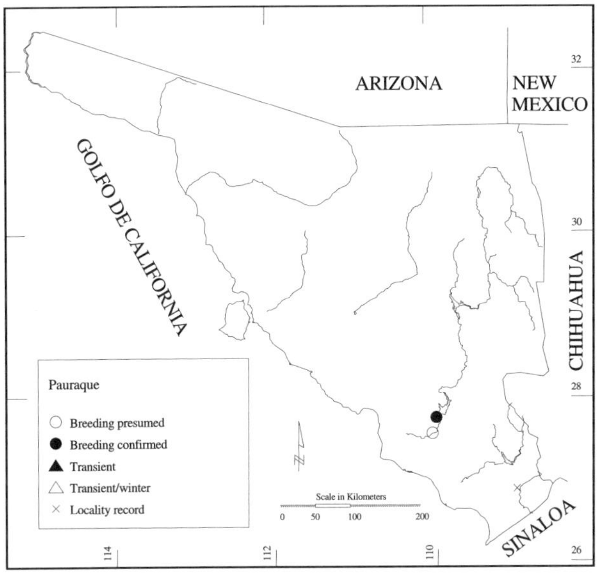

COMMON POORWILL

Tapacamino Tevíi

Phalaenoptilus nuttallii

The distinctive call of the Common Poorwill may be heard from February into October in coastal desertscrub and upward to about 1800 m in Sonora. We have not found it south of Guásimas, although van Rossem (1945) reported it at Agiabampo on 15 and 16 May 1937.

Some individuals remain all winter at least occasionally in southern Arizona, and the same may be true of Sonora, although most of our records fall between 20 February and 19 November. One was collected on 25 November 1956 between San José de Batuc and Villa Pesqueira (J. Marshall); one was seen on 2 December 1898 at San Luis Río Colorado (van Rossem 1945); two were noted on 28 December 1982 on the Alamos Christmas Bird Count, the only record from the lower Río Mayo drainage; and one was seen in the Christmas Bird Count period at Puerto Peñasco in 1979. We have heard them calling at the edge of mangroves near Punta Chueca on 19 January 1974 and in desertscrub north of Puerto Lobos (where it may nest) on 18 February 1989.

Although most individuals of the Sonoran population, along with those of the western United States, are believed to migrate, it is perplexing that no movements of transients have been detected in Sonora. Perhaps, unlike nighthawks that are diurnal migrants, poorwills travel and feed at night during migration, thus avoiding observation. Common Poorwills are known to hibernate, a most unusual behavior for a bird. It is possible that poorwills do not migrate at all in the warmer part of their range and are unseen in winter because of their inactivity and absence of calling.

The Common Poorwill is associated with rocky outcrops and rimrocks. Even when found in the open, rocky habitat is close at hand. In the drier parts of the desertscrub, as in the Sierra el Pinacate region, it is relatively scarce and may be entirely absent during periods of drought. The nesting season is from late March into July.

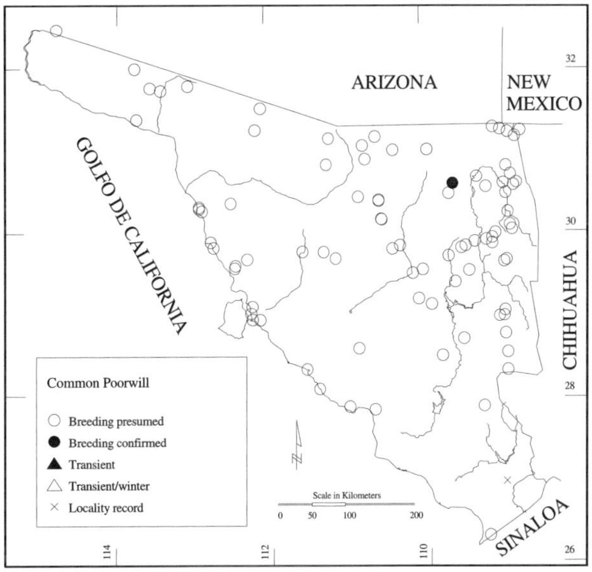

EARED POORWILL
Tapacamino Prío
Nyctiphrynus mcleodii

Termed one of the least-known and elusive birds of the North American continent by Schaldach and Phillips (1961), the Eared Poorwill was unknown in Sonora until four specimens were taken in 1946 by W. Sheffler and E. Harrison in the mountains east of Güirocoba (Miller 1948). A specimen (LSUMZ) taken by Sheffler on 27 May 1947 about 40 km northeast of Güirocoba was labeled "breeding male."

On 17 July 1987, Monson and J. Hoffman flushed an Eared Poorwill from the overhang of a large boulder in a side canyon of the Barranca las Colas below Sahuarivo. The bird alighted on a smaller boulder nearby in full sunlight, where it rested perhaps 15 seconds before it flew off with a loud *quark* note. The elevation was 1500 m in pine-oak with some Arizona cypress and alder.

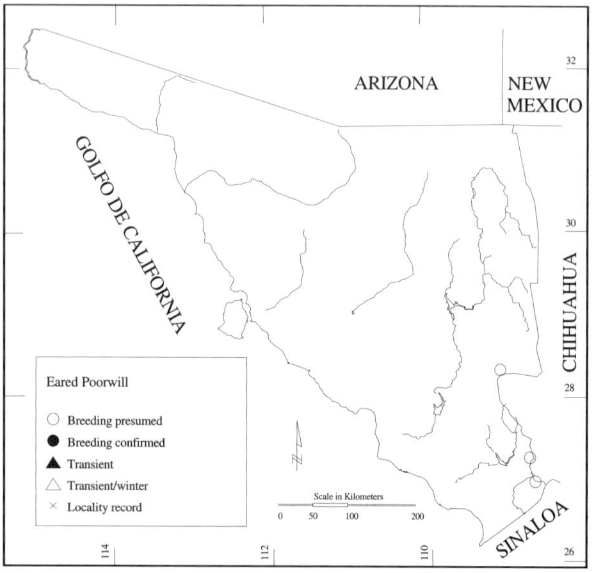

On 16 May 1988, Monson and R. Dorrance found another Eared Poorwill perched head-high close to the trunk of a pine on Mesa de Enmedio west of Yécora. It flew and alighted on a branch of another pine, again next to the trunk, where it was photographed. On close approach, it flew off to perch even higher in a third pine. The elevation of this pine-oak woodland with an understory of mainly manzanita and bracken fern was about 2140 m.

BUFF-COLLARED NIGHTJAR
Tapacamino Tu-cuchillo
Caprimulgus ridgwayi

At nightfall in the eastern half of Sonora, mostly in the spring and early summer, one hears the *cu-cu-cu-cu-ca-cuchillo* call of the Buff-collared Nightjar, especially in riparian sites bordered by open pasture or desertscrub or thornscrub. It is a fairly common to common bird from elevations of 50 to 1200 m in regions of dense thornscrub and tropical deciduous forest. Locally, it may be abundant, as it was on the night of 27 April 1977 around the corrals and buildings of Rancho las Guásimas about 14 km east of Presa el Novillo. The mostly bare ground was literally peppered with the birds, so many and so often calling simultaneously that numbers were difficult to estimate (perhaps 20 to 30). Van Rossem (1945) described it as abundant at Güirocoba.

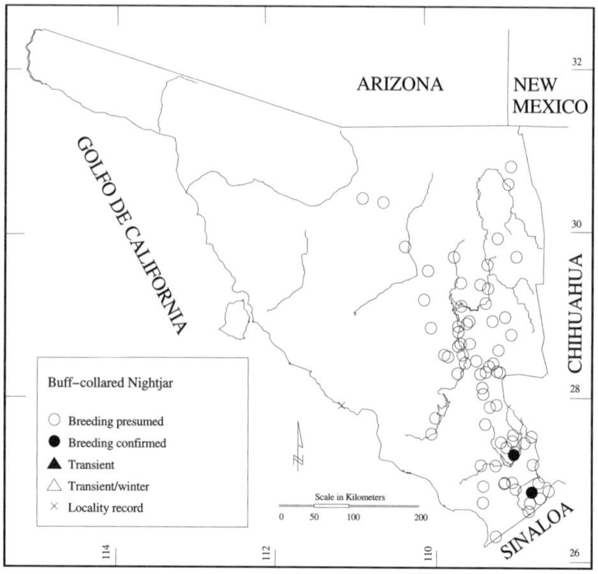

This nightjar appears to be chiefly migratory in Sonora, with dates from about 8 March to about 15 October at Lo de Campa. There are some midwinter dates: one was collected on the Río Yaqui north of Cd. Obregon on 2 January 1953 (A. Phillips) and another (UA) at Lo de Campa on 9 December 1983; it has appeared on the Alamos Christmas Bird Count in 1981, 1983, and 1990; we captured one in a mist net on 6 January 1983 at 780 m a few kilometers west of Tepoca and heard one at Lo de Campa the same evening; we heard two on the Río Carrizo north of Tonichi on 9 January 1983; and one was heard along the Río Yaqui at the mouth of Arroyo Bacadehuachi on 14 December 1983 (B. Brown). Eggs (WFVZ) have been taken in May and June, but this nightjar probably nests from April into July.

On the night of 4–5 June 1984 at Lo de Campa, Monson was awake all night because of uncomfortable heat and humidity, bawling cattle, and an onslaught of various insects. He took advantage of this involuntary insomnia by trying to count the times a nearby Buff-collared Nightjar called from 2000 to 0430. The approximate total was an amazing 5100 calls!

WHIP-POOR-WILL

Tapacamino Cuerporruín-norteño

Caprimulgus vociferus

In Sonora, the Whip-poor-will is seldom found away from pines, but we heard two birds calling from steep canyon slopes (1270 m) on the night of 11 May 1983 north of the village of El Trigo. The area lacked pines and was dominated by oaks. Marshall (1957:80) stated that the Whip-poor-will is not found on mesa tops, since it requires "dense growth of tall trees on canyon walls." In the absence of any certain winter records, it appears that the Whip-poor-will is a summer resident in Sonora. One recorded 27 February at Alamos (van Rossem 1945) was probably a migrant. In northern Sonora, we heard one calling 2 March 1984 in Cañon Evans in the Sierra los Ajos. The latest date is 19 September, when one was heard out of habitat at 780 m a few kilometers west of Tepoca. We must consider the possibility that some Whip-poor-wills winter in the eastern lowlands of Sonora, as they do rarely in adjacent southern Arizona.

J. Marshall found nests and young in June and July. He (1957) provides a perceptive and highly interesting account of Whip-poor-will habits.

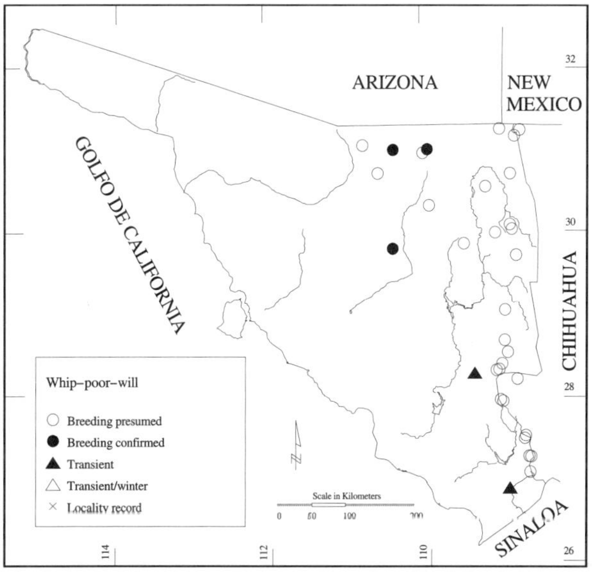

FAMILY APODIDAE

Swifts

BLACK SWIFT
Vencejo Negro
Cypseloides niger

The status of the Black Swift in Sonora is undetermined, as it has been found only three times. Monson and O. Allers saw two at close range from the highway a short distance south of Navojoa on 27 March 1982. Three were observed 6 km northeast of Yécora on 12 June 1986 on a ridgetop at 1680 m by A. Moorhouse. We saw as many as 60 flying daily with other swifts over and into Barranca las Colas near Sahuarivo 15 to 18 July 1987. It is possible that Black Swifts may breed in the mountains near Sahuarivo.

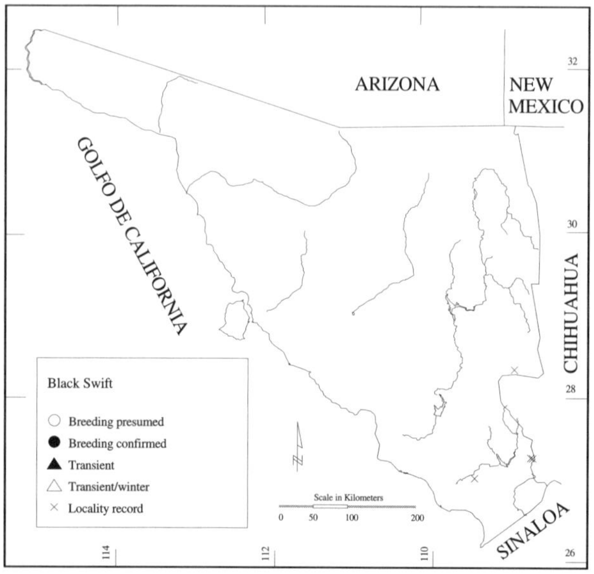

CHESTNUT-COLLARED SWIFT
Vencejo Cuello Castaño
Streptoprocne rutila

The Chestnut-collared Swift strays into southern Sonora from farther south or from the mountains of Chihuahua. We have two records: approximately 125 flying over El Naranjo canyon (460 m) of the Río Taymuco on 6 April 1987 (D. Stejskal, R. Bowers, J. Coons); and two flying with White-throated and Black Swifts over Barranca las Colas at Rancho San Rafael (1600 m) near Sahuarivo on 15 July 1987 (J. Bates, R. Bowers, G. Monson, A. Moorhouse, S. Russell, S. Suter, et al.).

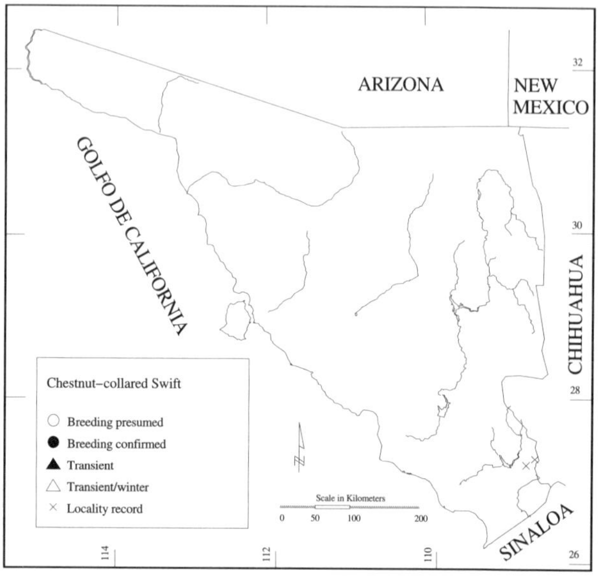

WHITE-NAPED SWIFT
Vencejo Nuca Blanca
Streptoprocne semicollaris

This swift has been recorded in Sonora three times. On 27 March 1959, W. Rook collected a female (WFVZ) from a flock foraging over a field on the outskirts of Navojoa. A flock of about 12 was observed and one was photographed a few kilometers east of Navojoa on the road to Alamos on 22 February 1974 (B. and D.

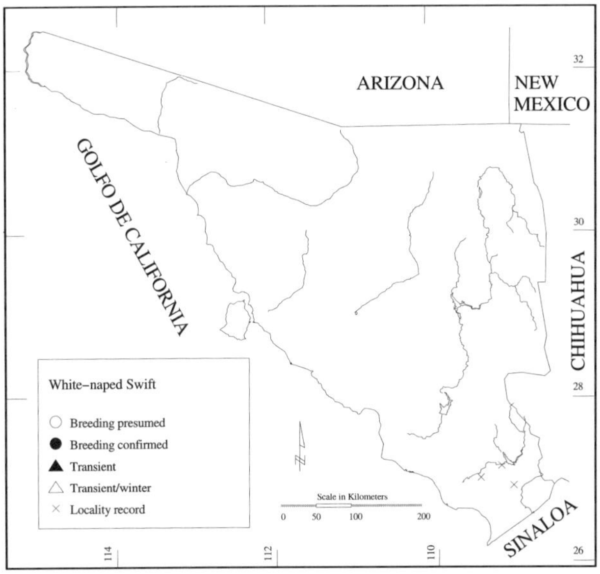

McKnight). The identification was substantiated by A. Phillips, but the photograph has been lost. On an Alamos Christmas Bird Count on 29 December 1987, ten were seen high on the north side of the Sierra Alamos by D. Stejskal and N. Krug.

VAUX'S SWIFT
Vencejo de Vaux
Chaetura vauxi

This small swift is a rare transient through Sonora in both spring and fall, usually as single birds or in small flocks. Migration periods seem short. Spring records are from 20 April (1925, in flocks at Magdalena; van Rossem 1945) to 15 May (1945, also at Magdalena; van Rossem 1945). Fall records are from 17 September (1974, four in hills about 16 km north of Bahía Kino; D. Lamm) to 6 October (1979, two at Puerto Peñasco; D. Stejskal). Most reports are from along and near the coast of El Golfo de California. The only report of large numbers is that of R. Taylor, who saw hundreds over fields southwest of Hermosillo on 2 May 1988.

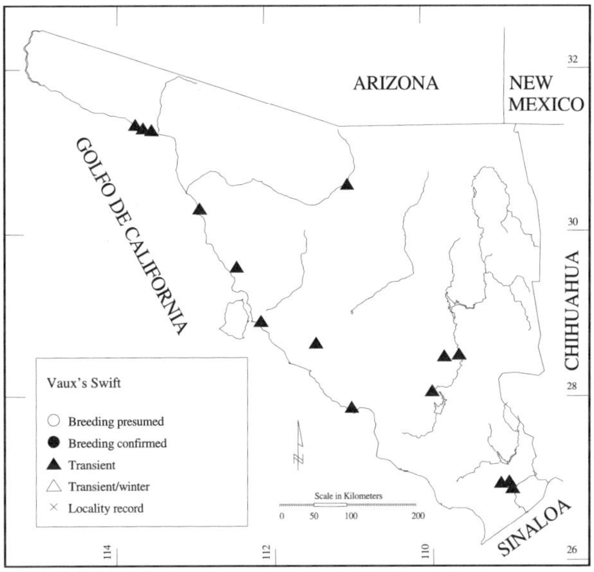

WHITE-THROATED SWIFT
Vencejo Pecho Blanco
Aeronautes saxatalis

White-throated Swifts are fairly common in the mountains of eastern Sonora. Because they are capable of long daily flights in search of insects, they may be expected virtually anyplace. Often they fly so high they are invisible except through binoculars; yet they will descend to feed low over fields. They have been recorded in Sonora in all months of the year, but there are fewer observations from the colder months (November through March). It is likely that some birds winter to the south. They often wander more widely in the rainy years; an example of this is ten at Volcan el Elegante in the Sierra el Pinacate on 15 September 1979.

There are few records from coastal areas, but we have noted birds along the coast (Punta Chueca, Caleta

Venecia, Bahía San Pedro) during May and June when they might be expected to be breeding. We have never seen more than 100 at once, but larger aggregations would not be unexpected. Only one nest has been found, by J. Marshall in the Sierra Azul on 13 July 1953.

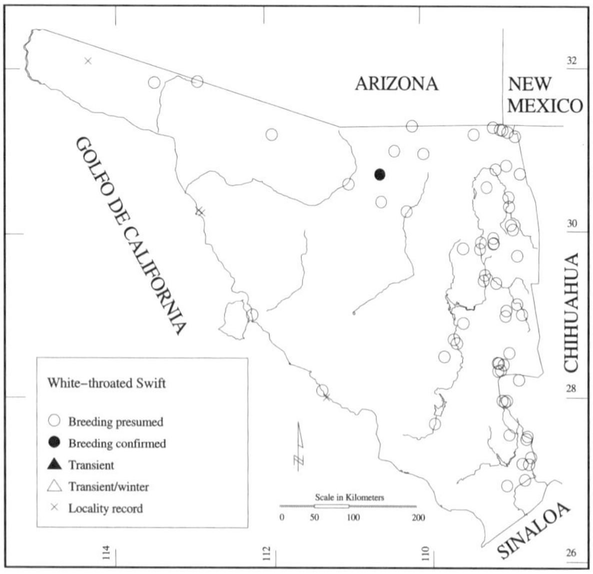

FAMILY TROCHILIDAE

Hummingbirds

BROAD-BILLED HUMMINGBIRD
Colibrí Pico Ancho
Cynanthus latirostris

This is the most common hummingbird in Sonora. It breeds from the coastal plains (north to Caleta Venecia) to 1250 m in the eastern mountains, and sometimes it wanders much higher (2000 m). It is common throughout the year in the tropical thornscrub and tropical deciduous forest northward to the latitude of Tecoripa and Lo de Campa, although greater numbers are present from March to October. We have estimated very large numbers on occasion: 280 in cactus forest at Agiabampo on 11 July 1987; 200 at Lo de Campa on 5 June 1984; and 150 at Rancho Jacalón northeast of Rosario on 11 June 1984. Spring migration peaks in the first half of March. Southbound migration is over by November. The presence of some individuals in northern Sonora in winter is not unusual: for example, one at Puerto Lobos on 4 January 1985 (J. Bates), one at Bahía Kino on 25 February 1987 (R. Taylor), one between Magdalena and Imuris on 16 and 27 January 1980 (Terrill 1981), and one on the Arroyo Santo Domingo above Rancho Saracachi on 21 February 1982.

Nesting may begin very early: R. Russell found a nest with eggs at the early date of 11 January 1981 in a small ravine (580 m) near Taymuco. The peak of nesting activity is in March, but at Lo de Campa and Arroyo Cajón Bonito we have noted that it may persist into mid-July. Breeding numbers are very high in the cactus forest along the lower coast of El Golfo de California, but we do not know the extent of the breeding season. In the north, nesting is confined to vegetation along permanent streams or water (e.g., Arroyo Cajón Bonito, Río Magdalena). A few birds breed into oak-pine woodland. A juvenile was collected in maple–Douglas-fir at 1920 m in the Sierra los Ajos on 1 August 1952 (J. Marshall).

The widespread introduced tree tobacco, which produces flowers yearlong that are utilized extensively by hummingbirds, may have increased the numbers and range of the Broad-billed, Anna's, and Violet-crowned Hummingbirds.

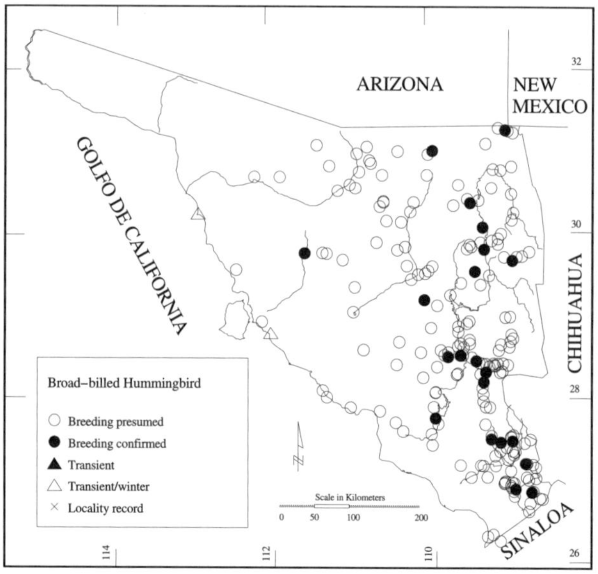

WHITE-EARED HUMMINGBIRD

Zafiro Oreja Blanca

Hylocharis leucotis

The White-eared Hummingbird is an uncommon summer resident of riparian areas above 1400 m in the pine-oak woodland and pine forest. Only in the Sierra Oscura (2000 m) and the vicinity of Sahuarivo (1600 m) have we noted more than 12 in a day. We have found it on the breeding range only from 13 March to 18 September, indicating that it is migratory.

The species is a rare winter visitant and transient at various southern foothills localities. It is regularly reported on Alamos Christmas Bird Counts, when it is usually found in the Sierra Alamos. The presence of six White-eared Hummingbirds on 5 January 1993 in less than a kilometer of Arroyo el Palmarito (670 m) northeast of Alamos suggests this and other lowland mesic canyons may be important wintering areas. We found several in sycamore-dominated habitat (920 m) in Arroyo el Reparo in March and mid-May. We do not know if the species breeds there, just below pine-oak habitat, or if these were late migrants. We saw no courtship displays.

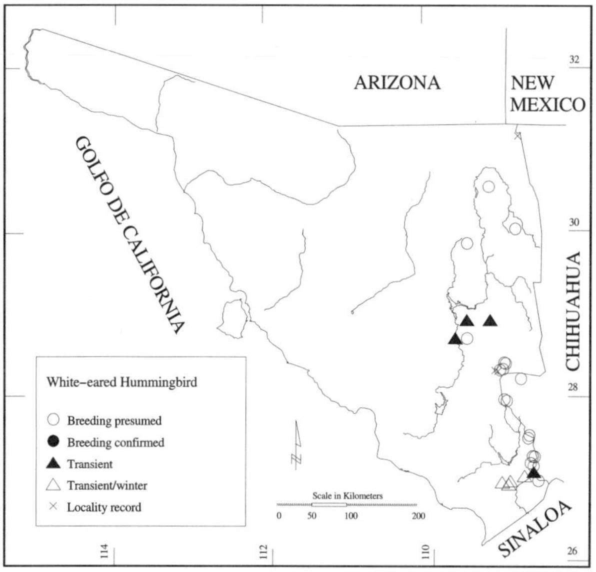

BERYLLINE HUMMINGBIRD

Colibrí Berilo

Amazilia beryllina

The Berylline Hummingbird is a summer resident above 1000 m in the mountains that fringe the eastern border of Sonora, being fairly common locally north to the Yécora region and rare farther north. Records extend from 12 March to 17 September, but most individuals probably arrive in May. It is most frequently

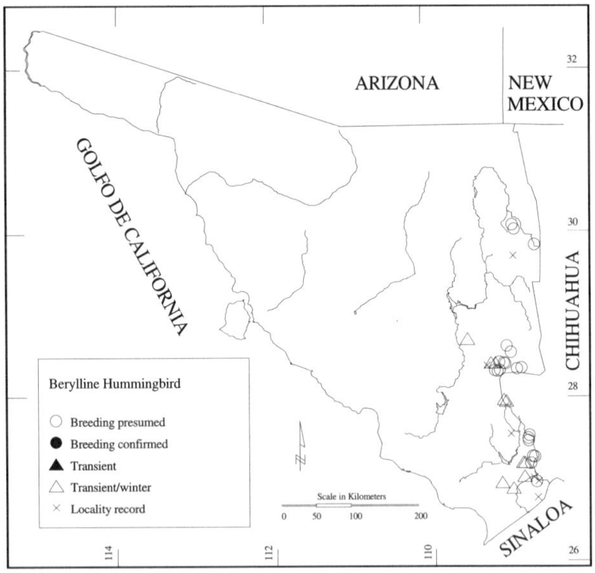

encountered in moist areas in the pine-oak woodland, where it is found in canyons and open woods.

In winter, Berylline Hummingbirds are infrequently found in the southern foothills and are recorded on Christmas Bird Counts in the Alamos area, places where they do not nest. We think most leave their breeding areas in Sonora to winter from Alamos southward. They return in spring through the foothills before ascending the mountains, and they may remain in the lowlands until mid-May.

No nests have been found in Sonora, and we can only speculate about the nesting area and season on the basis of breeding behavior. Successful nesting, as with all hummingbirds, is dependent upon an adequate supply of the appropriate wildflowers. Depending upon climatic conditions, there may be substantial shifts of breeding areas and season. In late July 1988 we saw numbers of territorial and singing Berylline Hummingbirds (and a female carrying nest material), in the company of Violet-crowned Hummingbirds. They fed avidly at the yellow flowers of *Mandevilla foliosa*, which in places was the principal understory plant in oaks (1500 m) on the northwest slope of Mesa de Enmedio west of Yécora.

We are uncertain of the status of several birds seen in mid-May at Guajaray (270 m) and Tepoca (580 m). We presume they were late migrants because none were singing or displaying. One at La Tinaja (195 m) on 13 July 1987 may have been nonbreeding or possibly an early migrant.

[CINNAMON HUMMINGBIRD
Colibrí Canela
Amazilia rutila

Three were seen at Los Algodones in extreme southeastern Sonora from 27 February to 1 March 1948 (A. van Rossem notes). This hummingbird is common north to central Sinaloa, and additional observations from Sonora should be anticipated.]

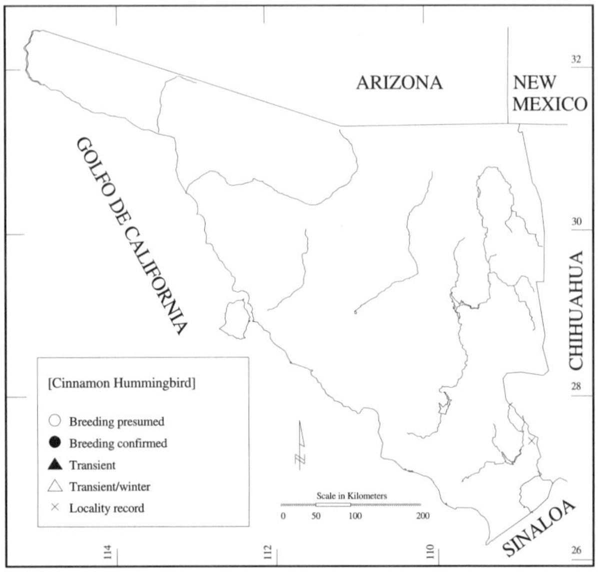

VIOLET-CROWNED HUMMINGBIRD
Colibrí Corona Violeta
Amazilia violiceps

The medium-sized Violet-crowned Hummingbird is a common resident of southern Sonora (200–900 m) north to the Tonichi–Lo de Campa region in the Río Yaqui valley. In northern Sonora, it is a fairly common summer resident, present in the drainages of the Ríos Bavispe, Sonora, and Magdalena (800–1300 m), from early April until mid-September. Three observed east of Moctezuma on 4 February 1983 indicate that either the species winters that far north or spring migration had begun very early.

Nests have been found from mid-March to 21 August. Principal breeding habitat is mainly riparian vegetation in tropical thornscrub and deciduous forest in the south and within gallery riparian woodlands in northern Sonora. In Arroyo Cajón Bonito (1240 m), Violet-crowns are fairly common nesting birds in July, yet they are not present there earlier in the year. Occasional birds seen in pine-oak woodland as high as 2000 m in the Sierra Oscura in September and one in the Sierra Huachinera at 2200 m on 2 September 1983 (Mark Larson) are regarded as postbreeding individuals. One netted in mangroves at Punta Chueca on 21 June 1978 was far from known breeding localities.

The Violet-crowned Hummingbird can be extremely common at times. We counted 60 at Agiabampo on 11 July 1987, where they were feeding on flowers of columnar cacti and a euphorb. The date is suggestive of breeding, but we observed no courtship. We saw many birds in oak woodland (1500 m) in late July on the slopes of Mesa de Enmedio. Individuals defended masses of flowering *Mandevilla foliosa* against other Violet-crowns and were dominant over the slightly smaller Berylline Hummingbirds. There was no song or display to suggest nesting, and we assume that

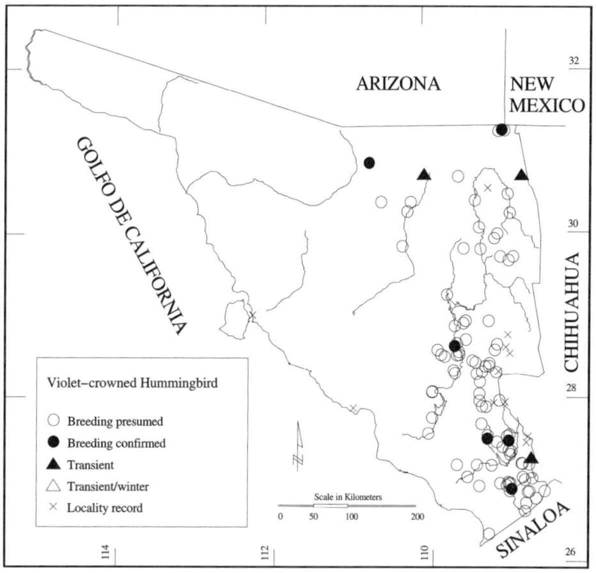

the birds had moved into the area because of the abundance of food. Forty-five were at Rancho Tesia north of Quiriego on 5 January 1986. The observation of 110 individuals at Lo de Campa on 17 July 1985 is exceptionally high and reflects a very high density of nesting birds. Numbers decline there rapidly in August, and only six to eight birds are present in the same area through the winter. There is much to be learned about the movements and breeding biology of this and other hummingbirds.

BLUE-THROATED HUMMINGBIRD
Colibrí Garganta Azul
Lampornis clemenciae

This species, the largest (8 g) of the hummingbirds in Sonora, is an uncommon summer resident of a few narrow moist canyons in coniferous forest (and oak-pine woodland) of the higher mountains of the east. Arroyo Cajón Bonito (at 1250 m) and Arroyo el Riito (980 m) northeast of Nacori Chico are the lowest elevations it is regularly found; these localities are characterized by permanent streams and gallery riparian forests including sycamores. Birds arrive in the breeding range in mid-March and remain until mid-September. There are no winter records in Sonora.

Only two nests in Sonora are known to us (found by D. Lamm and A. Moorhouse, respectively, both in May). Nests are usually constructed in a protected site, such as under an overhanging rock near a stream bed. Unlike most species, Blue-throated Hummingbirds tend to use the same nest in successive years. Marshall (1957) noted a reduction in numbers in the Sierra Huachinera from 1953 to 1954; he attributed the drop to drought and the resulting absence of flowering *Lobelia laxiflora*, upon which they fed.

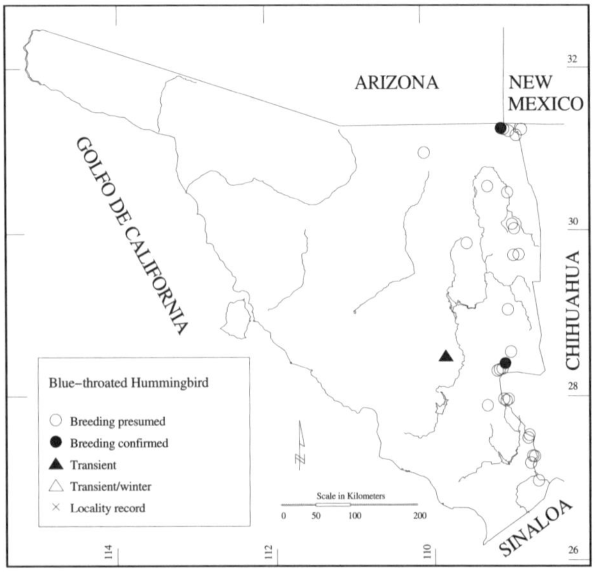

We know of only one record of a possible migrant at low elevations. R. Morse noted one at Lo de Campa (490 m) on 20 August 1984. The paucity of lowland observations suggests that Blue-throated Hummingbirds migrate south at high elevations in the mountains. Alternatively, because this is an uncommon species, individuals could migrate through the lowlands undetected.

MAGNIFICENT HUMMINGBIRD
Colibrí Magnifico

Eugenes fulgens

The Magnificent Hummingbird is a rare summer resident in the mountains of eastern Sonora, present from mid-March into the first week of September. Most birds probably do not arrive before early May. We know of no winter records in Sonora and know of only two probable migrants: two males we saw 3 km northeast (820 m) of Nacori Chico on 20 March 1985, and a specimen (OC) taken on 18 May 1933 northeast of Güirocoba at an elevation of 880 m. A male at coral-bean flowers on Rancho los Pozos southeast of Mazatán (660 m) on 25 June 1983 (S. and J. Levy) may have been an early migrant or a nonbreeding wanderer. J. Marshall found a nest on a horizontal alder branch 7 m above a small stream in the Sierra los Ajos on 29 July 1952; the female was apparently incubating. The species winters occasionally at feeders in southeastern Arizona.

Marshall (1957) considered the Magnificent Hummingbird chiefly a captor of flying insects and thus less dependent upon flowers than other hummingbirds. But it does feed extensively upon nectar, probably including that of the agaves that are plentiful in the mountains of Sonora.

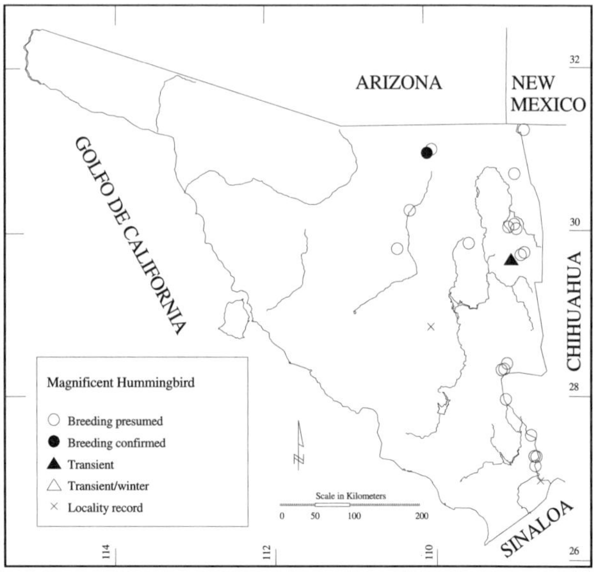

PLAIN-CAPPED STARTHROAT
Colibrí Picudo

Heliomaster constantii

The Plain-capped Starthroat is an uncommon summer resident in tropical deciduous forest between 60 and 500 m elevation and is present from mid-March to August. The species occasionally winters along the Río Cuchujaqui, along Arroyo el Palmarito, and at Güirocoba; up to seven have been noted on the Río Cuchujaqui on Alamos Christmas Bird Counts. We were surprised to see several of these hummingbirds in mangroves at Agiabampo on 11 July 1987. They may have been southbound migrants. By far the most northerly record is of one on 15 May 1988 in Arroyo Cajón Bonito (1280 m; M. Johnson, R. Taylor). The species is a straggler in southeastern Arizona, where there are nearly as many records as for all of Sonora, undoubtedly reflecting the lower number of bird observers in Sonora. The only evidence of breeding is a statement by van Rossem (1945:124) describing "young of the year on the wing as early as the middle of May" at Güirocoba.

Plain-capped Starthroat

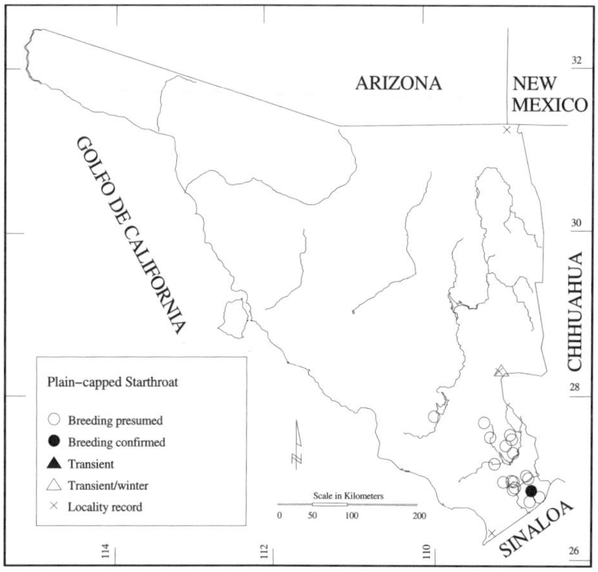

LUCIFER HUMMINGBIRD
Colibrí Lucifer

Calothorax lucifer

The Lucifer Hummingbird is rare in Sonora. Marshall (1957) collected (LSUMZ) a nonbreeding female and saw other females in the Sierra los Ojos Azules in rather open Apache pine, Arizona oak, grass, and *Ceanothus* on the ridge (2380 m) on 6 August 1952. On 16 May and 14 July 1976 we found several along Arroyo Cajón Bonito at 1240 m, with four nests located on the first date (Russell and Lamm 1978), in sycamore, ash, and netleaf hackberry. The birds foraged on the canyon slopes, presumably feeding extensively on the nectar of agave flowers. The Lucifer Hummingbird appears to be a recent invader to southeastern Arizona, hence it may be a similarly recent arrival in northeastern Sonora (Scott 1994).

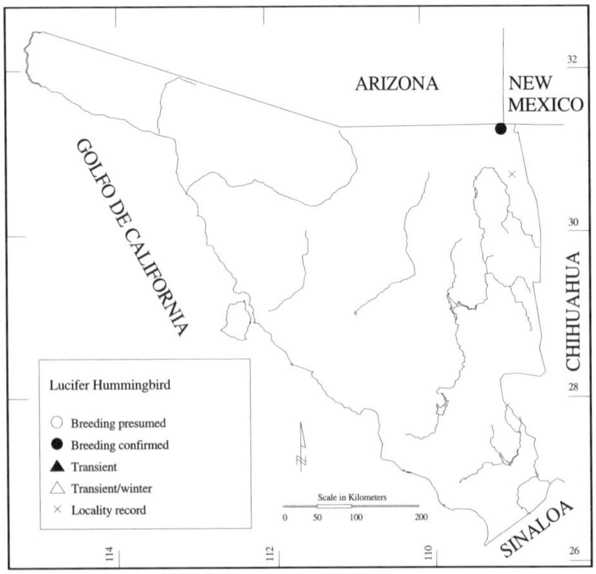

BLACK-CHINNED HUMMINGBIRD
Colibrí Barba Negra
Archilochus alexandri

Black-chinned Hummingbirds are uncommon spring migrants in March through mid-April through most of eastern Sonora. They are rare on the coast, from which there are only two records, both in April. Fall migrants appear in August. The species is fairly common all across eastern and central Sonora in September until the migration ends abruptly about 21 September. There is only one October record, and one in November (specimen collected by J. Marshall on 24 or 25 November 1956, 24 km west of San José de Batuc). Documented winter records are wanting.

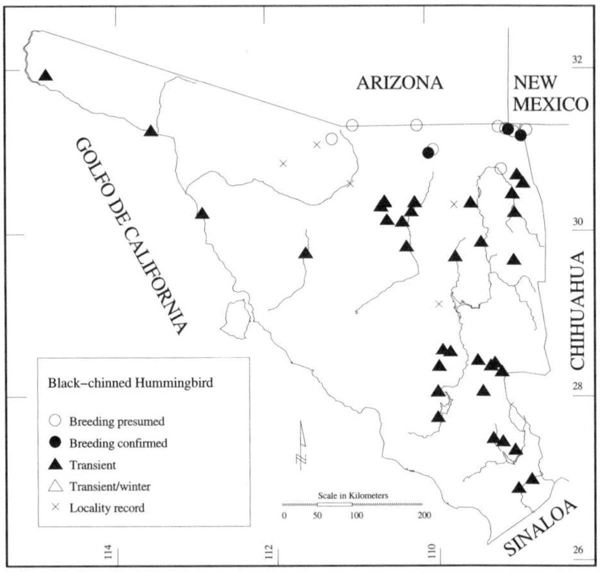

Breeding Black-chinned Hummingbirds from mid-April into early summer exploit permanent streambeds with cottonwoods and willows or sycamores.

ANNA'S HUMMINGBIRD
Colibrí Cabeza Roja
Calypte anna

Anna's Hummingbird is expanding its range in Sonora. Historically the species has been known only in fall and winter, as a migrant from its primary breeding range in California. Van Rossem (1945) cited only three records: Sierra de San Jose, 19 August 1893; Sierra el Pinito, 14 October 1890; and Puerto Peñasco, 21 February 1934. Phillips and Amadon (1952) found several in the Caborca-Sonoita area in late October and early November 1948. On 18 February 1960, A. Gardner collected an adult male (WFVZ) near Bahía Kino. In the 1980s, many additional records to the east and south were noted, and 9 to 23 individuals were regularly tallied on the Puerto Peñasco Christmas Bird Counts.

The species arrives in Sonora in numbers in early September and is present through the winter to at least as late as 18 April. It has not been found nesting. In southern Arizona, a similar migration pattern exists, but Anna's Hummingbirds were found nesting in 1962

(Yuma) and subsequently nesting from December to April at many localities close to the Arizona-Sonora border.

We think the species probably now nests in Sonora. Territorial, singing males have been noted at several locations in northeastern Sonora. On 5 January 1983 at least 12 singing males were distributed through an area of green, dense tropical thornscrub (dominated by flowering tree morningglories) 12 km east (520 m) of Tecoripa. It is highly probable that nesting occurred. Many searches have been made in the same area in subsequent years without finding them again. Thus, Anna's Hummingbird might be opportunistic by staying in an area if conditions are suitable for nesting or moving elsewhere when conditions are not good. Whether the birds nest in Sonora and then return to the Pacific Coast and breed again is unknown. There are no Sonoran records from 18 April to 19 August.

What has happened in the last 40 years to contribute to their range expansion? Anna's Hummingbird is a very successful urban species and has done well where many flowering exotic plants have been cultivated. Possibly the increase in introduced flowering plants such as eucalyptus and tree tobacco have provided a corridor. It is also possible that this hummingbird has become a more successful competitor or that it has undergone genetic change. It is a species well worth continued study.

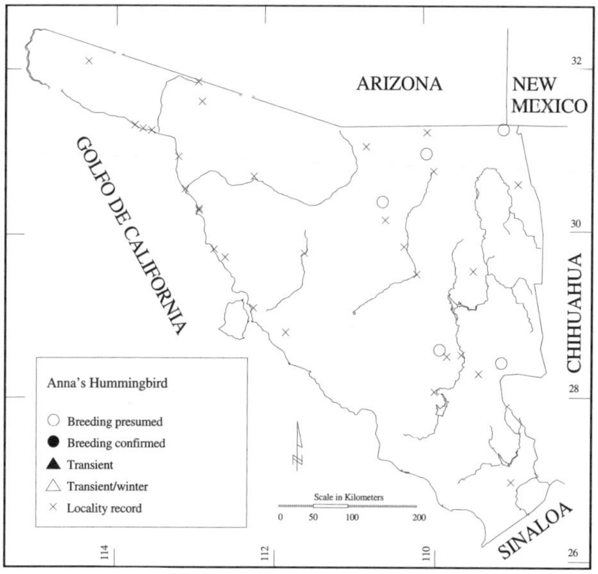

COSTA'S HUMMINGBIRD
Colibrí Cabeza Violeta

Calypte costae

This tiny (2.8 g) bird is characteristic of desertscrub, especially in coastal areas south to Guásimas. We have found it throughout most of the year at coastal localities (Puerto Lobos, Punta Chueca), but numbers vary seasonally and from year to year. Because its diet depends upon the availability of tiny insects and flower nectar, it may leave poor areas and search for better ones. Costa's Hummingbirds are most common at Puerto Lobos from February into April, when they nest there. In northeastern Sonora, they are uncommon late spring breeding birds up to elevations of about 1200 m. At locations east of the coastal plains and south of the latitude of Nacozari, they are fairly common but present only from late September to mid-March; no nesting behavior has been observed, and we believe these are wintering and transient birds. In some years, they may be common wintering birds in tropical deciduous forest in the Sierra Alamos.

The presence of a calling and displaying male before a female is indicative of nesting; on this basis the breeding season may be from January to June in years having substantial rainfall. Some male displays are used to defend a feeding area and are not indicative of nesting.

The movements of the Costa's Hummingbird in Sonora and Arizona are still poorly known. There is only one record from Sonora from mid-July to mid-

September. Where do they go? We don't know. Perhaps they are in places we have not visited; it is the time of summer rains and flowers, and they could be widely but thinly dispersed. Until someone conducts an intensive long-term study, including the marking of individuals with uniquely numbered bands, the Costa's Hummingbird, although sometimes common, must be considered a poorly known species.

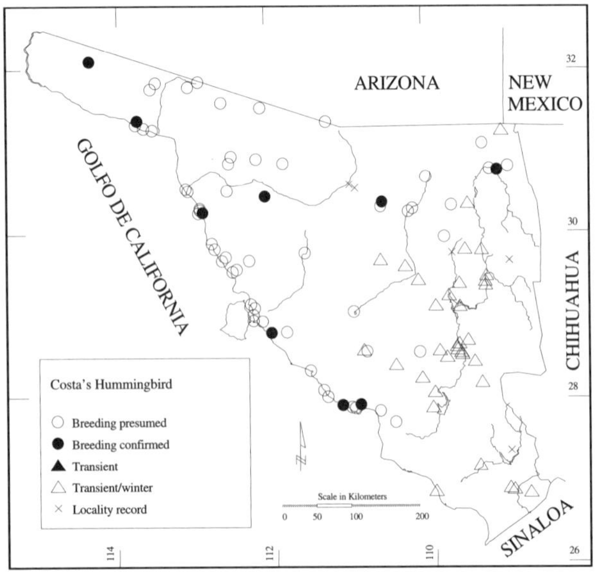

CALLIOPE HUMMINGBIRD
Colibrí Garganta Rayada
Stellula calliope

The smallest (2.0 g) of all Sonoran birds, the Calliope Hummingbird is a rare transient, known from seven spring and nine fall observations. Spring records fall in the period of 24 March to 16 May and include five western or coastal sites. All fall birds were noted from Naco and Arizpe east into the mountains (elevation range, 860 to 2000 m); dates are from 13 July to 1 October. Given that the Calliope Hummingbird is a regular late summer migrant in southeastern Arizona, it must be overlooked to some extent in Sonora.

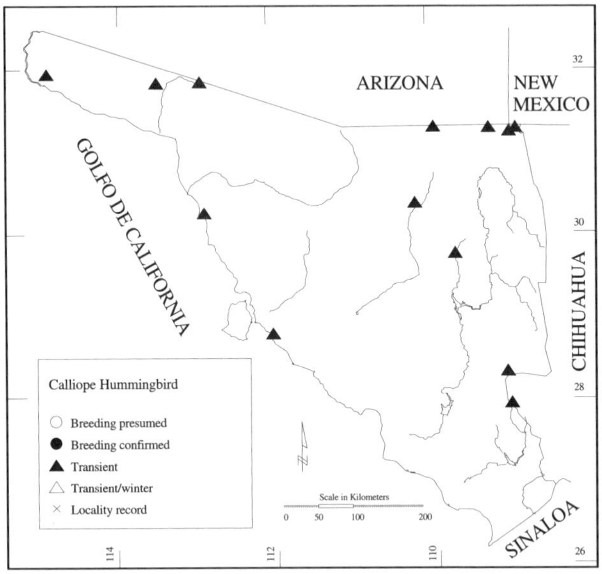

BROAD-TAILED HUMMINGBIRD
Zumbador Cola Ancha
Selasphorus platycercus

While other hummingbirds may not be noticed because they are silent at the time, the male Broad-tailed Hummingbird may make its presence known by a "distinctive trilling rattle" (Marshall 1957:82) as it flies past, a rattle produced by the wing feathers. The Broad-tail is a fairly common transient through the eastern mountains and an uncommon and local summer resident in the highest coniferous forests of northeastern Sonora. The earliest bird in spring was noted on 3 March, and the last fall record is on 6 September. Many birds that nest in more northerly areas migrate through Sonora between these dates. Three winter records, the only ones for Sonora, are of single males observed on two Alamos Christmas Bird Counts (R. Bowers and D. Stejskal, respectively) and one male on the Río Cuchujaqui on 30 January 1993 (W. Leitner).

Marshall (1957) stated that the Broad-tailed Hummingbird nests in spring, exclusively in pine forest at elevations well over 1830 m. However, he found that these hummingbirds may forage down into the pine-oak woodland, especially once the nesting time is past.

He noted one in pine forest at the summit (2100 m) of the Sierra Azul on 13 July 1953. The mountains have not been surveyed earlier in the year, and it is not known whether Broad-tails nest there.

The lowest elevation of a migrant was at Lo de Campa (490 m) on 12 March 1986 (A. Moorhouse). A notable record as being out of place altitudinally and seasonally is that of a male collected on the upper Río Bavispe north of San Miguelito on 21 June 1953 (A. Phillips) at 950 m.

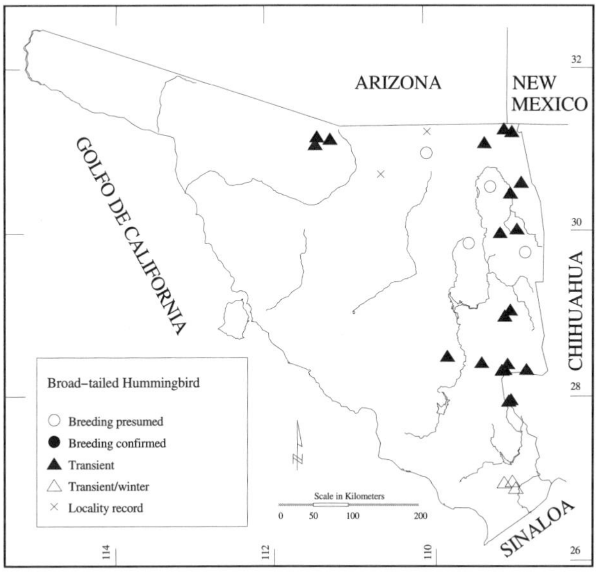

RUFOUS HUMMINGBIRD
Zumbador Rufo
Selasphorus rufus

Rufous Hummingbirds are transients through Sonora, migrating to and from their breeding grounds in northwestern North America. Southbound adult males, all cinnamon with an iridescent orange gorget, may pause temporarily at rich flowering food sources, chasing off larger hummingbirds with a vigor that belies their small (3.5 g) size. Fall migration begins in late July, peaks in early September, and is over by late September or early October. Fall migrants follow the eastern mountains (usually below 1200 m) and north-south valleys. There are only two coastal records in fall, 3 September and 6 October, both at Puerto Peñasco. A single bird taken and a second one seen on 31 October (Phillips and Amadon 1952) near Caborca represent the latest fall migrants.

In spring, northbound migrants move along the coast in the period of 18 January to 14 April—most from mid-February to the end of March. If winter rains have produced ample flowers, one may see up to 12 or 15 Rufous Hummingbirds in a few hectares. In dry years the birds may be rare, presumably making their flight with few stops. Fewer birds migrate along inland routes.

The largest number of Rufous Hummingbirds we have noted in one morning is 25. These were near Bacoachi, 18 September 1981, in a riparian area (1080 m) with many flowering tree tobaccos. The birds may have been following the Río Sonora southward.

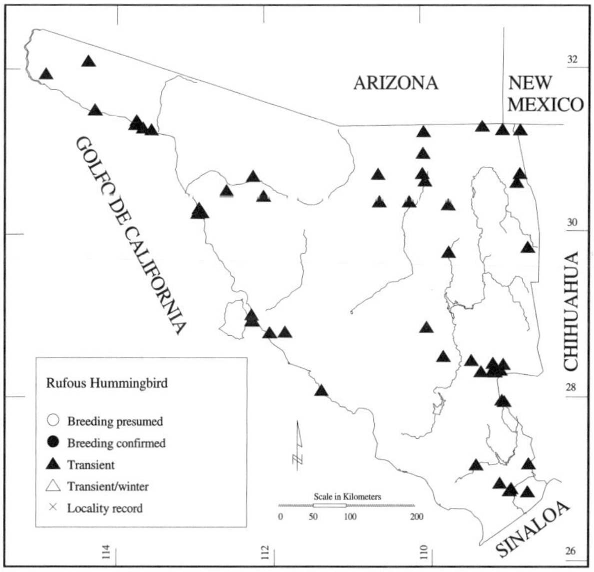

ALLEN'S HUMMINGBIRD

Zumbador de Allen

Selasphorus sasin

Two specimens of Allen's Hummingbird have been collected in Sonora: a male (LSUMZ) on 22 February 1955 in desertscrub near Sonoita (J. Marshall) and a female (UA) on 28 February 1971 in mangroves at Punta Santa Rosa (Russell and Lamm 1978). Van Rossem (1945) did not list the species. Typical males were seen on 18 January 1982 at Alamos (Monson) and on 12 February 1982 at Puerto Peñasco (K. Kaufman, D. Stejskal). Alden's (1969) statement that the species is common in winter at Guaymas is unfounded.

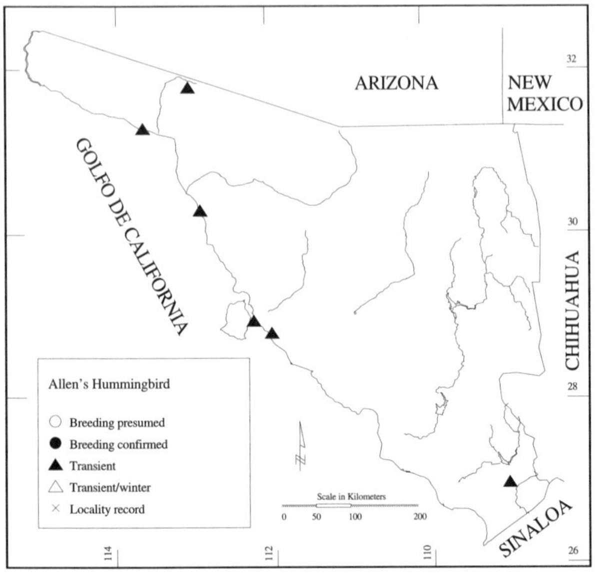

Based on the Russells' recent banding studies in southeastern Arizona, some Allen's Hummingbirds must migrate through mountainous eastern Sonora in July and August. It is also likely that some of the *Selasphorus* hummingbirds migrating northward along the coast in spring and identified as "Rufous" must be Allen's, especially in January and February. The similarity of female and immature Allen's and Rufous Hummingbirds is notorious; they can be identified reliably only in the hand.

FAMILY TROGONIDAE

Trogons

MOUNTAIN TROGON

Trogón Mexicano

Trogon mexicanus

We have found about ten pairs of Mountain Trogons in each of three mountain ranges of central eastern Sonora, all in coniferous forest at elevations above 1900 m. They would appear to be migratory; the earliest date is 10 May, and the latest is 18 September. We have seen paired birds but have found no nests.

In the Sierra Metate, southeast of Yécora, Mountain Trogons were on a north-facing slope in a mesic pine forest with some oaks, alders, and large madrones but no firs. The habitat in the Sierra Oscura was similar but contained many Douglas-firs. On Mesa de Enmedio the forest was drier, primarily open pine with a manzanita–bracken fern understory. Elegant Trogons are in the same places and are more common.

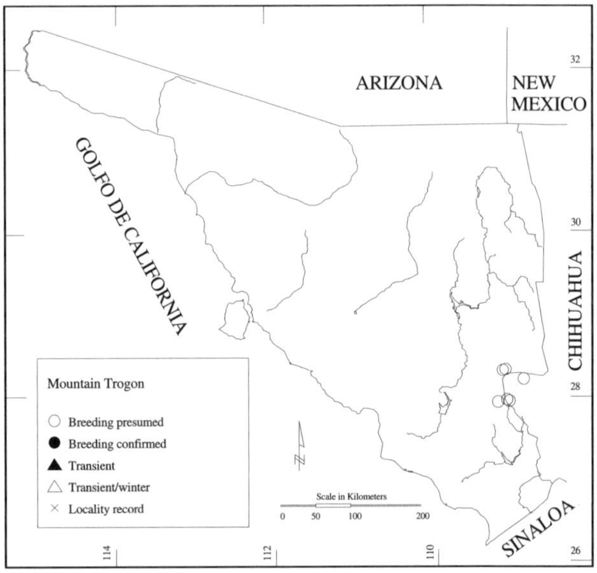

Mountain Trogon

ELEGANT TROGON
Trogón Elegante
Trogon elegans

In northern Sonora, the Elegant Trogon is associated with the pine-oak woodland, mainly in riparian situations with sycamore, Arizona cypress, and Chihuahua pine, from which it forages up canyon and mountain sides. In this part of its range it is uncommon, and the population migrates south for the winter. One we noted near Félix Gómez on 10 March 1989 was exceptionally far from its usual range.

South of latitude 28°30' N, which coincides roughly with the northern limits of the tropical thornscrub, many birds are found year-round, and the population is perhaps increased by wintering individuals from the north. Here the Elegant Trogon is common not only in the pine-oak woodland up to 2000 m (where it may not winter), but also at elevations down to 200 m or even lower, such as the Río Yaqui at Tonichi. In much of this region it is found in the tropical deciduous forest and tropical thornscrub.

In dense tropical thornscrub it is at times abundant. This is apparent at the beginning of the summer rainy season, when it is highly vocal in establishing territories and courting. At La Tinaja in the extreme southeast near the Río Fuerte, in mid-July of a dry year (1987) when nesting was postponed, we counted 75 along about 5 km of dusty road. We noted 50 birds at Lo de Campa in a morning in mid-July, when calling birds were well dispersed through the hills.

The Elegant Trogon nests in cavities in a variety of trees, including mesquites, sabal palms, feather trees, and chinos. Pairs may begin to investigate potential nest cavities in late May, but nesting invariably coincides with the onset of the summer rainy season. Once we saw a male attempt to enter a cavity occupied by nestling flickers! Young trogons fledge in July, and then the territorial calling of the adults abates rapidly. In September at Lo de Campa, groups of four to six birds forage together and there is no song; once (16 October 1983) we saw 11 in a single flock. Trogons eat buds and fruits of many plants but also forage on insects. We have seen them consume large green caterpillars and feed cicadas to nestlings.

In Sonora, the Elegant Trogon is known as "Coa," after its call.

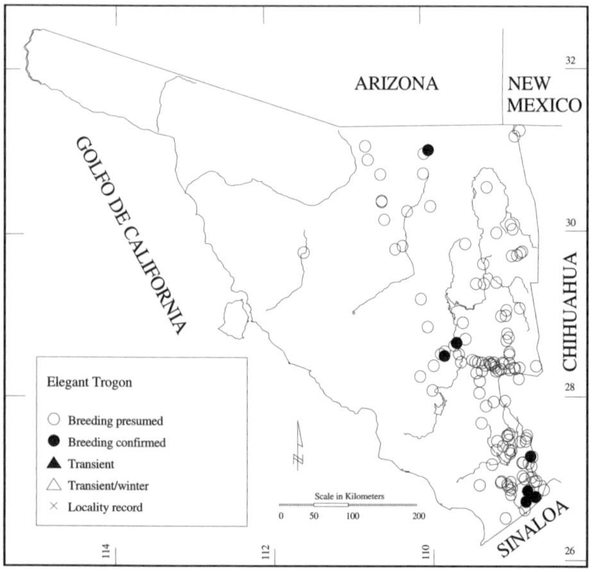

EARED TROGON

Trogón Orejón

Euptilotus neoxenus

Van Rossem (1945) reported a specimen of Eared Trogon collected on 16 December 1890 at a locality given as "El Puerto" in the Sierra Madre not far from Mesa Tres Ríos and close to the Chihuahua-Sonora boundary. The collector was F. Robinette, the zoologist for the famous Lumholtz expedition of 1890–1891 (Lumholtz 1902).

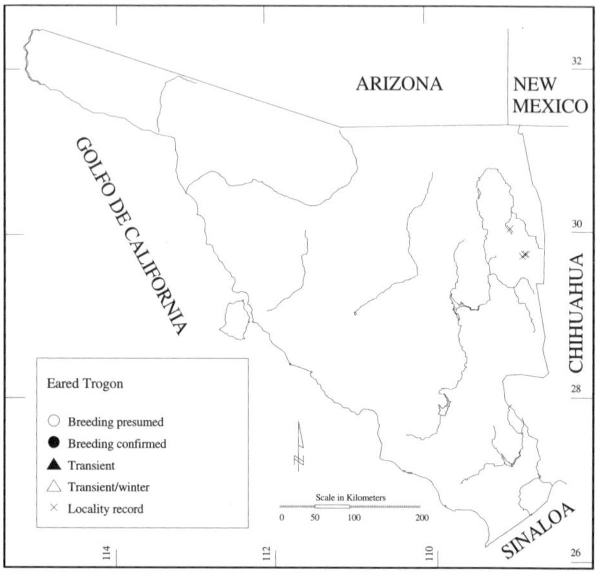

J. Marshall noted a pair of Eared Trogons on 10 June 1953 in the mountains northeast of Nacori Chico, very near the Robinette locality. On 1 June 1956, he and W. Lanyon saw a pair in the Sierra Huachinera. The species is found through much of the Sierra Madre Occidental in Chihuahua and southward, where it prefers undisturbed pine forest, and apparently it is uncommon and not easily found.

FAMILY MOMOTIDAE

Motmots

RUSSET-CROWNED MOTMOT

Momoto Corona Café

Momotus mexicanus

For such a large and striking bird, the Russet-crowned Motmot is surprisingly unobtrusive. It moves deliberately, sometimes only after sitting motionless on a branch for a long time. It is an uncommon and local summer resident of the tropical deciduous forest across a narrow belt of the southern tip of Sonora, between elevations of 100 and 1000 m. A half-dozen records from December, January, and February indicate that some individuals winter in Sonora. The species may be resident, but its small numbers and general reclusiveness may have prevented it from being seen more often in winter. Typically it is found in small arroyos, where it may excavate nesting burrows in precipitous banks. Nesting takes place from late May to mid-July.

Russet-crowned Motmot

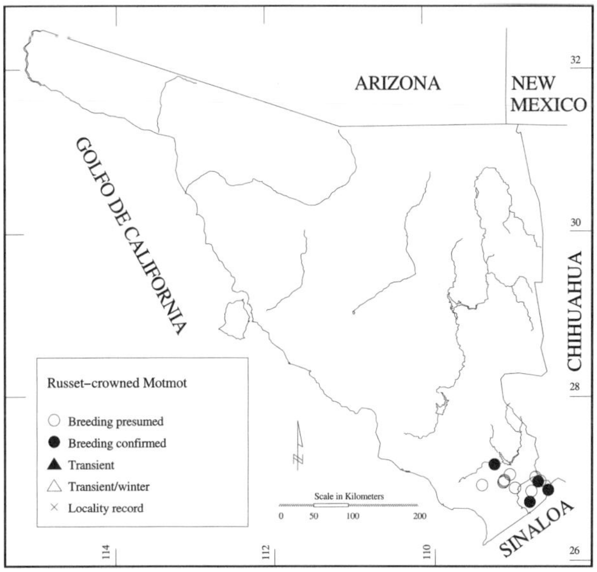

FAMILY ALCEDINIDAE

Kingfishers

BELTED KINGFISHER

Martín-pescador Norteño

Ceryle alcyon

This large blue kingfisher is a fairly common transient and winter visitant throughout Sonora at elevations below 1000 m, wherever bodies of water can provide its diet of small fish and crustaceans. Its distribution includes coastal areas (where it prefers mangroves), streams, rivers, ponds, and lake edges. These kingfishers are spaced well apart, and observers rarely see more than one in a kilometer of favorable habitat. The earliest fall date is 1 September, and the latest in spring is 9

May; most arrive after mid-September and have gone northward by mid-April. As would be expected, the extreme dates are from the most northerly localities. We can recognize a slight peak in numbers in September, indicating the passage of migrants, but numbers in other months are rather uniform.

There is no evidence that Belted Kingfishers nest in Sonora. Two birds on 10 July 1987 on the Río Yaqui 2 km below Presa Obregon were probably nonbreeding individuals.

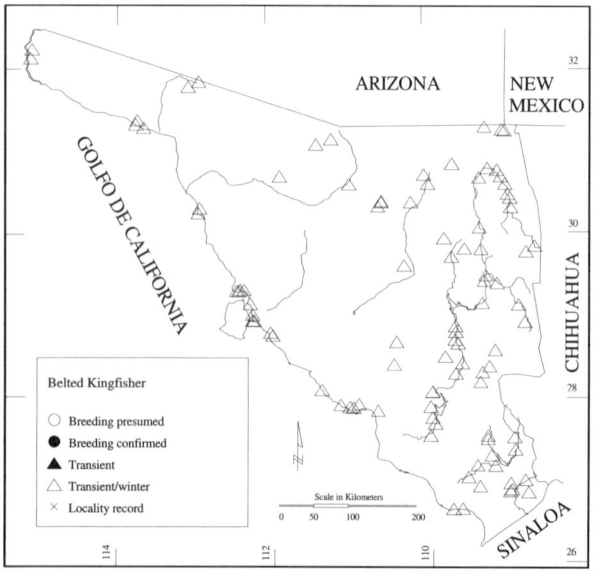

GREEN KINGFISHER
Martín-pescador Verde
Chloroceryle americana

This small kingfisher, with its rapid low flight and amicable chattering call, may be encountered on almost any stream, including very small ones, up to an elevation of about 1100 m. It is a fairly common resident in southern Sonora and uncommon in the north. Unlike the much larger Belted Kingfisher, it does not regularly frequent mangroves, although one was banded in mangroves near Punta Chueca on 18 September 1974. There are only four coastal records in all, in January, April, July, and September.

The highest elevation at which we have found this species is near Tarachi (1160 m), close to the village in a tiny wet canyon, a paradise where the walls were draped with masses of maidenhair fern, yellow columbines, yellow monkey-flowers, coral-bells, and *Epipactis* orchids. E. Mearns collected a Green Kingfisher in Arroyo Cajón Bonito (about 1280 m) in July 1892 or September 1893, but we have never seen the species there.

The only nest we have found was on the Río Tepoca (460 m) on 15 March. The adults were apparently alternating incubation duties, as they flew to a hole in the bank 38 cm above the water but largely concealed under a large tree root. Observations of behavior and data from specimens indicate a breeding season from February through April. Although a pair may normally forage along a kilometer of flowing water, we have seen pairs occupying no more than 500 m of stream. They typically forage by sitting 30 to 60 cm above the water, often concealed by overhanging vegetation. An abrupt and almost vertical dive into the water follows a period of intense downward peering.

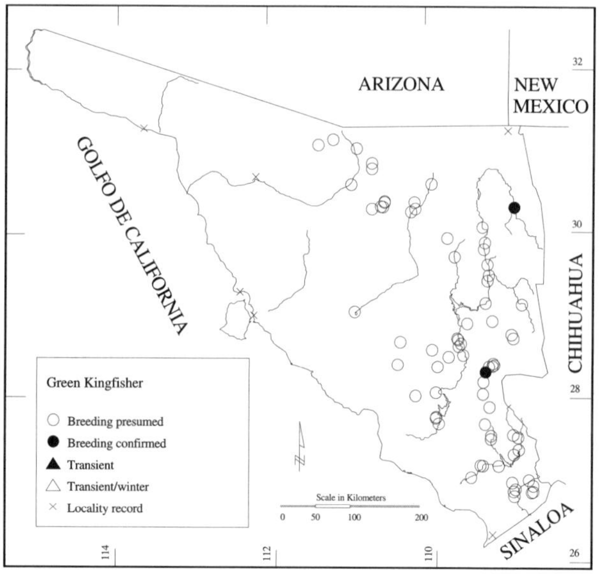

Green Kingfisher

FAMILY PICIDAE

Woodpeckers and Allies

LEWIS' WOODPECKER

Carpintero de Lewis

Melanerpes lewis

In so-called flight years, Lewis' Woodpeckers deviate from their more usual short migrations and wander extensively, often far from breeding grounds in open coniferous woodlands of the western United States and Canada. A few have moved as far south as Sonora in flight years. We know of records in 1928 (January), 1948 (February, October, November), 1953 (November), and 1984 (September). The most recent record (29 September 1984) is of a bird feeding on the fruit of tescalama, or rock fig, at El Capulin north of Tonichi (N. Davis).

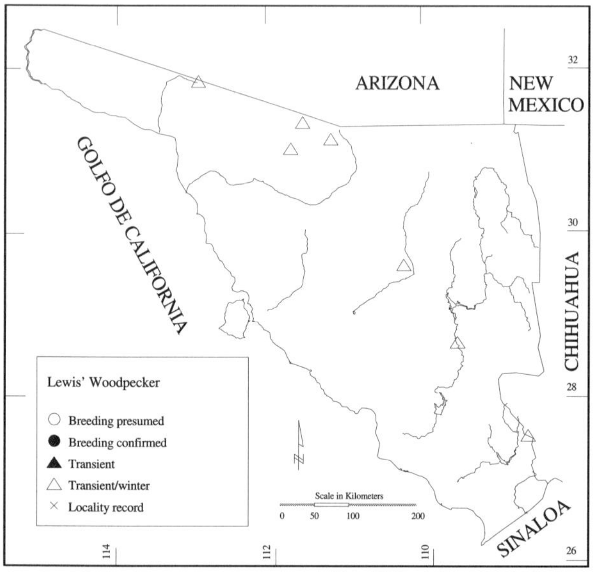

ACORN WOODPECKER

Carpintero Bellotero

Melanerpes formicivorus

The Acorn Woodpecker is a common resident in oak and pine-oak woodland (700–2100 m). As Marshall (1957) observed, it is most abundant where oaks are available for acorns and roost cavities and tall pine snags provide soft wood in which to store acorns and dig nests. In some riparian situations it substitutes cottonwood or sycamore snags for those of pines. At least a dozen Acorn Woodpeckers in the vicinity of El Naranjo northeast of Alamos on 6 April 1987 (R. Bowers) were a few kilometers from oaks; the locality is in the tropical deciduous forest at 460 m. The species is rarely found far from oaks.

Acorn Woodpeckers must have snags in which to nest. Marshall (1957) thought a shortage of snags limited nesting densities of the woodpeckers and probably other species that depend upon woodpecker-excavated cavities (such as Elf Owl, Purple Martin, Violet-green Swallow). We have found active nests in June and July. Birds store acorns in hundreds of small excavations in a single snag. The stored food is utilized

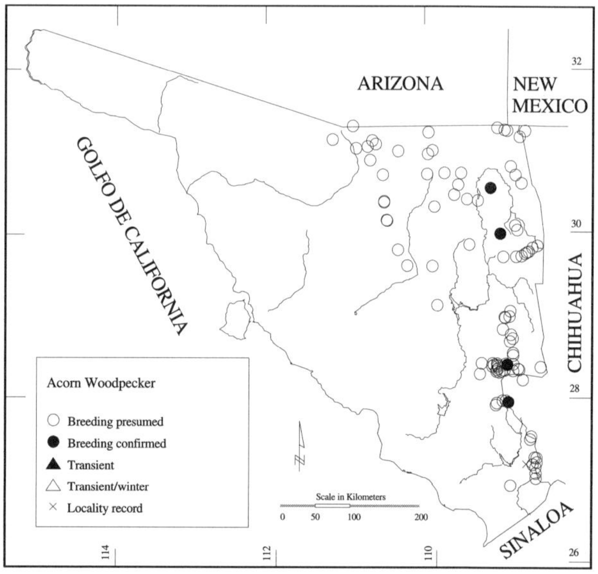

later by the woodpeckers, but also by thieving squirrels (which the birds chase). Near Los Aserraderos, we saw three small boys remove stored acorns from the deep crevices in pine bark and eat them. In addition to acorns, the birds eat insects that they pick from tree surfaces or dig from decaying wood; sometimes they catch large flying insects on the wing.

GILA WOODPECKER
Carpintero del Desierto
Melanerpes uropygialis

The resident Gila Woodpecker is the most common and conspicuous woodpecker throughout the desertscrub, thornscrub, and deciduous forest of Sonora. It reaches its highest numbers where there are large cacti—saguaro, cardón, organ pipe, and hecho—which it utilizes for food and lodging. In the north, it is fairly common in riparian woodlands and extends its range upward in canyons to about 1200 m. It forages upon adult insects and their larvae, which may be picked up from tree surfaces or dug from the wood with considerable pecking. Cactus fruit is a staple food; the soft fruit of cardón and saguaro is consumed as rapidly as it ripens. The tougher, more persistent barrel cactus fruit is used in colder months. Once in July near Agiabampo we watched a Gila Woodpecker on a tidal mudflat near mangroves as it ate tiny crabs.

Typically, nest cavities are 3 to 7 m above the ground in a living cactus or dead woody snag. Nesting begins as early as mid-March in coastal and southern areas and continues into July. In most of Sonora, nesting is most noticeable in June to early August, when three or four noisy young call raucously from cavities. Cactus fruit provides an abundance of food for fledglings, and the initial summer rains bring a flush of insects. Dispersal of young occurs in fall and winter, and Gila Woodpeckers are then seen occasionally at slightly

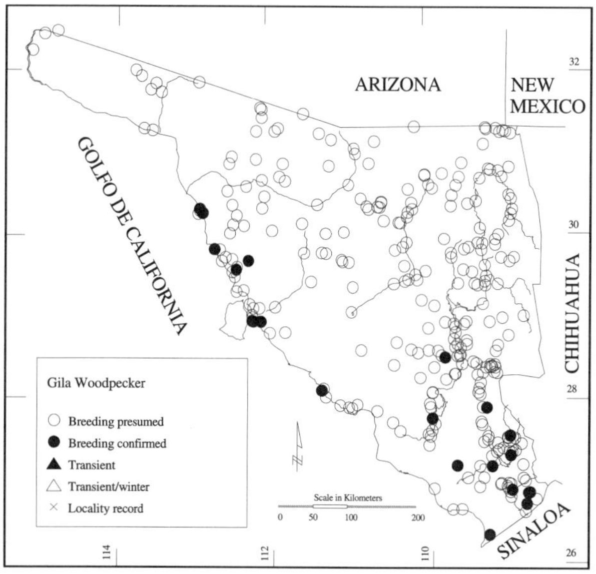

higher elevations or in sparse desert. On 20 December 1973, Lamm and Russell watched nine Gila Woodpeckers foraging in chollas, in palo verdes, and on the ground. They moved as a flock in desertscrub south of Pitiquito; it is plausible that all were young birds.

YELLOW-BELLIED SAPSUCKER
Chupasavia Maculado
Sphyrapicus varius

We know of only three unequivocal records of this species from Sonora, although it might be expected in extremely small numbers every winter. A. Phillips collected a specimen near Granados on 22 November 1953. One was carefully identified at the Río

Cuchujaqui south of Alamos on 24 November 1984 (C. Hunter et al.), and another was seen there on 3 January 1987, for which field notes were carefully recorded (D. Stejskal, D. Jones, et al.). The latter bird was working at sap holes in a Montezuma baldcypress. Until 1985, many field observers did not distinguish between Yellow-bellied and Red-naped Sapsuckers. We are uncertain of the status of the Yellow-bellied Sapsucker in Sonora.

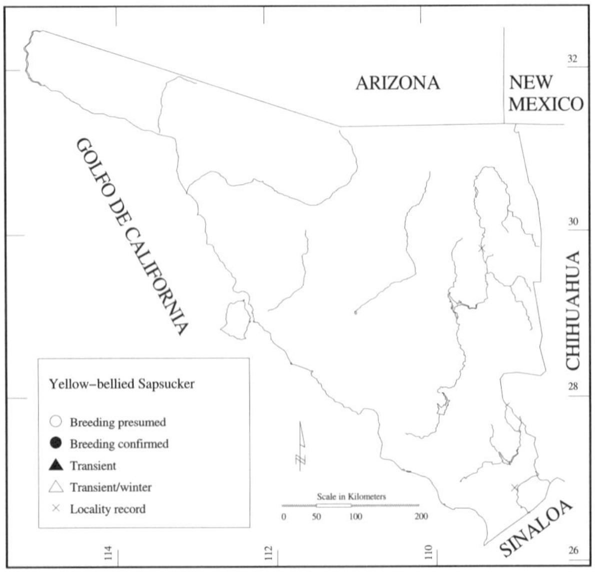

RED-NAPED SAPSUCKER

Chupasavia Nuca Roja

Sphyrapicus nuchalis

The Red-naped Sapsucker is an uncommon winter visitant and migrant in Sonora, being found chiefly in riparian situations at low elevations, but also occasionally in pine-oak woodland. It is found sometimes around human habitations where there are large woody plantings. Extreme arrival and departure dates are 17 September (Mututicache) and 30 April (near Tarachi, three in oaks). The maximum seen in one day was ten, counted about 40 km southwest of Nogales on 28 November 1953 (J. Marshall). Sapsuckers drill holes into the soft bark of willows and other trees. They feed upon the exuding sap and the insects attracted to it, as do a number of other bird species.

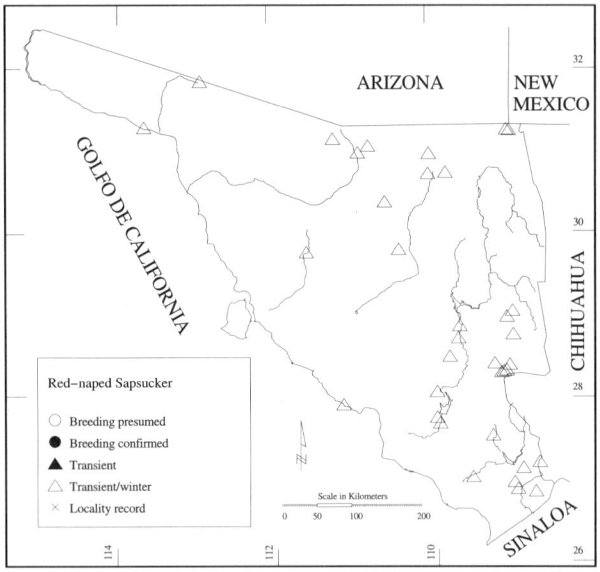

WILLIAMSON'S SAPSUCKER

Chupasavia Oscuro

Sphyrapicus thyroideus

This sapsucker is a rare transient and winter visitant in Sonora. The nine records are as follows: a specimen collected 22 December 1890 on the upper Río Bavispe (van Rossem 1945); a male at Guaymas on 26 November 1961 (D. Carter); two males and a female collected in the Sierra los Ajos on 29 December 1963 (UA, SWC; R. Crossin); a male in the Sierra Cananea on 13 October 1971 (E. L. Smith, S. Speich); a male near Las Chinacas (1500 m) on the Sonora-Chihuahua border on 10 January 1981; a pair near El Ocote (1480 m) east of Sahuaripa on 16 March 1984; a female in an ocotillo near Desemboque de Río San Ignacio on 6 March 1985 (R. Taylor); a male on the Río Cuchujaqui south of Alamos on 3 January 1988 (D. Jones *fide* D. Stejskal); and two males on Mesa de Enmedio on 13 March 1989. One of the latter males foraged on the ground with 12 Chipping Sparrows at the edge of a roughly plowed field in a pine-dominated area.

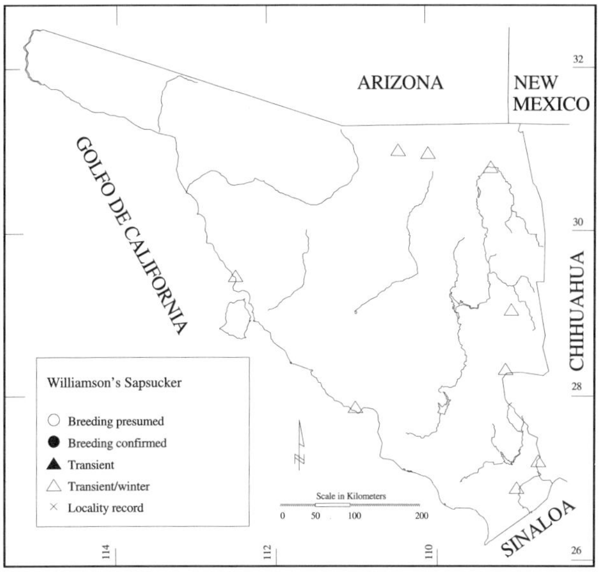

LADDER-BACKED WOODPECKER

Carpintero Mexicano

Picoides scalaris

This diminutive and unobtrusive woodpecker is a widespread and fairly common resident in Sonora at elevations up to 1200 m in the south and 1500 m in the far north. It is a characteristic species of desertscrub, thornscrub, and tropical deciduous forest, and its range extends upward in riparian corridors where cottonwoods dominate. At a few localities (such as along the road from Rosario to the Sierra Oscura and also at El Reparo), it is sympatric with Acorn Woodpeckers and Strickland's Woodpeckers, but this is unusual. Typically, the Ladder-backed Woodpecker does not extend far into oak-dominated habitats. It feeds upon insects, which it finds by probing and gleaning, sometimes from vegetation as formidable in appearance as chollas and ocotillos.

Nest cavities are excavated in the flinty wood of desert leguminous trees, as well as in desert willows, agave stalks, willows, and cottonwoods. Usually the nests are situated in the underside of a leaning branch or trunk. We have found only one nest in a tree cactus; a male fed young in a hole 0.3 m from the top of a 5 m cardón at Punta Cirio. Another nest was placed in an acacia, sheltered by a projecting bracket fungus. We have noted nests with young in May, June, and July. Ladder-backed Woodpeckers sleep in cavities at night throughout the year, as is typical of other woodpeckers. Male and female of a pair generally occupy different cavities.

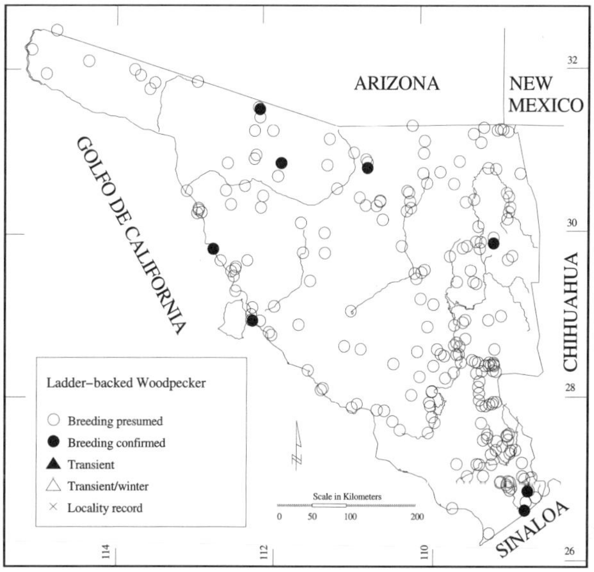

HAIRY WOODPECKER

Carpintero Velloso-mayor

Picoides villosus

This woodpecker is rare in Sonora but may be a resident of pine forest in most highlands from the Sierra los Ajos east and south as far as the Sierra Oscura. Van Rossem (1945) referred to two specimens taken (31 May and 10 June 1887) in the Sierra la Madera. It has not been found there subsequently, although J. Marshall visited suitable habitat. According to Marshall (1957:85), "The Hairy Woodpecker requires coniferous forest for its early nesting season. . . . Then families move down the mountain, especially along streams."

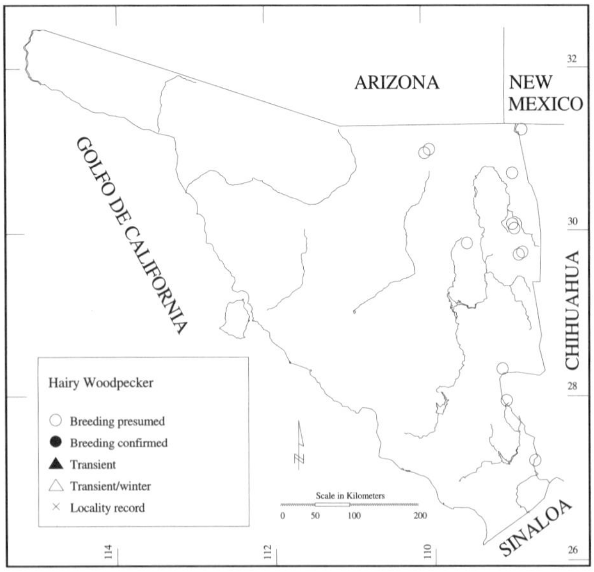

STRICKLAND'S WOODPECKER

Carpintero de Strickland

Picoides stricklandi

This woodpecker is a fairly common resident of the pine-oak association throughout Sonora, preferring sites where oaks are most dense. Intermediate in size between the Ladder-backed Woodpecker and the Hairy Woodpecker, it also occupies intermediate elevations (900 to 2100 m) and a different habitat. Strickland's Woodpecker pecks, pries, and probes into tree bark for its insect prey, frequently close to the ground.

Although we have not observed nesting birds, we have seen eight pairs with recent fledglings, all in the first half of June. Thus, the nesting season appears to be much more compressed than for other Sonoran woodpeckers. By July, adults are in heavy molt.

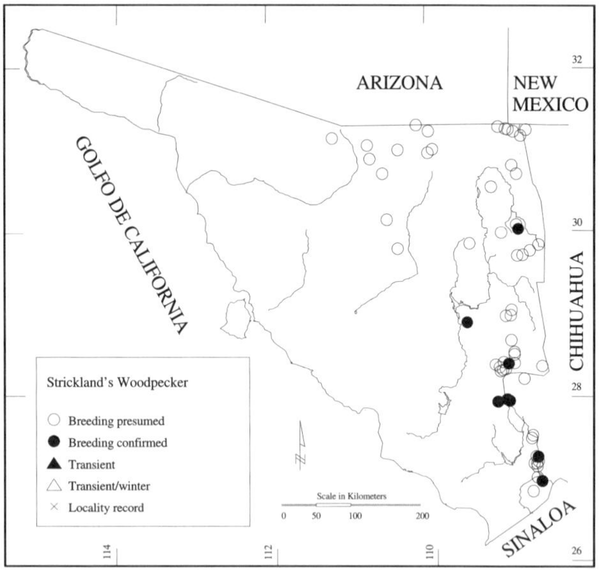

GRAY-CROWNED WOODPECKER

Carpintero Corona Gris

Piculus auricularis

The Gray-crowned Woodpecker is known from five localities in Sonora between the elevations of 900 and 1600 m. Our records are all in March through July, but it may be resident. It is definitely rare. Our few observations are of birds in mesic canyons and adjacent drier oaks and pine-oak woodland. Foraging birds probe small diameter (about 10–20 cm) trunks and branches of both living and dead trees for insects and their larvae. One near Sahuarivo was feeding on the fruit of a wild cherry.

At least three pairs were present in the lower portion of Arroyo el Reparo (920 m) in mid-May 1988. One pair had a nest about 10 m above the ground in a dead snag of a willow. We believe the adults were incubating.

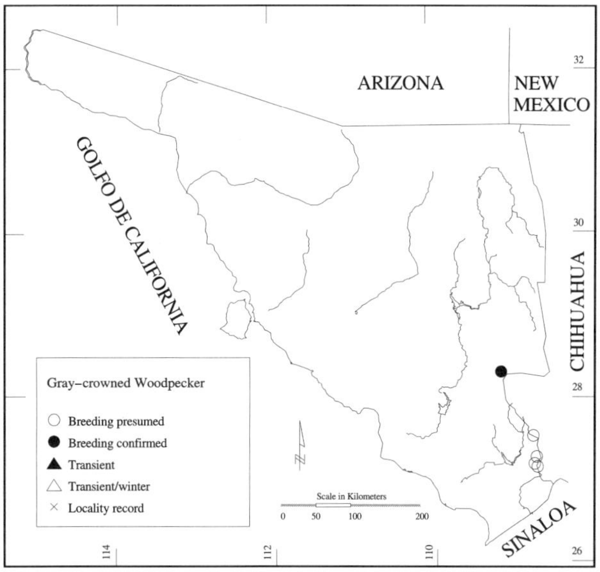

NORTHERN FLICKER
Carpintero de Pechera
Colaptes auratus

The western form of the Northern Flicker is the "Red-shafted Flicker" (*C. a. collaris*), called in Spanish "Carpintero Collarejo Alirrojo." It is a common to fairly common resident (some birds may be migratory) along permanent streams in the north and in pine-oak woodland upward into the pine forest. It requires large trees for its nesting cavities, usually selecting cottonwoods and pines. It breeds from late April through early June. Numbers are augmented in winter by visitants from the north. Many Red-shafted Flickers occupy the lowlands of the desertscrub, and rarely the tropical thornscrub and the tropical deciduous forest, in winter; these may be individuals from the mountains of Sonora or more northern localities. Extreme records in the lowlands are from 10 November to 19 March.

A second race, the "Yellow-shafted Flicker", *C. a. auratus*, has been found in winter throughout Arizona, hence it can be expected in Sonora.

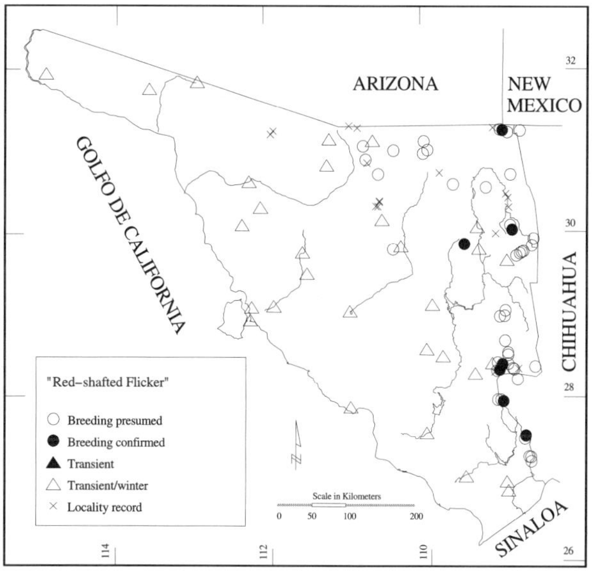

GILDED FLICKER
Carpintero Collarejo Desértico
Colaptes chrysoides

The Gilded Flicker is a common resident wherever large cacti grow, and upward along drainages to about 1200 m. It sometimes does range higher than 1200 m, as at La Pera, 10 km south of Esqueda, where it was found in riparian woodlands. The nest is customarily built in a large cactus, but large deciduous trees, usually cottonwoods, are also used. Nesting commences in March and may extend into July. Fourteen clutches of eggs (WFVZ) taken in southern Sonora between 23 April and 12 June consisted of two to six eggs (median, four). Cavities excavated by Northern and Gilded Flickers are much used by a variety of other birds.

Northern and Gilded Flickers intergrade wherever their ranges overlap in southern Arizona, as they probably do in Sonora. Both species have been seen together in summer a few kilometers south of Esqueda and in Arroyo Santo Domingo.

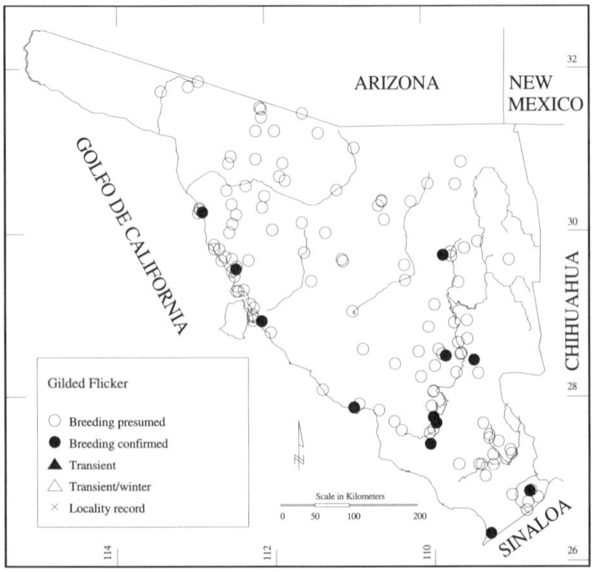

Lineated Woodpecker

LINEATED WOODPECKER

Carpintero Lineado

Dryocopus lineatus

This magnificent large woodpecker barely reaches Sonora at the northern end of a range extending south into Argentina. It is a rare resident in Sonora and is confined to regions of tropical deciduous forest in mountainous areas (200 to 1000 m) of the extreme southeast. We are aware of eight records since 1982, four of them single birds in the Sierra Alamos on Christmas Bird Counts. Fragmentary information from specimens (LSUMZ, WFVZ) indicate egg-laying in late May and early June. The "song" resembles that of the Northern Flicker, and its tattoo is much louder.

This fine woodpecker has probably always been rare in Sonora. But the infrequency of observations of so conspicuous a bird, combined with the growing popularity of the Alamos area by people watching birds, leads us to conclude that its numbers are declining. This species and the Pale-billed Woodpecker need in-depth studies in their northern ranges.

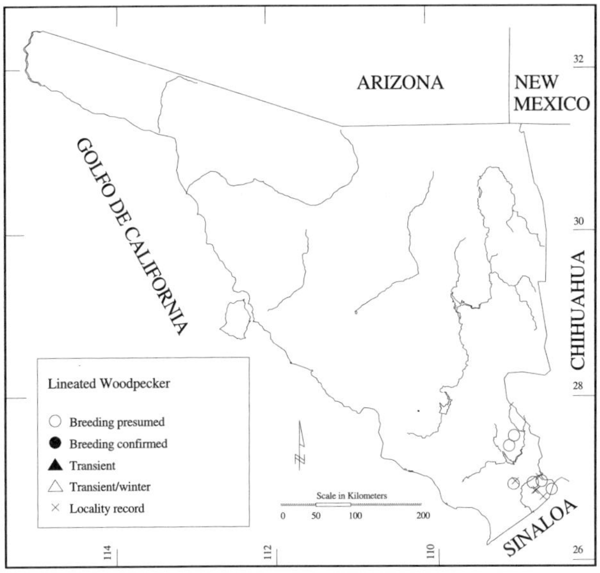

PALE-BILLED WOODPECKER

Carpintero Pico Plata

Campephilus guatemalensis

The Pale-billed Woodpecker, called "Sonora Ivory-billed Woodpecker" by van Rossem (1945), is probably no longer present in Sonora. It has not been found there since 1950, when four specimens (LSUMZ, OC, WFVZ) were taken near Güirocoba, the last of about 16 collected in the region since 1932. Before 1932, the Pale-billed Woodpecker was known in Sonora only from three specimens taken near Alamos in March 1888 (van Rossem 1934). Van Rossem (1945:137) described the Cañon San Francisco habitat, where some birds were found, as "giant cactus association in open gallery forest at 1,500 feet [460 m]." The cutting of large trees is often cited as a contributing factor in the decline of hole-nesting populations of large birds. This is a likely factor in Sonora, although we lack accurate information. We do urge that this species be studied in northern Mexico before complete extirpation occurs, as in the case of the Imperial Woodpecker.

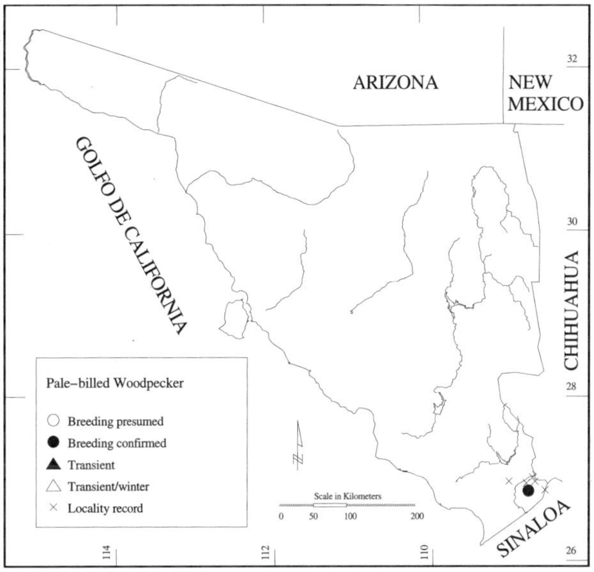

IMPERIAL WOODPECKER

Carpintero Imperial

Campephilus imperialis

This, the largest woodpecker in the entire world, is extinct. Its entire range was in the Sierra Madre Occidental of Mexico. It was too much to expect that such a large bird with specialized habitat requirements could survive very long once the mountains were invaded by tree-cutters and firearms were available with which to easily bring down these birds for food and ornamentation. When the last survivor met its demise is unknown, but apparently it was before 1960.

The Imperial Woodpecker was first reported from Sonora by Bendire (1895:45). He stated, "Lieut. Harry C. Benson, Fourth Cavalry, U.S. Army, found it to be common in the pine forests of the Sierra Madre in northern Sonora, Mexico, in 1887, and shot a specimen within 50 miles of the boundary line." The location was probably close to the Sonora-Chihuahua border, possibly in the Sierra Huachinera (or Sierra los Ojos Azules).

The next mention of the Imperial Woodpecker in Sonora is of one by Lumholtz (1902). In early December 1890 his expedition had advanced some 45 km from Nacori Chico into the Sierra Nacori but was still definitely in Sonora. He wrote, "Thus we . . . made camp, at an elevation of 6,300 feet [1920 m], on some old trincheras, with a fine view over the vast country we had left below. Large flocks of gray pigeons of remarkable size squatted on the pine trees nearby, and two specimens of the gigantic woodpecker we here observed for the first time" (Lumholtz 1902:36).

By mid-January 1891 Lumholtz had progressed into the Sierra Madre until he camped, from his account, near the confluence of the Río Gavilan and the Río Bavispe. It is difficult to tell if he was still in Sonora but it would appear that he was. While at this camp, he noted, "One morning our best marksman, a Mexican named Figueroa, brought in three specimens of that superb bird, *Campephilus imperialis*, the largest woodpecker in the world. This splendid member of the feathered tribe is two feet long; its plumage is white and black, and the male is ornamented with a gorgeous scarlet crest, which seemed especially brilliant against the winter snow. The birds go in pairs and are not very shy, but are difficult to kill and have to be shot with a rifle. One of their peculiarities is that they feed on one tree for as long as a fortnight at a time, at last causing the decayed tree to fall" (Lumholtz 1902:54). This closed the history of the Imperial Woodpecker in Sonora.

There are four specimens (AMNH) of this species taken by the Lumholtz expedition while apparently in Sonora, one on 24 December 1890 ("Bavispe River") and three on 10 January 1891 ("Rancheria de los Apache"). The first, not mentioned in his book, would seem to be almost certainly collected in Sonora. The other three are the ones mentioned above.

There is in the Los Angeles County Museum a specimen of a male Imperial Woodpecker collected by M. L. Neely in 1906 "west of Casa Grande, summit of Sierra Madre Mountains." The locality was almost unquestionably in Chihuahua.

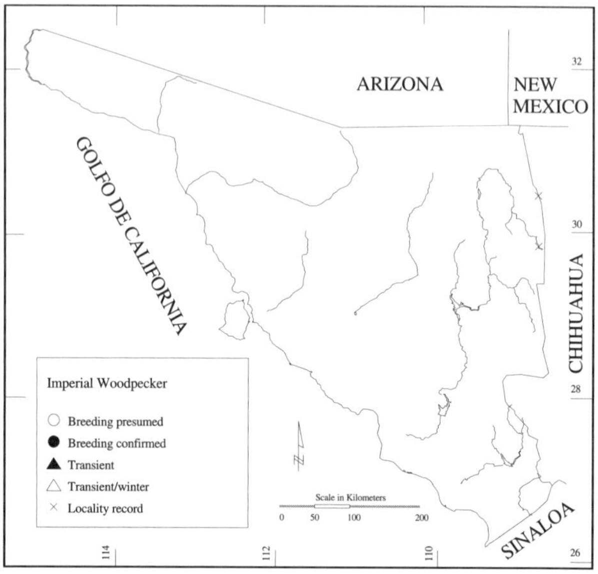

FAMILY DENDROCOLAPTIDAE

Woodcreepers

IVORY-BILLED WOODCREEPER
Trepatroncos Bigotudo
Xiphorhynchus flavigaster

The Ivory-billed Woodcreeper is now rare in Sonora; there are only four records since 1958. Short (1974:25) reported it "in hilly thorn forest N of Alamos." We

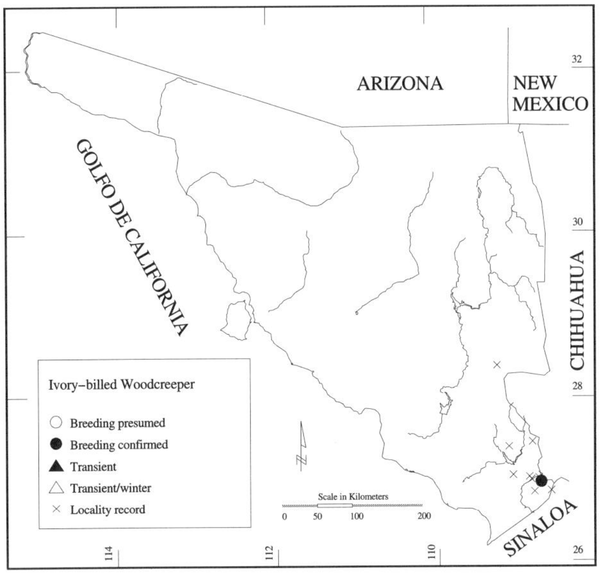

found one on 15 October 1983 approximately 2 km east (760 m) of San Nicolás, and A. Moorhouse and C. Green observed another on 16 May 1986 in the upper portion of Arroyo la Banderita (480 m). H. Marble watched one in upper Arroyo el Palmarito on 5 January 1993. A number of specimens were taken (OC, LSUMZ, WFVZ) between 1933 and 1958 in tropical deciduous forest at elevations between 400 and 1000 m. Several of the specimens were taken in December, January, and February, evidence that this species may be resident. W. Sheffler obtained a clutch (WFVZ) of three eggs near El Mirasol on 4 June 1947, the only nesting record for the state.

Van Rossem (1945) considered the species to be not uncommon and even fairly common in Sonora. We cannot explain its decline in numbers.

WHITE-STRIPED WOODCREEPER
Trepatroncos Escarchado
Lepidocolaptes leucogaster

This woodcreeper is a common summer resident in pine-oak woodlands and mixed conifers up to 2100 m in the southern two-thirds of Sonora from March into September. It is an uncommon summer resident in tropical deciduous forest above 400 m. We are aware of only one observation in October and no records from November. Although the species is not known to be migratory in other parts of its range, it is possible the northwestern Mexico population is not sedentary. We

White-striped Woodcreeper

have only 13 reports of its presence in December, January, and February, some at high elevations. The species becomes much less vocal after the end of the breeding season and may be easily overlooked, but year-round residency on the breeding grounds is yet to be established.

A. Moorhouse watched a single bird in tropical deciduous forest at Lo de Campa (490 m) on 28 September 1984. This may have been a postbreeding wanderer, for we have never seen another at that location. Another single bird at Presa Mocuzari (125 m) on 15 July 1986 was exceptional, as we have not otherwise found the species below 400 m.

The White-striped Woodcreeper typically forages by flaking bark from pines and by probing crevices in madrones and oaks. Pairs are regularly seen from March through July (and once in September). We watched a very vocal pair explore a knothole 15 m above the ground in a large sycamore in Arroyo el Reparo; the birds may have been selecting a nest site.

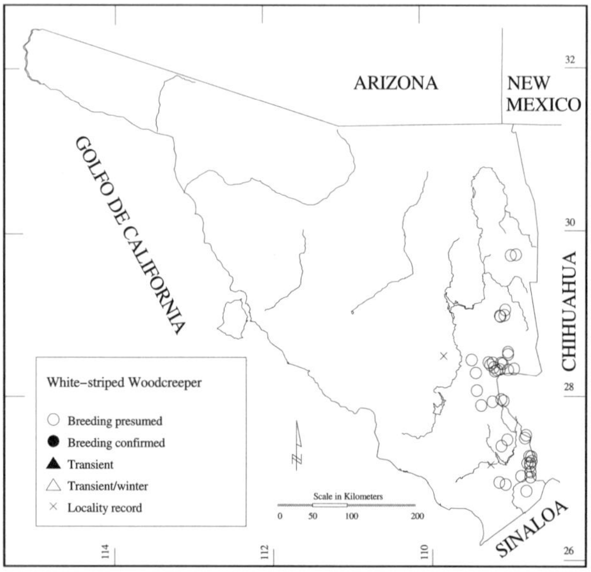

FAMILY TYRANNIDAE

Tyrant Flycatchers

NORTHERN BEARDLESS-TYRANNULET

Mosquero Lampiño

Camptostoma imberbe

A tiny drab bird, conspicuous only by its high, thin call, this flycatcher is a common resident of tropical deciduous forest and brushy riparian habitats with some trees at least 8 m tall. Permanent flowing water is often present, although the birds may be found where the only water is a stock tank or occasional pools along an intermittent streambed. During the winter some will wander to dry sites, especially on oak-grass hillsides. Some individuals winter at all locations where they have been found nesting; thus we surmise the species is resident. Often they are found in streamside brush next to the foot of bordering dry slopes or in adjacent mesquite bosques. Although they do not enter the coastal plain, they nest in the lower Río Yaqui valley (60 m); upper limits are at the lower edge of pine-oak woodland (1300 m).

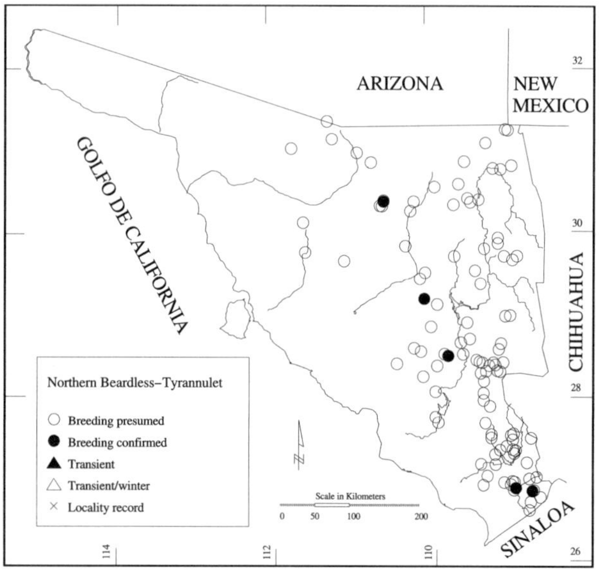

Song is most prevalent from early March through July, but some singing may be heard throughout the year. Nest building begins in mid-March, and incubating birds have been found as late as 7 August (Río Cuchujaqui; Short 1974). Nests are large for so small a bird, being globular with the entrance hole near the top. They may be well concealed in mistletoe, clumps of dead leaves, and old webworm masses and would be difficult to find were it not for the activity of the owners. The supporting plant may be a small shrub or even a tree as large as a mature Montezuma baldcypress.

Tufted Flycatcher

TUFTED FLYCATCHER

Mosquero Copetón

Mitrephanes phaeocercus

The distinctive Tufted Flycatcher is a resident from early March until October of moist canyons in the upper portions of pine-oak woodland (above 1400 m) in eastern Sonora north to the Yécora region. Its peculiar snapping call-note is persistent in spring and summer, especially from middle and higher perches in pines and oaks, where individuals pause between frequent sallying foraging flights. In fall, it departs its breeding grounds for well-wooded river bottoms (240 to 700 m) in winter, some birds perhaps going farther south. We have found it from mid-November to early April along streams in the lowlands. It is readily observed in winter along the Río Cuchujaqui southeast of Alamos.

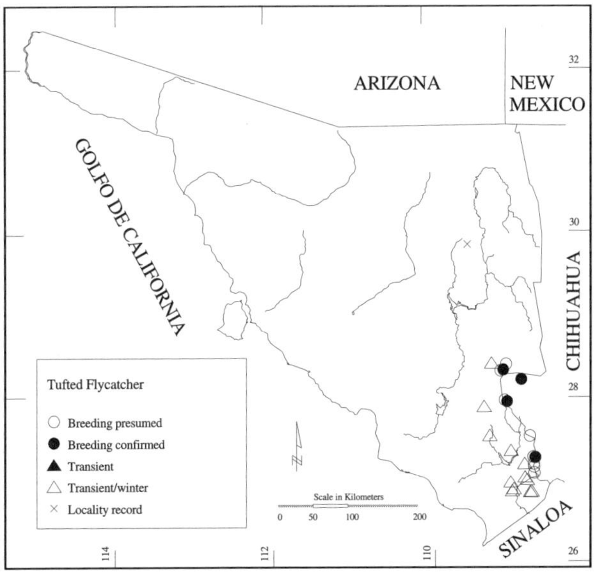

We have found four nests between 7 and 15 m above the ground on horizontal branches of oak, pine, and madrone. The inside and outside of the nests were lined with lichens and spiderwebs. One pair repeatedly made trips to the ground to take strands from a spider's funnel web. Nesting may commence in mid-March, but May and June are more normal.

It is remarkable that a Tufted Flycatcher was collected in the Sierra la Madera on 7 June 1887 by J. C. Cahoon, who was in the employ of the ornithologist William Brewster (van Rossem 1945). It has not been seen that far north since then. The Sierra la Madera is about 190 km northwest of the nearest presently known range.

OLIVE-SIDED FLYCATCHER
Pibí Boreal
Contopus borealis

The Olive-sided Flycatcher is a rare migrant in spring and fall at all elevations and in all habitats. Spring records are from the period 22 April to 7 June. Fall records are from 5 August to 1 October. Each of the 21 Sonoran records is of a single individual. Typically this species sits silently on high exposed perches to look for flying insects.

The scientific name of this species has been changed to *Contopus cooperi* (AOU 1997).

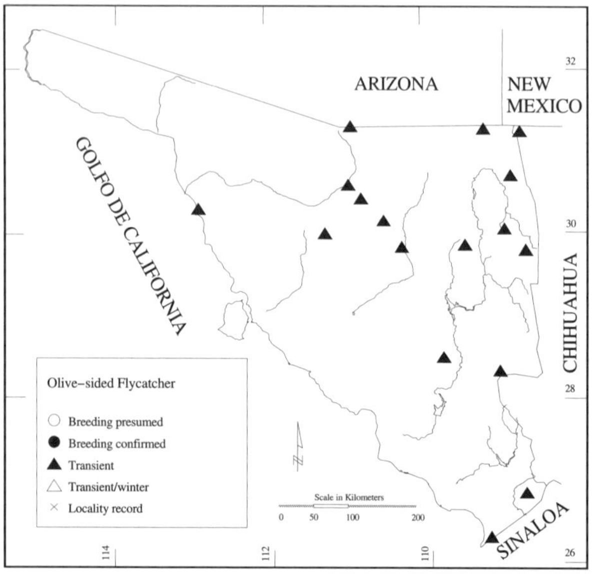

GREATER PEWEE
Pibí Tengofrío
Contopus pertinax

The Greater Pewee's most distinctive vocalization, its song, is a languorous "José María." It is a common summer resident of the pine-oak woodland, being most frequent in tall, open pines, from mid-March into September. Fall birds appear in the lowlands very early, the earliest being on the Río Cuchujaqui on 17 August (D. Stejskal).

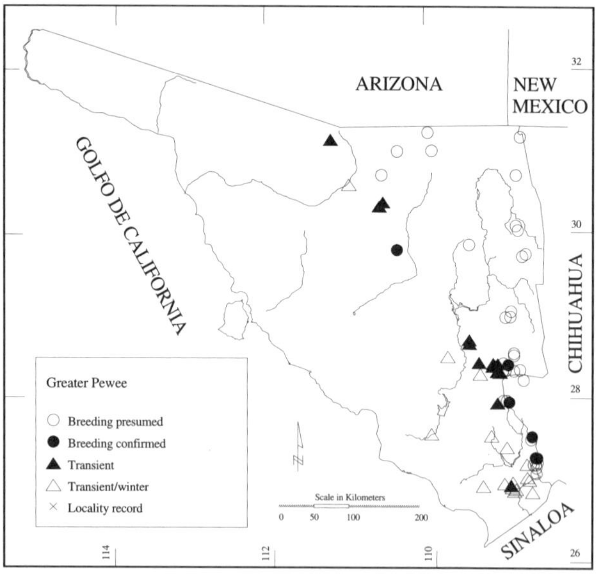

Although most Greater Pewees probably leave Sonora for the winter, a few remain in tall riparian woodlands as far north as the Río Cedros above Quiriego. One bird was observed in far northern Sonora in winter, along the Río Magdalena near San Ignacio on 16 and 27 January 1980 (Terrill 1981). A few also remain through the winter in the pines: for example, four on Mesa Grande (1820 m) near Yécora on 9 January 1983; one singing at the west base of Mesa de Enmedio (2000 m) on 9 January 1988; and two east (1450 m) of Santa Rosa on 8 January 1983.

Pairs begin nest building in early May, choosing a horizontal branch of a pine or oak about 8–10 m above the ground. Nests are lined with lichens and fine fibers. The adults are aggressive toward intruders, and we have watched them prevent the smaller Tufted Flycatcher from approaching closer than 15 m. These flycatchers are particularly aggressive toward Mexican Jays, which regularly consume eggs and young of any small birds they encounter. We once watched a pair of Greater Pewees harass a jay in late summer well after the end of the nesting season.

WESTERN WOOD-PEWEE
Pibí Occidental
Contopus sordidulus

The Western Wood-Pewee is a common transient almost statewide. It is also a common summer resident of cottonwood-willow-sycamore groves along permanent streams and of the pine-oak woodland, particularly in riparian sites and at high elevations. In southeastern Sonora, it is a common nesting bird where Chihuahuan oaks are along streams at relatively low elevation (above 600 m), such as at Rancho Jacalón (650 m) northeast of Rosario.

Early migrants may appear in the first week of April, but most pass through in May. One of the latest migrants was an individual captured and banded in the mangroves near Punta Chueca on 28 May. Fall migration is under way by 16 July and reaches its peak in late September. A lingerer has been noted near Alamos on 16 November, a very late date for a species that winters in South America.

In areas of optimal habitat, which includes large riparian trees and flowing water, this pewee may be abundant. It sits on exposed branches 8 to 20 m above the ground, peering constantly for flying insects that it takes on the wing. Its vocalizations create an impressive dawn chorus. Some individuals sing their coarse, short, descending *pzeeer* throughout the day, even during the

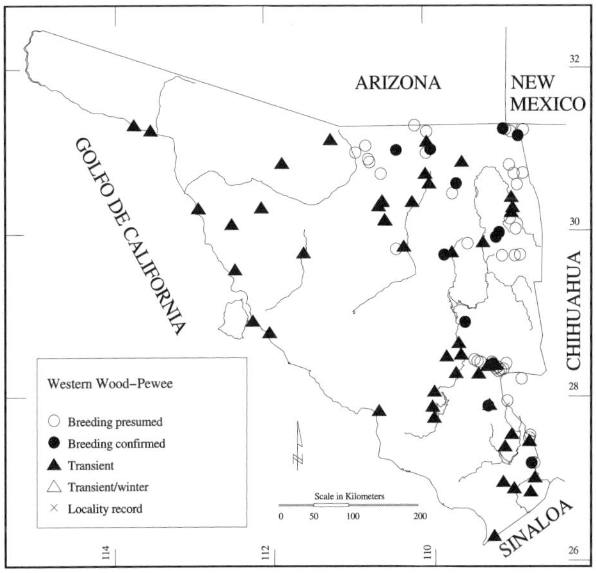

hot afternoon hours when other birds are silent. This pewee returns to its breeding areas well after most species have arrived. There are reports of May nests, but most have been found in June and July, coincidental with the onset of the summer rainy season. Large trees that are open beneath the canopy, such as cottonwoods and sycamores, are typically chosen for nest sites.

WILLOW FLYCATCHER
Mosquero Saucero
Empidonax traillii

The many species of the genus *Empidonax* share common features of appearance and behavior. Their vocalizations are distinctive, but they seldom call during migration. Because of the difficulties of field identification, we have been very cautious in accepting sight records. Fortunately enough specimens are available that we can describe the approximate status of each species in Sonora.

The Willow Flycatcher is an uncommon transient throughout Sonora below elevations of 1100 m. We know of only two records from higher localities: a bird seen in September at 1480 m northeast of San Miguelito and one collected (probably about 1700 m; van Rossem 1945) in the Sierra San Jose in October. The species may be common in the mangroves, especially in spring. In the Punta Chueca area during a banding study of mangrove birds in the early 1970s, Russell and D. Lamm captured 33 in mist nets on five spring dates, yet they saw only one individual in adjacent desert habitats. The largest numbers were taken during the last few days of May; only seven were captured in fall, all in mid-September. In general, the Willow Flycatcher is a late migrant compared with other *Empidonax* flycatchers; extreme spring dates are 9 April and 17 June. In fall, records extend from 3 August to 12 November. There are a few reports of this species in January, but because of the difficulty of positive field identification we have not accepted them. Most individuals winter south of Mexico.

There is no evidence that the Willow Flycatcher has ever nested in Sonora, although formerly it nested in southeastern Arizona in the Santa Cruz and San Pedro drainages. S. Howell found one on a small creek about 100 km east of Hermosillo on 21 June 1991. The date is late for a migrant, and he wondered if it might be breeding. Its numbers are declining throughout its range, perhaps due to its preference for open, wet areas, which are rapidly disappearing.

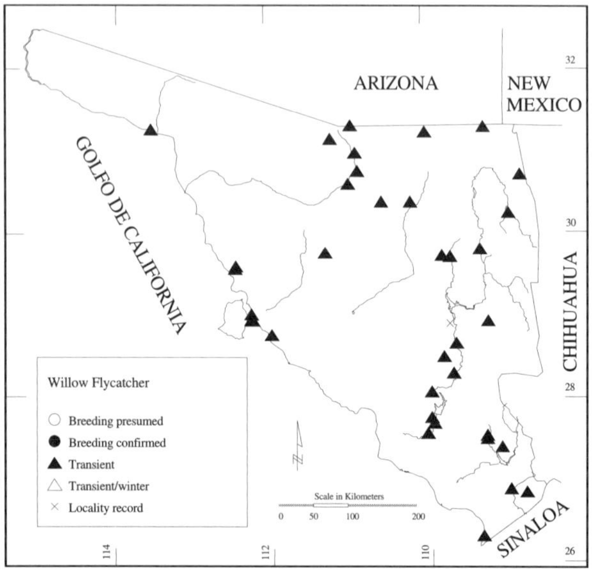

LEAST FLYCATCHER

Mosquero Mínimo

Empidonax minimus

We include this species on the basis of a specimen in MCZ taken at Alamos on 7 March 1888 (van Rossem 1945). A few sight records are indefinite, including one on an Alamos Christmas Bird Count on 28 December 1982 (D. Stejskal, N. Krug), one seen a few kilometers

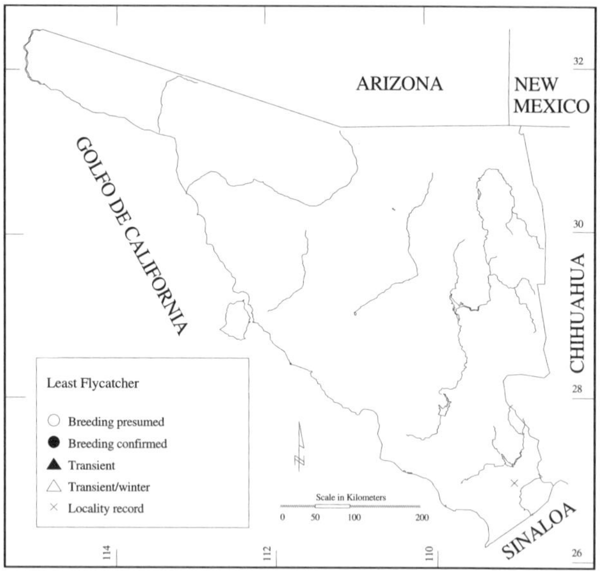

west of Ures on 14 January 1980 (Terrill 1981), and one watched by Russell as it foraged low over water in a Montezuma baldcypress grove near Rosario on 16 March 1982.

HAMMOND'S FLYCATCHER

Mosquero de Hammond

Empidonax hammondii

Hammond's Flycatcher is an uncommon transient through Sonora, appearing in fall as early as 11 August and present until at least 12 November. Northbound individuals have been noted from 11 March to 21 May. Most records are from the foothills and lower mountains.

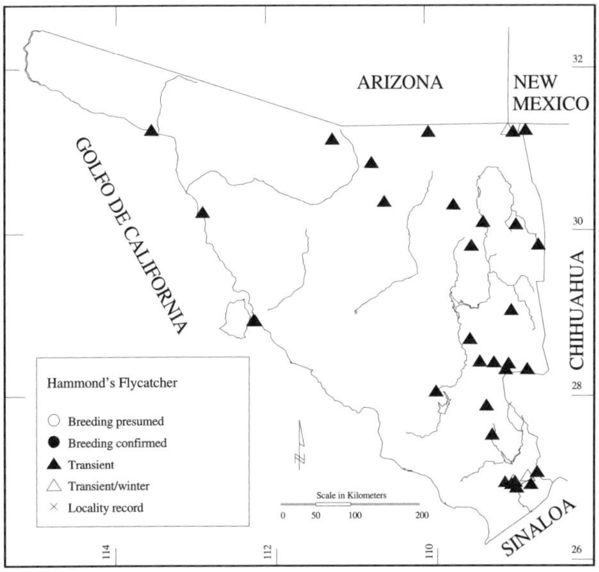

This species winters in Sonora, but we lack specimens and have few well-documented sight records. A. Moorhouse noted four in Arroyo Cajón Bonito on 29 January 1987, and Russell saw one in Arroyo el Palmarito northeast of Alamos on 5 January 1993. This flycatcher has been listed regularly on Alamos Christmas Bird Counts.

There are many *Empidonax* flycatchers in Sonora during migration and in winter. If the identification of each one were certain, we would undoubtedly raise our estimates of abundance for several species.

DUSKY FLYCATCHER
Mosquero Oscuro
Empidonax oberholseri

Based on specimens, birds handled, and well-documented sight records, the Dusky Flycatcher is a fairly common migrant and winter visitant in the lowlands in the vicinity of water, particularly in brush along streams. We have not found it above 1500 m in winter; most winter records are from below 1000 m. It is rare at any season in the desert and coastal plain.

Most spring birds are seen between mid-March and mid-April; the last date for spring is 29 May. The earliest fall record is 6 August, and peak numbers are present in the middle of September. One banded at Lo de Campa on 27 September 1984 was recaptured there on 15 March 1985. We think it wintered there.

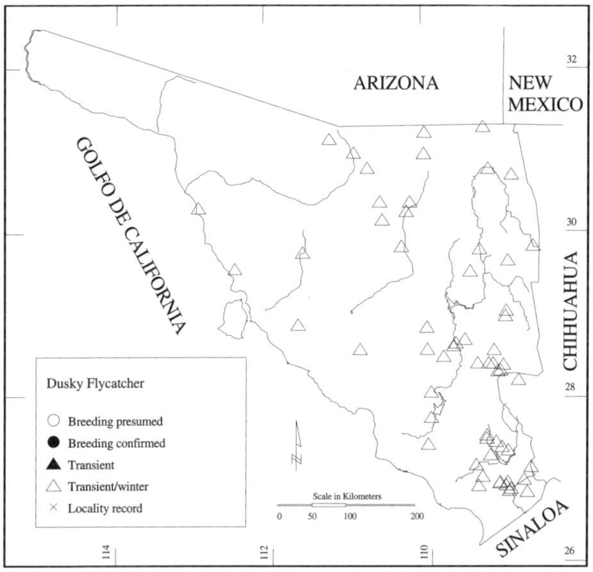

GRAY FLYCATCHER
Mosquero Gris
Empidonax wrightii

Except in extreme northwestern Sonora, the Gray Flycatcher is a common transient from the coastal plain (but not in mangroves) to the lower limits of the pine-oak woodland; it is rather uncommon in pine-oak woodland. It is common in winter at elevations below 1000 m and rare in higher pine-oak woodland. Mesquite bosques provide optimal habitat; one banded individual wintered in mesquites at Lo de Campa. As one measure of its abundance, about 40 were counted near Suaqui Grande on 31 October 1985. Outside migration dates are 27 April for departure in spring and 6 August for arrival in fall.

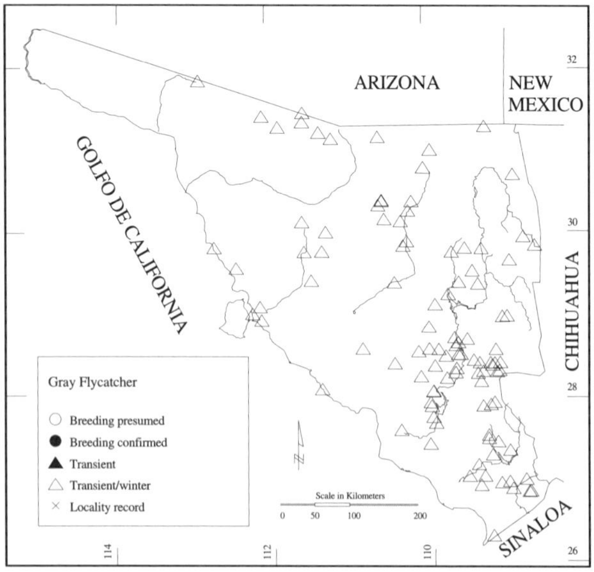

PINE FLYCATCHER
Mosquero Pinero
Empidonax affinis

A Pine Flycatcher was collected by W. Sheffler in pines at 1100 m in the extreme southeast on 24 February 1948 (LSUMZ). The specimen is labeled as having been taken 35 miles from San Bernardo "up Texas Canyon," a locality that we cannot define but believe is near Rancho Quemado, about 9 km north-northeast of Los Algodones. The species breeds in the extensive pine forests to the southeast in Chihuahua. Pinelands in Sonora are much less extensive than in Chihuahua and are perhaps too restricted to support this nondescript flycatcher.

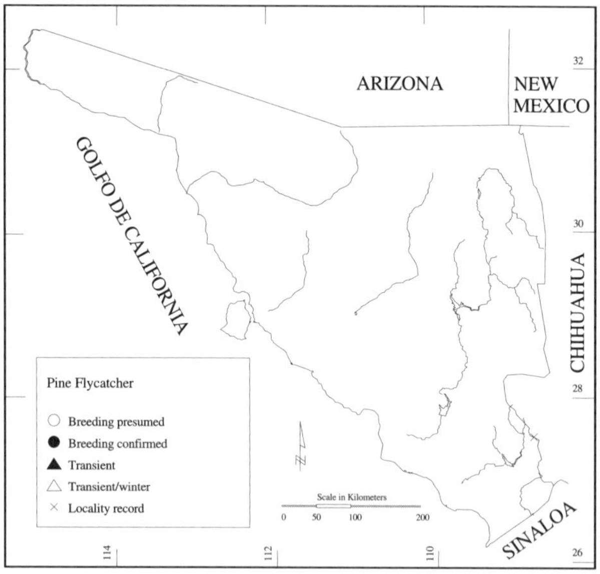

PACIFIC-SLOPE FLYCATCHER
Mosquero Californiano
Empidonax difficilis

In 1989, the AOU Check-list Committee, on the basis of recent genetic studies, split the Western Flycatcher (long known as *Empidonax difficilis*) into two species. The resultant new names became Pacific-slope Flycatcher (retaining the name *E. difficilis*) and Cordilleran Flycatcher (*E. occidentalis*). The two species are separable in the field only if the males are calling (which they do infrequently during winter and migration).

Our own files include more than 200 records of "Western Flycatcher" that cannot be correctly assigned to either Pacific-slope or Cordilleran Flycatcher. Specimens suggest that the Pacific-slope Flycatcher is a fairly common transient throughout Sonora in spring and fall and an uncommon winter resident, particularly in the southern portion. In winter, it is partial to thickets along streams and freshwater ponds, but in migration it may be found in any habitat, even arid portions of the desertscrub.

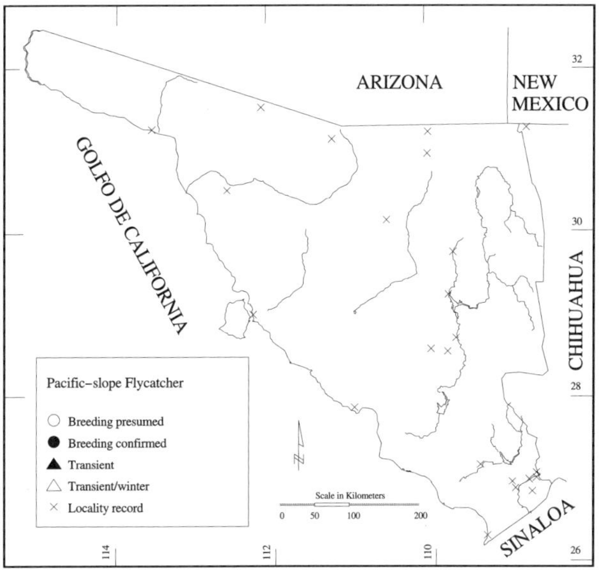

Field observers would do well to note the vocalizations of "Western"-type Flycatchers. Male Pacific-slope Flycatchers have a single note call and male Cordilleran Flycatchers utter a two-syllable call, whereas females of both species utter a single note (Kaufman 1990). On the morning of 22 August 1984, we counted 62 "Western" Flycatchers along Arroyo Carrizo north of Tonichi. The count indicates the magnitude of the migration and, because the birds were not identified to species, suggests how much is yet to be learned in Sonora.

CORDILLERAN FLYCATCHER

Mosquero Barranqueño

Empidonax occidentalis

The Cordilleran Flycatcher is documented by only a few specimens from Sonora and one breeding record (see comments on identification problems under the account for Pacific-slope Flycatcher). Specimens suggest that this species is an uncommon transient and rare winter visitant (in southern Sonora).

On 10 June 1984, J. Bock found a nest at 2100 m on the west slope of the Sierra Oscura. The nest was placed on a rock ledge in a roadcut shaded by pines and oaks about 1.5 m above the roadbed and contained four small young. Recorded vocalizations of the adults were the characteristic two-noted *whee-eet* of the Cordilleran Flycatcher. On 16 May 1988 we heard a Cordilleran Flycatcher singing this same vocalization on Mesa de Enmedio, and we think the species may breed there also.

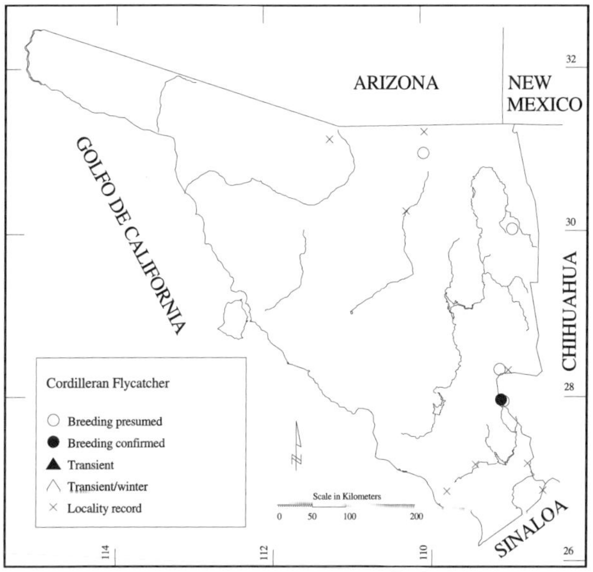

BUFF-BREASTED FLYCATCHER

Mosquero Pecho Leonado

Empidonax fulvifrons

The Buff-breasted Flycatcher, the smallest and most distinctly colored of the *Empidonax* flycatchers, breeds uncommonly and locally in open pine-oak woodland and ponderosa pine forest, usually near water where riparian vegetation is present. It winters in the lowlands of Sonora, uncommonly in the south and rarely in the north. We have seen as many as six in one day in winter near small streams in tropical deciduous forest. One wintering bird in a known breeding area (1 January

1983 on Mesa Grande) foraged in an old field and spent much of the time on small rocks on the ground. Over much of the wintering range the species is more frequent as a transient, especially in April.

We are aware of but one nest. On Mesa Grande on 20 July 1985, adults frequently fed small young in a nest 7 m above the ground on a horizontal branch of a slender Apache pine. Individuals are paired and calling by early May. Males sing a rather sharp *chee-bit* (Kaufman 1990) and often call incessantly at dawn.

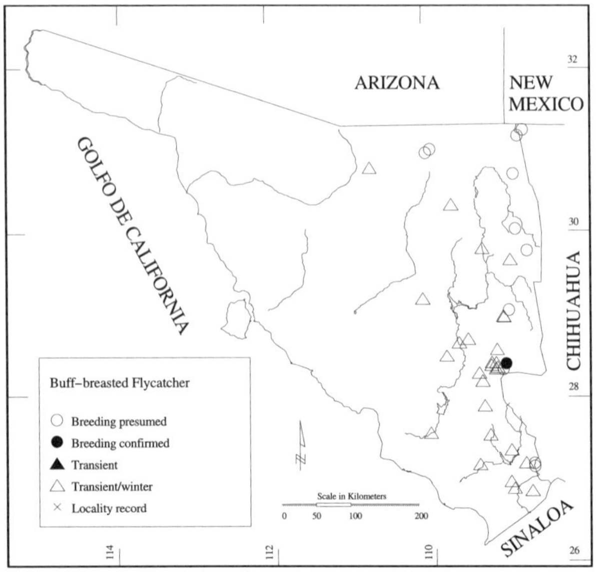

BLACK PHOEBE
Papamoscas Negro
Sayornis nigricans

Except along the coast of El Golfo de California, Black Phoebes are common and ubiquitous residents of watery habitats in Sonora. They are especially birds of streams, whether sluggish ones in the low desert or rapidly flowing mountain rivers. They also live along irrigation canals and drains and thus are becoming widespread in the coastal plains. A requisite is that nest

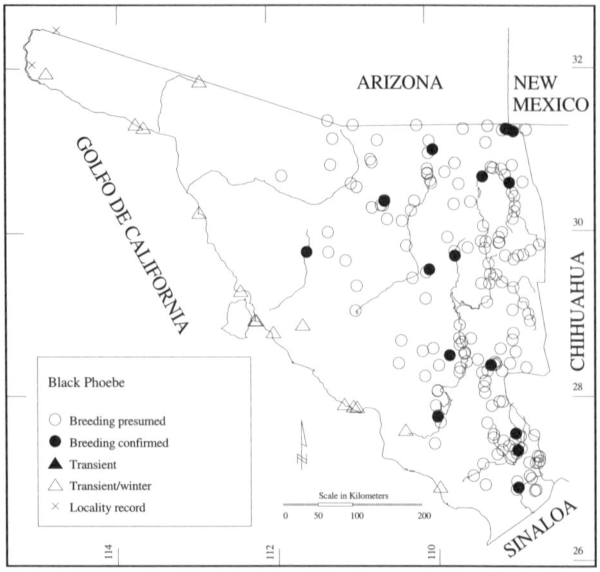

sites be nearby, whether a rocky cliff or a cut bank, or an abandoned building or bridge, to which they can attach their mud nests. The breeding season is late March into July. We have frequently encountered adults building a nest at the same time their neighbors are feeding fledglings. The greatest elevation at which breeding birds have been found is 1680 m in the Sierra los Ajos.

There is certainly some seasonal movement of Black Phoebes. Numbers are noticeably higher in September and October at Lo de Campa and Arroyo Santo Domingo. They are rare on the coast and appear there only from November to February. Some Black Phoebes may be seen at any time of the year in areas where they may nest. Breeding birds may migrate, only to be replaced by other migrants; this possibility awaits the type of evidence that would come from a long-term banding study.

EASTERN PHOEBE

Papamoscas Fibí

Sayornis phoebe

Sonora has six records of this bird: one collected near Granados on 18 November 1953 (A. Phillips); one collected near Santa Cruz on the Río Santa Cruz east of Nogales on 25 November 1954 (J. Marshall); one noted on 30 November to 2 December 1979 a few kilometers west of Ures (Terrill 1981); one seen near Magdalena on 9 February 1963 (W. Harrison, D. Carter); one seen in Arroyo Santo Domingo on 21 February 1982; and one at the Immigration and Customs check-point 21 km south of Nogales on 24 May 1991 (W. Leitner). In all cases, the phoebes were in the proximity of a stream. The bird in Arroyo Santo Domingo sang several times as it gleaned insects from a cliff face above the creek.

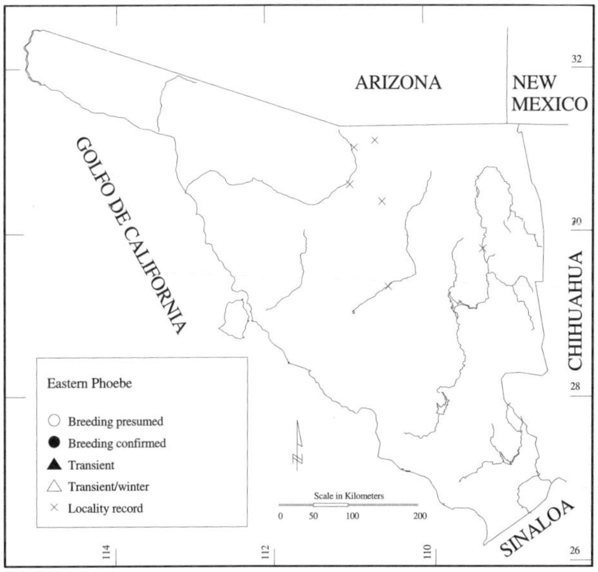

SAY'S PHOEBE

Papamoscas Llanero

Sayornis saya

The Say's Phoebe is an uncommon breeding species in northern Sonora. Individuals may be found there throughout the year, suggesting that they may be resident. Elsewhere, the species is a fairly common migrant and winter visitant, usually appearing in mid-September and remaining until early April. Although many are attracted to ponds, including small stock-watering tanks, others are found in open desertscrub areas where sufficient insects may be found. A single Say's Phoebe on an exposed perch in the vicinity of a corral or water tank is a familiar sight through much of the drier portions of Sonora. If successful in finding flying insects upon which to feed, one may remain several weeks or through the winter. More often a few days is the limit. Most observations are from below 1200 m, but there are a few from clearings in pine-oak woodland as high as 1900 m.

Nest building may begin as early as 2 March (under a bridge south of Naco). A nest north of Magdalena containing three eggs on 21 May is our latest record. There is circumstantial evidence that this phoebe may nest in very arid regions of the desert. We

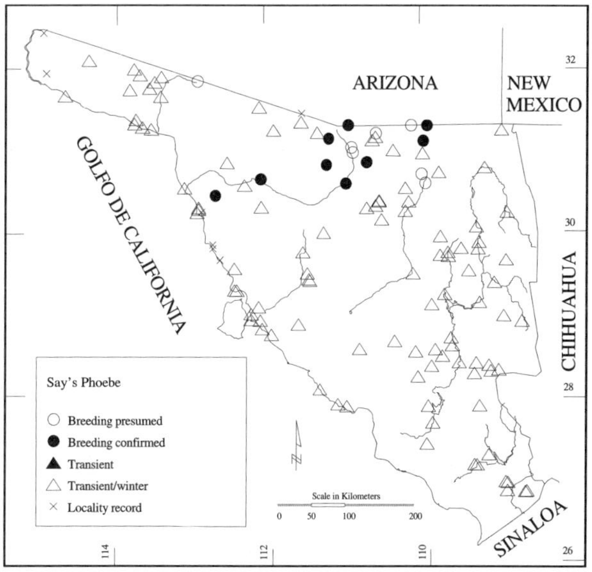

found an old and used nest near the coast at El Coloradito, south of Puerto Libertad. It was situated on a sandstone ledge above a dry arroyo. Near Puerto Libertad (June) one may have had a nest in a well, and at Puerto Lobos (April) one flew into a crevice on a cliff too high to be checked.

VERMILION FLYCATCHER
Mosquero Cardenal
Pyrocephalus rubinus

The male Vermilion Flycatcher is one of the most conspicuous and brilliantly colored of Sonoran birds. Seeing one perform its flight song high above the trees in bright sunshine creates a lasting impression. It is a common breeding bird in open areas with scattered trees and water. Especially favorable sites include open mesquite bosques after ample winter rains have produced a carpet of green ground cover, small irrigated pasture and farmland with some trees, and ranch buildings and corrals. Most records are from the foothills and lower mountains between 200 and 1200 m, although this flycatcher is present at lower elevations along the Río Yaqui to the vicinity of Esperanza and in suitable habitats up to 1500 m in the mountains. There are few records in any season in the coastal plain.

Breeding Vermilion Flycatchers in the northern half of Sonora are migratory, although some winter uncommonly in favorable localities. Typically individuals depart in late September and October and return in early March. Courtship begins almost immediately upon arrival, and nesting extends from mid-March into July. We speculate that Vermilion Flycatchers are opportunistic in choosing their breeding grounds. Some areas may have no breeding Vermilion Flycatchers in some years yet support high densities in years of unusually heavy winter rains. We encountered about 200 within 3 km of our camp in mixed farm- and pastureland northeast of Nacori Chico in late March 1985. The mesquites were beginning to show green buds, there was much early spring growth of annuals and other perennials, and several small streams were flowing. Many individuals may be seen in southern Sonora in winter. These could be migrants from the north that are augmenting a resident population.

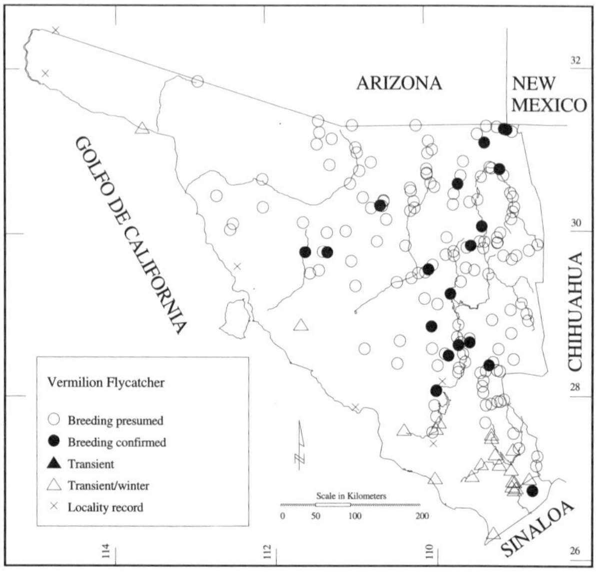

BRIGHT-RUMPED ATTILA
Atila
Attila spadiceus

This flycatcher is little known in Sonora. The first record from the state was a specimen (DVM) taken near Güirocoba in April 1957. Two additional specimens (SDMNH, LSUMZ) have also come from the Güirocoba area in April and May. The species has been found on the Río Cuchujaqui in April, August, and November.

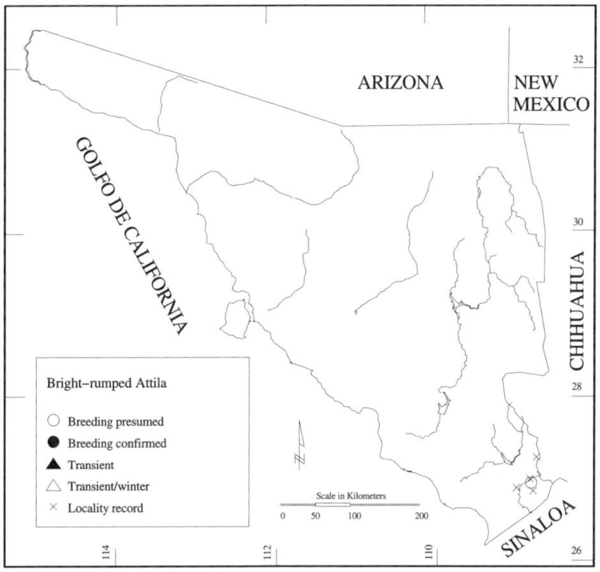

The most recent record is of a bird observed in July 1987 in the bottom of the Barranca las Colas at Sahuarivo. The preferred habitat appears to be riparian growth along narrow permanent canyon streams; the seven records have been from elevations between 260 and 1320 m. Nesting is indicated by a large brood patch on the female collected in May. The species is probably resident, although we have insufficient data to be certain.

DUSKY-CAPPED FLYCATCHER

Papamoscas Triste

Myiarchus tuberculifer

The Dusky-capped Flycatcher is an inconspicuous summer resident, haunting densely foliaged trees but betraying its presence by plaintive *peeur* notes. Typically it feeds by briefly hovering to pick an insect from a leaf or twig. In northern Sonora it is limited to pine-oak woodland up to 2100 m and to riparian woodland at elevations as low as 1000 m. It occupies more diverse habitats in central and southern Sonora and may be found in open pine forest, tropical thornscrub, tropical deciduous forest, and cottonwoods and willows in the lower river valleys (60 m). Individuals return to breeding areas in late March and April and remain until late August or September. It is uncommon in winter in southern Sonora, north to Lo de Campa. The coast is avoided in migration, although this species has been found in spring at Agiabampo and Guaymas.

Pairs form soon after the birds return in spring, with pairing often accompanied by much calling and chasing. As do other members of the genus *Myiarchus*, the Dusky-capped Flycatcher nests in cavities. It initiates nesting in late April in southern Sonora and in June in the north. Either a natural cavity or one excavated by a woodpecker in a tree or large cactus is used. According to notes accompanying clutches (WFVZ) from Güirocoba, the three or four eggs are laid on a mat of fur. Palm fibers or shed lizard skins may be used as alternative lining materials.

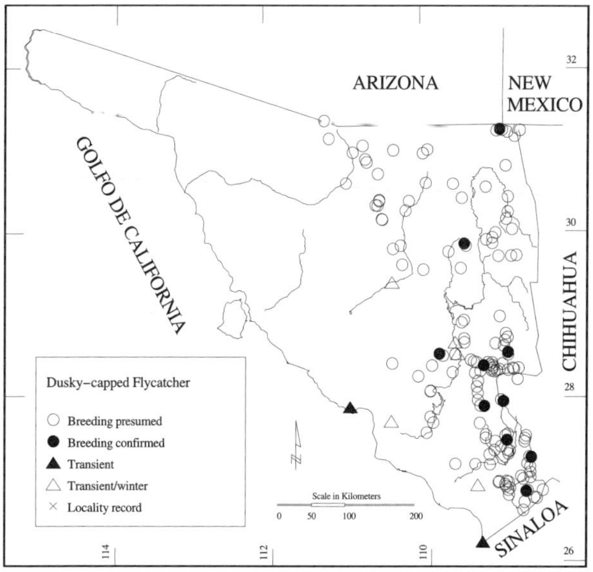

ASH-THROATED FLYCATCHER
Papamoscas Cenizo
Myiarchus cinerascens

The Ash-throated Flycatcher is a common breeding species in desertscrub habitats upward into Chihuahuan oak encinal and pine-oak woodlands. Along the coast of southern Sonora it is uncommon and local. Güirocoba (400 m) is the only known breeding locality in the extreme southeast, although this flycatcher is common in winter throughout the area.

The very similar Nutting's Flycatcher replaces it in tropical thornscrub and deciduous forest clearings, although both species occur together in many places. Field identification of the two species is difficult, and we are uncertain of the extent of overlap. In the Alamos area, the Ash-throated Flycatcher is present from late July to early April as a migrant and winter resident. At Lo de Campa both species are present throughout the year.

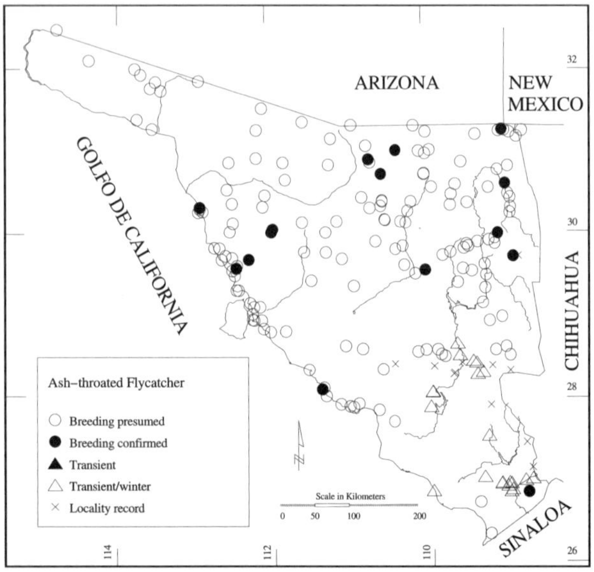

Banding studies in the mangroves of the Punta Chueca area in the 1970s by Russell and D. Lamm indicate that many Ash-throated Flycatchers there are year-round residents. The flycatchers nested in desertscrub but foraged and roosted in the mangroves. Some young birds were found to remain in the area and breed in subsequent years. One adult was retrapped 5.5 years later. Numbers increase in migration and in winter. Farther north, in the Puerto Lobos area, banding data also establish permanent residency. Elsewhere in northern Sonora, individuals migrate south in August and September and return the following March and April.

Nests are placed in cavities in cactus, oaks, mesquite snags, and even upright pipes. Three to five eggs are laid as early as late March; nesting continues until at least mid-June.

This flycatcher supplements its usual diet of insects with fruits of elephant tree and cardón, saguaro, and organ pipe cactus. In June, underparts of individuals that have been feeding on cardón fruit are stained pink. In November, December, and January, when ripe fruit is available and insect food is in low supply, the winter distribution of the species is tied closely to the availability of elephant tree fruit. Some but never all elephant trees have fruit every year. The flycatchers hover briefly before a ripe fruit, take it quickly with a snap of the bill, and fly to a nearby perch before swallowing it, pit included. They are often aggressive toward one another and Gray Vireos, which also eat this bitter fruit.

NUTTING'S FLYCATCHER
Papamoscas de Nutting
Myiarchus nuttingi

Nutting's Flycatcher is a common resident of tropical thornscrub and open portions of tropical deciduous forest, often coexisting with the Ash-throated Flycatcher. The two species are extremely similar in appearance and vocalizations, and our caution in accepting records has undoubtedly led to the exclusion of some valid observations. Elevational limits are from about 50 to about 1200 m. The species is present in thornscrub up to the lower oaks of the eastern foothills.

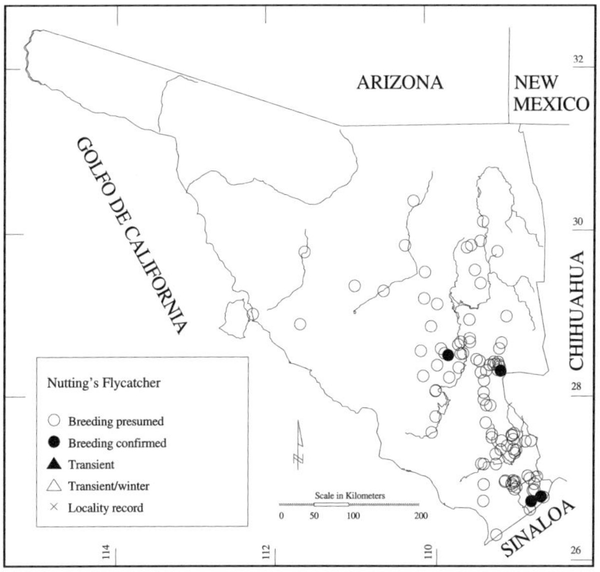

We are aware of nesting from mid-May to early July. Nests are in cavities, as is the case with all *Myiarchus* species, and are lined with fine fibers. Holes excavated in hechos and organ pipe cactus by woodpeckers provide many sites.

BROWN-CRESTED FLYCATCHER

Papamoscas Tirano

Myiarchus tyrannulus

This largest and most vocal of the Sonoran species of the genus *Myiarchus* breeds commonly over much of the state. Its distribution is from low coastal areas, where large cacti (cardón and saguaro) dominate, into high elevations in riparian communities supporting extensive large trees with a closed canopy. It is not common in tropical deciduous forest, which may have sizable trees, except in riparian areas. It regularly occurs up to elevations of 1700 m and rarely up to 2000 m in riparian areas of pine-oak woodland. On the Sierra Aconchi it also frequents encinal (Marshall 1957). It is a migratory species in Sonora, arriving in mid-April, with some birds departing early in August and most gone by mid-September. There are few records from winter months: one on 14 November (A. Phillips) on the lower Río Yaqui, one at Lo de Campa in January, and one near Moctezuma in February (both A. Moorhouse). There are several records in March that may represent very early migrants.

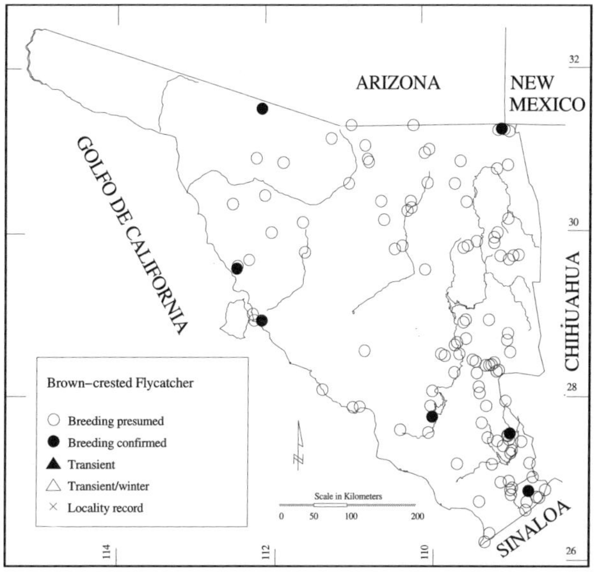

The species nests in cardón cactus in the vicinity of Punta Chueca but apparently does not return there before mid-May, a date much later than in the interior. It nests at Punta Chueca at the end of May and in early June. Some foraging is done in mangroves. One captured in mangroves for banding on 30 May laid an egg in the holding container. Although there are many large cacti along the coast north of Punta Chueca, we have not found this species there, even during migration periods. The coast is hotter and more arid to the north, and this large flycatcher may be food-limited, whereas the smaller Ash-throated Flycatcher finds more prey items and is widespread. Nearly all of our observations of fruit eating by the Brown-crested Flycatcher have been in the Punta Chueca area, where it consumes saguaro and cardón. The fruit was taken as the bird hovered in front of it. On the lower Río Yaqui, we have seen it eat guamachil fruit in May.

As is the case with other *Myiarchus* flycatchers, nests are in cavities. The large body size of this species requires a large cavity. We have not seen a Brown-crested Flycatcher nest in an hecho; perhaps that cactus is too small. Holes deemed to be those excavated by woodpeckers in saguaro, cardón, sycamore, cottonwood, and pine are often used. Egg laying in inland regions takes place from early May to early July.

GREAT KISKADEE
Luis Bienteveo
Pitangus sulphuratus

Large, colorful, and raucous, the Great Kiskadee has been found irregularly and locally in Sonora. It is an uncommon breeding bird along the lower portions of the Río Yaqui, frequenting large cottonwoods and willows of the riverbanks. We have observed it most frequently on the Río Yaqui below Presa Obregon. Individuals have been noted in the winter months at several localities not in the Yaqui or Mayo river systems. We have no way of knowing if these birds had nested earlier in Sonora or if there is a general northward postbreeding movement from the south. Birds noted in winter are also typically, but not exclusively, associated with freshwater habitats with trees.

In the gallery cottonwoods 2 km below Presa Obregon, kiskadees were common (perhaps 20 pairs) in 1986. On 11 April, we watched six pairs initiating nest construction in large cottonwoods. In May and July some pairs were adding material to nests; some birds in July were feeding nestlings. In July 1987, many birds were present, but we noted no nesting activity. We have not found kiskadees in some habitats in extreme southern Sonora (including the riparian growth of the Ríos Mayo and Fuerte) that appear suitable.

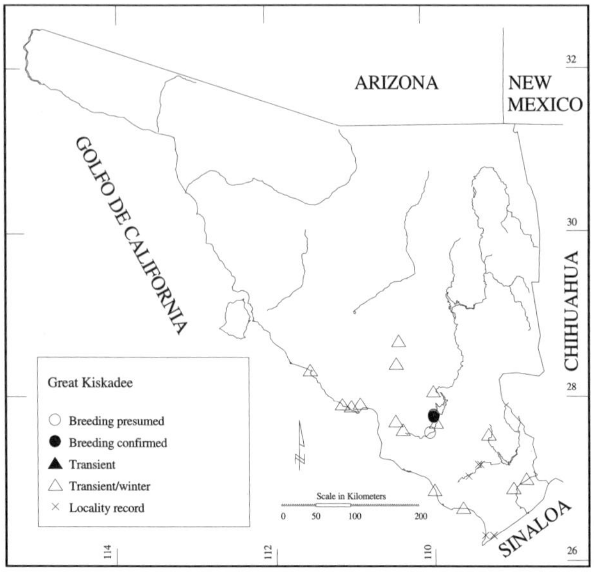

SOCIAL FLYCATCHER
Luis Gregario
Myiozetetes similis

Resident Social Flycatchers are fairly common in the tall riparian growth of the Río Yaqui downstream from Presa Obregon and are uncommon along the lower Río Mayo below Presa Mocuzari. As the name implies, they are gregarious, usually chattering noisily. With the exceptions of a single bird seen in July near El Chinal in extreme southern Sonora and flycatchers observed rarely in late July and early August on the Río Cuchujaqui southeast of Alamos (Short 1974), all recent records of this species away from the Río Yaqui have been in winter.

We once observed the initiation of nest building, on 11 April on the Río Yaqui below Presa Obregon. A

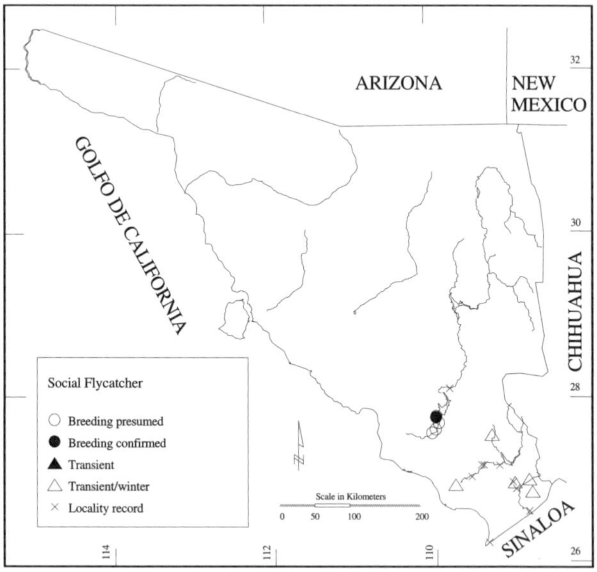

Social Flycatcher placed the first strands of grass at the base of small twigs high in a cottonwood, only to have the breeze immediately blow them away. The bird chased after them, caught them in the air, and replaced them more securely. Adults were still entering nests on 10 July, but we do not know what the nests contained.

SULPHUR-BELLIED FLYCATCHER
Papamoscas Atigrado
Myiodynastes luteiventris

This flycatcher is a common summer resident in the tropical deciduous forest and in riparian stretches of the tropical thornscrub. Farther north it is locally common in sycamore-walnut-cottonwood canyons with flowing streams, upward into pine-oak woodland at 2000 m. It arrives in late spring; the earliest date is 2 May at Güirocoba (van Rossem 1945). Some individuals do not arrive on the breeding grounds until the end of May. Birds begin to depart southward at the end of July, and few are present in September. The latest fall dates are 15 September (Arizpe) and an exceptionally late bird on 16 October (Lo de Campa).

We are puzzled by the absence of the Sulphur-bellied Flycatcher in apparently suitable habitat in the lower Río Yaqui valley. It is unreported below Presa Obregon and is uncommon on the lower Río Mayo and the Río Cuchujaqui. Yet it is abundant on the Río Fuerte at the Sonora-Sinaloa border.

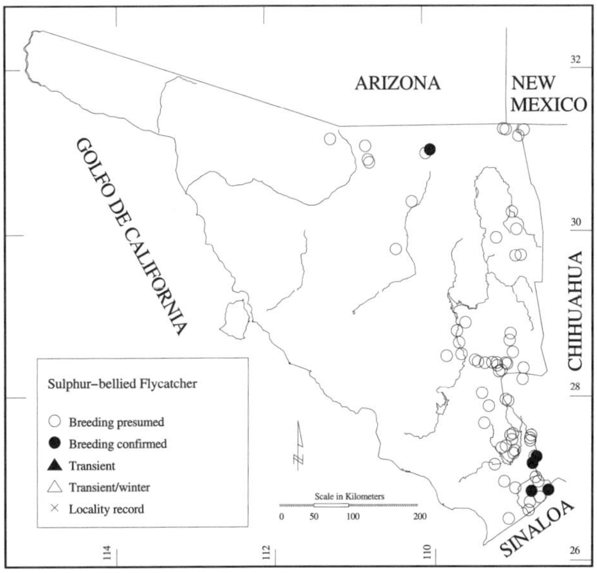

The Sulphur-bellied Flycatcher finds a natural cavity or a woodpecker hole in a large tree in which to nest. Cactus is infrequently used, but S. and J. Levy watched a pair repeatedly fly in and out of a hole in an hecho cactus a few kilometers east of Tonichi. This flycatcher is a summer nester, mostly in June and July after the summer rains have begun.

TROPICAL KINGBIRD
Tirano Tropical
Tyrannus melancholicus

The Tropical Kingbird is a summer resident of Sonora, ranging from abundant in the south to rare and local in the north and northwest. It is particularly numerous in mangroves and adjacent cactus "forests" of the Agiabampo area and in cottonwood groves along the lower Ríos Yaqui and Mayo. It ranges less commonly northward in mangroves to Punta Sargento (Russell and Lamm 1978). In the interior, it is uncommon in trees bordering farms in the tropical deciduous forest, in open situations along major streams in the tropical thornscrub, and along streams and in irrigated farmland of desertscrub.

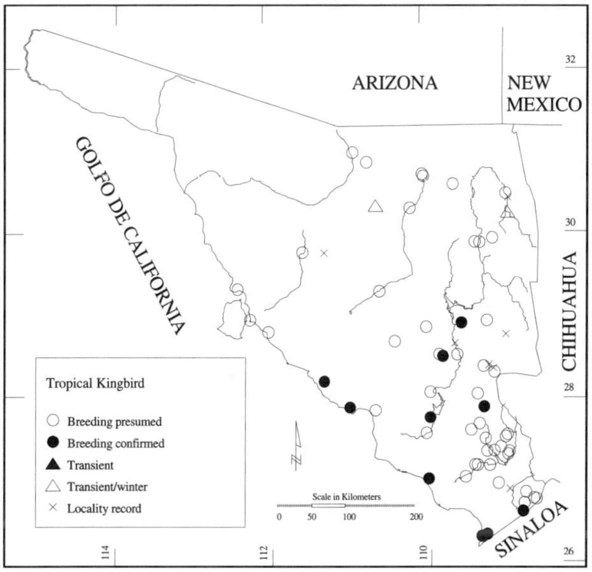

Its northward migration takes place in April and early May, the southward flight in late July and August. Extreme dates are 9 April (Piojillo at the north end of Presa Obregon) and 29 September (El Capulin north of Tonichi). One bird at the Río Cuchujaqui southeast of Alamos on 16 November 1986 constitutes our only well-documented record for the colder months.

Nesting is in June and early July. The nest is typically a flimsy structure, often including many root stems. Where there are mangroves, nests may be over water. In big trees along rivers, nests may be as high as 16 m. On the lower Río Yaqui, we watched a pair nest building on 10 July while their neighbor cared for fledged young. Although these kingbirds usually catch large insects by sallying out from a perch in pursuit, we have watched them flying swallowlike over the calm water of the river before sunrise, dipping every 10 m to pluck insects from the surface in surprisingly graceful flight. At Agiabampo in July, they moved from mangroves to forage on ripe cactus fruit.

CASSIN'S KINGBIRD
Tirano Gritón
Tyrannus vociferans

A strident *key-deer* announces the presence of a Cassin's Kingbird. It is a common summer resident of riparian areas, including cottonwood groves, and isolated large trees in stream bottoms, fields, and meadows from 700 to above 2000 m through the highlands of northern and eastern Sonora. It is a common wintering species in the lowlands of southern Sonora and a transient in all areas except the coast and arid west. The species moves northward slowly, beginning in mid-March and reaching a peak in mid-April. Some fall migrants may be noted as early as late July, but most individuals migrate in September and October, when they may be conspicuous on fence and telephone wires along roads. Numbers of migrants have been noted into November. But they are never as abundant as Western Kingbirds during migration.

A predawn chorus of noisy Cassin's Kingbirds characterizes the early breeding season in sycamore- and cottonwood-dominated canyons. Nesting begins in mid-May and continues through the summer. We have seen parents with short-tailed begging young in early September. Occasional individuals are reported from the tall cottonwoods of the lower Ríos Yaqui and Mayo, and at Nuri, in May and June. The dates suggest nesting, but nests have not been found. Cassin's Kingbirds often nest in areas with other species of kingbirds; we have found four species together near Bacerac and Esqueda.

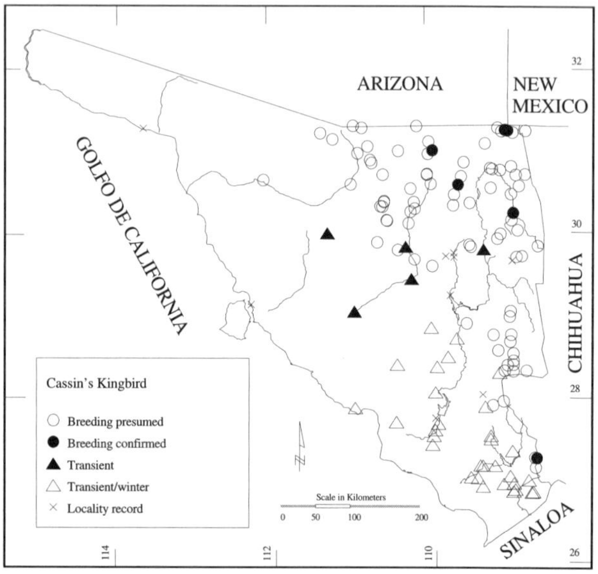

THICK-BILLED KINGBIRD
Tirano Pico Grueso
Tyrannus crassirostris

Thick-billed Kingbirds utilize large deciduous trees in open areas for foraging perches and nesting sites. In such places they are common in tropical deciduous forest. They are less common in tropical thornscrub and desertscrub, where they are found in large cotton-

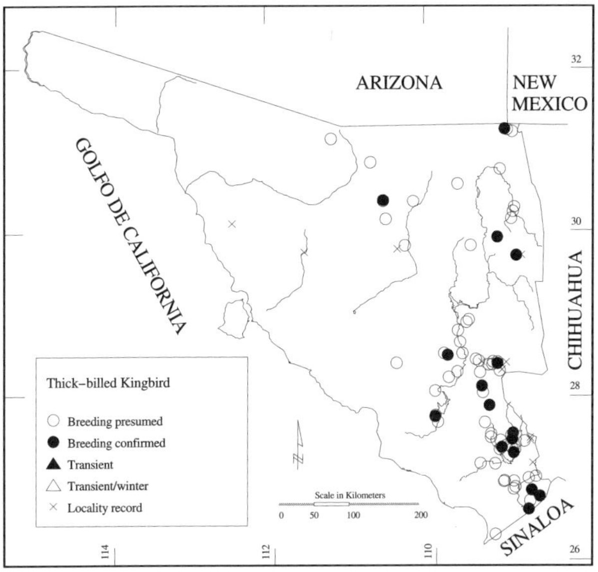

woods, willows, and walnuts along streams, and in oak country, where again they are found in tall riparian growth.

The migration northward begins about 10 April in the Alamos area; northern breeding areas may not be occupied until early May. Fall departure begins in early September and continues into October. A few individuals may winter in extreme southern Sonora. As many as ten have been noted on an Alamos Christmas Bird Count.

They forage from the topmost perches in trees and bushes and call loudly and frequently—habits that make them conspicuous. At Lo de Campa, birds return about 20 April, and nest construction commences in early May; we have seen family groups there in the first week of June. July nesting is common, usually involving the construction of new nests. It may be that nesting in July is the normal pattern, and May nests are infrequent and dependent upon above-average winter rains. Although Thick-billed Kingbirds typically make their nests more than 6 m above the ground in large trees, they will sometimes nest lower in medium-sized oaks. These are aggressive kingbirds and usually exclude other kingbirds from their territories. Cassin's Kingbirds usually avoid the larger Thick-billed Kingbird (Marshall 1957), but Tropical Kingbirds sometimes nest nearby. At Nuri, we watched a nesting Thick-billed Kingbird only 5 m from an active Streak-backed Oriole nest; the territory of a Tropical Kingbird overlapped that of the Thick-billed.

WESTERN KINGBIRD

Tirano Pálido

Tyrannus verticalis

The Western Kingbird is a fairly common spring migrant, a local and uncommon breeding species in the extreme northeast, and an abundant fall migrant. In spring individuals move northward from about 11 March to 18 May. One observed by W. Russell at Puerto Peñasco on 28 February 1988 was exceptionally early, especially for so far north. The southbound migration begins about 15 July, peaks from mid-August to mid-September, and ends about 15 October. Literally hundreds may rest on fence wires and power lines in places with some grass or in fields. On very hot August and September days, migrants spend much time sitting on power lines in the shade of supporting poles. In the drive from Hermosillo northeast to Ures, we have seen

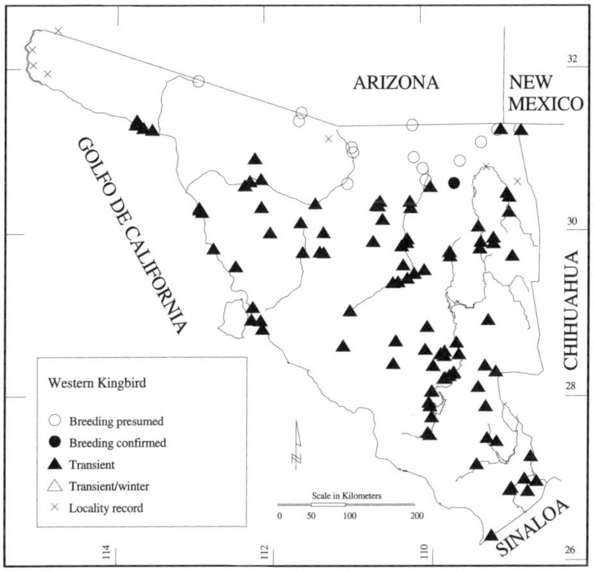

Western Kingbirds perched in the shade of about 80 percent of the poles in many miles of highway. When resting in regions without power lines or fences, they often group in treetops, sometimes 10 to a tree. The latest fall record is of two at Piojillo (upper Presa Obregon) on 11 November. There are no winter records.

The number of northbound migrants seen is far less than that of southbound migrants. Possibly southbound birds move slowly and feed extensively on the numbers of insects produced during the summer rains. Northbound migrants may move with few stops and without lingering, or they may use a different route.

We have established nesting only at La Pera (11 km south of Esqueda), where we watched a juvenile being fed by a parent on 22 July. The presence of Western Kingbirds along the upper Río Magdalena north of Imuris in late May suggests that nesting may occur there, and June records from Morelos (J. Marshall) and lower Arroyo Cajón Bonito also may represent nesting birds. Van Rossem (1945) reported breeding at Santa Rosa Tesia (on the lower Río Mayo) on 22 June 1937. Unless the breeding range has changed drastically during the last 50 years, we doubt his report. Although this kingbird is viewed as almost exclusively insectivorous, except in winter, Russell saw one near Presa Obregon eating the bitter berries of *Bursera laxiflora* on 22 September.

SCISSOR-TAILED FLYCATCHER

Tirano-tijereta Rosado

Tyrannus forficatus

Russell and Lamm (1978) watched a Scissor-tailed Flycatcher with kingbirds along the main highway about 26 km south of Nogales on 10 October 1973. R. Russell found one in a *Maytenus phyllanthodies* thicket adjacent to the beach at Caleta Venecia on 19 May 1993. The silent bird was seen by many people the same day but could not be found the next day.

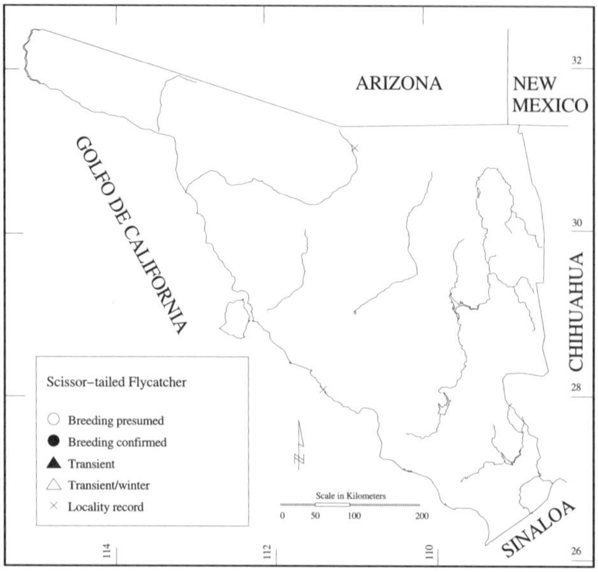

GRAY-COLLARED BECARD

Mosquero-cabezón Mexicano

Pachyramphus major

The Gray-collared Becard was not detected in Sonora until 1983 (Monson 1986). The first bird, a male, was discovered about 11 km west of Yécora on 9 January 1983 (J. Bates, C. Green, et al.). Subsequently we saw a pair about 24 km north of Yécora on 11 May 1983 and a male about 26 km east of Sahuaripa on 17 March 1984.

On 19 May 1986 three males were found at El Terreito (north of Los Algodones), and another singing male was noted at Rancho Quemado on 21 May 1986. On 16–17 July 1987 two nests were discovered in separate canyons west of Sahuarivo by Moorhouse, Monson, and Suter. The nests were globular masses about 24 cm in diameter, placed in crotches of slender oaks (*Quercus epileuca* or similar species) 6 and 9 m above the ground, respectively. Both had many pine needles incorporated in the exterior, and at one the female had collected *Tillandsia* flower stalks and frag-

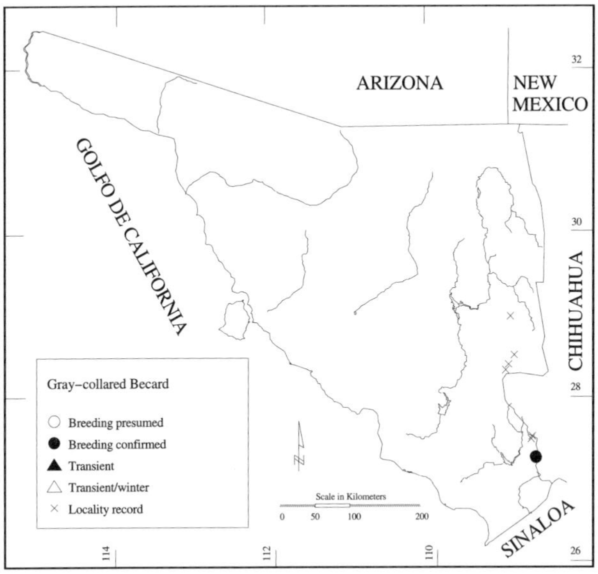

ments of green *Usnea* for the nest. We are unaware of any other descriptions of the nest of this species.

All Gray-collared Becards were seen in oak-dominated pine-oak woodland at elevations varying from 1320 to 1820 m. The seasonal status is unknown. A male at El Terreito foraged and sang in an area about 300 m across, his song consisting of four or five loud, slurred notes: *wheop, wheop, wheop, wheop*.

ROSE-THROATED BECARD
Mosquero-cabezón Degollado
Pachyramphus aglaiae

The Rose-throated Becard is a fairly common summer resident in northern Sonora, occupying riparian habitats dominated by cottonwoods and sycamores. In the south it is more common and inhabits tropical thornscrub and tropical deciduous forest. Although chiefly a bird of lower valleys, it occasionally extends upward to the lower limit of the pines at 1600 m, as in the Sierra Aconchi (Marshall 1957) and at Rancho Quemado northeast of San Bernardo.

The bulky hanging nests are often suspended from branches varying from 5 m above the ground to the tops of large trees, including Montezuma baldcypresses, cottonwoods, and sycamores. At Rancho Jacalón northeast of Rosario, nests were built in Chihuahuan oaks. Construction begins in May, although eggs may not be laid before June (WFVZ). Both male and female may continue to add materials to the nest during the time eggs are being incubated (Skutch 1969). Monson saw an adult and a young bird still close to a nest near Bacerac on the late date of 4 September.

The migration is protracted, with birds appearing gradually rather than in large numbers in a short period. In spring, migration commences by mid-March in the south and not until early May in the north. Two birds at Lo de Campa on 16 March perhaps were somewhat early for that locality in central Sonora. The fall migration begins in August and continues into early October, with one individual near La Casita south of Nogales on 15 October (A. Phillips and R. Dickerman) exemplifying a late record. A few Rose-throated Becards winter in extreme southern Sonora in the tropical deciduous forest. A male was found in Arroyo Santo Domingo, far to the north of the wintering range, on 20 February 1982.

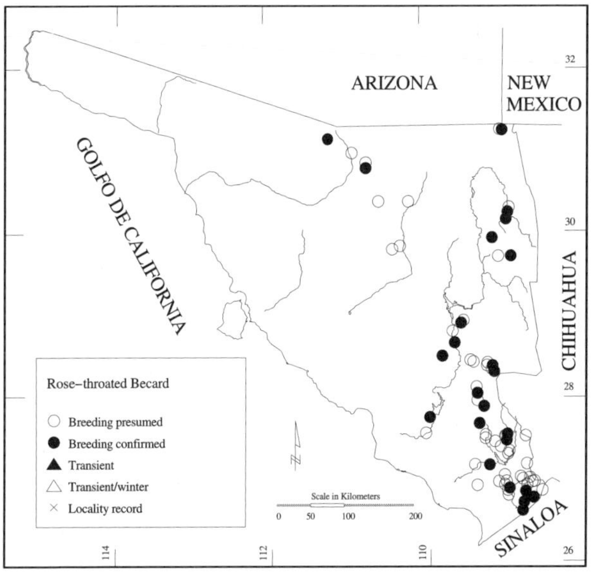

MASKED TITYRA
Titira Enmascarada
Tityra semifasciata

Van Rossem's (1945:143) record, stated as follows, is still the only one of this species for Sonora: "Known only from the type locality [of the subspecies *hannumi*], where two specimens, a pair in breeding condition, were taken and four other individuals seen on May 29, 1937. All were in semi-deciduous woodland in the Tropical zone at altitudes of from 1400 to 2000 feet [426 to 610 m]." The type locality is Arroyo San Francisco, close to the common point shared by the states of Sonora, Chihuahua, and Sinaloa.

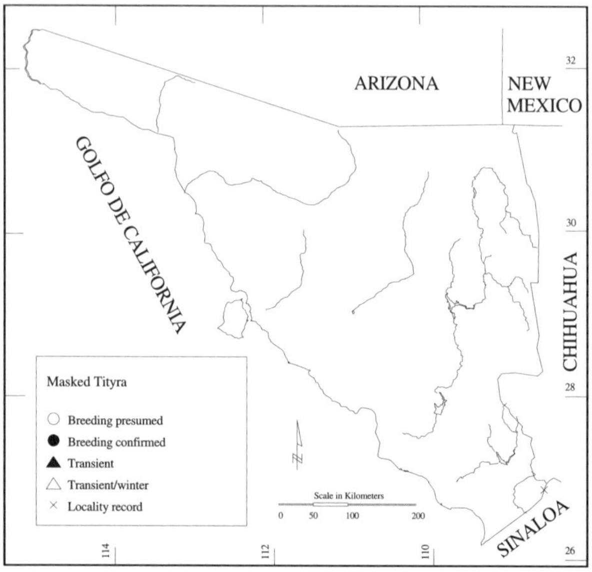

There has been much clearing of tropical deciduous forest in southeastern Sonora, but that change probably does not account for the absence of recent tityra records. More likely the northern limits of the range fluctuate over time, a normal phenomenon. If the species has become scarce in Sinaloa, there is cause for concern.

FAMILY ALAUDIDAE
Larks

HORNED LARK
Alondra Cornuda
Eremophila alpestris

Horned Larks breed in two different regions in Sonora. In northeastern and north-central Sonora they occupy rather level short-grass areas, especially overgrazed rangeland. In the lowlands of the northwest, they may be found in coastal dunes and other sandy places, basins, and abandoned or fallow fields. Nonbreeding birds from other populations frequently occur in these same areas in winter in loose flocks of up to 30 individuals.

We can say little about the seasonal distribution of Sonoran larks. They move about unpredictably, a reflection of our inadequate knowledge of their responses to varying environmental conditions. Numbers vary within seasons and between years. The presence of these larks in the very arid coastal plain is dependent upon rainfall. Although we have found them nesting at Puerto Peñasco and present there in each month, in some years there are few or none. At Bahía Sargento, where they sometimes nest in *Frankenia-Salicornia* flats near the beach at the southern end of

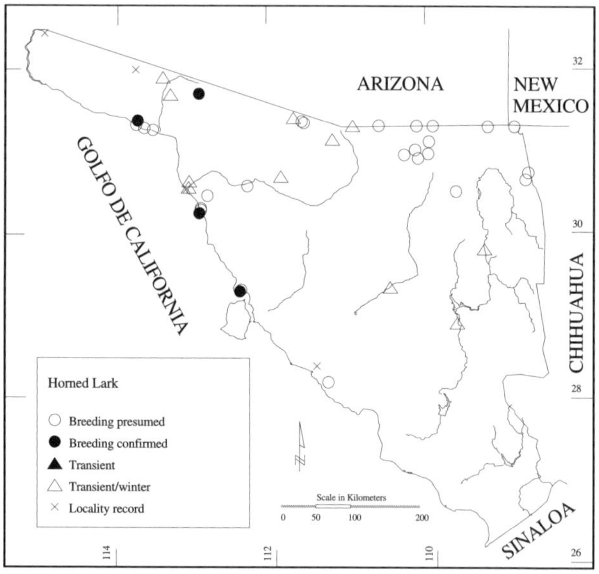

their coastal range (Russell and Lamm 1978), they are absent in more years than they are present. Paired Horned Larks seen in fields near San Miguel north of Caleta Venecia on 22 May 1993 were the southernmost recorded in Sonora and may have been nesting.

In the arid northern coastal areas following winters of above average rainfall, we have found singing birds and pairs in late February. Successful breeding does not always follow pair formation, and we know of fledged young in only four years, all in late March and the first half of April. In years without adequate precipitation, few or no larks are present, and if pairs are formed, nesting is unsuccessful.

FAMILY HIRUNDINIDAE

Swallows

PURPLE MARTIN

Golondrina Azulnegra

Progne subis

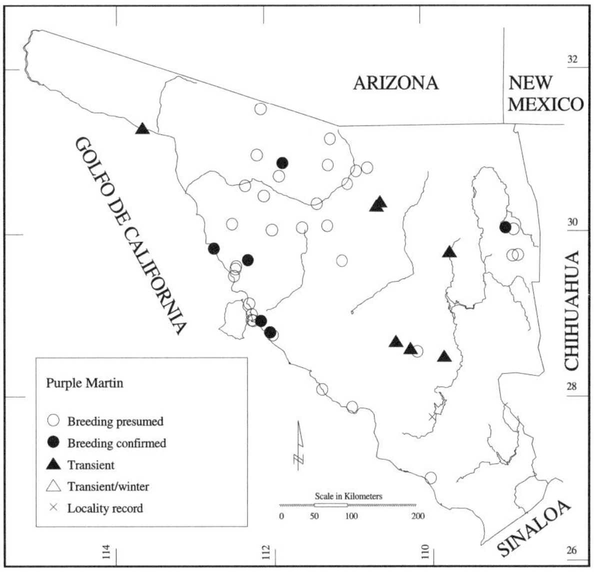

Although J. Marshall saw Purple Martins as early as 18 April in Moctezuma, most individuals appear in mid to late May and are uncommon summer visitants in cardón- or saguaro-dominated desertscrub. Along the coast they are local but widely distributed in small numbers. Almost immediately upon arrival they begin exploring cavities in the large cacti for nest sites. They appear to use such sites exclusively. Nests may be separated by a kilometer or more, or the birds may nest in loose colonies with up to 25 pairs (e.g., just north of Punta Cirio) in places where the cacti are clumped and surrounded by areas supporting few large cacti. Egg laying and incubation take place in June and early July. By the time young are fledging, the summer rains have begun and flying insects are abundant.

At the end of the summer, large assemblages occur at a few roosting sites. Once, on 17 September, we watched many small loose groups gather over the mangroves at Punta Santa Rosa in the last hour of daylight. As darkness approached, larger flocks arrived and all coalesced into a single massive flock of an estimated 15 000 birds. After much swirling, they dropped suddenly with a roar of their wings to roost over the water in mangroves. They dispersed early the next morning before sunrise. This roosting aggregation probably included both local birds and others migrating south from a more northern part of their range. We know of no spring or summer roosts in Sonora. Our latest fall date in desert areas is 21 October at San Carlos.

On 15 and 16 July 1975, while camped in mountains north of Bahía Kino among the cardóns, we heard at about 0400 both mornings the beginning of a veritable din of martin noise that seemed to be coming from birds flying overhead in the blackness. This kept up for about an hour and a half, by which time it was daybreak and the birds had dispersed. Two mornings later north of Caborca, the martins began their performance at 0330, singing in pitch dark, and so far as we could ascertain, from saguaro tops and hole entrances. The most intense calling and singing occurred while it was completely dark, and the sounds ceased with the birds' disappearance at the arrival of dawn.

Away from saguaro and cardón, we know reports

of nesting only in pine-oak woodland (Marshall 1957; Phillips 1986) in the Sierra Huachinera and the Sierra Nacori. We have never seen these martins in the mountains of eastern Sonora, and we suspect they must be very local. There are records of small numbers of fall migrants from several northeastern localities. The largest group, more than 100, foraged over a marsh and drank there on 11 September at Saracachi. At Lo de Campa, where they do not nest, we saw late migrants on 12 October.

SINALOA MARTIN
Golondrina Sinaloense
Progne sinaloae

We found Sinaloa Martins in mid-July 1987 at 1600 m on the mesa west of Sahuarivo. On each of four days we observed up to 15 individuals foraging over the pine-oak woodlands, pastures, orchards, and the barranca edge, usually with swifts but sometimes with Barn Swallows or alone. Six sat and preened in a dead pine; two were adult males with the white breast, belly, and crissum in striking contrast to the glossy blue-black of the chest. The other birds were females or immatures. Over a period of an hour they would exchange places with others in flight. A shrike once supplanted two of the martins. We saw no Purple Martins in the area.

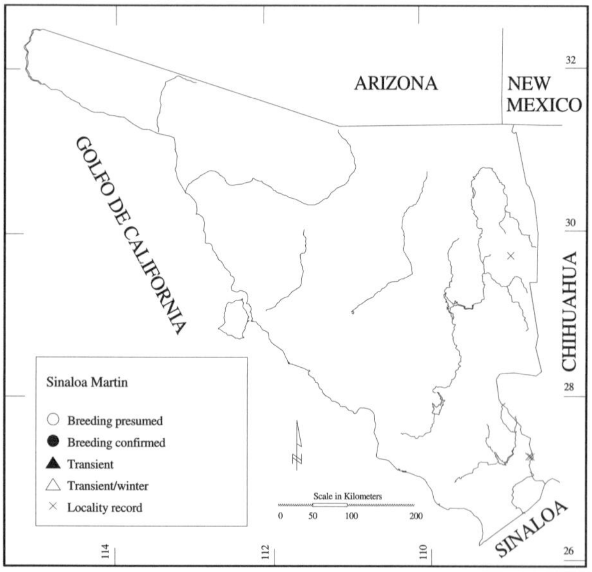

A. Phillips collected two specimens on 9 and 10 June 1953 at El Riito (980 m) in the Sierra Nacori. J. Marshall, who was with Phillips, noted that the martins foraged over a pond near the stream and associated with Northern Rough-winged Swallows along a nearby cliff. In the Sierra Nacori, the Sinaloa Martin was found on the same slope as the Purple Martin but at a lower elevation, both evidently breeding (Phillips 1986). Phillips (1986) did not recognize the Sinaloa Martin as a species and merged it with the Caribbean Martin, *Progne dominicensis*.

The Sinaloa Martin is both local and rare throughout its range, thus vulnerable to abrupt population declines due to habitat change. Its wintering area is unknown. It should be considered a species of concern, and efforts should be made to learn more about its natural history.

TREE SWALLOW
Golondrina Bicolor
Tachycineta bicolor

Tree Swallows do not nest in Sonora, but they migrate through the state in large numbers, and a few individuals are present in winter. Northward flights, typically of fewer than 50 individuals, are in late March and early April (latest date, 6 April). Fall migrants appear as early as July 15 and reach a peak in mid-September, when hundreds may be seen along the Río Yaqui and in the rolling desertscrub of central Sonora. At that time, summer rains have resulted in an abundance of winged insects, and swallows (diurnal migrants) seem not at all out of place in the desert. On 16 September 1974 we estimated more than 20 000 (including one flock of about 3200) on wires along a few kilometers of highway about 90 km north of Hermosillo. These large flights do not linger but progress steadily southward. By October, few individuals remain anywhere, most being concentrated at favorable feeding areas such as the head of Presa Obregon. Most winter records are from coastal areas.

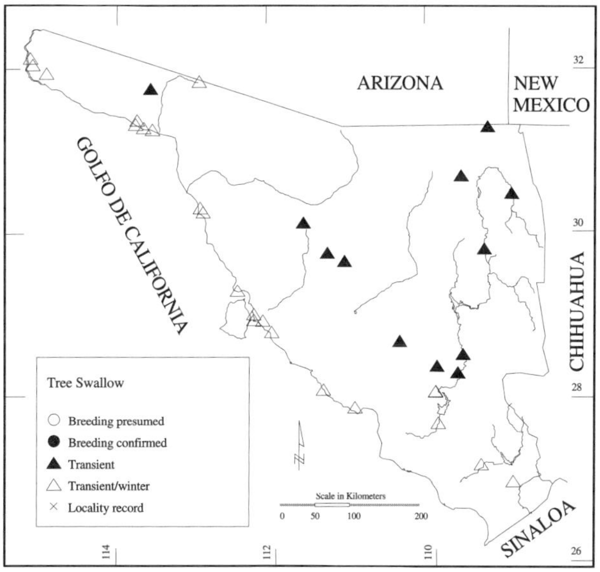

MANGROVE SWALLOW

Golondrina Manglera

Tachycineta albilinea

This swallow, most common in tropical Latin America, may be found almost year-round in small numbers locally along the coast north to Punta Cirio and inland along the Río Yaqui (to the head of Presa Obregon) and Río Mayo (Presa Mocuzari). We lack records from 16 July to 21 October. Most reports from winter months have been coastal, but six individuals seen on 7 January at the head of Presa Obregon suggest wintering along the Río Yaqui.

Along the rivers, Mangrove Swallows are most frequently seen where vegetation is densest and insects most abundant. On the coast they feed primarily on insects flying over tide-flushed esteros where mangroves occur. They may fly several miles inland from mangroves to find suitable nest sites. We once saw them feeding low over the beach at Guásimas (9 January) where there were no mangroves; the birds sat on baked mud 50 m from the beach when not foraging.

The absence of records from 16 July to 21 October is difficult to explain. It is possible that Mangrove Swallows move south and out of Sonora during this period, but it is more likely that the birds are widely dispersed throughout central and southern lowland Sonora. This is the period of summer rainfall, and insects are plentiful away from the coast and rivers. The species is local and occurs in small numbers; if widely dispersed through thornscrub and tropical deciduous forest areas it could be undetected. In

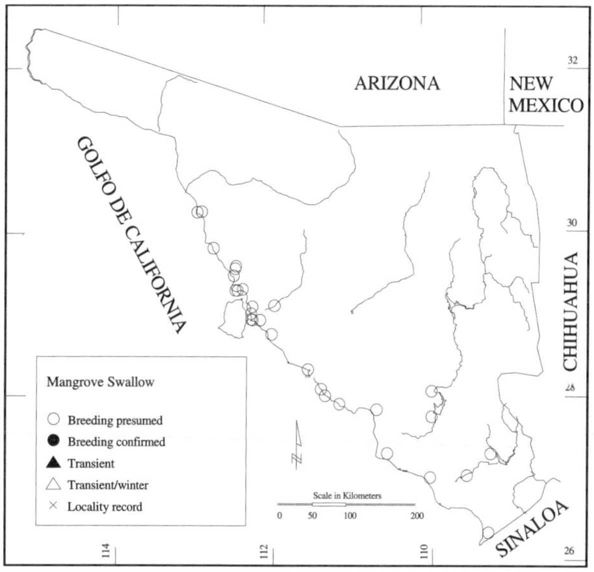

November, as terrestrial flying insects become less abundant, the swallows may return to areas with shorelines.

Cavities in cardón cacti are preferred as nesting sites. We have seen pairs (and once, a trio) investigating cavities as early as 18 March, about 8 km inland near

Punta Chueca. By the first week of April, birds were gathering nest material. A pair at Punta Sargento included eel grass in lining a cardón cavity. At Puerto Lobos, students in an ornithology class watched a pair carry string and other material into a pipe 7.5 cm in diameter that projected a meter beyond the eaves of a house, where the pipe was used as a roof support. Southwest of Cd. Obregon, at Liliba and the nearby Isla Lobos, G. Bancroft collected four nests on 27 May 1930 (WFVZ). Included were a fresh and probably incomplete clutch of two eggs and a clutch of five eggs in an advanced stage of development. Two of the nests were in cavities in driftwood, less than a meter above the beach. Another nest was in a crack in the lighthouse wall, and one was on a beam in the light-keeper's cottage. All nests were lined with large feathers. Mangrove Swallows are regularly present from early April to mid-July along the Río Yaqui, where they probably nest. A pair investigated a cavity in a dead cottonwood flooded at the head of Presa Obregon on 10 April, and a cottonwood cavity below the dam contained a pair on 11 April and also on 15 May.

VIOLET-GREEN SWALLOW
Golondrina Verdemar

Tachycineta thalassina

Violet-green Swallows nest in the giant cacti of the central and southern coastal regions and possibly in the coniferous forests of higher elevations. Large numbers move through Sonora on migrations to and from their more northerly breeding range, which extends all the way to Alaska and northern Canada. There are records from all months, although the paucity of November and December records suggests that most individuals winter to the south. There are several late December records from the Alamos area.

Coastal birds are flying among the cardóns from Puerto Lobos south at least to Guaymas as early as mid-February. They are not common, and usually only a few pairs are found in an area. Nesting begins in early March (large young were seen in a cardón cavity at Puerto Lobos on 4 April 1986) and continues through April. We do not know if these swallows also nest in summer, but we did see a pair investigate a cavity on 15 July 1975, only to be repulsed by resident Purple Martins. A male taken 21 June was in breeding condition.

There are no breeding records from the mountains of eastern Sonora, although Marshall (1957) found Violet-green Swallows nesting in the drainage of the Río Gavilan in the Sierra Madre just across the border in Chihuahua. Several individuals were flying among the pines at Rancho Quemado on 8 May 1986 and near Yécora on 12 June 1986; these may have been local birds. Between the coast and the high sierras, Violet-green Swallows have been observed at many localities at all elevations in all months from January into November. Most of these birds were undoubtedly transients, but it is possible that some were locally breeding birds. Substantial flights of hundreds of migrating birds have been seen in the interior river

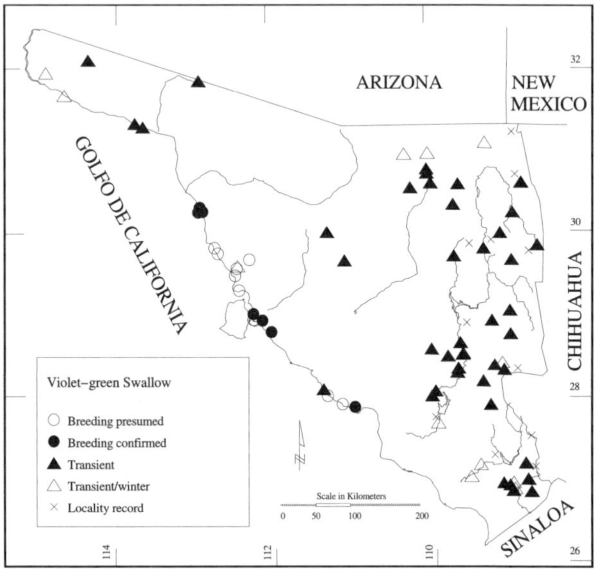

valleys and mountains in March, August, September, and October. There were at least 10 000 along Highway 15 near El Oasis on 18 August 1984; there were none there on 20 August (S. and J. Levy).

NORTHERN ROUGH-WINGED SWALLOW
Golondrina Ala Aserrada

Stelgidopteryx serripennis

This nondescript brownish swallow breeds in Sonora and also migrates through the state en route to and from more northerly nesting areas. Most migrants move northward during the last half of March and in April; southbound birds occur from mid-July through September but in small numbers. Only once have we ever seen more than 100 birds in a day; on 19 March 1985, we counted nearly 400 migrating over the fields at the edge of the Río Bavispe at Granados. On 1 August 1961, J. Marshall saw thousands near Bácum. Very few individuals (maximum of five seen in a day) migrate along the coast and at higher elevations in the mountains; most follow the north-south river valleys and the foothills. The species is rarely seen away from water, even during migration.

There are very few observations of the species from October to February and none in December. The only birds reported in January are ten foraging over Presa Agua Caliente on the 4th and two seen in Alamos on the 17th. There are four February dates, but these are of only one or two birds. The only October records are from Puerto Peñasco and Lo de Campa. In late autumn, only one bird has been seen, at Navojoa on 28 November.

Despite widespread occurrence of the Northern Rough-winged Swallow during the period of April through July, there is very little evidence of nesting. A pair on 11 April copulating in flight below Presa Obregon fell into the water (R. Chapin). Individuals were still in the area in mid-July. Van Rossem (1931) referred to birds taken near Guaymas on 7 May as breeding. Monson and J. Stevenson found three pairs apparently nesting in ruins of an adobe building at Guaymas on 12 April 1983. A specimen from the Cd. Obregon area had enlarged gonads on 3 June. A bird was incubating in a cavity in an earthen bank about a meter above a small wash in oak woodlands north of Nacozari on 31 May. R. Cunningham (Wilt 1976) found nests on the Río Sonoita in March 1986. On the basis of multiple observations during the period of May to mid-July, other localities are indicated on the range map as "breeding presumed."

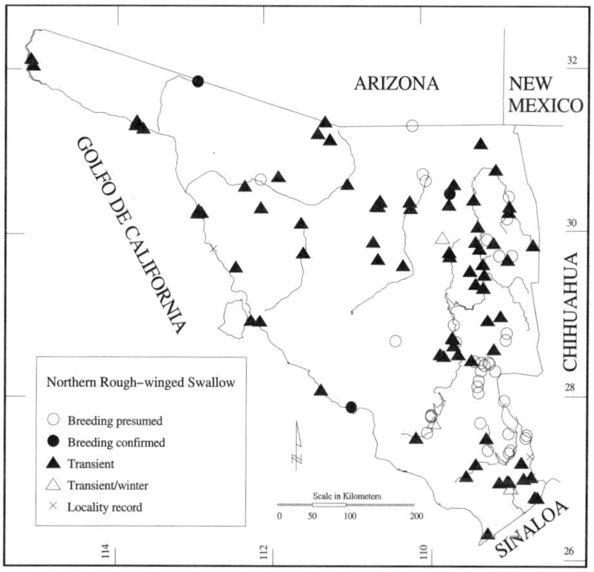

BANK SWALLOW
Golondrina Ribereña

Riparia riparia

The Bank Swallow, a breeding species of high northern latitudes around the world, is an infrequently recorded migrant through Sonora. We are aware of only eight records: three in spring (Bahía Kino, 7 April 1976; Ciénega de Santa Clara, 9 April 1994; Tepoca, 12 May 1988) and five in fall (Granados, 12 August 1954; northwest of El Oasis, 17 August 1976; Puerto Peñasco,

3 September 1979; Huachinera, 4 September 1983; Bahía la Choya near Puerto Peñasco, 5 October 1979). There are many records from Arizona (Monson and Phillips 1981), and the species should be expected regularly in Sonora.

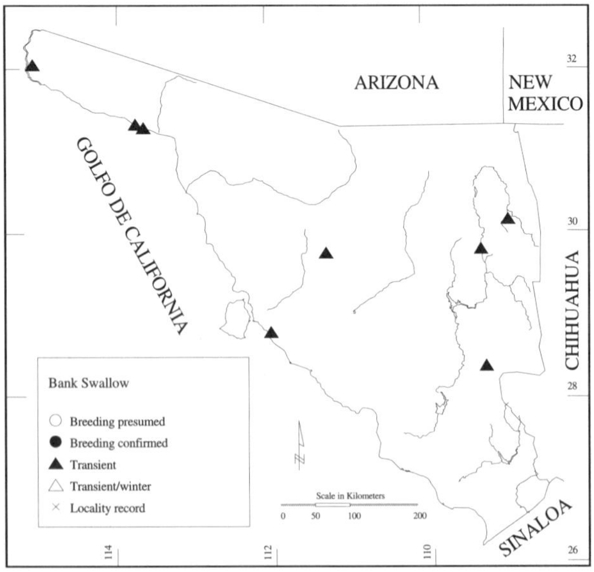

CLIFF SWALLOW
Golondrina Risquera
Hirundo pyrrhonota

Cliff Swallows breed locally from the coastal farmlands into the mountains, usually on cliffs overhanging water and under bridges but occasionally on buildings (such as the church on the plaza in Yécora). They winter far to the south, perhaps in Brazil and Argentina, return in March and April, and nest into July and occasionally August. The species ranges over most of the United States and Canada, and transient birds have been noted in Sonora as early as 16 March and as late as 5 October. More than 1000 were seen over fields east of Hermosillo on 19 March 1982, a mass spring movement of migrants. A flock of more than 1000 birds foraged over fields north of Navojoa on 19 July 1987; probably all were southbound migrants.

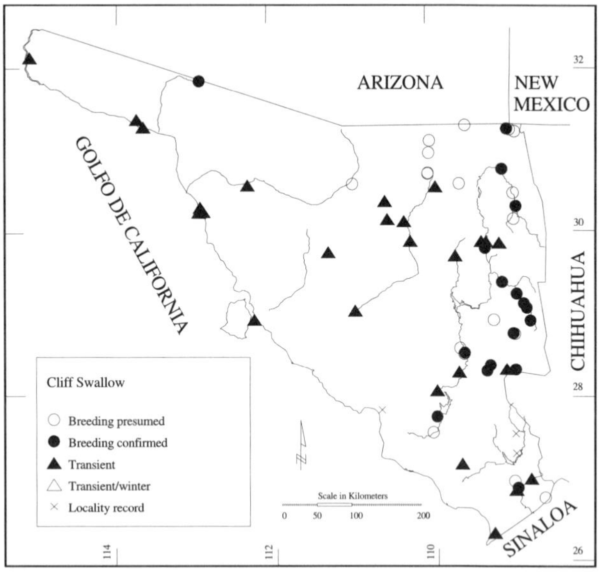

The Sonoran population has a dark chestnut forehead, in contrast to the light forehead of some migratory Cliff Swallows that may occur as transients. Because some migrants also have a dark forehead, it is not possible to identify birds in the field as local or transient, unless they are observed nesting. Nesting activity could be difficult to establish because migrants sometimes pause in their travels to investigate nests of local birds and have even been seen entering nests not their own (Phillips 1986).

This species, like many other swallows, is highly social. Cliff Swallows live in small groups that cluster their mud nests rather closely together under some sheltering overhang. Each pair enters the nest through an opening on the side; incubating or brooding birds often peer out from the nest. The more exposed nests may weather during the nonbreeding season and require extensive repair before reuse. A source of mud is a prerequisite for their nesting. It is probable that the exact timing of nesting may be tuned to the summer rains, but little supporting evidence is available. Clustered mud nests on rock faces along streams in Sonora are probably those of Cliff Swallows. Barn Swallows also build a mud nest, but theirs is an open platform and is not retort shaped. Cliff Swallows prey upon large numbers of flying insects, usually over water or crop and grazing lands.

The Cliff Swallow has recently been moved to the genus *Petrochelidon* (AOU 1997).

BARN SWALLOW
Golondrina Tijereta
Hirundo rustica

Barn Swallows are fairly common breeding birds and migrants. Northbound migrants pass through Sonora from 22 February to 29 May; fall migrants arrive late (3 September), and nearly all are gone by 12 October, although there are three November records (latest date, 23 November, at Presa Mocuzari). We have noted nesting from early May into September and from the foothills (210 m) into the southern mountains (1600 m).

As the name of this species implies, Barn Swallows regularly nest in buildings. They are present in many villages and towns and at ranch buildings, where their pleasant chattering and graceful flight in search of flying insects make them appealing. In 1972 Russell saw one sitting on its mud nest inside a tienda in Sinoquipe

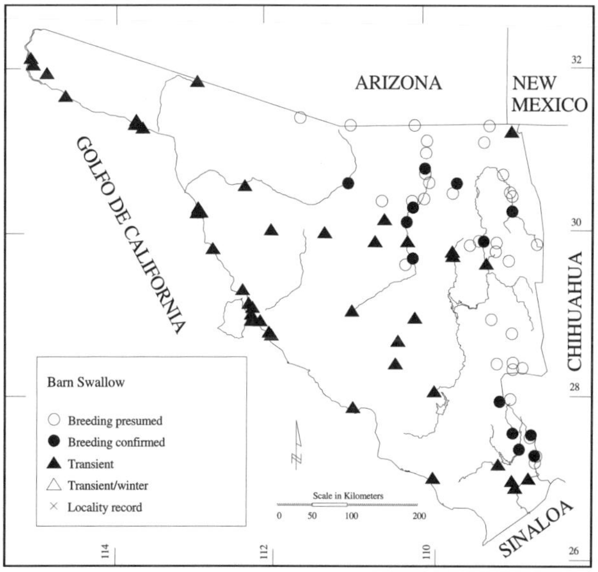

where the only access was through the front door opened by customers. The Barn Swallow is loosely social, and usually several nests may be scattered among several buildings. We have not observed large flocks in Sonora; 100 with other migrating swallows on the Río Moctezuma on 25 September is the largest number noted. Migrating flocks numbering into the thousands are noted along the Colorado River in Arizona and California in fall. The absence of observations of large flocks in Sonora is unexpected.

FAMILY CORVIDAE
Jays, Magpies, and Crows

STELLER'S JAY
Chara Crestada
Cyanocitta stelleri

Breeding Steller's Jays are limited to extensive coniferous forests, but their territories frequently overlap those of Mexican Jays in pine-oak woodland at high elevations. In winter they occasionally move to lower elevations and in irruptive years even into the deserts and lowlands (e.g., one collected in January at Sonoita in 1894 and four at Alamos in March 1888). We have seen Steller's Jays in March in a number of pine-oak localities at 1400–1700 m where they do not nest.

We have encountered this species in pairs and in flocks of usually less than ten individuals; a group of 22 was exceptional. Nesting usually occurs in May and June. Steller's Jays are omnivorous; although they forage

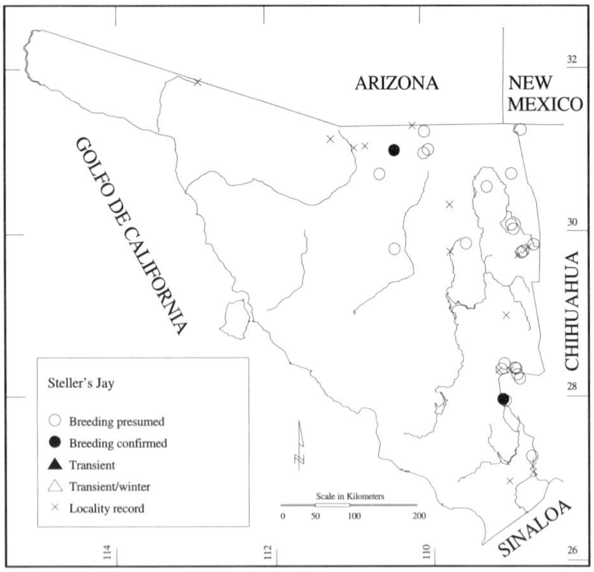

most often in pines, they also consume acorns, seeds, berries, nuts, beetles, and other adult and larval insects (Marshall 1957).

BLACK-THROATED MAGPIE-JAY
Urraca-hermosa Cara Negra
Calocitta colliei

This striking long-tailed jay is a moderately common resident of tropical deciduous forest, typically at elevations from 100 to 900 m. It becomes less common in the northern part of its range, where thornscrub plants become more dominant. It is usually seen in pairs or flocks of less than ten birds. Loud and raucous calling may attract one's attention to these jays, but they often forage silently for some time and can be remarkably difficult to detect. One could spend a day within the territory, which may be greater than 100 ha, of a flock of Black-throated Magpie-Jays and not be aware of their presence.

They consume fruit, seeds, and insects; we once saw one carry a baby bird taken from the nest of another species, presumably to feed to its own young. Nests are made of bulky sticks and are placed in trees (figs and *Bursera* among them) about 8 m off the ground; one near Rosario was made of thorny sticks but the lining lacked thorns. In the extreme southeastern portion of Sonora, nesting takes place in late May, June, and early July. Of 32 clutches of eggs in the WFVZ from the Güirocoba area, 15 had six eggs, 9 had five eggs, 5 had seven eggs, and the remainder had four eggs.

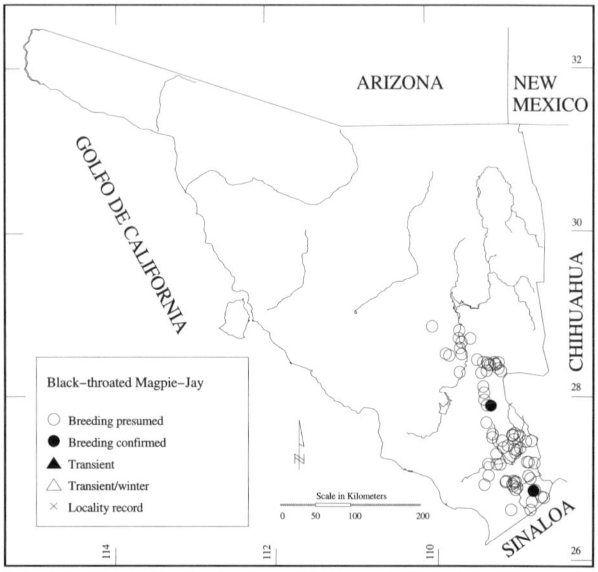

PURPLISH-BACKED JAY
Chara de Beechy
Cyanocorax beecheii

Small flocks of Purplish-backed Jays are resident in tropical deciduous forest north to Quiriego and Los Algodones. They are uncommon everywhere but are especially local and difficult to find north of Alamos, although apparently suitable habitat is widespread. Their range extends upward to an elevation of about 600 m and westward to the edge of the coastal plain (100 m). One individual collected (WFVZ) in thorn forest at Agiabampo was far from its usual habitat.

All nests were found in the period of 4–15 June but in three different years. Nine of 11 clutches (WFVZ) contained nine eggs; 2 clutches had four eggs. Nests are of sticks, twigs, and vine stems. Flocks usually have fewer than six birds and contain a pair of breeding adults. Some of the additional birds may help by providing food for nestlings and by watching for predators. We watched a group near Quiriego shaded by a Montezuma baldcypress in mid-April during an especially harsh dry season drink and bathe in the Río Cedros.

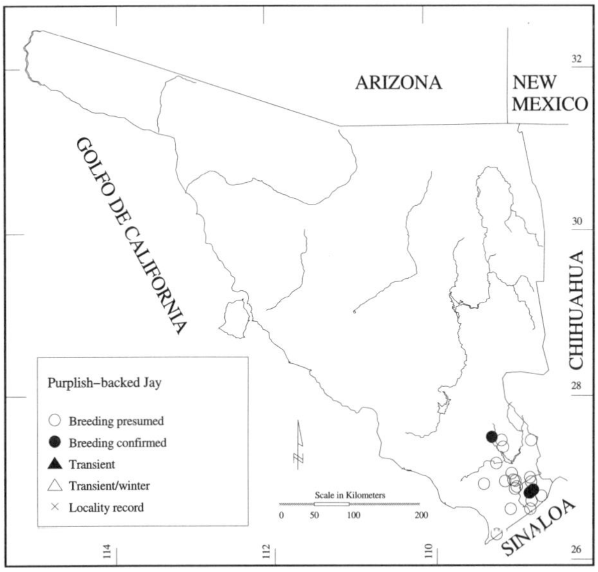

WESTERN SCRUB-JAY
Chara Pecho Rayado
Aphelocoma californica

Western Scrub-Jays, widespread in chaparral elsewhere, are rare and local in Sonora, where there are few suitable areas of low oak thickets. Although the species has been recorded from several mountain ranges in the north and northeast, evidence of nesting is known only

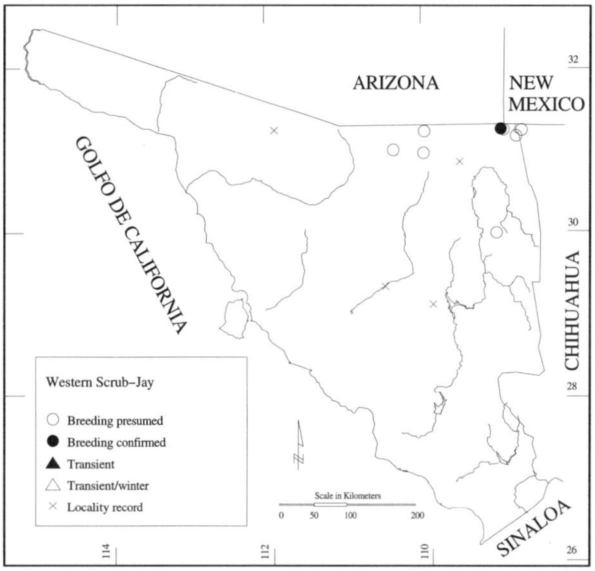

from the hills south of Arroyo Cajón Bonito. There on 14 July 1976, Russell saw a young bird attended by adults remove an acorn from a tree and eat it. J. Bates watched a pair in the dense oaks near Aribabi on 19 July 1986 but could not find a nest. Although we found two pairs in the Sierra el Humo in November 1975, we do not know if they are resident there. Western Scrub-Jays wander from their breeding range in some years and are likely to be found almost anywhere in northeastern Sonora (e.g., Ures; Terrill 1981).

MEXICAN JAY
Chara Pecho Gris
Aphelocoma ultramarina

Mexican Jays are conspicuous and common residents of encinal and pine-oak woodland, from the Sierra el Humo east and south through the length of Sonora. They occur at elevations with oaks from about 900 m up into pine-oak habitat above 2000 m. Where oaks are scarce in conifer forests, Mexican Jays are replaced by Steller's Jays. They are not present in areas of deciduous oaks (e.g., *Quercus chihuahuensis, tuberculata, albocincta*).

In spring, these jays are among the first birds to vocalize at daybreak; they add their calls to the singing of American Robins, Brown-backed Solitaires, and Greater Pewees, as Whip-poor-wills and Whiskered Screech-Owls retire for the day. They breed cooperatively (J. and E. Brown 1985). Young of a previous brood may assist in the rearing of subsequent generations. They live in noisy territorial flocks that may occasionally contain as many as 16 individuals but usually have only six to eight members. Only fully adult birds have solid black bills; young may have some light coloration in the bill for two years. Each flock has at least a pair of breeding birds; some flocks may have as many as three pairs of nesting birds. Nesting begins in early spring when breeding pairs construct a stick nest about 30 cm in diameter and line it with fine grasses, often at a height of 3 to 4 m in an oak tree. April and May are peak months for breeding, and if all goes well four or five young may fledge.

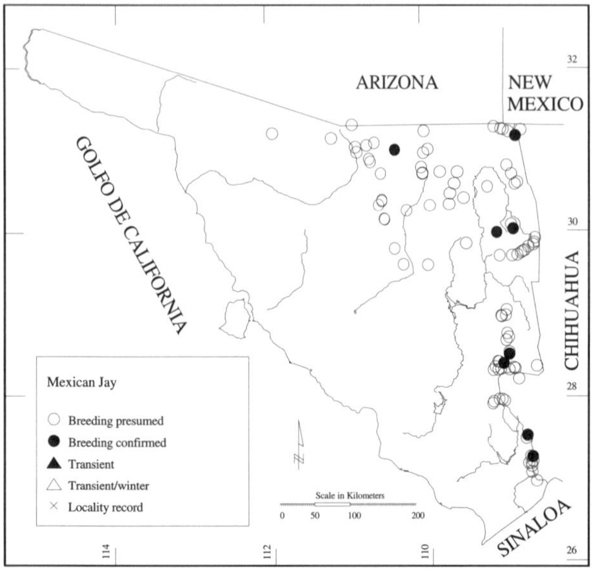

PINYON JAY

Chara Piñonera

Gymnorhinus cyanocephalus

Pinyon Jays, as their name suggests, are normally associated with pinyon pines at more northern latitudes. In Mexico they are known to breed only in Baja California Norte. Large flocks do occasionally leave the breeding areas and wander extensively. J. Marshall saw hundreds moving in a loose group in an area of about 32 ha, in Emory oaks 25 km south of Esqueda, on 19 April 1953. In October 1961, three or four flocks totaling 200 birds were observed in the Sierra Cananea (P. Westcott). Monson saw and heard one on 1 May 1974 on the crest of the Sierra San Luis. We found about 80 individuals foraging in mesquite-grassland dotted with *Haplopappus* about 6 km south of El Sasabe on 11 November 1975.

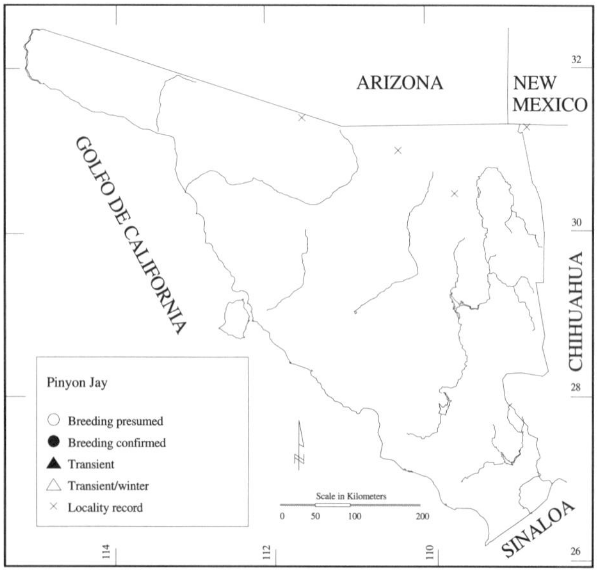

CLARK'S NUTCRACKER

Cascanueces Americano

Nucifraga columbiana

Only two records exist for this bird in Sonora: a skull was found at Bahía la Choya near Puerto Peñasco on 16 December 1972 (A. Rea); and a specimen (YU) was taken by P. Martin on 14 June 1956 at 2100 m in the Sierra el Tigre. Clark's Nutcrackers will occasionally wander from their breeding areas in coniferous forests. The winters of 1955–56 and 1972–73 were "invasion" years, when nutcrackers moved long distances south.

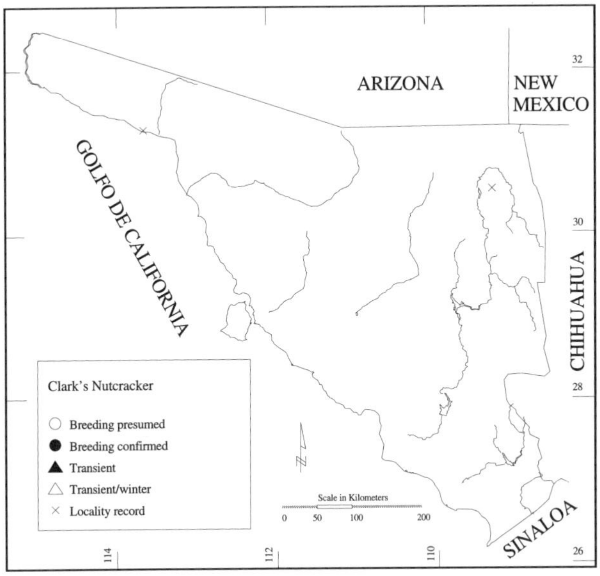

AMERICAN CROW
Cuervo Americano
Corvus brachyrhynchos

The single record cited by van Rossem (1945) continues to be the only one from Sonora: a specimen collected at Monument 204 on the Río Colorado on 14 March 1894. American Crows nest along the Mogollon Rim of central Arizona and in some winters wander to the south and west, yet the specimen from Sonora is from the subspecies that breeds on the west coast of the United States south to northwestern Baja

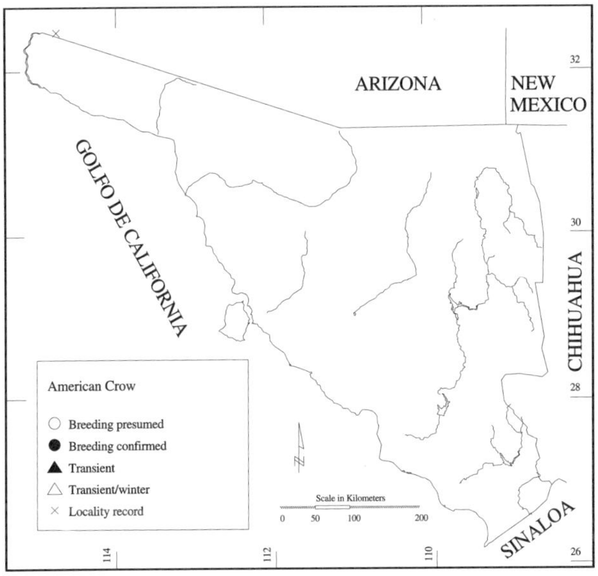

California Norte (A. Rea in Phillips 1986). Rosenberg and colleagues (1991) cite records of large flocks of crows in the lower Colorado River valley in Arizona.

SINALOA CROW
Cuervo Sinaloense
Corvus sinaloae

Sinaloa Crows are common residents north to the agricultural areas of the lower Ríos Mayo and Yaqui. They frequently occur in towns and villages, often in flocks in the nonbreeding season. One bird in September 1985 on the Río Yaqui at Piojillo constitutes the northernmost record. Accumulated records indicate that Sinaloa Crows have increased in numbers in the Cd. Obregon vicinity in recent years, probably attributable to an increase in the extent of agricultural lands as irrigation water has become more widely available.

Sinaloa Crows lay four or five eggs (WFVZ) in bulky nests of sticks from late May to mid-July. Their

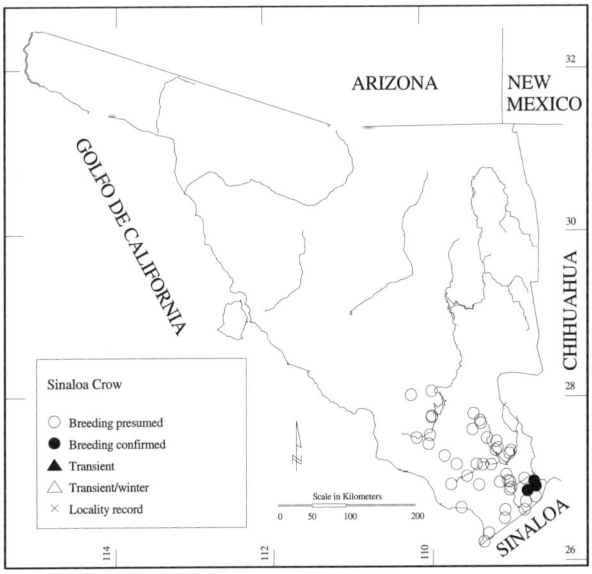

young fledge at the beginning of the rainy season and benefit from the increasing food supply resulting from the rain. Crows eat a wide range of foods, including seeds, insects, fruit, roadkills, and garbage. Figs are choice food items; we saw a flock of 125 feeding upon ripening figs near Cedros in April 1986.

CHIHUAHUAN RAVEN
Cuervo Llanero
Corvus cryptoleucus

Chihuahuan Ravens and Common Ravens are notoriously difficult to separate in the field. Both species occur in Sonora. Chihuahuan Ravens are presumably resident in northern Sonora but may wander extensively. Migrants may come to Sonora from other parts of their range. These ravens are most common in the shrub grasslands from Nogales to the Sierra San Luis. Moderate numbers occur south of Nogales, and we believe that some of the ravens reported between Santa Ana and Hermosillo are this species. The southernmost record of which we feel confident is from the Rosario area, where we watched a flock of 60 Chihuahuan and Common Ravens in March 1982. Despite considerable collecting in the southern third of Sonora in the period of 1930–1960, we know of no specimens taken south of Imuris.

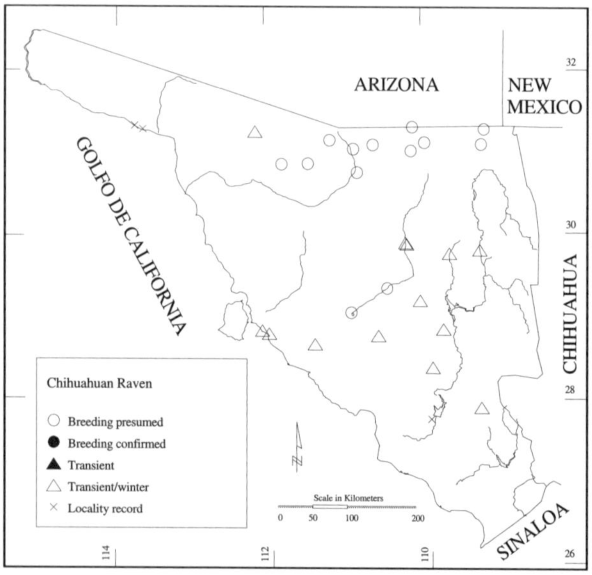

Many nesting ravens have been noted in Sonora, but none has been identified as the Chihuahuan Raven. Of all the more common species pairs of Sonoran birds, the two ravens are perhaps in greatest need of careful field study. We think the Chihuahuan Raven is much more widely distributed than our data indicate.

COMMON RAVEN
Cuervo Común
Corvus corax

The Common Raven is not only a common resident throughout the entire state of Sonora but is also distributed throughout much of the northern hemisphere. Although one is most likely to see single birds or pairs, groups of up to eight or so birds are not unusual. Large flocks of perhaps 10 to 50 or more may occur in the nonbreeding season, especially when abundant food sources are available. Sometimes ravens flock when air currents are favorable for soaring, and then they may swirl and dive upon each other as if at play. Common Ravens are very large birds, weighing about a kilogram; they are the largest of the passerines. Their wide distribution is due in part to their omnivorous diet. They feed upon anything from carrion to grain and fruit. We have seen them carrying small birds and rodents to feed their young. We have not seen them drink water. Harsh guttural notes carry over consider-

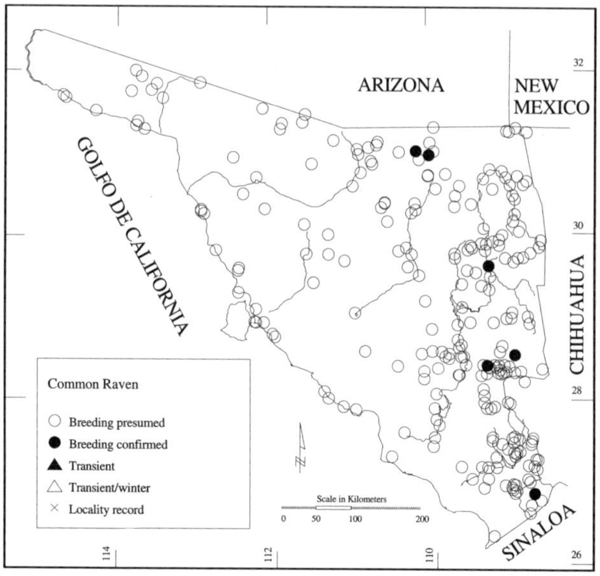

able distances, enabling the birds to remain in contact with mates and group members that share in food searching. We have often seen Common and Chihuahuan Ravens in the same flock.

Common Ravens typically select a rocky ledge as a nest site and only infrequently utilize trees. They may use the tall towers that support transmission lines. Nesting typically begins in April, and the several young may not leave the nest until mid-June.

FAMILY PARIDAE

Titmice

MEXICAN CHICKADEE
Carbonero Mexicano
Parus sclateri

The species has a spotty distribution in Sonora, reflecting the lack of high elevations supporting its preferred habitat of extensive tracts of Douglas-fir and other conifers. We have never found many individuals at a locality, and we consider it an uncommon resident of forests above 2000 m. Mexican Chickadees may move to slightly lower elevations in winter; we have records from Milpillas in Chihuahua and from Tres Rios, where one or two birds were seen in oaks at 1380 m in March. T. Huels watched a family group of five, including three fledglings, in the Sierra Oscura on 10 June 1985, the only direct indication of nesting in Sonora. When not nesting, chickadees often join mixed-species flocks including Bushtits, Bridled Titmice, Hutton's Vireos, and Olive Warblers.

The genus *Parus* has recently been divided (AOU 1997); the scientific name of this species becomes *Poecile sclateri*.

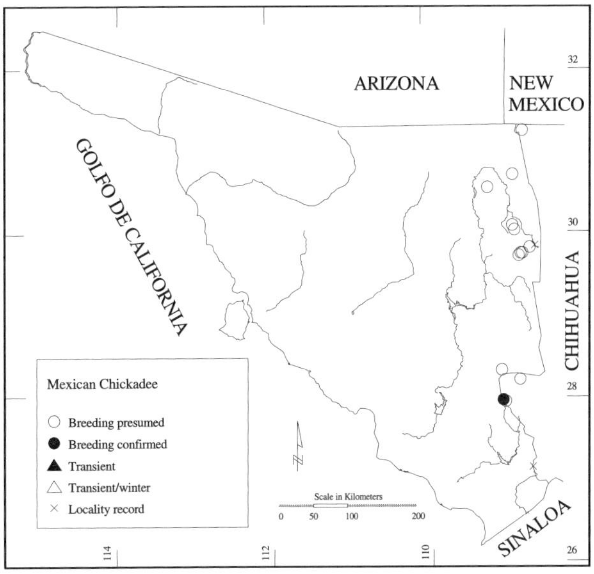

BRIDLED TITMOUSE
Carbonero Embridado
Parus wollweberi

Bridled Titmice are common residents of riparian habitats at elevations above 850 m with large trees (sycamores and cottonwoods are characteristic) and oak and pine-oak woodlands. Some tendency for winter wandering is indicated by three records from localities where we do not think they breed: in March in the north end of the Sierra Alamos and in February and November on the lower Bavispe (Villa Hidalgo and Granados).

Some pairs have been noted in February, when other titmice are still in flocks. Nesting begins in late March and continues into May. Nests of Bridled Titmice are in cavities of trees, often within a few meters of the ground. Young birds, some still begging, are seen into July. Titmice form small flocks except when nesting. In late summer, several family groups may merge into a flock of as many as 16 individuals. Winter groups are usually smaller. The activity and almost constant vocalizations of Bridled Titmice attract other species, thus forming a mixed-species flock of small insectivorous birds. The titmice move about actively, picking insects from leaves and twigs of leafy plants a few meters above the ground.

The scientific name of this species has been changed to *Baeolophus wollweberi* (AOU 1997).

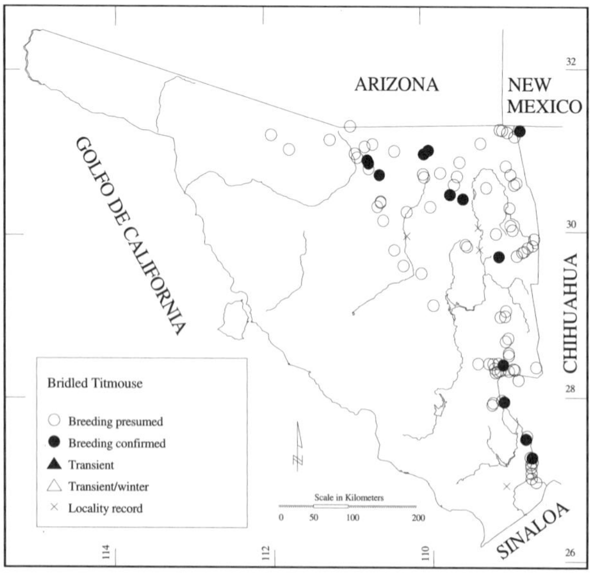

PLAIN TITMOUSE

Carbonero Sencillo

Parus inornatus

The Plain Titmouse is known from mainland Mexico only from two places. These titmice were first found in November 1951 by A. Phillips (Marshall 1957) on the west slope of the Sierra San Luis. Marshall collected two specimens in the same location on 2 September 1952. Russell watched a pair foraging in scrubby oaks on north-facing slopes at 1620 m in Arroyo Cajón Bonito, above the riparian community occupied by Bridled Titmice, on 22 May 1982. The date suggests that the pair may have nested.

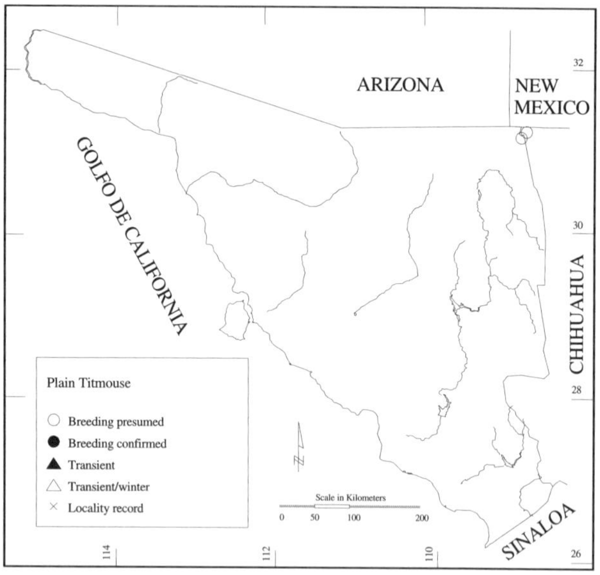

The Plain Titmouse has been divided into two species (AOU 1997). The Juniper Titmouse, *Baeolophus ridgwayi*, is the only one of the two newly recognized species in Sonora.

FAMILY REMIZIDAE

Penduline Tits and Verdins

VERDIN

Baloncillo

Auriparus flaviceps

The tiny Verdin, with a distinctive yellowish head as the scientific name states, is one of the most widely distributed birds in Sonora, resident from sea level up to 1200 m. It prefers open thornscrub and desertscrub habitats with many spiny or thorny shrubs, including cholla cacti, which it uses as nest sites. Although Verdins may occur to the edge of tropical deciduous forest and upward to oaks, they do not occur where spiny shrubs are lacking or where the vegetation is very dense.

Verdins make their presence known by high thin *pink* calls, repeated frequently, that aid members of the pair in locating one another as they forage rapidly for small insects in the leaves and twigs of desert shrubs. In the most arid regions of Sonora, such as the Gran Desierto and Sierra el Pinacate, they are especially noticeable, because few other birds live there. Where mangroves occur along the coast, Verdins nest in nearby shrubs in the desert but forage extensively in the dense foliage of the mangroves. They consume nectar if it is available, often puncturing the bases of long tubular red flowers (such as chuparosa) to gain access to the nectar. We have seen them eat the fruit of *Bursera hindsiana* and *B. laxiflora* in years in which the summer rains failed.

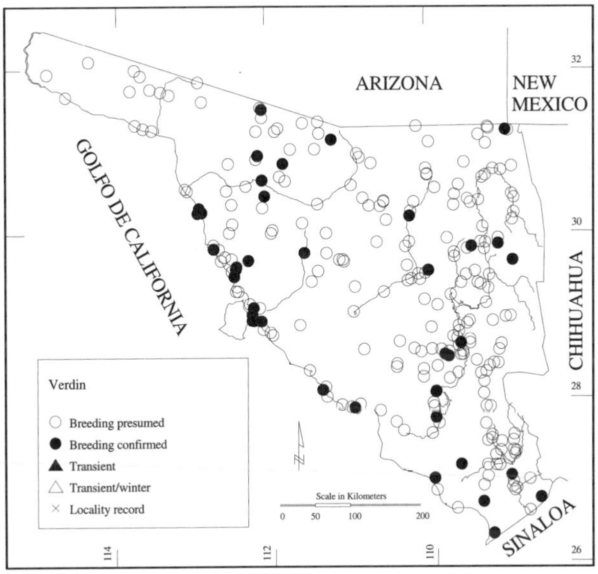

The nests are rounded masses about 18 cm in diameter of small but stiff spiny twigs and are usually placed in a spiny shrub; acacias, palo verdes, and even cholla cacti are used. A. van Rossem found a nest placed in *Salicornia* on Isla Alcatraz at Bahía Kino. The adults enter the nests through an opening directed downward and out from the shrub in which it is placed. Verdins utilize nests for roosting, as well as for rearing young; thus more than one nest is to be expected in a territory. The nesting season is long, beginning in early March and extending into July. The peak nesting time may come early in the season. We noted one very late nest, at which the adults were feeding nestlings near the upper Río Sonora on 14 August.

FAMILY AEGITHALIDAE

Long-tailed Tits and Bushtits

BUSHTIT

Sastrecillo

Psaltriparus minimus

Flocks of as many as 30 of these small resident birds characterize encinal (woodland dominated by oaks) and pine-oak associations, except in the early spring when they breed. Bushtits feed by picking insects from small leaves and twigs, often adopting acrobatic positions while feeding. Occasionally some plant material is consumed. Because of their association with evergreen oaks, they are rarely found below 1000 m or above 2200 m. A. Phillips found them near Arizpe (820 m) on 27 December 1952, indicating that the species does occasionally wander to areas where it does not nest.

Flocks break down into pairs in midwinter, and nest building may begin in early February. The nest is a long (about 42 cm), pendulous structure, suspended from the end of a drooping branch. It includes many fine fibrous materials and serves as a well-insulated chamber for eggs and nestlings. Sometimes the parents

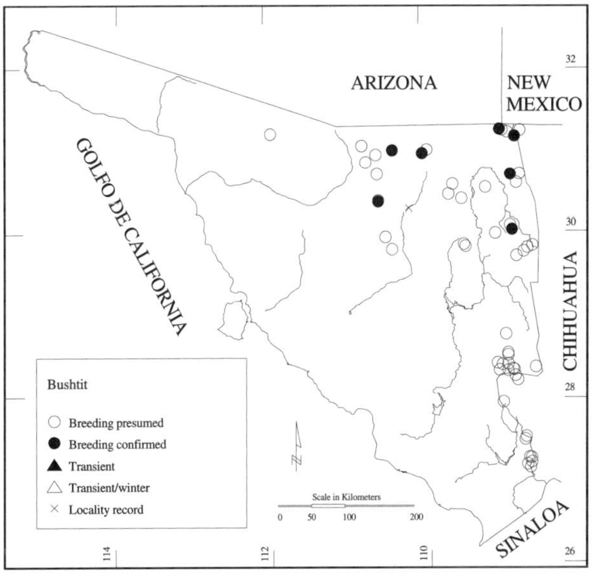

are assisted by one or two nonbreeding adult helpers. After fledging, the family moves as a small flock before joining others. Bushtits sometimes join other small insectivorous birds in flocks moving through oaks and associated trees.

Geographically, there are populations of brown-headed or gray-headed Bushtits. Sonoran Bushtits are all gray-headed. In northern Sonora, immature males have a bold black patch extending from the bill posteriorly through the eye and ear. Adults lack the black patch. From central Sonora southward, males of all ages have the black patches (Phillips 1986).

FAMILY SITTIDAE

Nuthatches

RED-BREASTED NUTHATCH
Sita Canadiense
Sitta canadensis

Four winter records constitute the basis for the inclusion of this species as a Sonoran bird. Red-breasted Nuthatches typically occupy damp and dense boreal forests at northern latitudes. They nest as close to Sonora as the high mountains of southeastern Arizona. Lamb collected one (LSUMZ) near Güirocoba on 2

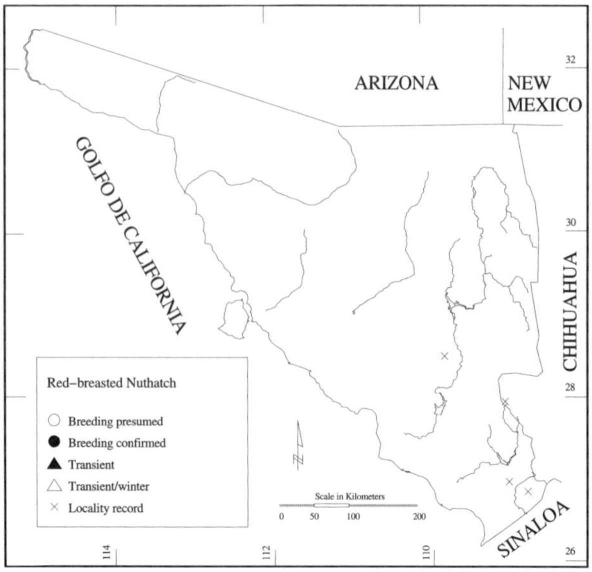

February 1958. S. and R. Russell watched one in a large fig tree at Lo de Campa on 6 January 1983, and several people observed one at El Osa in the Sierra Oscura on 14 March 1982. Two were seen in the Sierra Alamos on 21 December 1989 (D. Stejskal et al.). In some but not all winters, this species is known to wander away from the cool forests in which it nests. The winters involved in these Sonoran records are all ones that had substantial flights of this nuthatch to the lowlands of the southwestern United States.

WHITE-BREASTED NUTHATCH
Sita Pecho Blanco
Sitta carolinensis

White-breasted Nuthatches are present in wooded and forested regions of mountains, most predictably where there are large trees above 1200 m. They are resident in encinal and pine-oak woodland and in coniferous forests and are rather typical of riparian woodlands of canyons. Although in most localities they are fairly common, they are common on Mesa Grande northwest of Yécora. Marshall (1957) found them to be absent or rare in mountains lacking snags. We are aware of lowland winter records at five localities where they do not nest: the Río Yaqui above Esperanza; Huepac; Opodepe; north of Quiriego at Santa Rosa Tesia; and near San Nicolás. Some populations of this species north of Mexico are migratory, so it is surprising that there are not more lowland records in Sonora. In some years there are sizable flights of nuthatches in the southwestern United States.

The foraging motions of a nuthatch are unmistakable. Nuthatches often cling upside down on a tree

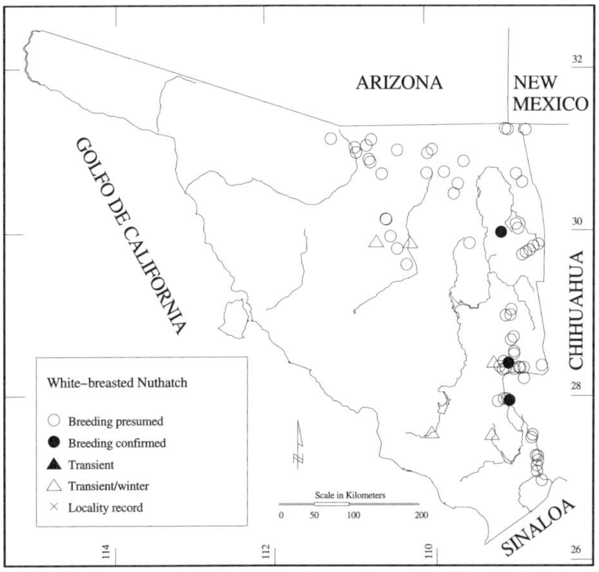

trunk or large limb and work their way downward as they probe crevices in the bark for insects. Occasionally they pry bark loose with their bill in search of insects or pound in the fashion of a woodpecker. At intervals they hesitate long enough to look around them, holding their head and neck away from the tree as they do so. They consume some seeds, including those of pines, and acorn fragments.

White-breasted Nuthatches are often seen in pairs, even in winter. Nests are in natural cavities or crevices in trees and also sometimes in woodpecker holes. Nesting apparently begins in May and continues into July, but we have little data on breeding.

PYGMY NUTHATCH
Sita Enana

Sitta pygmaea

In Sonora, the Pygmy Nuthatch is an uncommon resident in fewer than a dozen mountainous places. This is in sharp contrast to large numbers in the ponderosa pine forests of southern Arizona. The species forages with head downward much like other nuthatches but utilizes pines almost exclusively, preferring to probe twigs and leaves. The pine forests of Mesa Grande near Yécora appear well suited for it, yet it has never been reported there.

We have no evidence that this species nests in Sonora, other than its presence in all seasons. It nests in tree cavities, as do other nuthatch species.

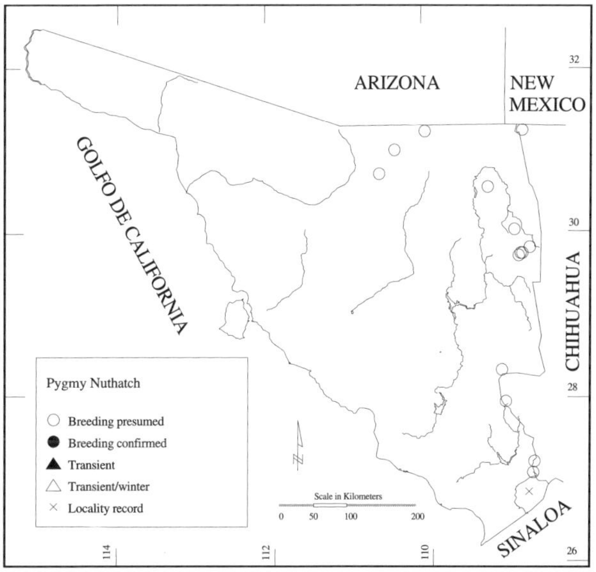

FAMILY CERTHIIDAE

Creepers

BROWN CREEPER
Trepador Americano

Certhia americana

The Brown Creeper is a resident of pine-oak woodland and pine forest and riparian woodland at elevations above 1400 m. These creepers are common in areas with tall pines and oaks, such as the Sierra Oscura and mountains near Yécora. Some populations north of Mexico are migratory, and individuals from those populations may account for the three lowland winter records in Sonora: one near Ures (December 1952, A. Phillips) and on Puerto Peñasco Christmas Bird Counts in 1981 and 1982.

Creepers forage by a methodical upward movement on tree trunks and large branches, keeping their body very close to the bark and using their tail as a brace as they take short hops. As they spiral around a tree, they probe for insects and larvae in the bark with their slender and slightly decurved bill. At some uppermost point the bird drops to the base of another tree and proceeds to work upward again. Creepers are difficult for humans or predators to see, for their color blends well with that of the bark. Creepers are not gregarious birds, and more than two at once are likely to be seen only in family groups. In the nonbreeding season, one or two creepers may join mixed-species flocks of Bridled Titmice, Hutton's Vireos, kinglets, and various warblers.

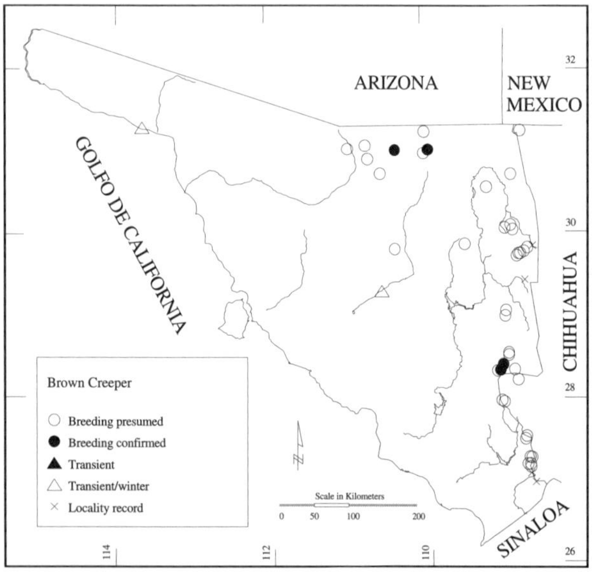

Males begin to sing and pairs form in early March, followed by nesting in late March into May. Creepers typically place their nest under a piece of loose bark, but we found one nest near Yécora in a crevice in the base of a broken-off oak branch that hung 2 m above the ground. Four eggs seem usual. A family unit of six on 20 July is the latest group we have seen in summer.

FAMILY TROGLODYTIDAE

Wrens

SPOTTED WREN

Matraca Serrana

Campylorhynchus gularis

The resident Spotted Wren is dispersed locally through the mountains and foothills of the southeastern portion of the state. Groups (presumed to be family groups) of as many as half a dozen birds live in rocky and scrubby situations, usually on steep slopes from 650 to 2000 m where oaks grow. Along the upper Río Cuchujaqui (560 m) we have noted them on rocky slopes where palms are present. We have found both this species and the Cactus Wren together only at Güirocoba (400 m) and Rancho Jacalón (650 m; northeast of Rosario). In our experience Spotted Wrens live in much denser habitats than Cactus Wrens. Even at Güirocoba Cactus Wrens are found in the more open thornscrub and in or near fields and pastures.

The similarities between Spotted and Cactus Wrens are striking. Although about 20% smaller in body weight, the Spotted Wren shares behavioral and nest building characteristics with the Cactus Wren. Individuals of a Spotted Wren group may forage close to one another, frequently chattering and loudly calling

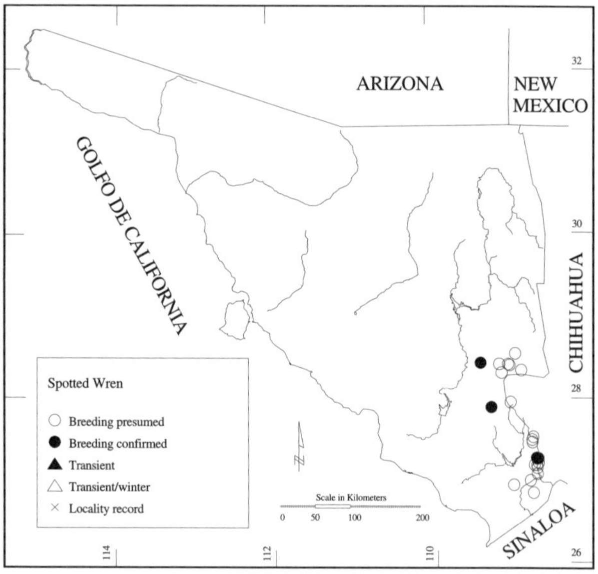

whit-purr. They search for food on the ground, on and under rocks, and in the bark of trees. Most foraging is close to the ground, but we have seen them moving through oaks and Douglas-fir at 12 to 15 m above the ground. We have found nests and fledglings (up to four) with adults in June and July. An active nest in July near Sahuarivo had been placed on the underside of a dead palm frond, but its contents could not be examined. Although their enclosed nests (resembling those of the Cactus Wren) would presumably be difficult for a cowbird to enter and lay an egg, we have seen one adult Spotted Wren feeding a larger-bodied young Brown-headed Cowbird.

CACTUS WREN
Matraca del Desierto
Campylorhynchus brunneicapillus

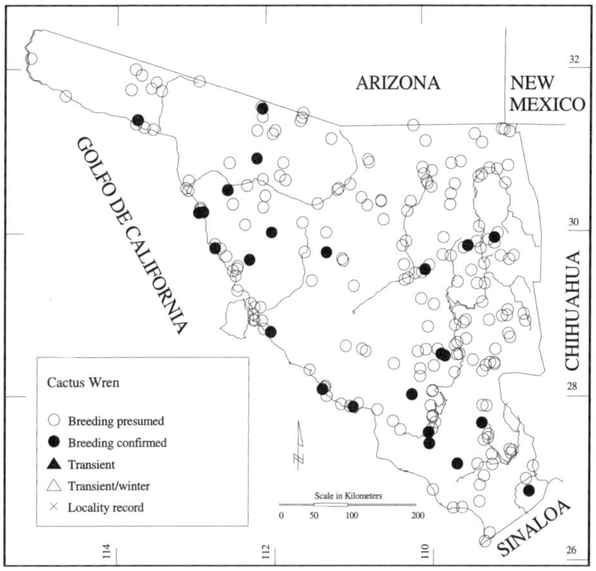

Cactus Wrens are well named, for they typically live where cacti are present, particularly chollas. They are abundant and conspicuous residents of all open scrub communities up to 1400 m. They have been found as high as 1550 m in open pine woodland near El Ocote and in oak woodland in the Sierra el Humo.

Their nests are prominent elongated structures about 40 cm long and 20 cm in diameter, rounded on one end and somewhat pointed with an entrance about 5 cm in diameter at the other. Grass and other plant fibers are used in nest construction, and feathers are used for the inner lining, to provide insulation against the sun or nocturnal cold. Cholla cacti are used for nest sites where they are available. If there are no cacti, the birds usually place their nests in other spiny shrubs. Several nests are built within each territory, but only one is used for the eggs. The others are used by the adults and young for roosting year-round. Three or four eggs are laid, sometimes as early as February or as late in the season as August. Seasonal temperatures and precipitation may determine local nesting dates. Young remain with their parents for many months and like the adults are curious and noisy. Although Cactus Wrens are basically insectivorous, they do consume some fruit. D. Danforth told us of watching an adult feed young frogs (genus *Rana*), 2.5 cm long, to its fledglings near Tarachi. Adults sing *char char char* persistently from an exposed perch during the breeding season. Cactus Wrens and their nests are so noticeable that the species is not likely to be overlooked wherever it occurs.

ROCK WREN
Chivirín Saltarroca
Salpinctes obsoletus

The cheerful melodious song and buzzy calls of the Rock Wren are familiar sounds in Sonora up to about 1600 m, particularly where there are rocky outcrops supporting only a little vegetation, often in the most barren places. This wren searches for insects and spiders in cracks and crevices, frequently pausing to bob deeply. Although most inland populations north of Sonora are predominantly migratory, some breeding birds of Sonora may be resident. Migrants from the north winter throughout the state at low and middle elevations, where they may be found in barren deserts, abandoned fields, and other sites away from rocky areas.

The Rock Wren carries small twigs into a crevice in rocks, or even earthen banks, where it builds its nest upon a platform of small pebbles. A trail of small rocks several centimeters long often leads to the nest cavity. In the rocky coastal hills of northwestern Sonora, Rock Wrens begin to nest in mid-February. At Puerto Peñasco, where we have found nests, and perhaps in some other very hot lowland areas, there are no summer records, and we think these wrens may go elsewhere. In the hills and mountains up to about 1800 m, nesting begins in late March or April and may continue into August. We do not think Rock Wrens nest in extreme southeastern Sonora, even in the mountains. There is a specimen (UK) of a nonbreeding male taken on 12 June 1958 about 14 km southwest of Alamos. Short (1974) did not find the species in the Alamos area in the summer. They do winter in small numbers in the southeastern foothills.

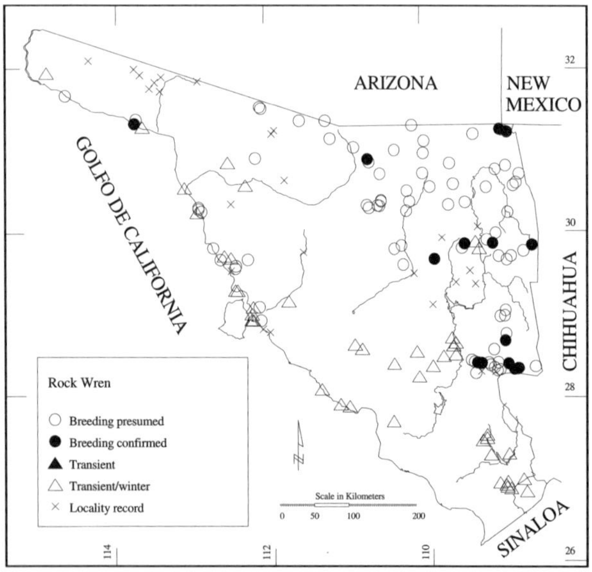

Although this is a familiar bird to all who go into the field in Sonora, remarkably little has been learned about its nesting habits and movements. Breeding in extreme desert areas may not occur every year but may depend upon unusual rains. Conceivably, the individuals that attempt nesting in late winter in the desert may go elsewhere and nest again later in the year, a behavior rarely documented for any bird species.

CANYON WREN
Chivirín Barranqueño

Catherpes mexicanus

Few birds are as well named as this small wren that lives in canyons and other steep rocky places in much of Sonora. It is resident from sea level to 2000 m, spanning a wide range of vegetational types but requiring cliffs or precipitous rocky slopes that offer ledges and crevices in which to feed. Its descending series of deliberate whistled notes, ending abruptly with a buzzy *zee zee*, are among the most appealing sounds of nature from mid-March into early September. The notes are usually sung from the top of a cliff or high exposed rock and carry for several hundred meters. Although characteristic of wild and remote areas, Canyon Wrens live in some towns (Alamos, Arizpe, and Moctezuma

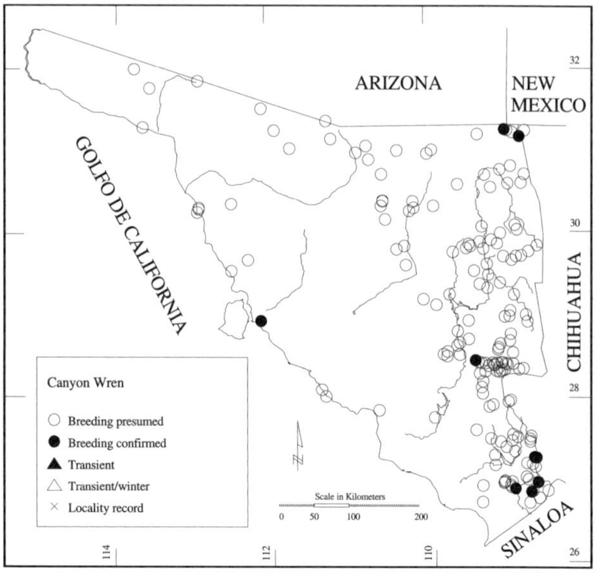

among them), where they call from old buildings. We have found no evidence of migratory movements, either of Sonoran birds or of other populations moving into Sonora seasonally.

Nests are constructed in crevices of rock faces. Occasionally abandoned mud nests of Cliff Swallows are utilized. First clutches of four or five eggs may be laid as early as mid-April, and second nestings are not unusual. At one nest in Arroyo Cajón Bonito, the adults fed young on 16 May and young of a second brood on July 13. Nesting is generally completed by early August. The species, like other wrens, is highly insectivorous. One bird carrying a damselfly in its bill paused to sing before taking it to the nest.

SINALOA WREN
Chivirín Sinaloense
Thryothorus sinaloa

Sinaloa Wrens are resident in thickets in thornscrub and tropical deciduous forest, particularly in areas that have large emergent trees. They are birds of arroyo banks and adjacent lower slopes where undergrowth is prevalent. Most localities of record are between 200 and 1000 m; there are no records from the coastal plain. The species may be extending its range northward. R. Russell noted the Sinaloa Wren's characteristic small retort-shaped nest in Arroyo Santo Domingo (980 m) on 11 September 1986 in an area we had visited several times previously and where we subsequently found a pair. We are not certain the pair had nested, for the nest may have been used only for roosting.

Often common but difficult to see, Sinaloa Wrens sing loudly and emphatically from March into October. The birds forage by searching ground litter, dead leaves and stems, and crevices in bark for insects. Rarely do they emerge from their shaded habitat or ascend higher than 3 or 4 m in the vegetation.

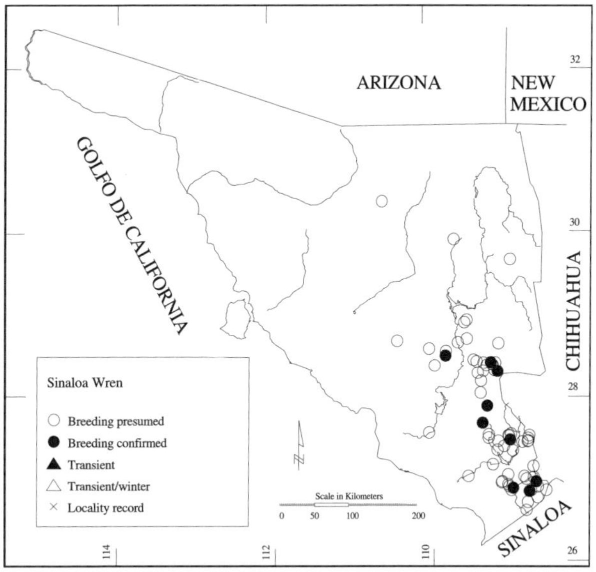

The clutch of four or five eggs is laid in the period from late April to early August. Nests are conspicuous closed structures of grass and other plant fibers, often placed near the tips of branches of leguminous trees within 2 or 3 m of the ground. We failed to find evidence of nesting in May of 1988, a particularly dry year, which would suggest that there may be some dependency upon rainfall.

HAPPY WREN
Chivirín Feliz
Thryothorus felix

This wren is rather local and generally uncommon in dense thickets of southern Sonora. It occurs typically in arroyos where water is present and trees are large. The highest densities we have found are in riparian thickets adjacent to the Río Yaqui below Presa Obregon, where the Sinaloa Wren is absent. The Happy Wren ranges northward along the Río Yaqui drainage, interspersed with Sinaloa Wrens but always in smaller numbers and in denser, more moist situations. It is resident and has been found only at localities below 600 m.

Songs of the Happy Wren—loud, melodious, and often repeated—are the best indication of the bird's presence. It is secretive and forages on insects in the undergrowth, rarely appearing in the open. We have seen fledged young attended by their parents from May into July but have never found an active nest.

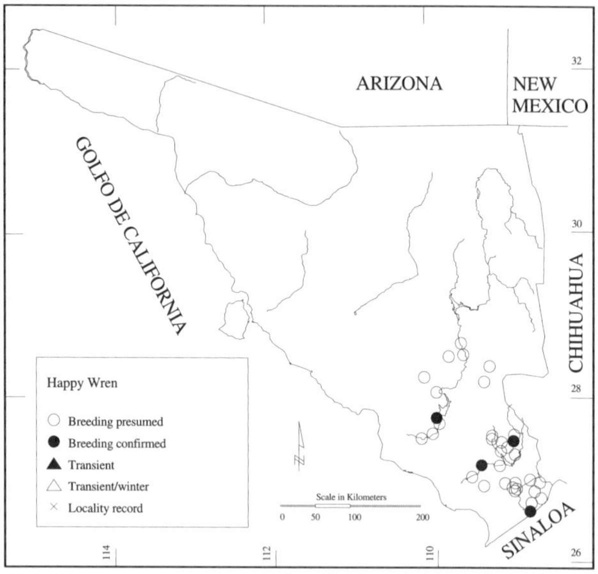

Happy Wren

BEWICK'S WREN
Chivirín Cola Oscura
Thryomanes bewickii

The Bewick's Wren is resident in mountains and foothills from the Sierra Oscura northwest to the El Sasabe area at elevations above 800 m. It is characteristic of oak woodlands (encinal) but nests in some areas without oaks, as in cottonwoods along the upper Río Bavispe. We also found them along the Río Bacoachito (600 m) south of Félix Gómez on 4 August 1986, when young birds were being fed by their parents. The area is considerably west of any other breeding locations, has permanent water, and is characterized by mesquite and willow. Phillips (1986) implied that these wrens have nested at Granados (520 m); we have no evidence of breeding (although we have seen them there in early spring).

Bewick's Wrens are uncommon migrants and winter visitants, recognizable as such in areas below 800 m where they do not breed. There are records from only three coastal localities: Punta Chueca, Puerto Peñasco, and Golfo de Santa Clara. It has been found twice in the vicinity of Alamos (28 December 1982, D. Stejskal; 3 April 1987, R. Bowers).

Bewick's Wrens have loud and melodious songs that vary considerably from one geographic area to another. If present, these wrens are not easily overlooked during the breeding season (March into August), when they sing frequently. They prey upon insects (Marshall 1957). Nests are placed in cavities or crevices within a few meters of the ground.

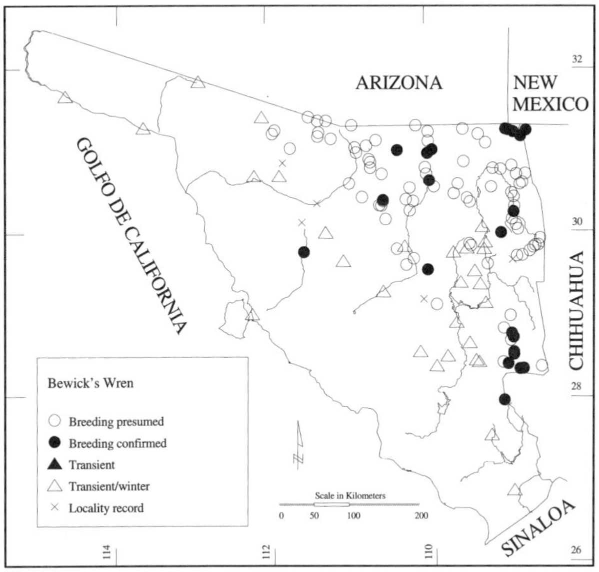

HOUSE WREN

Chivirín Saltapared

Troglodytes aedon

The House Wren nests locally above 1600 m in eastern Sonora. Contrary to its common name, it does not choose to live near dwellings in Sonora. Its most typical habitat is coniferous forest, where it frequents dense undergrowth and mesic arroyos, sometimes as low as the upper edge of pine-oak woodland. But it does not occur everywhere the habitat seems suitable. It does not nest on Mesa Grande northwest of Yécora, for example, yet is common to the southwest on Mesa de Enmedio and in the Sierra Oscura. Marshall (1957) found it in only some of the mountain ranges that he visited in northeastern Sonora.

There also may be a disjunct lowland population. At two places in tropical deciduous forest near Tepoca at elevations just under 800 m, we have found House Wrens under circumstances suggesting nesting. On 13 March 1984 at a site (720 m) at the lower edge of oaks 3 km north of San Nicolás, several were singing and two were paired. On 20 September 1982, we saw a family group including still-begging young 7 km west (780 m) of Tepoca.

A subspecies of the House Wren was named from specimens collected by J. C. Cahoon in the Sierra la Madera in 1877. No one has found the species in that range subsequently. Perhaps the specimens were taken not at higher elevations, but from lower sites where no one has recently encountered them. The Tepoca-area birds may be part of this population.

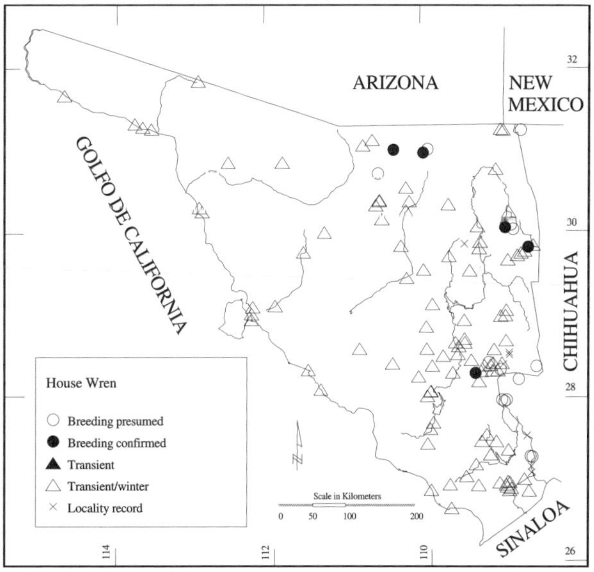

House Wrens are common transients and winter visitants in thickets throughout Sonora. The first individuals arrive from the north in mid-September, and the species is fairly common into November. Fewer birds are found in the winter months, suggesting that many House Wrens migrate to points south of Sonora.

Two individuals banded at Lo de Campa on 8 December 1983 were still present there on the following 9 March, indicating that wintering birds may be sedentary. Numbers of birds pass northward in March, and a few linger into April. Although the wrens have been seen in such open places as Puerto Peñasco and Golfo de Santa Clara, most occur inland where denser vegetation offers cover and foraging opportunities. These wrens call occasionally and may even sing during migration. They eat small insects they glean from rough plant and rock surfaces in well-shaded places.

WINTER WREN

Chivirín Chochín

Troglodytes troglodytes

There are two recent observations of the Winter Wren made by competent observers under optimal conditions. J. Bock and J. Bates studied one at Cerro Prieto, just north of Puerto Lobos, on 4 January 1988. D. Danforth watched one in Cañon Evans in the Sierra los Ajos on 16 November 1991.

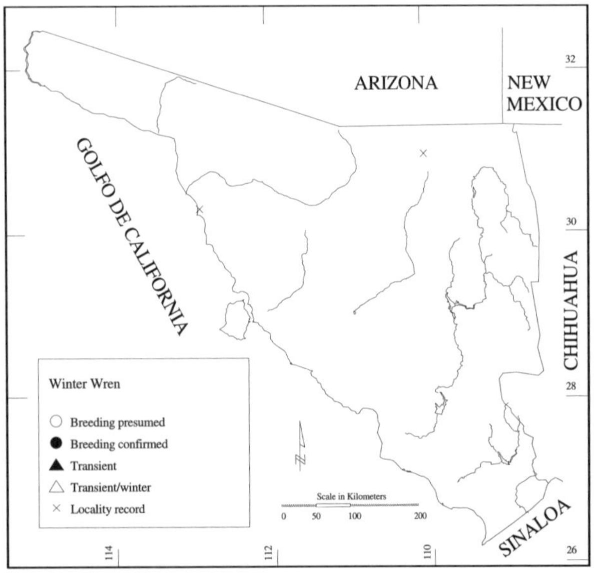

[SEDGE WREN

Chivirín Sabanero

Cistothorus platensis

We accept a single well-documented sight record as the basis for inclusion of the Sedge Wren. P. Pyle and S. Howell made careful observations of a single bird in a small marsh in the Río Colorado delta about 16 km northwest of Golfo de Santa Clara on 4 December 1983.]

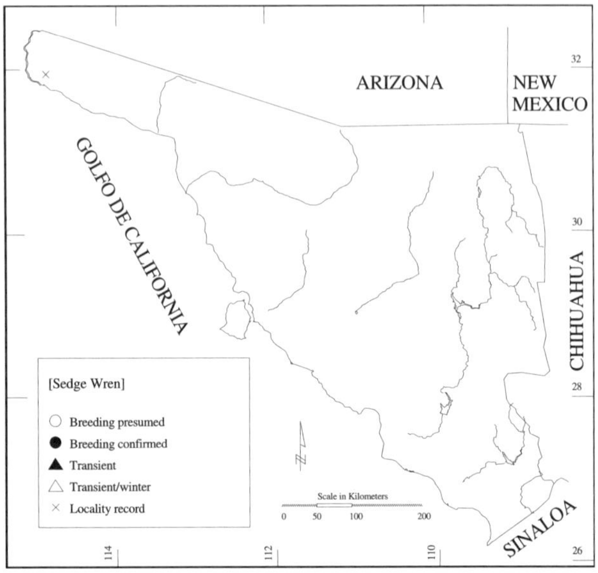

MARSH WREN

Chivirín Pantanero

Cistothorus palustris

The Marsh Wren is found in a variety of wet and marshy areas throughout the state during migration and in winter; extreme dates are 19 September and 15 April. It is uncommon, and usually no more than one is found at a location. The lack of records is attributable to its reclusiveness and choice of cattails or other dense marsh vegetation. We have seen it only once in mangroves (at Punta Chueca). Where marshes are unavailable during migration, it may occur in tall grass and

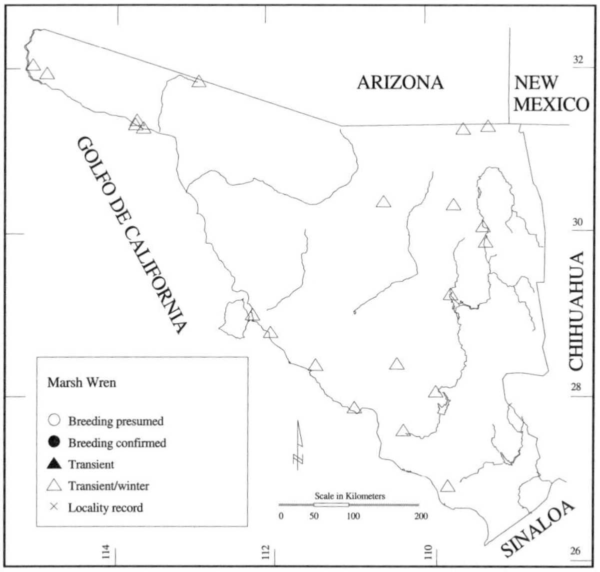

weeds and in *Senecio* thickets but always over water. Occasionally it does search for food in the open. Once we watched a bird at the headwaters of Presa Novillo searching for insects in a dead flooded palo verde while a Vermilion Flycatcher nesting in the same tree tried unsuccessfully to drive it away.

Van Rossem (1945) presumed the species to be resident in marshes along the Río Colorado, where E. Mearns and F. Holzner took three specimens in late March 1894. These dates are well within the migration period, and the birds could have been individuals from farther north. It is possible that this wren nests in Ciénega de Santa Clara.

[FAMILY CINCLIDAE

Dippers

AMERICAN DIPPER

Mirlo-acuático Norteamericano

Cinclus mexicanus

We have only one record of the American Dipper in Sonora. On 12 July 1986 David Barker of Brownsville, Texas, and a companion were climbing up the Cañon las Chimeneas on the west slope of the northern part of the Sierra San Luis. They found a lone dipper at about 1500 m where there was a stream with many small waterfalls. It was going from stone to stone and would walk under the surface of the water as they watched it for fully five minutes.]

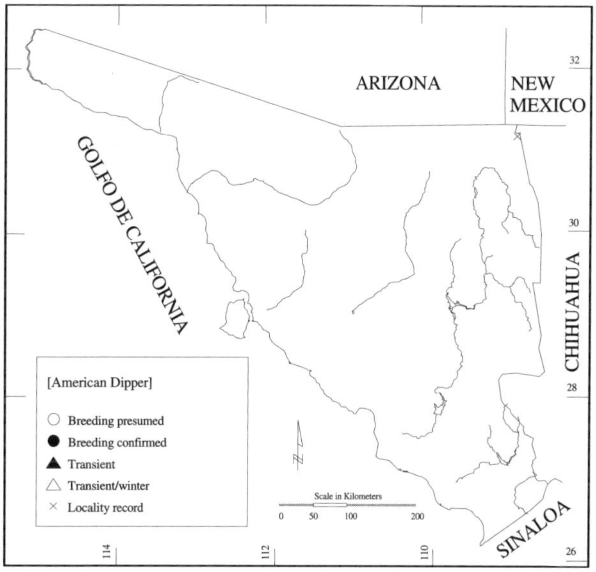

FAMILY MUSCICAPIDAE

Muscicapids

GOLDEN-CROWNED KINGLET

Reyezuelo de Oro

Regulus satrapa

Although the diminutive Golden-crowned Kinglet breeds from the arctic southward in coniferous forests to Guatemala, only seven individuals have been found in Sonora, which lacks the extensive high fir forests that would provide suitable breeding habitat. The records are as follows: one in the Sierra el Pinito (1500 m), 27 November 1954 (J. Marshall); one banded near Imuris

(820 m), 5 February 1964 (A. Collister; Monson and Phillips 1981); two in cottonwoods on the Río Magdalena near San Ignacio, 22 January 1977 (K. Kaufman et al.); one in Montezuma baldcypress on the Río Cuchujaqui (420 m) east of Alamos, 1 November 1981 (D. Stejskal and N. Krug); one near (1600 m) Mesa Tres Ríos, 21 March 1985 (J. Bates); and one in the middle section of Arroyo Cajón Bonito (1280 m), 29 January 1987 (A. Moorhouse, R. Taylor).

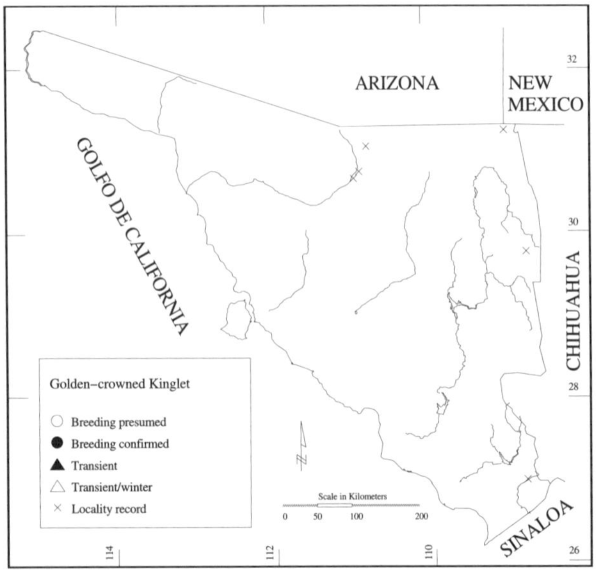

RUBY-CROWNED KINGLET
Reyezuelo de Rojo
Regulus calendula

This kinglet breeds north of Mexico but migrates though Sonora in large numbers and winters commonly. The first fall birds arrive in early September. During the peak of fall migration in November, dozens may be seen in dense vegetation. Spring migration is concentrated into the last half of March, and most depart by 10 April, although a few birds linger into mid-May (once until 8 June). Ruby-crowned Kinglets

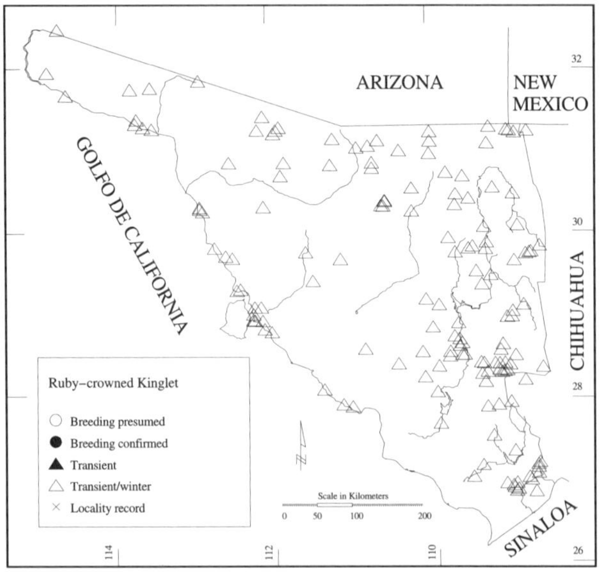

occur in all habitats in Sonora, from mangrove thickets to high-elevation coniferous forests. They are most common where trees are dense and large. Kinglets consume small insects that they glean from leaves and twigs, often in a hovering flight. We saw nine in a single flowering ash tree in mid-February, preying upon the myriad of tiny insects. Although we have banded only 13 of these kinglets in Sonora, one male banded near Puerto Lobos in January of 1985 returned the following winter. Ten of the banded birds were females, but they were banded over several months, and we do not know the actual sex ratio.

BLUE-GRAY GNATCATCHER
Perlita Azulgris
Polioptila caerulea

The insectivorous Blue-gray Gnatcatcher, a small long-tailed slender bird of undergrowth and thickets, is a summer breeding bird of pine-oak woodland in Sonora. In migration and in winter, when its numbers are augmented by birds from farther north, it may be found almost anywhere, including arid coastal communities. Migrants arrive in the lowlands in late August and are widespread in September. Northbound migrants are uncommon, and most move through in March and early April.

These gnatcatchers are frequently found in twos, often identifiable as male and female, during the winter and migration periods, leading to speculation of unconfirmed breeding areas away from oaks. Perhaps pairs are formed on the wintering grounds (although that behavior is rare in small passerines), and the two birds migrate together to the nesting area.

We saw pairs of closely associating birds in spring in dense shrub communities at Lo de Campa and below Presa Obregon. These pairs had singing males, suggesting that pair bonds had been formed. Neither area is typical of known breeding sites in Sonora. However, farther south in Mexico, *Polioptila caerulea* does nest in lowland communities. Females and winter males of *P. caerulea* and *P. nigriceps* are very similar to one another, posing difficult field identification problems. We have discarded many of our field records of these two species because of our uncertainty of the species involved.

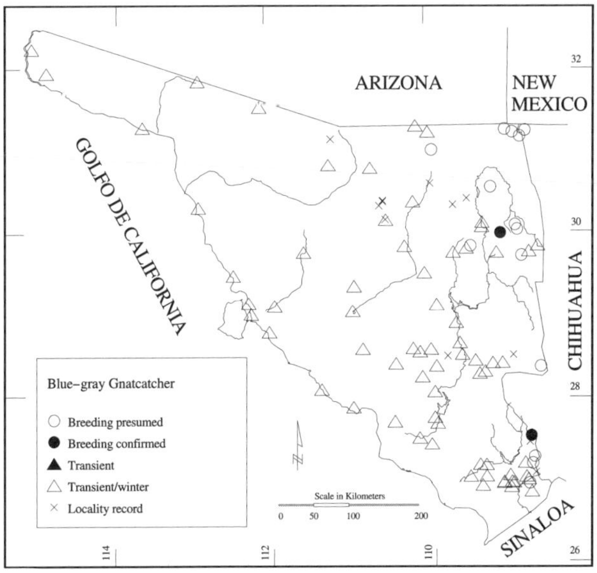

BLACK-TAILED GNATCATCHER

Perlita del Desierto

Polioptila melanura

Black-tailed Gnatcatchers are permanent and characteristic residents of desertscrub, common south along the coast to Huatabampito and inland toward Navojoa and north on the Río Yaqui. In southern Sonora, they are present locally on open dry slopes dominated by tropical thornscrub. Black-tailed Gnatcatchers in the northern half of Sonora range east to the foothills of the Sierra Madre, reaching a maximum elevation of about 1200 m. They are commonly in contact with Black-capped Gnatcatchers. Where both occur, *P. melanura* typically frequents the more open scrubby and often overgrazed slopes, and *P. nigriceps* the taller and more dense thornscrub habitats.

Banding at Puerto Lobos and Lo de Campa indicates that the species is sedentary throughout the year. Pairs of this gnatcatcher are encountered at all seasons, suggesting that they mate permanently. The nesting season begins in early March and extends through August. Nesting may be opportunistic, with breeding postponed in the spring of unusually dry years. When summer rains are persistent, southern birds may still be feeding fledglings in late September. Their nests are tiny cups consisting of innumerable bits of fibers, thin bark, tiny leaves, and spiderwebs. Rather than collect materials anew for a second nesting attempt, adults frequently dismantle a nest and reuse the materials at a new site, sometimes more than 100 m distant. Brown-headed Cowbirds frequently lay their eggs in gnatcatcher nests, typically resulting in the failure of the gnatcatchers to fledge their own young.

This gnatcatcher has the distinction of being the smallest bird, other than hummingbirds, to nest in Sonora. It maintains its 5 g body through rather

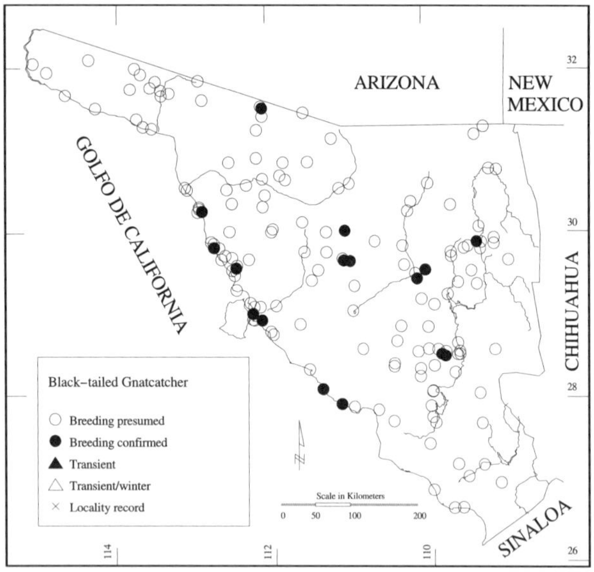

constant foraging for tiny insects, most picked from vegetation but occasionally taken in the air in a short flight. Black-tailed Gnatcatchers call frequently, perhaps using vocalizations as a primary means of avoiding interactions with Black-capped Gnatcatchers (Atwood 1988) where both occur.

BLACK-CAPPED GNATCATCHER
Perlita Sinaloense
Polioptila nigriceps

The Black-capped Gnatcatcher is a characteristic species of dense thornscrub, where it is a common and conspicuous resident. It is less common in tropical deciduous forest. Most records are from elevations below 800 m, but it has been found locally in brushy habitats up to about 1400 m. We have few records from winter months at latitudes north of Hermosillo, and we think the northern populations may be migratory. Where both this species and the Black-tailed Gnatcatcher are resident, as for instance they are in the sparse and open thornscrub north of Navojoa, the Black-capped Gnatcatcher occupies the denser vegetation and thickets along the washes.

As is the case with other Sonoran *Polioptila*, pairs may be seen in every month. Males acquire their distinctive partial black caps in February, and breeding usually commences in March. The last broods usually have fledged by July, but K. Kaufman saw a pair feeding a recently fledged Brown-headed Cowbird on 31 August 1975 along the Río Cuchujaqui near Alamos.

P. nigriceps males, without black caps in winter, are very similar in appearance to *P. caerulea* individuals in winter. Females and young of the two species are also similar. Great care must be used in identifying "white-tailed" *Polioptila* except for distinctive males. There are subtle differences between the two species, particularly in vocalizations (Atwood 1988) and in the appearance of the females (*P. nigriceps* is more brown).

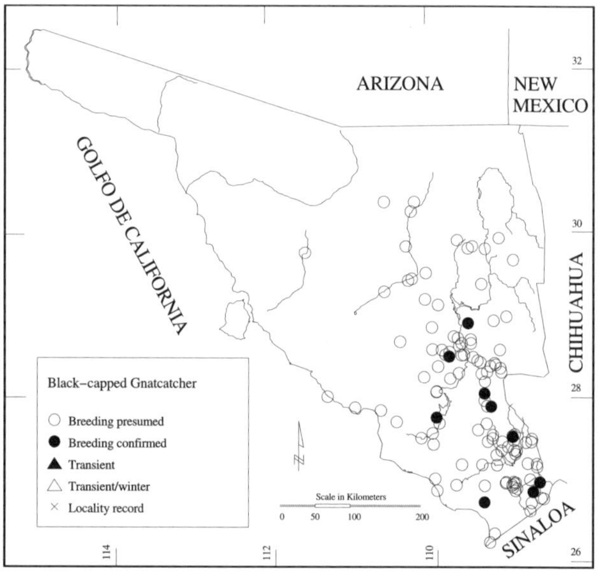

EASTERN BLUEBIRD
Azulejo Garganta Canela
Sialia sialis

The Eastern Bluebird is a common breeding resident of pine-oak woodland of eastern Sonora, where it nests at elevations from as low as 1000 m (in encinal) upward to more than 2000 m (in pine forests).

Bluebirds typically nest in cavities in dead trees, often within 3 m of the ground. Most pairs nest in May, but some occasionally begin in April. Many birds are nesting in July after summer rains have begun. The young retain their distinctive spotted plumage for several weeks after fledging, and family groups of parents with their young may be seen into September. Insects are extensively preyed upon during the warmer months, often being captured on the wing. In winter, bluebirds shift their diet to fruit, and they form flocks and wander from breeding areas but remain in the mountains.

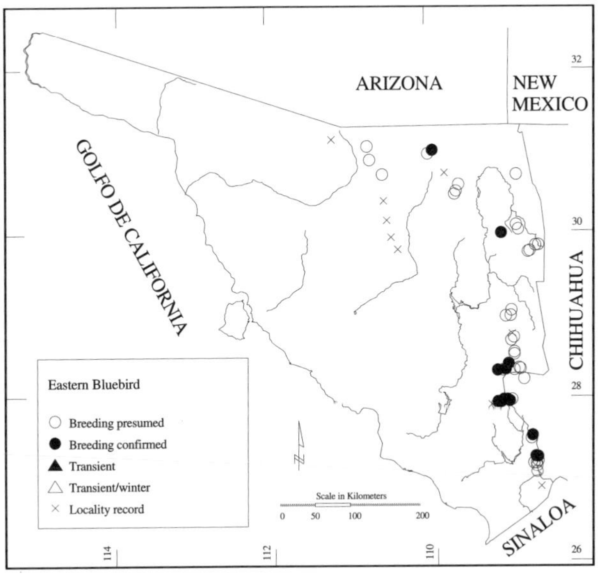

WESTERN BLUEBIRD

Azulejo Garganta Azul

Sialia mexicana

This bluebird species is probably a rare and local summer visitant in open pine and oak grassland of northeastern Sonora. Although no nests have been found, spotted young birds have been noted in the Sierra San Luis (July), in the Sierra Hachita Hueca (early September), and near Mesa Tres Ríos (August). J. Marshall found six adults in Sierra los Ojos Azules in August. About 20 additional records are all from the months November through March, indicating that Western Bluebirds are uncommon but widespread winter visitants. Western Bluebirds breed commonly in the higher mixed-conifer forests in Chihuahua but barely extend west into Sonora. Marshall (1957) commented on the habitat preferences of Eastern and Western Bluebirds.

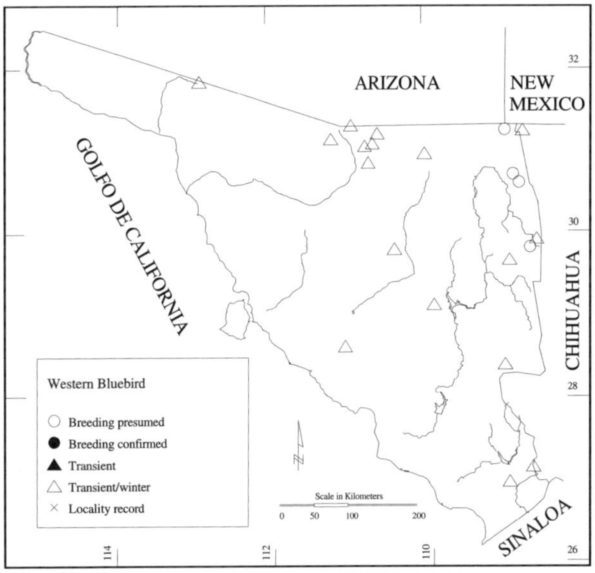

MOUNTAIN BLUEBIRD
Azulejo Pálido
Sialia currucoides

Mountain Bluebirds are irregular winter visitants to Sonora. Dates range from 7 November to 25 February. Most of the 14 records are of only a few individuals, but A. Moorhouse noted 140 in oaks south of Santa Cruz on 31 January 1987. They can be expected almost anywhere and in any habitat in the state. All species of bluebirds tend to wander in the winter, in search of small fruit on which they prefer to feed; they are also adept at flying to the ground to capture insect prey.

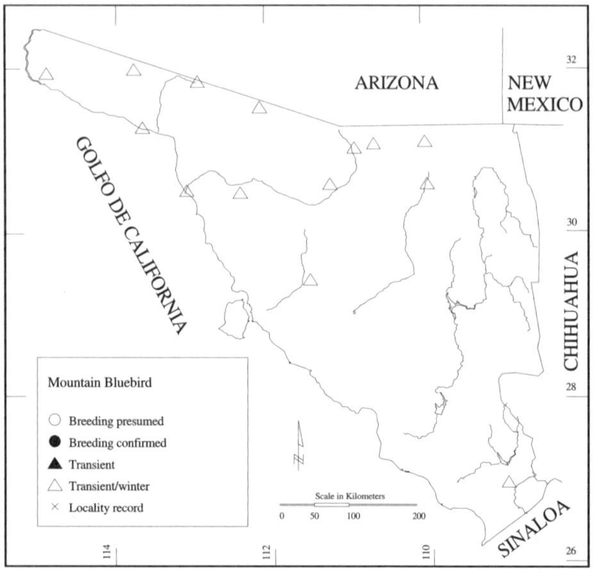

TOWNSEND'S SOLITAIRE
Clarín Norteño
Myadestes townsendi

This solitaire, widely distributed in western North America, has a discontinuous distribution south of central Arizona and New Mexico. It is known to be a nesting species in Sonora only in the Sierra Huachinera. There we found spotted juveniles with adults on 3 September 1983 in pine forest at 1750 m. On 16 May 1988, we observed a persistently singing paired male at 2000 m in pine forest on a mesic northwestern slope of Mesa de Enmedio; the species probably nests there also. We do not know where these populations winter.

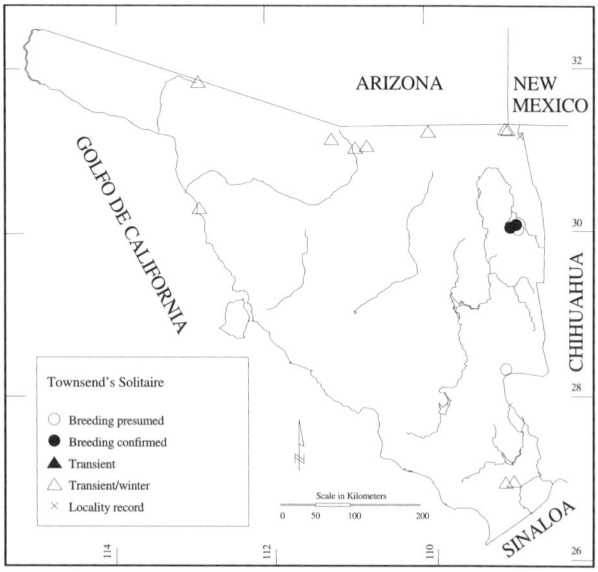

The Townsend's Solitaire is a rare wintering species with about a dozen records at various localities between 1 November and 2 April. One individual was found in coastal desertscrub at Puerto Lobos on 11 November 1986.

BROWN-BACKED SOLITAIRE
Clarín Jilguero
Myadestes occidentalis

This solitaire, nondescript in appearance but endowed with a remarkably beautiful song, is known to breed in Sonora only on the mesic northwestern slopes of Mesa de Enmedio, southwest of Yécora. There we found adults feeding spotted juveniles on 30 July 1988; nearby, another pair copulated. On the basis of singing territo-

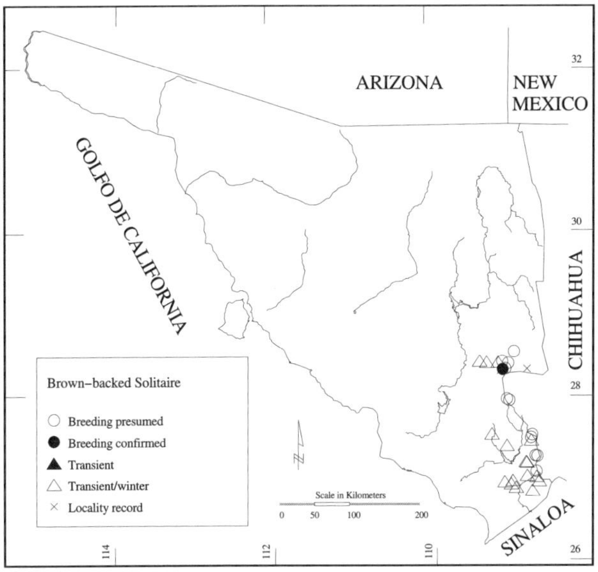

rial or paired birds, some found as early as 14 May, we believe that this species nests in additional localities between 1440 and 2100 m. We have not seen solitaires in breeding areas between August and March; they migrate to the lowland foothills, and perhaps some go south. On 14 May 1988, we found five singing and apparently territorial birds (one paired) in Arroyo el Reparo (920 m). Their behavior suggested nesting. The tall green riparian trees were in striking contrast to the drought-ravaged slopes of that spring. The birds may have been waiting for summer rains before moving to usual breeding grounds at higher elevations. The young birds seen on the slopes of Mesa de Enmedio 2.5 months later were only 4 km away but 600 m higher.

There are many records of Brown-backed Solitaires from moist lowland areas (240–900 m) between 18 August and 15 May. Some of these solitaires sing full songs. Fruit is often eaten; they appear to be fond of *Bumelia* and *Prunus*.

ORANGE-BILLED NIGHTINGALE-THRUSH
Zorzal Pico Naranja

Catharus aurantiirostris

This species was first found in Sonora on 25 November 1965, when a specimen (UA) was collected on the Río Cuchujaqui 10 km southeast of Alamos. Subsequently, about 15 individuals have been seen in the period of 15 August to 15 May in the foothills of southern Sonora, most of them at the Río Cuchujaqui location in December.

Several singing birds (one paired) were found in mid-July 1987 in damp situations in the Barranca las Colas (1320 m) immediately west of Sahuarivo. All were near flowing water and could be approached very closely while they were vocalizing. If the birds nest at this site, the location is only a few kilometers west of known points of record in Chihuahua (Moore 1938). Two birds in mid-May in Arroyo el Reparo were in a box canyon where nesting may have been plausible.

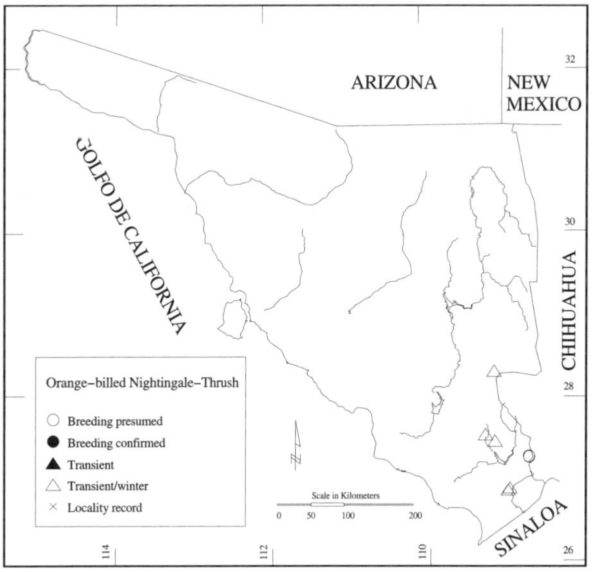

[RUSSET NIGHTINGALE-THRUSH
Zorzal Mexicano
Catharus occidentalis

One Russet Nightingale-Thrush was found with Rufous-capped and Orange-crowned Warblers, Rusty Sparrows, and a House Wren on 27 November 1993 by C. Green and A. Moorhouse. The birds were in shrubby growth of a steep canyon (1600 m) draining the northwest escarpment of Mesa de Enmedio, just west of the divide on the Hermosillo-Yécora highway. Excellent notes provided for this observation eliminate the possible confusion with other species of *Catharus*.]

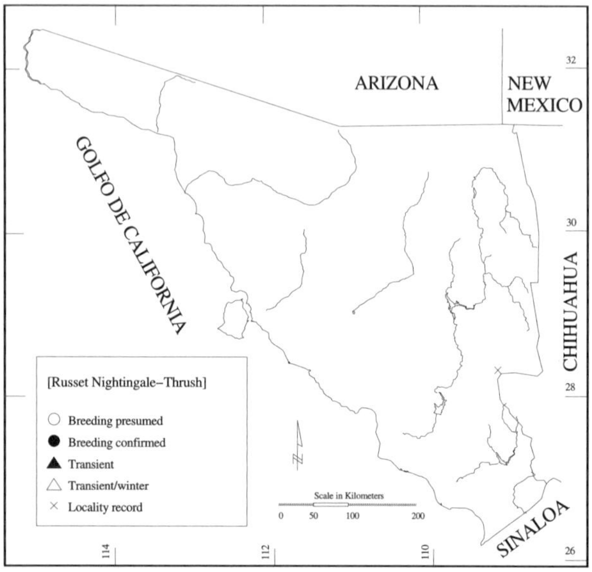

SWAINSON'S THRUSH
Zorzal de Swainson
Catharus ustulatus

Swainson's Thrush is a migrant through Sonora. Silent individuals seem secretive and typically forage on the ground in shady moist areas under dense trees. This thrush has been recorded on fewer than 20 dates in fall (4 September–24 October), on the northern coast and inland to the central mountains at elevations under 1000 m. An intense thunderstorm may have accounted for several individuals netted on the coast on 17 September 1974 (Russell and Lamm 1978). Five were netted on 11 September 1986 in Arroyo Santo Domingo following the passage of a weak frontal system from the north; they were without subcutaneous fat reserves and were associated with an exceptional flush of fall migrants that included hundreds of warblers.

In spring, small numbers have been noted on about 30 occasions between 20 April and 1 June (most near mid-May). Most spring records are from the mountains and foothills of eastern Sonora, as high as 1820 m; they have not been found on the coast in spring. We saw 26 in Arroyo el Reparo on 14 May 1988; most were attracted to a fruiting *Bumelia*. They also eat the fruit of *Bursera laxiflora*.

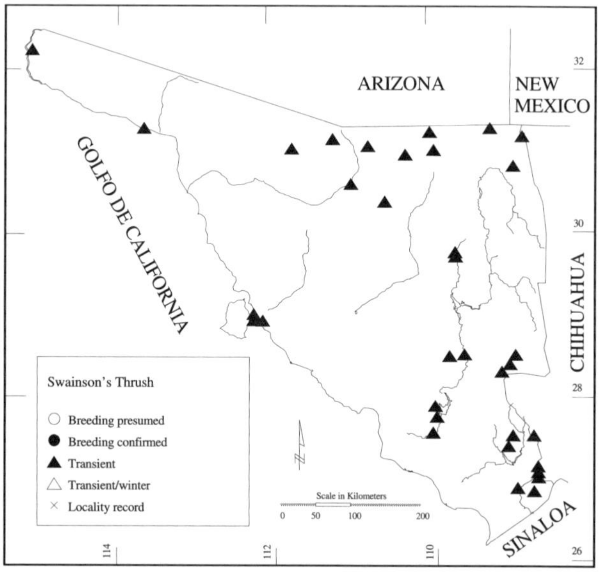

HERMIT THRUSH
Zorzal Cola Rufa

Catharus guttatus

Hermit Thrushes are widely distributed and common as migrants and winter visitants in Sonora. They usually forage singly on the ground for insects in damp places under dense growth, where they call occasionally but do not sing. They are fond of most wild fruits, including hackberries. Although the species nests in the mountains just across the border in Arizona, it has not been found breeding in Mexico, although it may nest in Baja California (Grinnell 1928; Howell and Webb 1992).

During migration and in winter Hermit Thrushes may be found anywhere in the state. Fall migrants have been seen as early as 17 September, but most pass through in November. Northbound migrants are most numerous from mid-March to mid-April, and small numbers continue as late as 21 May. Most wintering birds have been found in the eastern foothills and mountains to 1800 m, but there are a few coastal winter records.

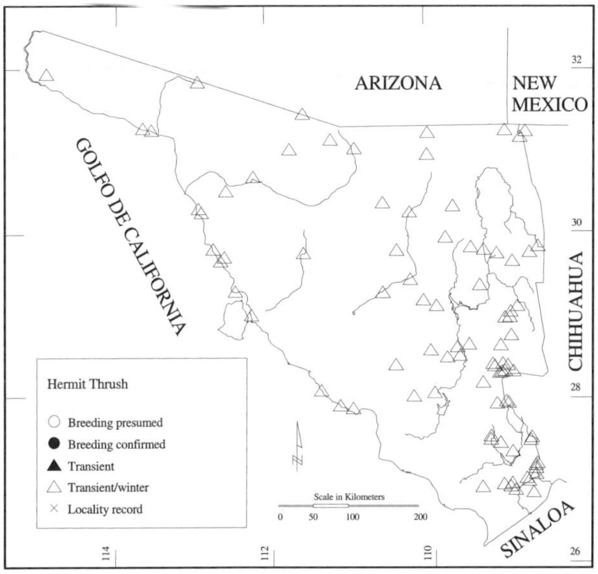

[WOOD THRUSH
Zorzal Maculado

Hylocichla mustelina

There is a catalogue entry in the Delaware Valley Museum of a female Wood Thrush collected by A. J. van Rossem southwest of Sonoita on 25 March 1947. The specimen cannot be located, and this species must be considered hypothetical in Sonora.]

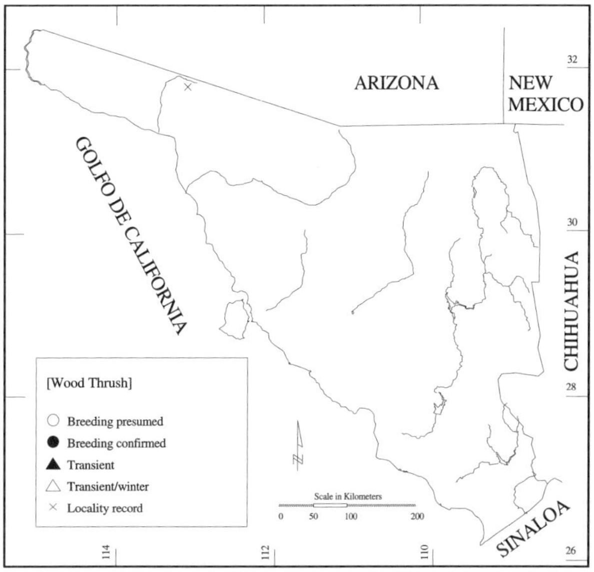

WHITE-THROATED ROBIN
Mirlo Garganta Blanca

Turdus assimilis

Southeastern Sonora is the northernmost portion of the range of this robin. It is uncommon and breeds locally from May into July in deep canyon bottoms and along wooded mountain streams between 900 and 2000 m, where gallery riparian forest shades moist undergrowth. In winter, White-throated Robins have been found only in the Alamos and Güirocoba areas, where they are uncommon from November into May. Twenty birds, an unusually large number, fed upon small figs in a single large tree in Arroyo el Palmarito (670 m) on 5 January 1993.

Several singing birds and at least one copulating pair were in the lush growth along the stream in Arroyo el Reparo in mid-May 1988. Another ten fed upon the ripe fruit of *Bumelia*. W. Sheffler found a nest 3 m up in a small tree on a stream bank on 31 May 1947 near El Mirasol. He noted that it was a typical robin's nest of mud but heavily lined with fibrous rootlets. The bird ultimately incubated a clutch of two eggs (WFVZ). J. Hoffman watched an adult, apparently incubating, in a nest 9 m above the ground in an Arizona cypress on 17 July 1987 in the barranca west of Sahuarivo.

This robin sings rather infrequently in Sonora, forages inconspicuously, and generally remains within dense woodland. Consequently, it is easily overlooked. Many canyons in eastern Sonora are difficult of access and unvisited by students of birds. White-throated Robins may be more widespread than we have found them.

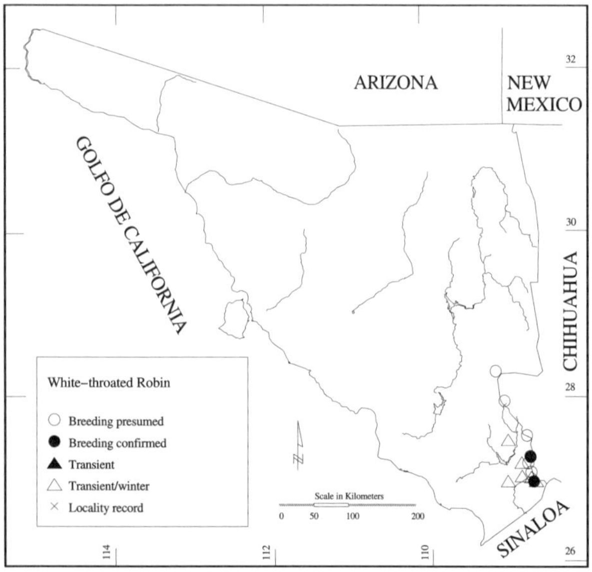

RUFOUS-BACKED ROBIN
Mirlo Dorso Rufo
Turdus rufopalliatus

The Rufous-backed Robin is a moderately common resident species of inland riparian woodland below an elevation of about 500 m. Its range extends northward in the Río Yaqui drainage to the vicinity of Tonichi. We have considered the species resident on the basis of year-round records at breeding localities, but individuals are known to wander widely. There are many records from Arizona and California, most in fall and winter. This robin is regularly reported from Guaymas, where we have no evidence of nesting. As with other robins and thrushes, fruit is a dietary staple. Several birds may be attracted to a fruiting tree in the nonbreeding season.

Short (1974) speculated that the Rufous-backed Robin breeds in the Alamos area mostly in late June to August. We have found that these dates apply to the species throughout Sonora; this robin nests coincidentally with the onset of summer rains. It is not a conspicuous bird at any season. Song is concentrated into a brief period before nesting commences and is not obvious. Rufous-backed Robins are not often seen, in part because they frequent thick vegetation and tall trees.

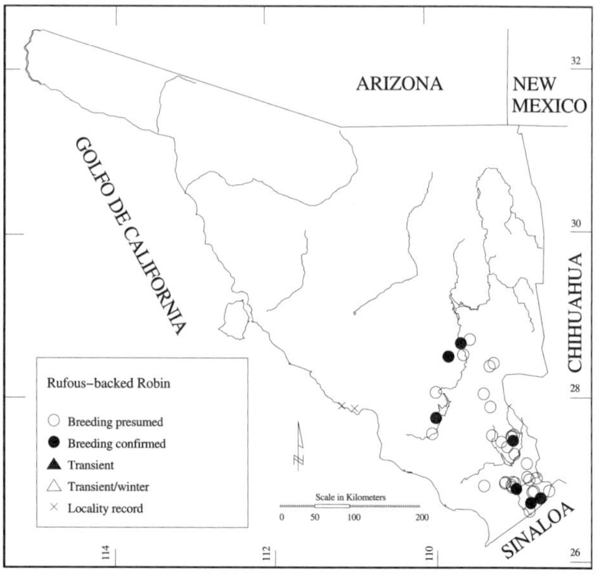

AMERICAN ROBIN

Mirlo Primavera

Turdus migratorius

The American Robin is a common summer resident of the eastern mountains, where its repetitive though melodious singing is characteristic of most pine-oak and coniferous habitats from the end of April well into the summer. It prefers areas with open and rather level ground, where its characteristic run, pause, and peer feeding actions can be effectively utilized. Nesting commences in mid-April and peaks in early June. We have seen nest building as late as 15 July and juveniles with spotted breasts as late as mid-September.

When not nesting, robins tend to wander widely. Sonoran birds are joined by migrants from the north, and it is possible to see robins anywhere in the state. One in January in Caborca fed with a large flock of cowbirds in a cattle feedlot. They are often seen in wet and grassy areas, and small ripe fruits, including mistletoe berries, attract them. There is no pronounced migration period; migrants have been found from late October to mid-April. Their numbers vary greatly between years, and their movements are erratic.

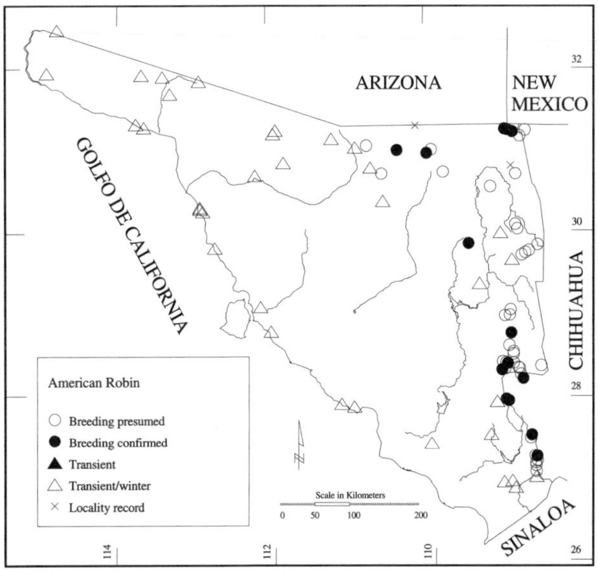

[VARIED THRUSH

Mirlo Pecho Cinchado

Ixoreus naevius

The only observation of this species from mainland Mexico is of a female carefully observed at Bahía Kino on 2 November 1985 by R. Taylor. Varied Thrushes nest from northern Alaska south to northern California and have been recorded from Arizona and northern Baja California in winter.]

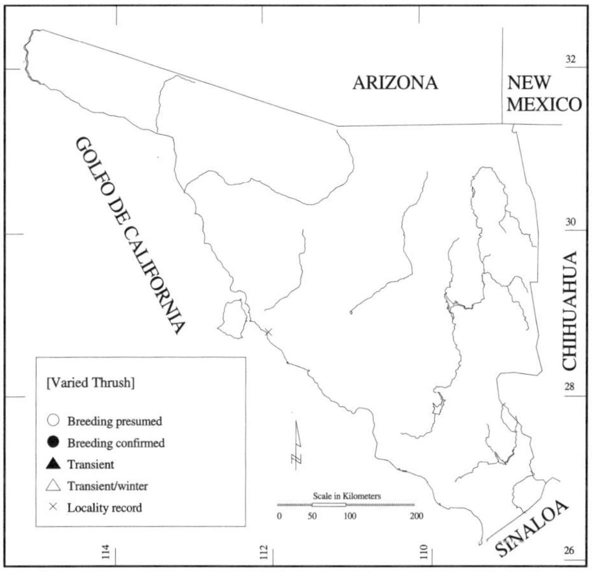

AZTEC THRUSH

Mirlo Pinto

Ridgwayia pinicola

Two observations, both in small canyon bottoms in coniferous forest at 2000 m at El Osa in the Sierra Oscura, are the only records. Three young birds capable of only short flights were accompanied by a female-plumaged adult, and they foraged on steep slopes above a small stream on 17 September 1984 (R. and S. Russell). We presume that this was a family group and that the nest had been somewhere nearby in this mesic small ravine. J. Bock observed a female-plumaged bird briefly on 9 June 1984.

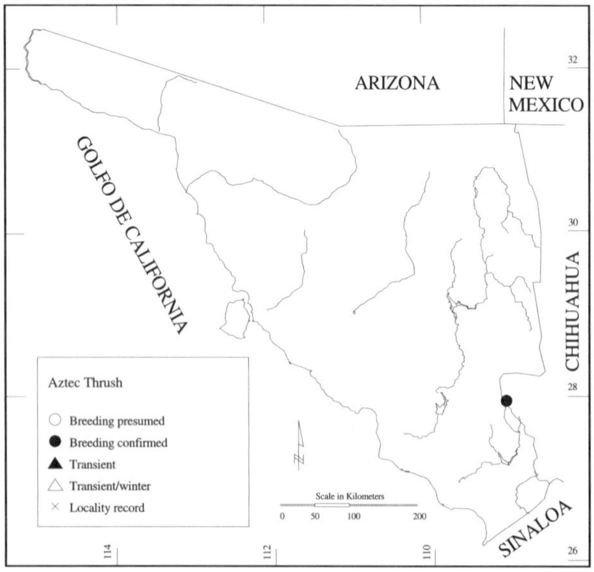

FAMILY MIMIDAE

Mockingbirds, Thrashers, and Allies

[GRAY CATBIRD
Maullador Gris

Dumetella carolinensis

On 28 November 1992, D. Jones and T. Corman found a silent Gray Catbird in a thicket on the bank of the Río Cuchujaqui near its confluence with Arroyo Mentidero. It fed upon the blue berries of a vine growing in a Montezuma baldcypress.]

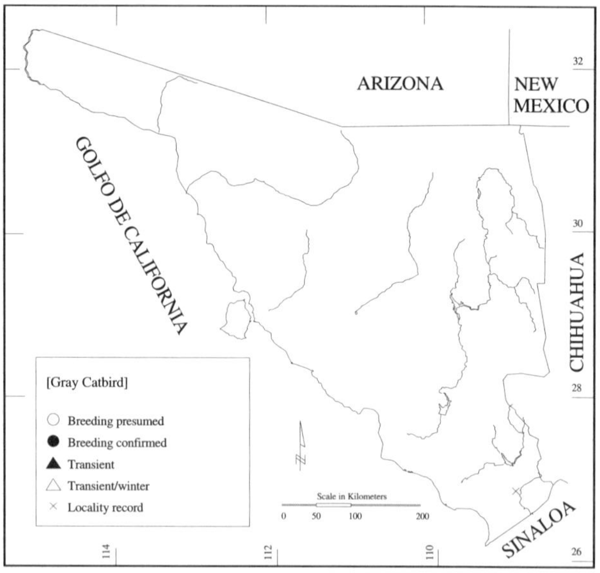

NORTHERN MOCKINGBIRD
Centzontle Norteño

Mimus polyglottos

The Northern Mockingbird is one of the most conspicuous roadside birds in Sonora. With a flash of white in wings and tail, it attracts attention as it abruptly flies from a low perch in almost any open scrubland. It prefers areas with thickets and some grass, where it preys predominantly upon insects and feeds upon small fruits. Perhaps no Sonoran mockingbirds are year-round residents, except possibly in irrigated areas of the Ríos Yaqui and Mayo. In winter, most of the population consists of migrants from the north.

Mockingbirds winter widely in small numbers in coastal regions. In March, small groups of silent mockingbirds migrate in the daytime in loose flocks over the coastal dunes and bluffs. Inland, spring migrants are rather evenly dispersed through all open areas below about 1500 m elevation. In late August and in September, southbound migrants occasionally form small loose flocks in central Sonora.

We have found mockingbirds in more than 200 localities in the course of our studies in Sonora, but we have confirmed nesting in only seven places. Mockingbirds may sing and be territorial and paired, yet not nest. They nest when conditions have produced an abundance of green vegetation and numerous insects. A shortage of rain may terminate nesting attempts and cause the birds to move in search of potentially better nesting areas.

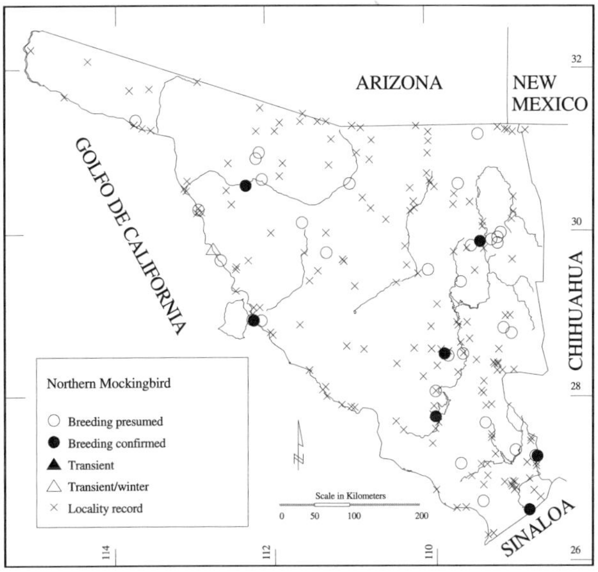

All of our nesting records are for summer months after the onset of the rains, except for one in May in an irrigated green area along the Río Yaqui below Presa Obregon. Under suitable wet conditions, mockingbirds may be expected to nest in open areas almost anywhere in Sonora below 1600 m. In very dry years, no mockingbirds will be found where they have been known to nest previously.

In desertscrub regions the species is normally absent from mid-April until July, and nesting may occur only in those years when above-average amounts of rain have fallen. In northern coastal areas, winter rains are usually insufficient to provide suitable conditions for spring nesting. But in March 1985 mockingbirds were singing, territorial, and paired (but not yet nesting) at Puerto Lobos when the vegetation was very green from heavy winter rains. We think they probably nested.

Mockingbirds are excellent songsters and incorporate phrases of other birds into their repertoire. Singing is at a peak in March and again in July, when the birds are defending a territory and nesting is most likely. They rarely sing in Sonora in the months from September through February.

SAGE THRASHER

Cuitlacoche de Chías

Oreoscoptes montanus

This species, a breeding bird in the Great Basin of western North America, is an uncommon winter visitant in northwestern Sonora but rare south of Puerto Lobos. Most records are from coastal areas, particularly dunes. The Sage Thrasher spends much time on the ground, running rapidly from concealing plant to concealing plant. Although it has been found as early in the fall as 15 September, nearly all of our records are from January and February. In winters that receive especially heavy rainfall, these thrashers may occur in numbers in the Puerto Peñasco district. There, in February and March of the wet 1978–79 winter, we noted about 40 in lush growth north of Cerro Prieto; once we saw nine in a single ocotillo. Some individuals sang extensively. The latest spring date is 5 April.

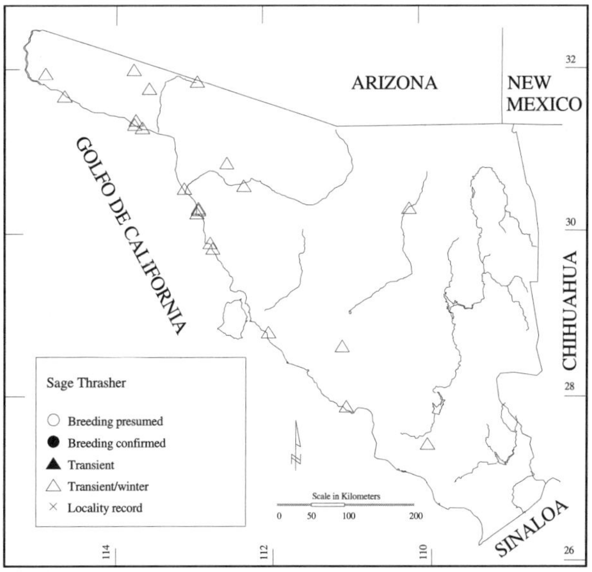

BROWN THRASHER
Cuitlacoche Rojizo
Toxostoma rufum

In winter, some Brown Thrashers occasionally wander from their usual range in the southeastern United States. There are four records from Sonora: one collected at Huasabas on 15 November 1953 (A. Phillips); one seen at Guaymas in December 1968 (Alden 1969); one observed at San Ignacio near Magdalena on 16 and 27 January 1980 (Terrill 1981); and one noted at Puerto Peñasco on 19 December 1981 (T. Gatz, D. Stejskal).

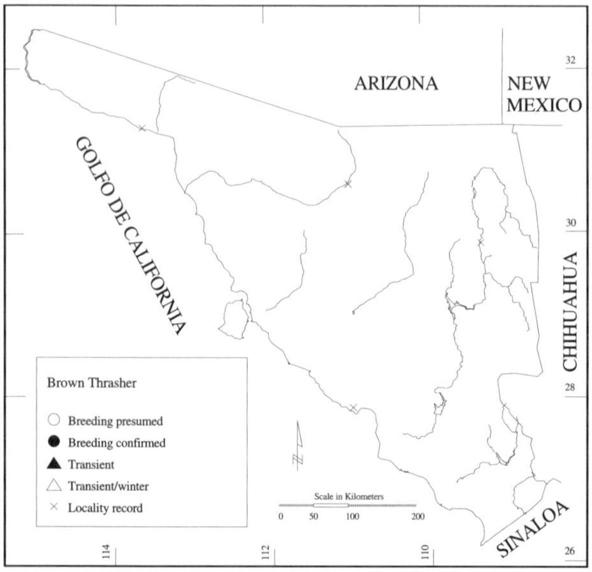

BENDIRE'S THRASHER
Cuitlacoche Pico Corto
Toxostoma bendirei

Bendire's Thrashers are inconspicuous and uncommon residents south to Guaymas. They prefer level bare ground with scattered low bushes, small trees, and some grass. Migrants from the north first appear in early September, and many winter throughout Sonora (but rarely east of the Río Sonora and its tributaries). The northward migration begins in February, and most have departed by the end of March. A bird was collected (WFVZ) at Güirocoba on the very late date of 25 May 1958. We think the bird was a delayed migrant.

We are aware of five nests spanning dates from 28 February to 18 July.

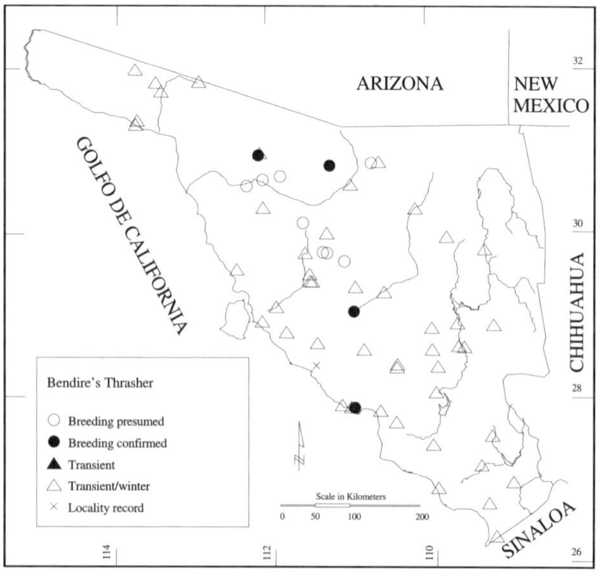

CURVE-BILLED THRASHER
Cuitlacoche Pico Curvo
Toxostoma curvirostre

One of the most widely distributed bird species in Sonora, the Curve-billed Thrasher is resident in open areas in all habitats up to elevations of about 1600 m. It is conspicuous and common in all desertscrub regions except in the extreme northwest. In the denser thornscrub and deciduous forest communities, it selects places where the ground is bare. Even at higher elevations in oaks and conifers, this thrasher may be found in natural openings or man-caused clearings. Villages and towns provide appropriate combinations of foraging and nest sites. The Curve-billed Thrasher searches for food primarily on the ground by using its bill to toss litter aside or probe deeply, but it also takes fruit, particularly the fruit of many species of cacti. Over short distances it runs rapidly, but over longer

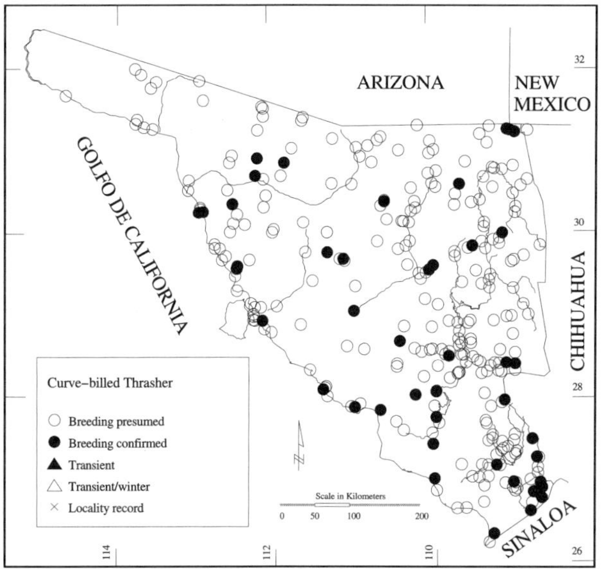

ones it flies. The song is rather attractive, and the thrasher's distinctive loud double-call notes alert one to its presence.

Nests are obvious bulky structures of sticks most frequently placed in *Opuntia* cacti, usually in large round-stemmed chollas but occasionally in dense prickly pears. Trees and shrubs are also used; one nest at Los Aserraderos was 12 m up in a large oak. Inland, most active nests have been found from early April to mid-July; the large number in July probably reflects a second breeding season, as suggested by Short (1974). Coastal populations typically nest in March and April, with some pairs perhaps continuing into May and June; there seems to be almost no nesting coincidental with summer rains. We have summarized breeding over a large geographic area, but local populations are very responsive to current weather conditions. If plants are green and growing in late winter, thrashers begin to nest in March. If conditions remain favorable because periodic rainfall maintains a supply of insects, thrashers may continue their efforts into July. If rains do not occur, most thrashers do not nest, although some make an attempt in late spring that is rarely successful.

The species is generally considered to be nonmigratory. But young birds disperse from the nest area and may move considerable distances. Consequently, thrashers are found (especially in fall and winter) in areas where they are not likely to nest.

Near the Agua Caliente arm of Presa Obregon, C. Green saw a 1.2 m long Mexican Boa (*Boa constrictor*) coiled around a Curve-billed Thrasher.

CRISSAL THRASHER
Cuitlacoche Crisal
Toxostoma crissale

Crissal Thrashers are generally uncommon residents in northern Sonora, with most records from elevations above 600 m. The distribution is not continuous, and they are not found everywhere in their preferred habitat of dense brushy thickets. We have found them to be common only near Arizpe, Benjamin Hill, El Oasis, Trincheras, and Félix Gómez. We have never found the species south of Bahía Kino, but van Rossem (1931) reported collecting a nest and eggs between Empalme and Guaymas. We think birds from northern populations may occur in Sonora and account for points of record in winter, contrary to the AOU Check-list (1983), which indicates that the species is resident throughout its range. These winter localities appear unsuitable for nesting, and we lack observations of Crissal Thrashers in summer from them.

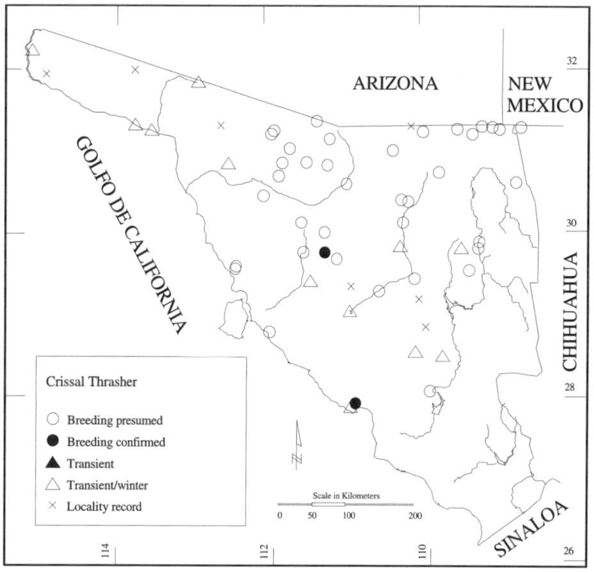

LE CONTE'S THRASHER

Cuitlacoche Pálido

Toxostoma lecontei

The inconspicuous and elusive Le Conte's Thrasher is resident from sandy northwestern Sonora south to Bahía Sargento, a region characterized by annual precipitation of less than 7 cm. It is found in coastal sand dunes and in sparsely vegetated desert areas dominated by *Frankenia*, *Opuntia*, or creosote bush. The most terrestrial of the thrashers, it runs long distances over the sand, or out of sight in small washes, often with the tail up at an angle.

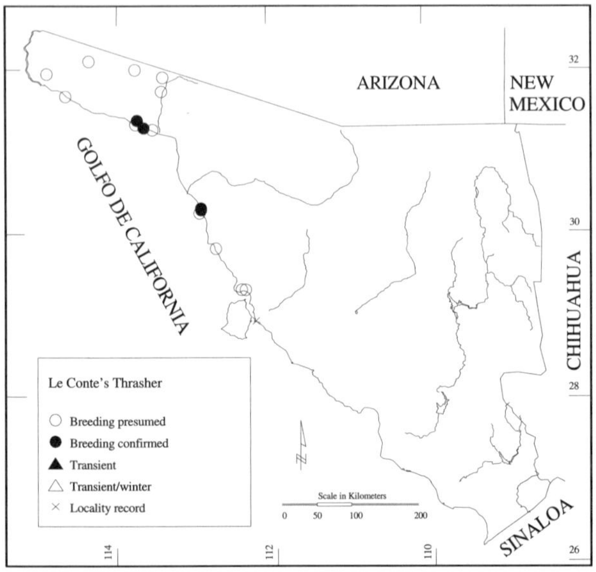

Le Conte's Thrashers nest in the winter months of December into early April; the exact timing depends upon the extent of the winter rainfall. The stick nests are placed low in the densest shrubs available, which may be *Lycium* or *Cercidium*. Song is infrequent, especially outside the breeding season. We once saw three birds involved in building a nest in early April at Puerto Lobos. Although all individuals appeared to be adults, it is probable that one was a juvenile from an earlier nesting.

BLUE MOCKINGBIRD

Mulato Azul

Melanotis caerulescens

The Blue Mockingbird is uncommon but widely distributed during the late spring and summer breeding season between 900 and 1700 m in the southeastern portion of the state. At that time these mockingbirds are somewhat less secretive than in other seasons, and their varied and melodious song issues unexpectedly from more open vegetation. Birds foraging in the ground litter are more readily observed. All of our summer observations have been of birds in riparian situations, some near tumbling mountain streams and others in floodplain thickets. In a kilometer along the stream in Arroyo el Reparo in mid-May, we found 20 birds, most in obvious pairs. We have never encountered such a concentration elsewhere. Our only records of nesting are in June and July. We are uncertain of the status of paired birds found at Sejaqui (400 m) in summer, near San Nicolás in March (720 m), near Santo Tomás (460 m) in May (Clark 1984), and in the Sierra Alamos (440 m) in November. It is possible that Blue Mockingbirds nest at these locations and are resident (except perhaps at Santo Tomás).

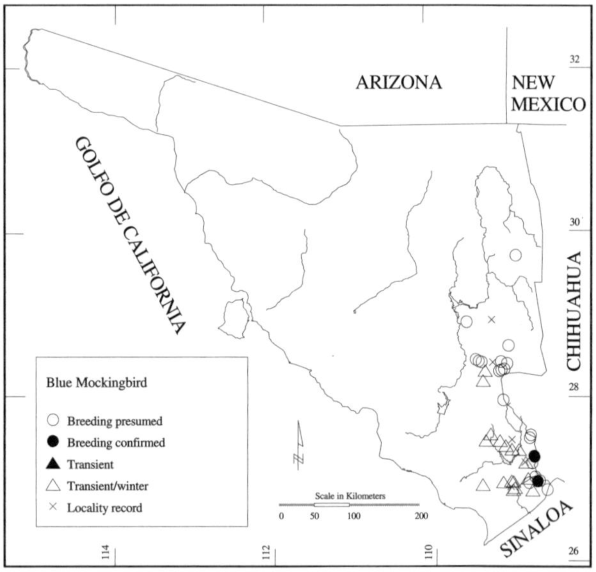

Some or most of the breeding birds may spend the winter at lower elevations, where there are many records ranging from the edge of the coastal plain at Chinobampo (110 m) to about 500 m in the foothills. Birds in the lowlands during the winter are generally quiet and forage in the densest thickets in riparian areas, where they remain almost invisible. We have only two reports of the species from September and October, the time when presumably it moves from breeding to wintering areas. The latest spring date at a low elevation is 13 April.

FAMILY MOTACILLIDAE

Wagtails and Pipits

[WHITE WAGTAIL
Lavandera Blanca
Motacilla alba

A vagrant of this primarily Old World species was observed at length on 30 April 1974 near El Diablo (1240 m) in Arroyo Cajón Bonito by Monson, C. McMoran, and V. Roth (Morlan 1981).]

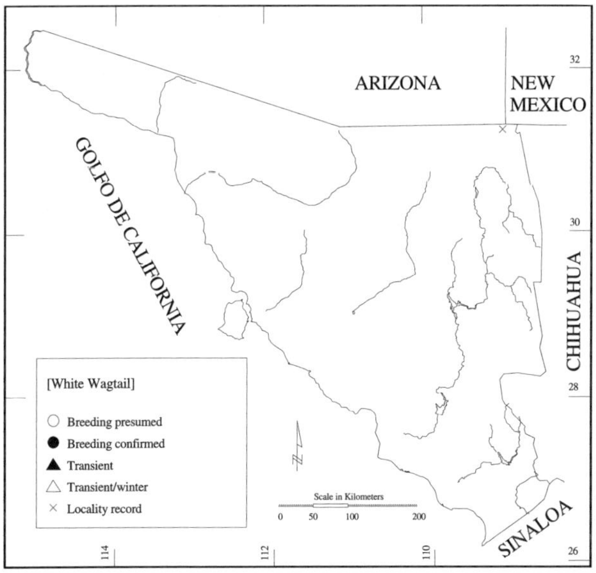

AMERICAN PIPIT
Bisbita de Agua
Anthus rubescens

The American Pipit occurs in Sonora in small numbers from mid-October to late March; some individuals winter, whereas others migrate through the state. Extreme dates are 21 September and 6 June, the latter represented by a specimen (DVM) of the eastern Asiatic race (Phillips 1991). This species is most often found foraging for insects in wet places such as irrigated fields and pastures, river banks, and the edges of ponds and lakes. It may frequent grasslands or coastal sand dunes in the occasional years when there is a profusion of winter annuals. Numbers vary from year to year, and rarely flocks of a hundred birds may be seen.

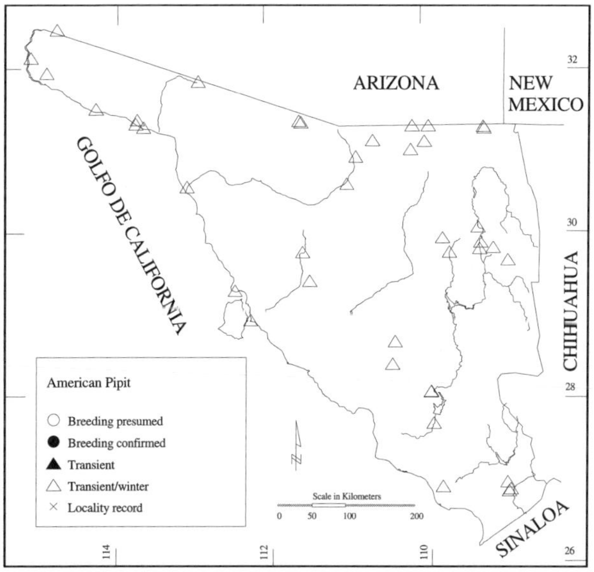

SPRAGUE'S PIPIT
Bisbita Llanera
Anthus spragueii

This pipit is a rare winter visitant in Sonora. Phillips and Amadon (1952) collected three in the plains southwest of El Sasabe on 21–22 October 1948.

L. Goldman found one on the shore of Presa Rodriguez at Hermosillo on 6 April 1961, and R. Tweit observed one on the beach at San Carlos on 28 September 1984. The April and September records, although well documented, do not exclude the possibility of vagrant Asiatic species. Sprague's Pipit is to be expected in the grasslands of the area between Cananea and Agua Prieta, as it has been found often in adjacent areas of Arizona.

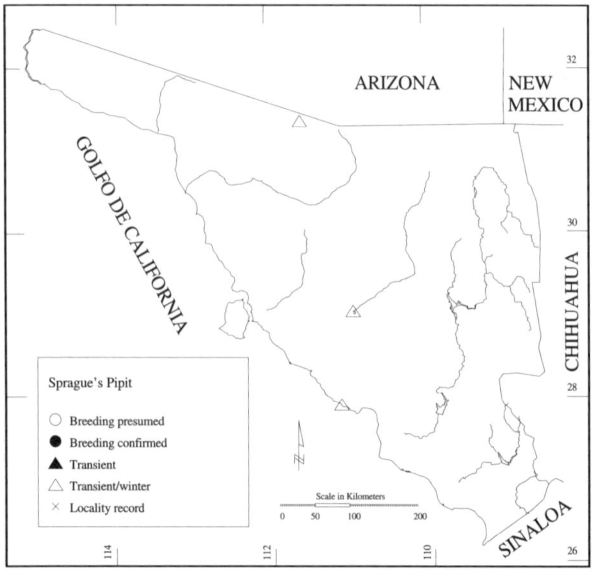

FAMILY BOMBYCILLIDAE

Waxwings

CEDAR WAXWING

Ampelis Chinito

Bombycilla cedrorum

Cedar Waxwings breed across the more northern latitudes of the continent and, like other species that utilize fruit extensively, fluctuate markedly in numbers from year to year in southern wintering areas. In Sonora, in years in which they are present, they are seen

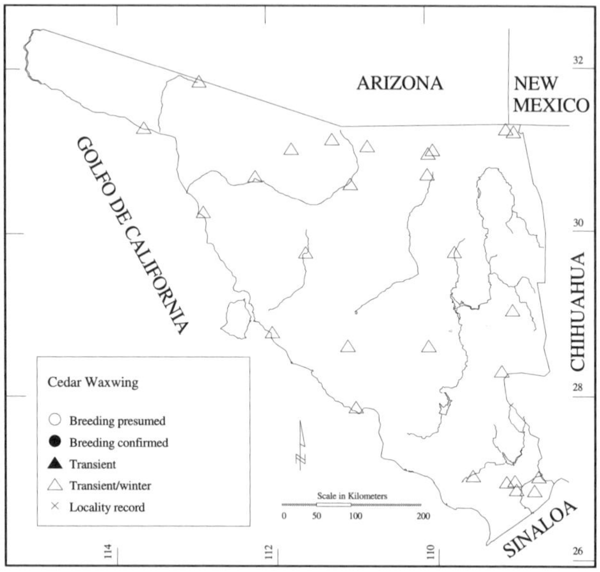

from November (earliest date, 14 October) to late May (latest date, 31 May) but most often in winter and spring. Typically, small flocks of fewer than a dozen birds are found in fruiting trees or resting in close proximity in a treetop, their presence marked by distinctive high-pitched *seet* notes.

FAMILY PTILOGONATIDAE

Silky-flycatchers

GRAY SILKY-FLYCATCHER

Capulinero Gris

Ptilogonys cinereus

Gray Silky-flycatchers have been found seven times in pine-oak habitats above 1500 m and twice in lowland areas. The species was first found in Sonora in March 1982, when we noted seven feeding on mistletoe fruit in the Sierra Oscura. On 19 July 1985, R. Russell found

Gray Silky-flycatcher

a nest 8 m above the ground on a horizontal branch of an Apache pine on Mesa Grande. The parents alternated nest attentiveness, each arriving with food for the young and then brooding them until replaced by its mate after five to eight minutes. The silent adults perched on dead branches of oaks or in the tops of pines and sallied out to seize insects. A. Moorhouse noted a pair at El Terreito on 19 May 1986 and three individuals at La Cueva north of Yécora on 12 June 1986. In mid-July 1987 we observed at least 12 at two localities near Sahuarivo. The behavior of a pair at a nest 20 m in a pine suggested that they were incubating. W. Leitner watched a pair on the northwest slope of Mesa de Enmedio on 27 May 1991. A. Moorhouse found six on the Sonora-Chihuahua border east of Maycoba on 29 December 1993.

R. Taylor saw a flock of six on 8 April 1983 at the Alamos airport, and R. Bowers and colleagues found a flock of 25 to 30 at Aduana on 4 April 1987.

We have visited the Mesa Grande site in years other than 1985 without seeing the species. These few observations suggest that Gray Silky-flycatchers are unpredictable with respect to their movements. Sonora is at the extreme northern limit of the species' range, and perhaps conditions suitable for their presence do not occur every year.

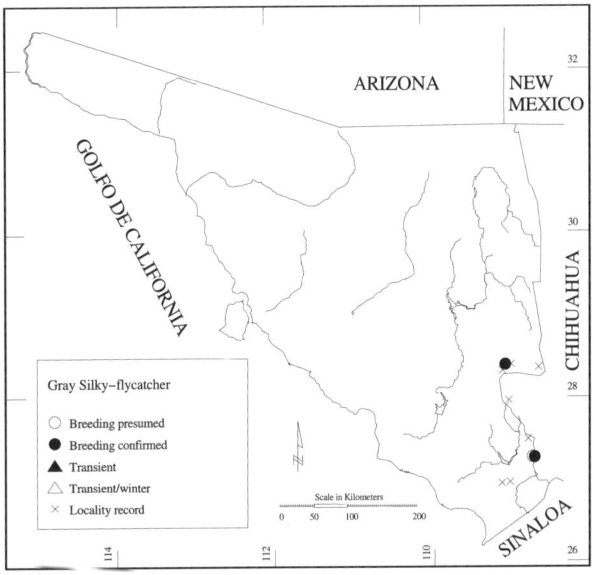

PHAINOPEPLA
Capulinero Negro
Phainopepla nitens

Phainopeplas are conspicuous wherever they occur. They sit in treetops or on exposed dead branches and frequently dart out in pursuit of an insect or to confront another Phainopepla. They feed almost exclusively upon fruit of mistletoe and are absent from mistletoe-infested regions if there are no berries. They are also fond of elderberries. Insects caught upon the wing supplement their diet. They may be found in almost any habitat if mistletoe fruit is abundant, up to about 1300 m at some time of year, but they do not remain in the same place throughout the year. Small flocks of presumably migrating birds have been seen on inland river systems in September. Although no intensive studies have been conducted in Sonora, our own field records indicate a seasonal shift in elevation from the lower coastal regions in the cooler months to higher elevations in summer.

Nesting in the coastal plain occurs in March, April, and early May and at inland localities from mid-May to early August. We have no evidence to suggest that birds that have nested along the coast move inland and nest again. Much could be contributed to the knowledge of this species simply by recording its numbers and nesting activity month by month at various places in Sonora.

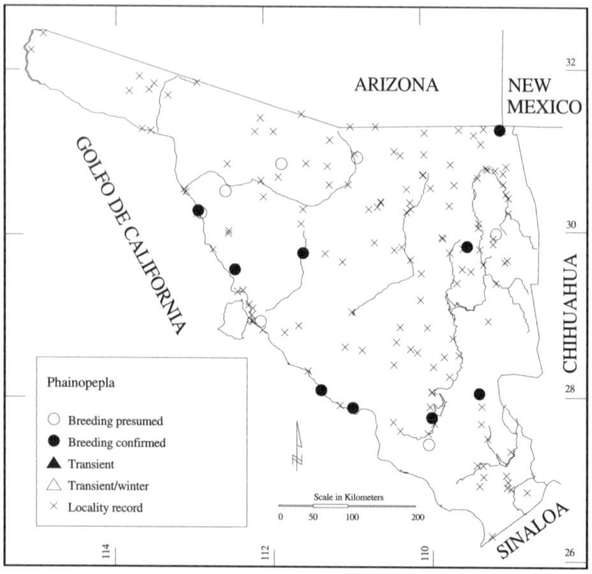

FAMILY LANIIDAE

Shrikes

LOGGERHEAD SHRIKE

Alcaudón Verdugo

Lanius ludovicianus

The Loggerhead Shrike is characteristic of open and preferably grassy habitats from sea level into the mountains. It perches conspicuously on snags, fences, and power and telephone lines as it scans for prey such as large grasshoppers and even small lizards, rodents, and birds. The Loggerhead Shrike is an uncommon breeding species; the breeding population presumably may or may not be resident. Many shrikes from breeding ranges to the north winter in or migrate through Sonora. Fall migrants are most visible from late September through November. In spring, shrikes are widespread in March, but rarely are more than two or three noted in an area.

The Loggerhead Shrike has been found nesting in ten places in Sonora. Our limited data indicate a seasonal progression from late winter breeding in coastal areas to July nesting in the high mountains. We found it nesting in mid-February, March, and April near the coast. It has nested near Mazatán (660 m) in March and Santa Ana de Magdalena (660 m) in April. In the mountains we found fledglings in the oaks at Aribabi (1300 m) on 20 July and in a clearing (1900 m) in pine forest near Sahuarivo on 18 July. J. Marshall saw two fledglings in the Sierra Cananea on 17 July. A pair in an open area (1630 m) above Los Algodones behaved in such a way that we suspected they were nesting on 19 May.

We have summer records of shrikes from Puerto Peñasco and think they may be resident in some but not all years. We do not know whether shrikes are present at Puerto Lobos in the summer, nor do we know whether the inland breeding populations are resident. At Lo de Campa, where we have visited

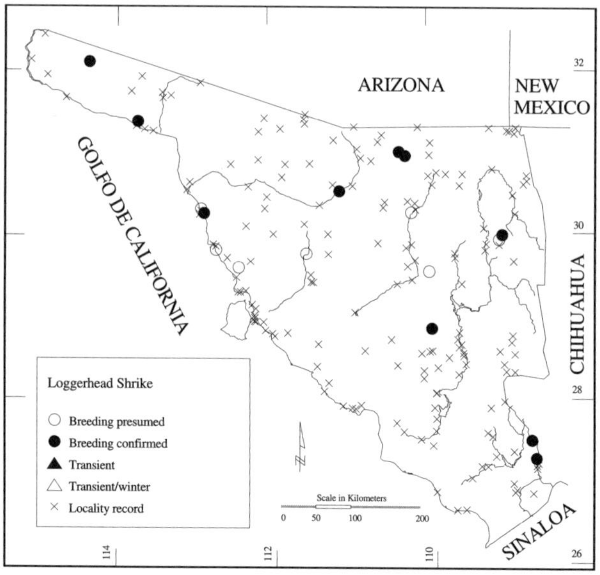

throughout the year, shrikes are present only from late August to mid-March. In the coastal Punta Chueca area we have not found shrikes between 27 February and 7 September, although we have been there often in that period.

The Loggerhead Shrike in Sonora is another poorly known species, even though there are hundreds of observations. We think that breeding in desert areas may be dependent upon precipitation. Further, we believe that shrikes may leave desert areas under unusually dry conditions. At Punta Cirio, one or two shrikes are present throughout nearly every winter, but we have found them there only once in summer. They probably nest there when conditions are favorable and stay through the summer in some years. That pattern is perhaps typical in other desert areas.

FAMILY STURNIDAE

Starlings and Allies

EUROPEAN STARLING

Estornino Pinto

Sturnus vulgaris

This species successfully became established in North America after an introduction in New York City in 1890. It spread across the continent and nested in Arizona for the first time in 1954 (Phillips et al. 1964). In 1968 it was noted in Sonora (Sonoita, Puerto Peñasco, and 20 km south of Nogales), in 1973 it was found at Empalme, and in 1974 it was seen investigating potential nest sites at Navojoa (Russell and Lamm 1978). The first documented nest with young was southeast of Rancho San Bernardino in a cottonwood on 30 April 1977. The species is now widely distributed and apparently resident in Sonora. The largest concentrations (of about 300 birds) have been seen in the agricultural regions of the southern coastal area. There is no evidence of any seasonal migration of birds from the United States into Sonora.

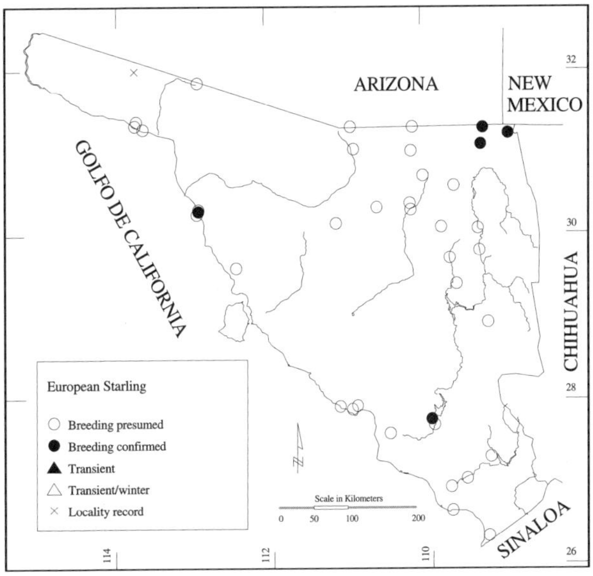

Starlings were first noted at Puerto Lobos on 30 March 1985, when parents were feeding noisy young in a cardón cavity. Seven years later, starlings still foraged in the dump at the village, but no more than six have been seen there. The species has not been found in the agricultural areas near Caborca, and we wonder about the route followed by the individuals colonizing Puerto Lobos. They appear to be increasing in numbers more slowly in Sonora than in their range to the north. They are generally regarded as undesirable because they may crowd native birds from a limited supply of suitable cavity nest sites.

FAMILY VIREONIDAE

Vireos

[WHITE-EYED VIREO
Vireo Ojo Blanco

Vireo griseus

One was seen and heard singing repeatedly in a riparian locality just east of Alamos from 11 to 14 May 1992 by K. B. Ostyn.]

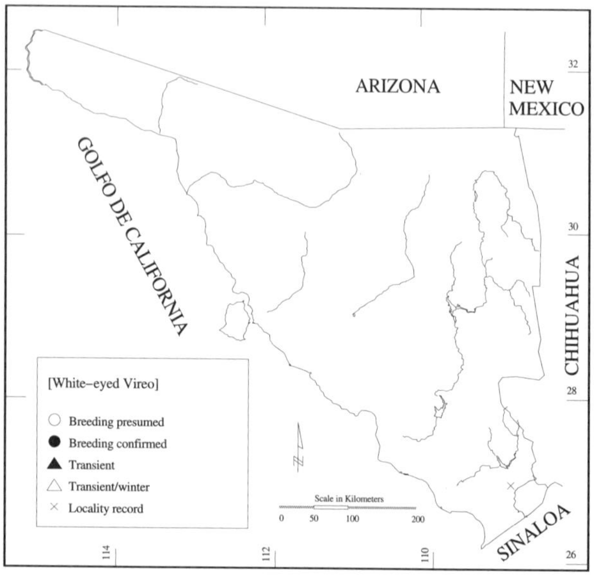

MANGROVE VIREO
Vireo Manglero

Vireo pallens

Van Rossem (1945) collected (UCLA) a male in breeding condition on Isla Masocarit on 14 May 1937. We are aware of no other records from Sonora. As its name implies, the species is an inhabitant of mangrove thickets. We did not find it in a search of mangroves near Agiabampo in 1987. Further efforts should be made to establish it as a Sonoran resident.

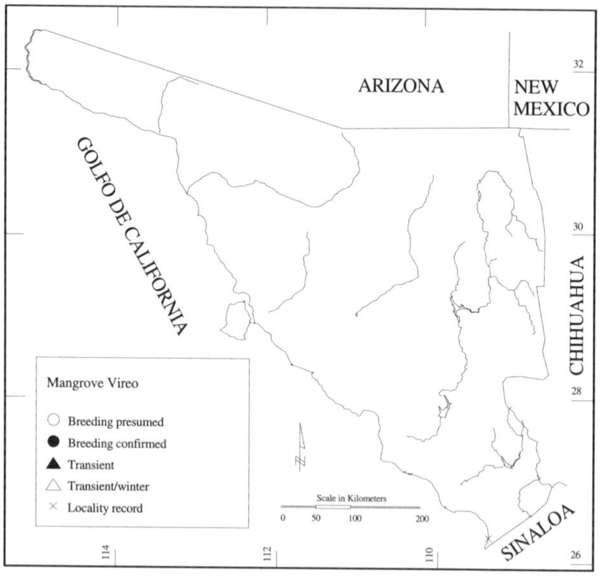

BELL'S VIREO
Vireo de Bell

Vireo bellii

The Bell's Vireo is a common breeding species in riparian thickets from near sea level to 1200 m, except in the extreme northwest and southeast. Most of these vireos arrive on the nesting areas about 10–15 March. The Bell's Vireo is partial to mesquite bosques, where its repetitive singing may be heard throughout the day from its arrival into September. In very mesic areas nesting is in full progress in early April, but in drier areas much (if not most) nesting activity is in the period from June into August. Most birds have departed by the end of September, but some lingering birds or northern migrants remain until early November.

Bell's Vireos winter in small numbers from the lower Río Yaqui southward, thus they may be there throughout the year. We do not know if the southern-

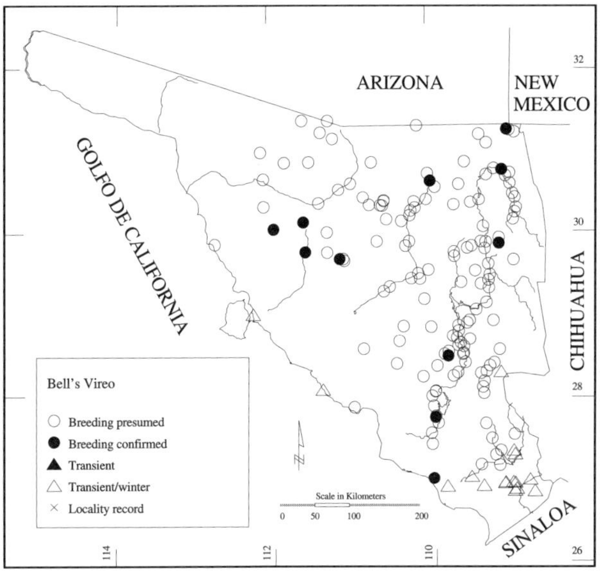

most breeding vireos migrate or are replaced by birds from the north. Transient Bell's Vireos move northward through lowland valleys from mid-March to mid-April and southward in late summer. The species is rare in northern Sonora in winter; it has been found near Ures (twice: December, January), near Magdalena (December), and at Granados (February).

This species has suffered serious population declines in parts of its range north of Sonora, attributable to habitat destruction and cowbird brood parasitism. Although this vireo is presently doing well in Sonora and is a widely distributed and common species, it can be expected to suffer as riparian lands are cleared.

BLACK-CAPPED VIREO
Vireo Gorra Negra
Vireo atricapillus

Two observations from extreme southern Sonora are the only ones from the state. On 29 November 1954, A. Phillips attempted to collect one on the Río Yaqui northwest of Cd. Obregon. D. Danforth and G. Gregg saw one on the Río Cuchujaqui southeast of Alamos on 29 November 1975. The species is known to winter in central Sinaloa. Its breeding population in Texas and Oklahoma is now so reduced that any observations of the species in Sonora will be highly fortuitous.

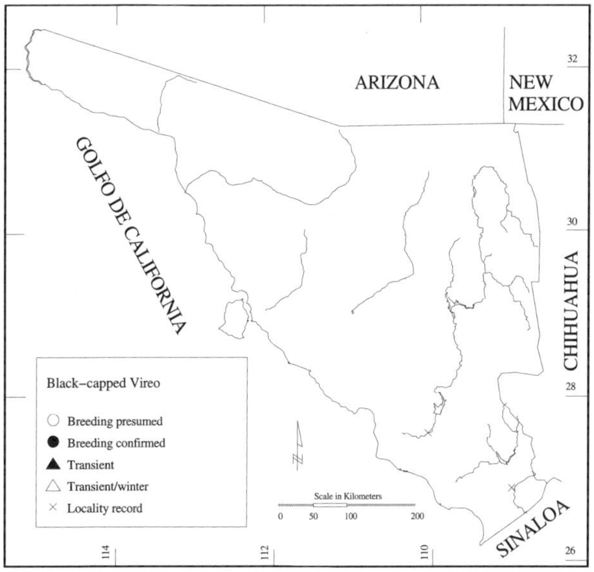

GRAY VIREO
Vireo Gris
Vireo viclnior

Gray Vireos winter in coastal and lowland desertscrub areas, most commonly where elephant tree is abundant, between Puerto Lobos and Guaymas. There during the dry winter months they may be the most obvious North American wintering species, and one may find four birds in a one-kilometer transect. The earliest date of arrival is 9 September 1973 (Puerto Los Mochos), but most birds appear on the wintering grounds in late September. In spring, departure for the north is in late March, and most vireos have gone by 5 April. The latest spring date is 12 April 1973 (Punta Chueca).

Banding indicates that individuals often return to the same wintering spot in successive years, where they defend a foraging area through the winter (Bates 1992a). One bird occupied the same elephant tree– and ironwood-dominated arroyo for six winters. The vireos use chases and vocalizations to ward off intruders. In addition to a simple song, a trill likened to that of a Harris Antelope Squirrel (*Ammospermophilus harrisii*) is used often in the winter but rarely on the breeding grounds. Bates (1992b) noted a remarkable overlap in the winter distribution of this vireo and elephant tree. The vireo, an insectivorous bird during the breeding season, consumes much fruit in winter, almost exclusively that of this species of *Bursera*. At Punta Chueca banded birds also returned in successive years to territories at the interface of scrub and mangroves. They foraged extensively on insects well into the mangrove swamps.

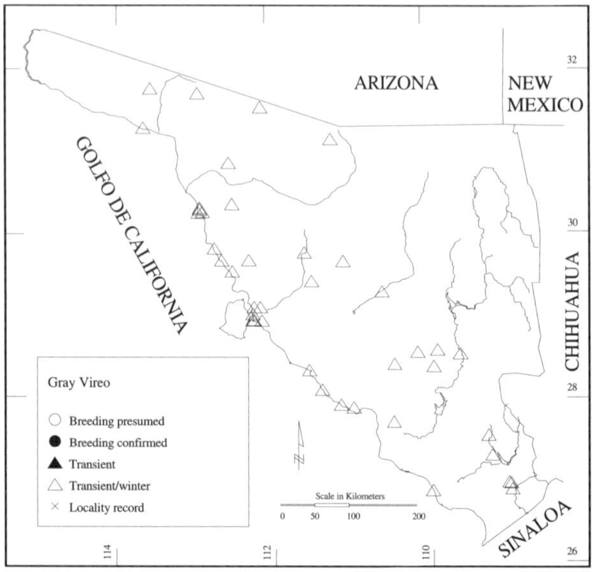

SOLITARY VIREO

Vireo Anteojillo

Vireo solitarius

The Solitary Vireo is a common summer breeding species in the pine-oak woodlands and pine forests of eastern Sonora at elevations between 1400 m and 2000 m. Many additional birds migrate through Sonora. An occasional individual may be expected in winter in almost any lowland wooded area, but at that season it is common only in the southern third of the state. The Solitary Vireo has a short song, which it repeats almost incessantly during the breeding season, sometimes even from the nest. Transient and wintering birds also sing but less frequently. The species is insectivorous, picking its small prey from leaf surfaces, pine needles, and twigs with a glancing movement of its bill. We have seen it consume fruit only once in Sonora, when one ate unidentified small red berries near San Nicolás on 13 March.

Nesting has been verified in relatively few mountain ranges in Sonora, although the species is widespread in pine and pine-oak during summer. Marshall (1957) noted its preference for tall riparian woodland in the northern mountains and for pine-dominated forests farther south. Nesting is in May and June. We have seen begging fledglings as late as mid-July.

The fall migration becomes apparent in mid-September and continues to mid-November, when we counted 30 near Félix Gómez in a 2 km section of the Río Bacoachito. In spring birds move northward from mid-March to mid-April. Most Solitary Vireos migrate and winter to the east of wintering Gray Vireos. The distinctively marked yellowish form, representing Pacific coast populations, is conspicuous during migration and winter but is generally less numerous than all-gray (including Sonoran-breeding) birds.

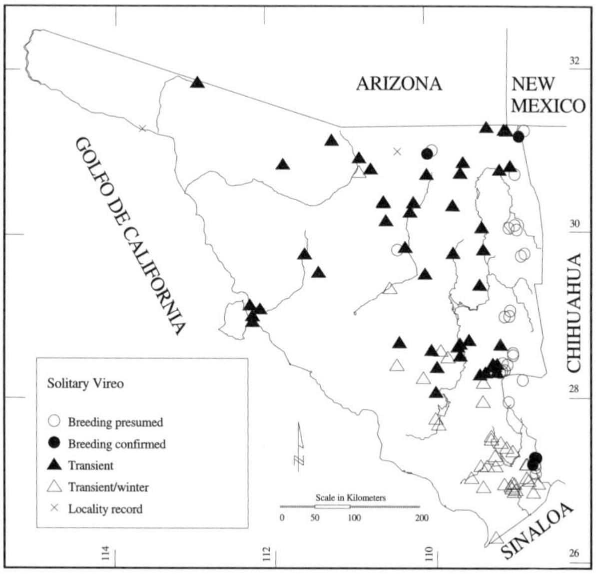

The Solitary Vireo has been divided into three species (AOU 1997). Publication schedules do not permit revision of this account, but both Cassin's Vireo (*V. cassinii*) and Plumbeous Vireo (*V. plumbeus*) are found in Sonora. The Plumbeous Vireo is the species breeding in Sonora; Cassin's Vireo is present as a migrant and in winter.

HUTTON'S VIREO
Vireo Reyezuelo
Vireo huttoni

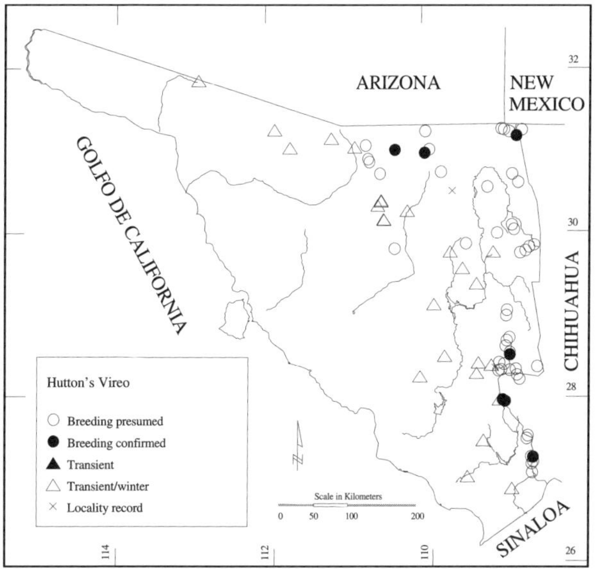

This diminutive vireo is a characteristic breeding bird of pine-oak woodlands of mountain canyons and also woodlands dominated by large oaks. From about 1200 m to the highest elevations in Sonora, the Hutton's Vireo is a retiring but moderately common bird from March into September; for the winter months many birds move to elevations below their breeding range. It has an almost unnoticeable song, calls infrequently, and quietly goes about its business of picking insects from foliage. Nest building begins in at least some years as early as 17 March, when we watched a bird gather spiderwebs for nest lining material in Arroyo Cajón Bonito (1240 m). The birds nest later at higher elevations, mostly in early June. We have seen parents feeding their begging but flying young in mid-July.

Individuals, perhaps birds from northern populations, appear in late July in places where they do not nest. Some birds remain in the higher breeding grounds until very cold conditions suppress the insects to the point that the vireos are forced to lower elevations. Hutton's Vireos are rare to uncommon anywhere in Sonora in winter but are present in the eastern foothills, typically at elevations above 600 m. Lo de Campa (490 m) is the lowest place we have regularly found them in the winter.

GOLDEN VIREO
Vireo Dorado
Vireo hypochryseus

Seven specimens of the Golden Vireo were collected between 2 and 8 May 1888 by M. Frazar at Hacienda de San Rafael. Another bird was taken by A. van Rossem at Cañon San Francisco on 29 May 1937 in a willow and cottonwood association. Except for one undocumented bird listed during the count period of the Alamos Christmas Bird Count in 1982, the species is not otherwise reported from Sonora.

Van Rossem (1945) described the habitat of this vireo as "Tropical zone foothills" and thought the elevation to be about 450 to 600 m. We have unsuccessfully searched many areas meeting this description in southeastern Sonora. The bird is strikingly marked and has a loud repetitive song that is likely to attract attention. The area where it has been found has experienced no unusual habitat modifications. We doubt that the species has become extinct in Sonora; rather, it likely is very local and probably present for only about three months of the year. The bird should be sought in extreme southeastern Sonora.

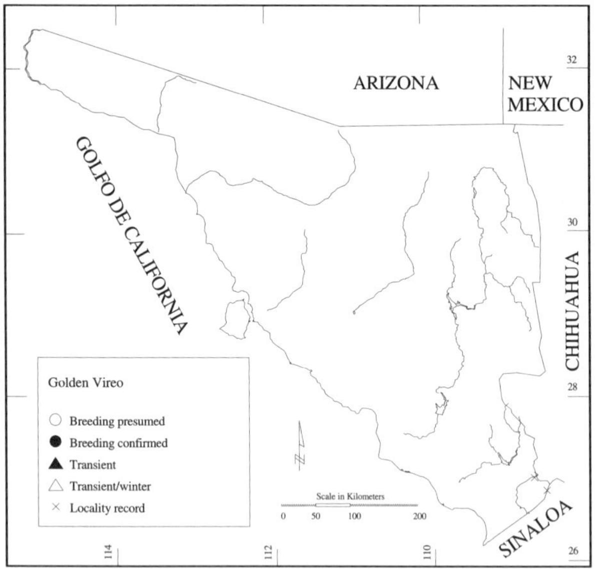

WARBLING VIREO
Vireo Gorjeador
Vireo gilvus

Warbling Vireos are fairly common spring and abundant fall migrants through the foothills and riparian areas of central and eastern Sonora. Some move along the coastal plain, but there are only two records for the far northwest. Spring migrants appear by mid-March and may be seen until early June, during which period one may expect to see one to three in a morning in the field. Only one has been found in mangroves (netted on 30 May 1971 at Punta Chueca).

Some individuals appear on their fall migration as early as 17 July in northeastern Sonora (Sierra los Ajos), and numbers increase greatly in August and September, when one may easily see five to ten in a morning. They are attracted to water for bathing, especially in September, when we have seen ten at once at a small pool, where they would quickly fly into the water, appear to bounce out, and then spend several seconds preening in nearby brush. On 16 September 1982 we observed 200 northeast of Rosario in a period of an hour in the early morning, most in the vicinity of a bathing spot. Fall migrations involve large numbers compared to spring. In spring, when hillsides are mostly bare, the birds concentrate in riparian areas. In fall, green vegetation is not limited to stream banks, and the birds may be anywhere, although they are most numerous along streams.

Van Rossem (1945) reported the species breeding uncommonly at Rancho Santa Barbara, but we have only circumstantial evidence of nesting elsewhere. We have found one or two singing birds in summer near Sahuarivo and El Trigo. Warbling Vireos regularly sing during migration and even occasionally in winter. Their migration period is unusually long. We do not conclude on the basis of song alone that they have been nesting. Marshall (1957) reported three pairs in a mapped one-mile census route at his site on the east slope of the Sierra Huachinera in June. There they preferred tall riparian trees.

Warbling Vireos winter uncommonly to fairly commonly in the lower Río Yaqui and Río Mayo valleys and in the extreme southeast, mainly in riparian situations. Elsewhere they are not to be found between early November and early to mid-March.

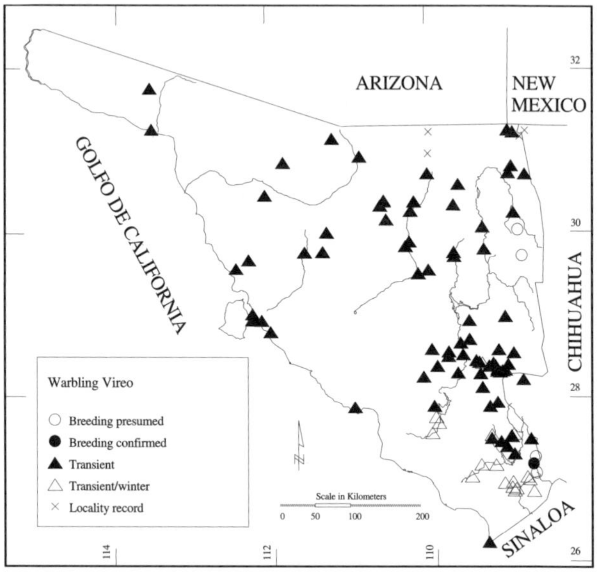

RED-EYED VIREO
Vireo Ojo Rojo
Vireo olivaceus

Four individuals of this common and widely distributed North American species were seen (two, and a supposed nest, were collected, DVM) at the north end of the Sierra los Ajos on 16 July 1952 by A. Phillips. Marshall (1957) stated that the birds in question were early fall migrants, but Phillips (1991) left open the possibility the species had nested. The Red-eyed Vireo is not known to nest in Arizona or Chihuahua, and we await new records before considering it a breeding species in Sonora.

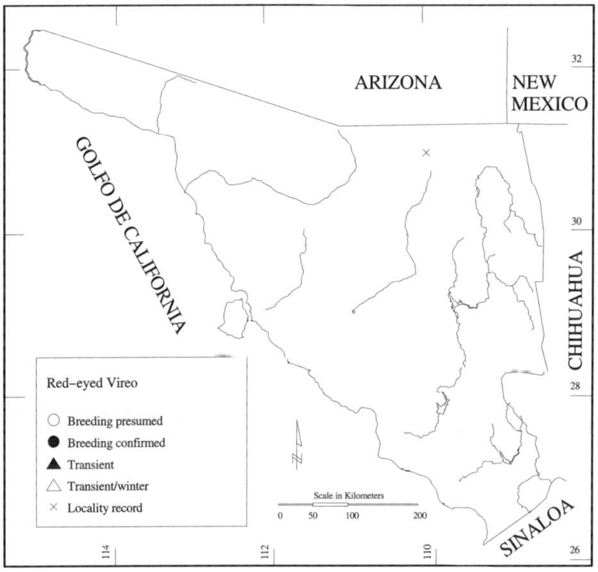

YELLOW-GREEN VIREO
Vireo Verdeamarillo
Vireo flavoviridis

This species is a locally distributed summer resident in southern Sonora. Its presence and nesting are perhaps more finely tuned to the summer wet season than those of any other migratory bird in the state. It usually arrives during the last week of May (earliest date, 8

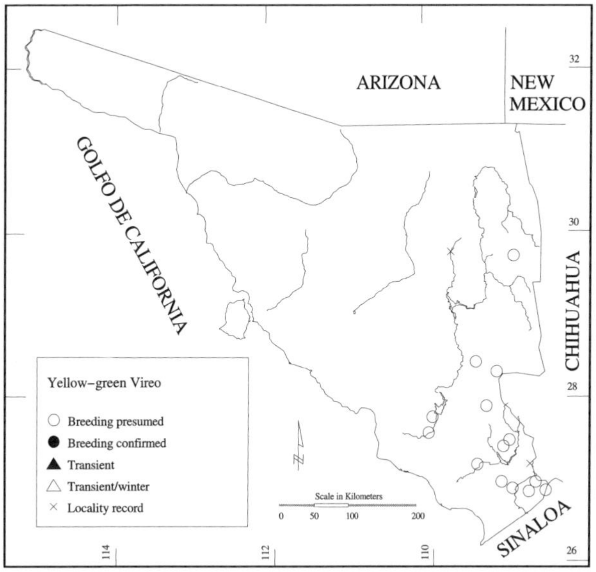

May, at Güirocoba), sings repetitiously and establishes territories in June, nests in July, and departs in August (latest date, 27 August, near Alamos). Although arrival dates may be rather consistent from year to year, we think the onset of breeding may vary according to the timing of initial summer rains.

Tropical deciduous forest provides the favored habitat, but the Yellow-green Vireo is not found everywhere the habitat appears suitable. Although van Rossem (1945:215) reported it from pine forests and "not uncommon to an altitude of 5,500 feet [1675 m]," no subsequent observations have been made where pines are present or higher than 1000 m.

The northernmost record is of a bird collected (van Rossem 1945) by J. Cahoon at a place he called "Oposura" on 6 June 1887. Although the locality is now known as Moctezuma, when Cahoon collected in the region (including the Sierra la Madera) he labeled his specimens as Oposura. Consequently, the exact locality is unknown.

FAMILY EMBERIZIDAE

Emberizids

SUBFAMILY PARULINAE

Wood-Warblers

TENNESSEE WARBLER

Chipe Peregrino

Vermivora peregrina

The Tennessee Warbler is casual in Sonora. Three have been found: one collected (DVM) near Cibuta south of Nogales on 14 October 1954 (A. Phillips); one seen at La Aduana near Alamos on 15 March 1982 (D. Stejskal); and one seen in Arroyo Cajón Bonito on 21 May 1982 (R. Morse).

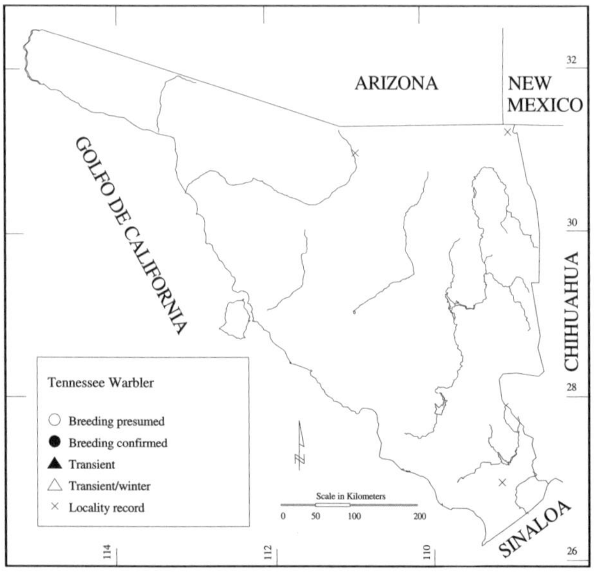

ORANGE-CROWNED WARBLER

Chipe Corona Naranja

Vermivora celata

As a transient, the Orange-crowned Warbler is common and found nearly everywhere in leafy vegetation in Sonora. In winter it is fairly common at elevations below 1000 m; it is present but uncommon in the higher mountains from the Yécora region south.

The first fall migrants appear in late August. Earliest fall dates in the north are 1 September in the lowlands (Bacerac) and 26 August in the highlands (Sierra Madre north of Río Bavispe; Marshall 1957). In the south, migrants appear by mid-September and continue to move through Sonora until late November. In wet years when coastal vegetation is especially lush and green, Orange-crowned Warblers linger there in numbers and may be the most abundant migrant. On 26 November 1977 near Punta Chueca, Russell captured and banded 26 individuals as they flew into mangroves at dusk to roost; another 38 were captured in half an hour the next morning as they left the mangroves. None were retrapped later in the winter, as they presumably had moved farther south. At Lo de Campa, individuals banded in mid-October were still present in December, suggesting that they may winter there. In the Alamos area, participants in the Christmas Bird Count have tallied as many as 150 Orange-crowned Warblers in one day (in 1990).

The spring migration commences in mid-February and extends to early May, with stragglers into the first days of June (2 June in the Sierra los Ajos; Phillips). Although the migration extends over a long period, it peaks in March and early April. We counted 200 in 3 km of coastal desertscrub at Puerto Lobos on 1 April 1973.

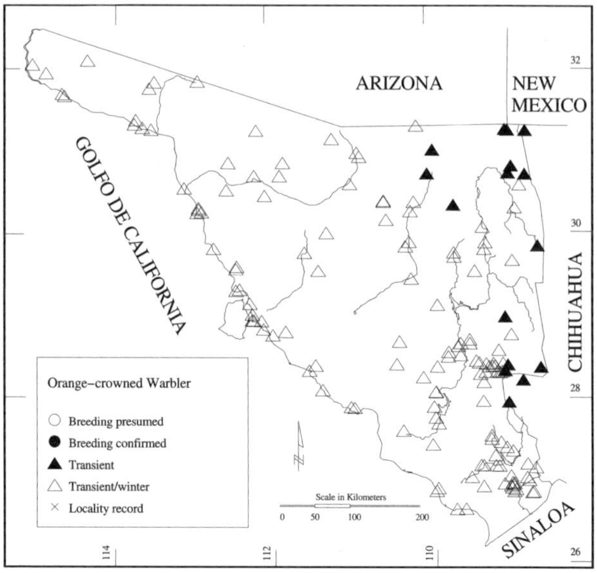

NASHVILLE WARBLER

Chipe de Coronilla

Vermivora ruficapilla

The Nashville Warbler is a fairly common transient in Sonora, except in the higher mountains and the northwest. A few winter from the Quiriego and San Bernardo areas south. We found 24 on Arroyo el Cobre near Güirocoba on 12 December 1983.

Migration is in March–April and September–October. The earliest spring date north of the wintering range is 9 March (Lo de Campa), and the latest spring record is 16 May (Arroyo Cajón Bonito). Fall migrations are sometimes spectacular and utilize the deciduous forests of the eastern foothills. We have counted from 150 to 300 birds in a morning at several locations in mid-September. Riparian areas are favored, and migrants on hot days frequently bathe in shallow trickles of water. We saw six in the water at once on 30 September north of Tonichi, some of many that bathed. They are readily attracted to an imitation of a pygmy-owl call, and we have had as many as 18 gathered in response to the whistled calls. The species rarely migrates through oak woodlands or in higher communities.

Our earliest fall date is 21 August, when we found 24 near San Nicolás. The latest fall observation is 16 October (40 km south of Nogales; A. Phillips). Two November records could represent wintering birds: five at Lo de Campa on 25 November, one at San Carlos on 13 November. There is an aberrant winter occurrence: one near Puerto Lobos on 4 January 1988 (J. Bates). The highest records are of one at El Ocote in the mountains east of Sahuaripa at 1480 m on 15 March and one collected at about 2100 m in the Sierra San Luis on 3 September (Marshall 1957).

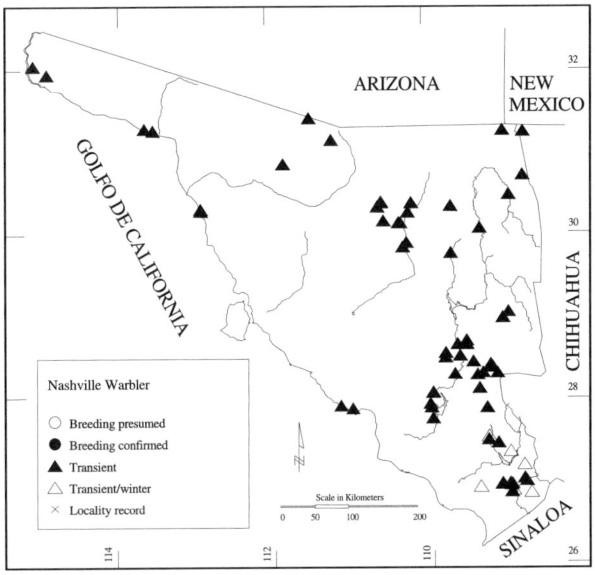

On 15 September 1986, Nashville Warblers appeared to be everywhere in the *Prosopis-Celtis-Pithecellobium*–dominated bosques 8 km north of Arizpe. They silently gleaned centimeter-long green lepidopteran larvae, probed upside down into crevices of trunks and branches, and pulled apart pieces of bark.

VIRGINIA'S WARBLER

Chipe de Virginia

Vermivora virginiae

Virginia's Warbler is an uncommon (rare in the southern half) migrant in Sonora, where it is usually found in deciduous trees of riparian areas at elevations under 1000 m. The many individuals that nest in the extensive area north and west of Sonora must migrate to the southeast, thus barely passing over the northeastern corner of Sonora. Marshall (1957:107) pointed out that "south of Arizona, the mountains are not high or moist enough to provide deciduous brush close to forest, which is the Virginia Warbler's preferred habitat," and thus it does not nest in Sonora.

There are only two records from near the coast: one banded at Puerto Lobos on 1 April 1973 and another seen on 15 April 1973 about 30 km northeast of El Desemboque de Río San Ignacio. Extreme fall dates are 21 August (north of San Nicolás) and 26 September (near Moctezuma), and for spring they are 17 March (Lo de Campa) and 14 May (Arroyo el Reparo).

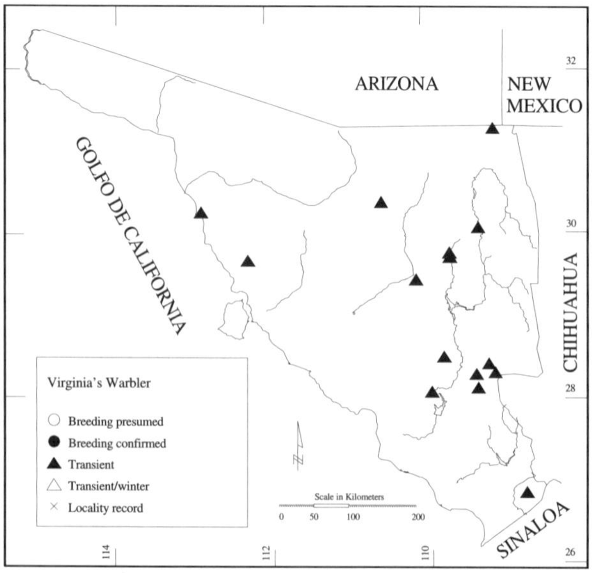

LUCY'S WARBLER

Chipe Rabadilla Rufa

Vermivora luciae

Lucy's Warbler might well be called the Desert Warbler, as it is the only wood-warbler that tolerates the xeric conditions of the desertscrub. It is a common summer breeding species in northeastern Sonora (formerly common in the Río Colorado delta, where its present status is uncertain). It is most common in mesquite bosques and in washes with large ironwoods, mesquites, palo verdes, and chinos. It is also found in willows, sycamores, and oaks at low elevations (including Chihuahuan oaks). Nesting has not been confirmed south of the latitude of Hermosillo, but it probably nests at Lo de Campa where plants of the tropical thornscrub have replaced desertscrub. A number of records for southern Sonora, including the border with Sinaloa, in the first half of July are thought to represent southbound migrants rather than breeding birds.

The spring migration is impressive in its abruptness. The first birds appear about 10 March, and within two weeks all of eastern Sonora has many birds. Most have reached their breeding grounds by early April. The fall migration begins in late June, and in July the species becomes widespread in southeastern Sonora. Most individuals have moved on south toward their wintering grounds by early September, but stragglers may linger into October. Our latest record not in the wintering range is 17 October near Tecoripa.

Lucy's Warbler winters sparingly in the extreme south of Sonora. The largest number found on an Alamos Christmas Bird Count has been six, and in some years these warblers are not seen at all. One noted at Lo de Campa on 8 December may have been an unusually late laggard.

Lucy's Warbler is an incessant songster from the arrival time in spring through the nesting season. It begins to nest only a few days after it arrives on the breeding grounds, and most nests contain fledged young by May. It does not place its nest among exposed twigs but rather puts the nest behind a loose piece of bark (often near the ground) or in a small cavity in the trunk of a large shrub or small tree.

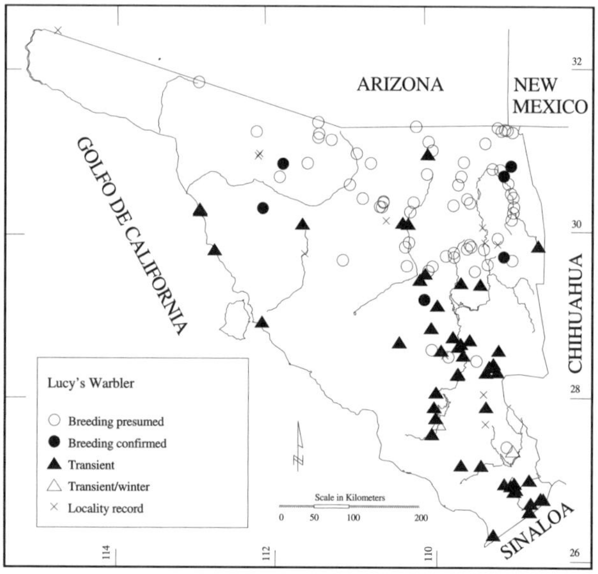

NORTHERN PARULA

Parula Norteña

Parula americana

A male Northern Parula with enlarged testes was collected in the Sierra los Ajos on 16 July 1952, and a female was taken in the same spot on 29 July 1952 (both Marshall 1957). The birds were foraging in riparian woodlands, including alders and Arizona white oak. The male, which was not singing, did have enlarged gonads, and the female had a regressing brood patch. The possibility exists that they constituted a nesting pair. Otherwise, the only Sonora record is of one seen along the Río Cuchujaqui near Los Llanos about 32 km by road east of Alamos on 23 October 1974 (Monson).

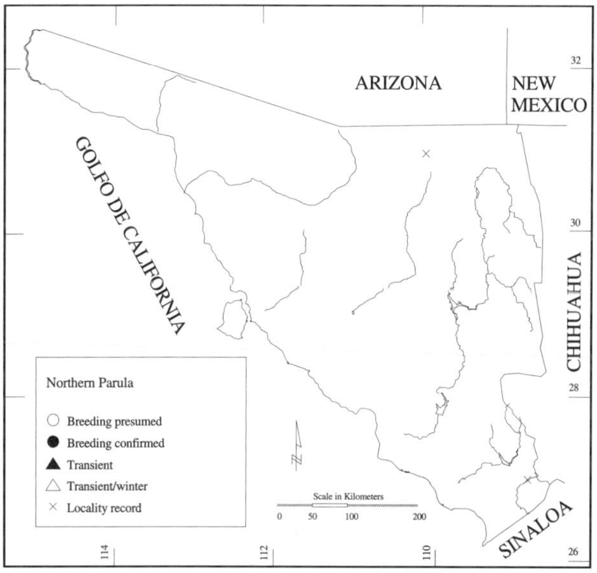

Tropical Parula

TROPICAL PARULA

Parula Tropical

Parula pitiayumi

The Tropical Parula is a local summer breeding species between 400 and 1400 m in foothill canyons with streams in southeastern Sonora. It is often common where it does occur and is made conspicuous by its extended singing throughout the day. Habitat requirements include deciduous trees and pools of water, if not streams, in canyon bottoms. At El Riito and Arroyo el Reparo oaks were also present. All records fall in the period of 20 April and 4 September. In mid-May we found 12 pairs in a kilometer along the intermittent stream in Arroyo el Reparo.

Evidence of nesting comes only from Güirocoba, where a nest containing three eggs was taken on 3 June (WFVZ). The species has been reported several times from the Río Cuchujaqui southeast of Alamos, but we do not think it breeds there. This warbler responds to imitations of the Northern Pygmy-Owl, and not surprisingly, J. Marshall once saw one harassing a pygmy-owl it had discovered.

Van Rossem (1945) refers to a specimen taken 26 May 1905 in the Sierra de San Antonio in northern Sonora. There is no reason to suspect an isolated colony there, as van Rossem suggested.

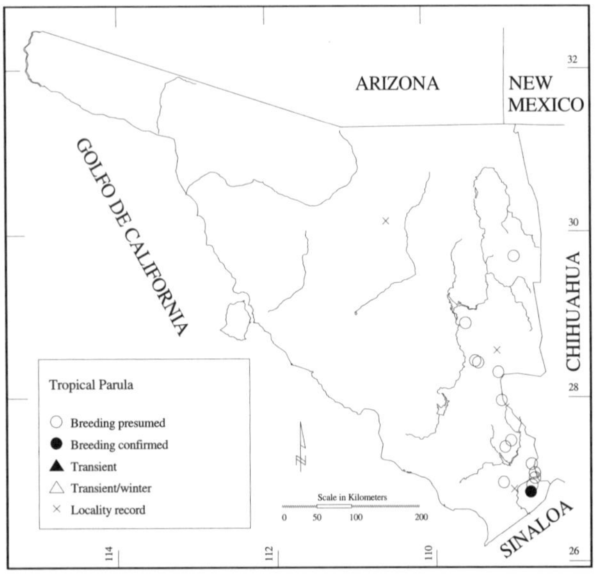

CRESCENT-CHESTED WARBLER
Parula Ceja Blanca
Parula superciliosa

The Crescent-chested Warbler, a rare summer resident, has been found in few places in Sonora. One was taken at San Antonio (Mina Abundancia) about 32 km northeast of Alamos on 9 April 1888 (van Rossem 1945). It was not noted again for nearly 100 years. We found it to be common in the Sierra Oscura on 14–15 March 1982, 8–9 June 1984, and 17 September 1982. It has been noted singing on and just below Mesa de Enmedio west of Yécora on various dates from 13 March to 30 July. The only other Sonoran record is of one observed carefully by S. Terrill north of Esperanza on the Río Yaqui on 24 December 1977.

Individuals forage predominantly in oaks, where they seek insects or insect larvae on leaves. They move rapidly, looking briefly at clusters of oak leaves and petioles and peering or probing upward into dead curled leaves before flitting a few inches or several feet to the next leaves. They sing frequently, often while feeding. Occasionally they forage with other species; two were in a mixed flock in June on Mesa Rodriguez (2100 m) in the Sierra Oscura that also included a White-striped Woodcreeper, four Mexican Chickadees, a pair of Brown Creepers, a pair of Eastern Bluebirds, a family of Hutton's Vireos (including two fledglings), and a pair of Olive Warblers. T. Huels found a bobtailed young bird on Mesa Rodriguez on 10 June, the only positive nesting record.

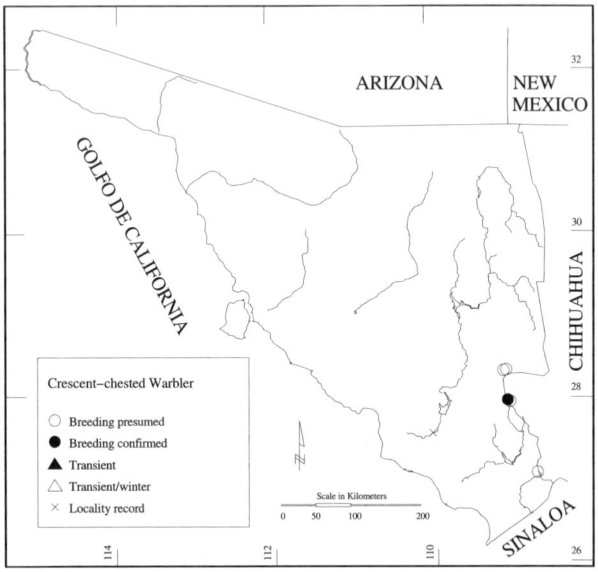

YELLOW WARBLER
Chipe Amarillo
Dendroica petechia

The familiar Yellow Warbler (excepting the Mangrove Warbler; see below) is a common transient throughout Sonora and a local summer resident in the major river valleys of northeastern Sonora.

The earliest spring migrants have been noted at Alamos on 9 March; birds on territories have been noted in Arroyo Cajón Bonito on 15 March. Migrants continue to move northward through the state as late as 30 May. Southbound transients appear in early August and continue to move through Sonora until late October. Transients are common in spring and fall over the entire state, barring the higher parts of the eastern mountains and including the shrub-lined washes of the desertscrub. Records from along the coast of El Golfo de California are scanty, and individuals have been netted and banded in the mangroves near Punta Chueca only in late May and late September. There is one spring (1 April) record, from Puerto Lobos.

Yellow Warblers nest in cottonwood groves along streams and marshes. In the upper Río Bavispe, nesting is underway by the end of April, if not earlier. We have seen fledglings as late as mid-July in Arroyo Cajón Bonito. The species was probably once a common summer resident in the Río Colorado delta, and van Rossem (1945) indicated that it had nested there, but it is not believed to breed there now.

Yellow (Mangrove) Warbler

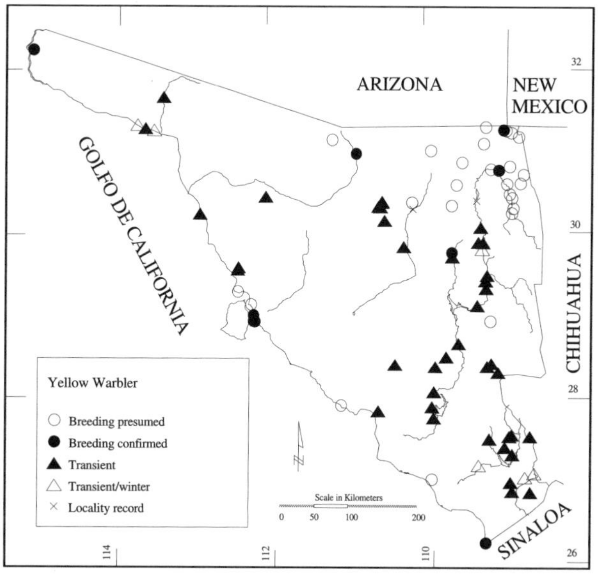

Five birds may have been wintering: one collected near Granados on 19 November 1952 (A. Phillips); two found on the Puerto Peñasco Christmas Bird Count on 3 January 1981 and another on 1 January 1992; and one in Arroyo el Palmarito east of Alamos on 5 January 1993.

A distinct ecological form of the Yellow Warbler, sometimes called the Mangrove Warbler, breeds commonly in coastal mangroves. The males have distinguishing chestnut heads not found in any other Yellow Warblers in Sonora. Extensive banding in the mangroves near Punta Chueca indicates that this population is a summer visitant. These warblers arrive in spring about 11 April but do not commence to nest until late May. Young fledge for a period of about four to six weeks in June and July. Departure is in late September. Several individuals were recaptured at later dates, one three years later. We have observed only one individual in these mangroves in winter, a female banded 18 January. Other winter records are one at Bahía Kino on 29 December, two at San Carlos on 15 January, and one at Guaymas on 26 November (last record by D. Stejskal).

[CHESTNUT-SIDED WARBLER
Chipe Flanco Castaño
Dendroica pensylvanica

A male was observed under excellent conditions on the Río Sonora west of Ures on 28 May 1980 by R. Dummer, G. Rosenberg, and B. Whitney.]

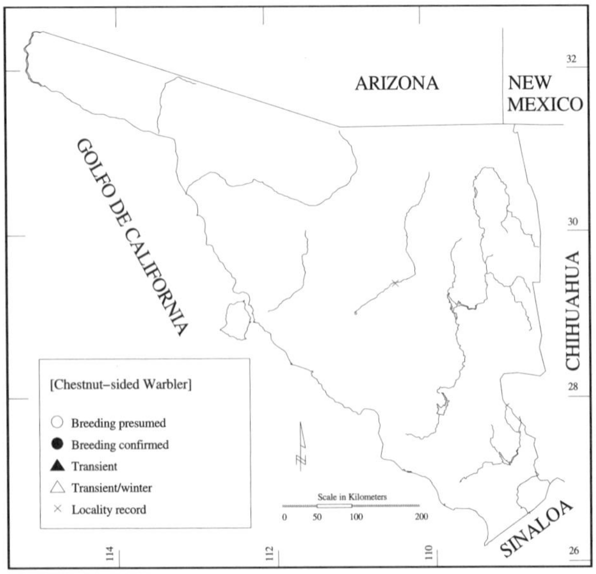

MAGNOLIA WARBLER
Chipe de Magnolia
Dendroica magnolia

One was collected at Alamos on 27 February 1888 by M. A. Frazar (van Rossem 1945). Another was observed on Arroyo el Cobre, 2 km east-northeast of Güirocoba, on 12 December 1983 (C. Green).

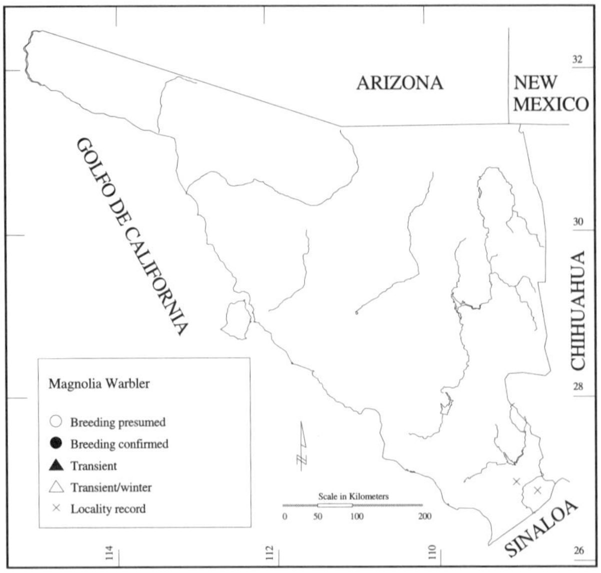

[CAPE MAY WARBLER
Chipe Atigrado
Dendroica tigrina

The only Sonora record for this eastern warbler is of one seen on a Christmas Bird Count, in the residential section of Puerto Peñasco on 16 December 1989 by D. Stejskal.]

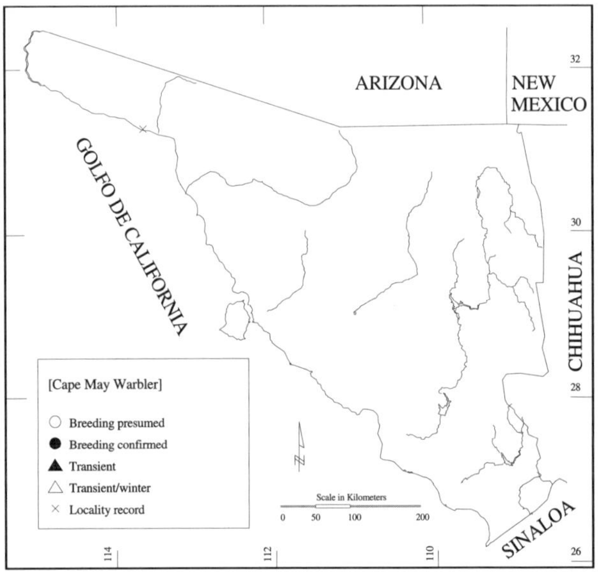

BLACK-THROATED BLUE WARBLER
Chipe Azulnegro
Dendroica caerulescens

R. Wilt observed one above the Río Sonoita dam at Sonoita on 17 October 1975 (Groschupf et al. 1988). A male was discovered about 5 km north-northeast of Guisamopa in dense thornscrub on 14 March 1984 by A. Moorhouse and was viewed by several observers.

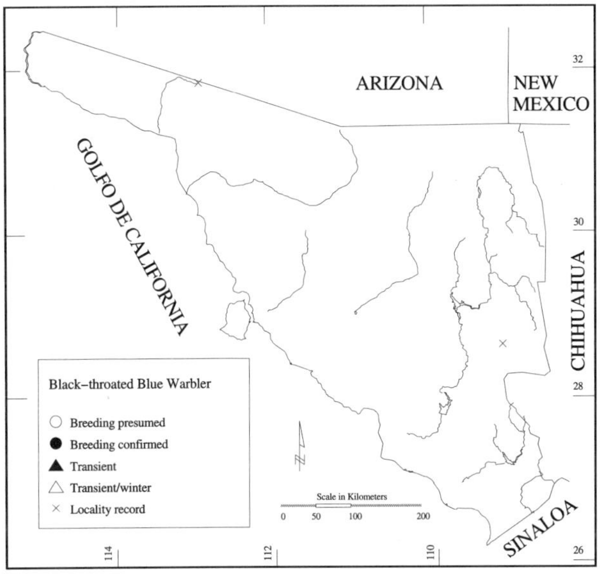

YELLOW-RUMPED WARBLER
Chipe Coronado
Dendroica coronata

The Yellow-rumped Warbler is a fairly common to sometimes abundant transient and winter resident all over Sonora, being most common along wooded stream bottoms and in agricultural areas and sometimes along the coast of El Golfo de California. Territorial individuals have not been found in the state, and this species probably does not nest there. Some examples of maximum abundance are nearly 600 on an Alamos Christmas Bird Count; 100 in the Seri village of El Desemboque in January; 180 at Granados in February; 100 at Arroyo Santo Domingo in November; and 200 at Presa Cuauhtémoc in April. These numbers are well below counts often made in southern Arizona.

Spring migration continues into mid-May and, exceptionally, to early June (Sierra los Ajos on 3 June, per Marshall 1957; and one in the Sierra Huachinera on 1 June, per J. Marshall). The earliest fall arrival dates are 3 September in the Sierra San Luis (Marshall 1957) and 9 September in Arroyo Cajón Bonito.

Yellow-rumped Warblers not only forage by gleaning at all levels in trees and bushes, but they often feed with sparrows on the ground in the open, and even among the rocks of the shoreline of El Golfo de California. Frequently, especially in late winter, they pick insects from the air over flowering trees and fields.

The name "Audubon's Warbler" was once used for the subspecies that commonly occur in Sonora. Occasionally, and possibly more often than is realized, individuals of the more eastern group of subspecies (once known as a separate species, the Myrtle Warbler) are found in Sonora. Van Rossem (1945) mentioned specimens taken in the Río Colorado delta on 3 February 1929 and at Tecoripa on 30 March 1929. A. Phillips collected a Myrtle Warbler on the Río Yaqui above Esperanza on 19 November 1954 and another south of Nogales on 19 December 1956. The Myrtle Warbler was identified at Alamos in late December 1985 (D. Stejskal) and near Ures in the winter of 1979–80 (Terrill 1981).

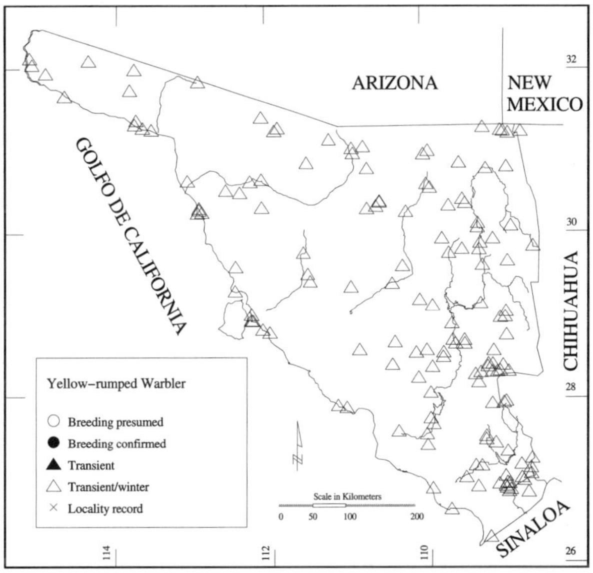

BLACK-THROATED GRAY WARBLER
Chipe Negrogris
Dendroica nigrescens

The southern breeding limit of the Black-throated Gray Warbler, with the possible exception of Baja California Norte, is found in northeastern Sonora, in the Sierra los Ajos and the Sierra San Luis (Marshall 1957). There it nests from May into July, usually in oak-juniper-pinyon but also in mixed pine-oak woodland.

The species is a transient statewide except along the coast from Bahía Kino south. In the driest part of the desertscrub it is usually limited to oases or large drainageways such as the Río Asunción. It is more common in the denser tropical thornscrub and the tropical deciduous forest. Noticeable migratory flights occur from about 10 March to 10 April and again from mid-September through October.

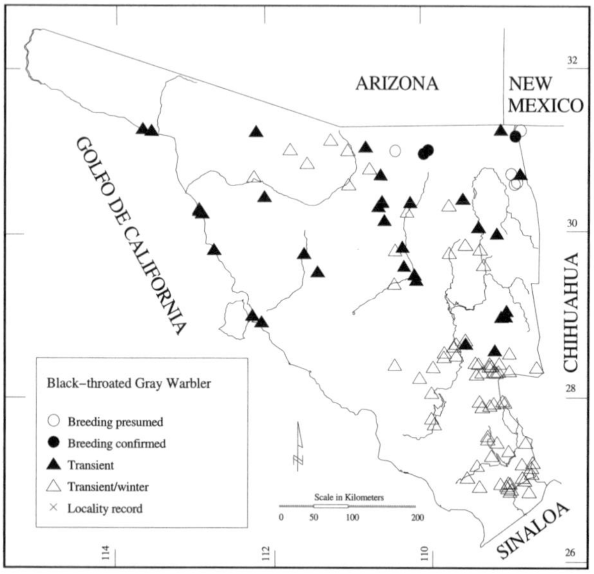

The Black-throated Gray Warbler winters commonly in Sonora north to the latitude of Hermosillo, chiefly in woodlands along streams and sometimes in well-vegetated waterless canyons. As many as 113 have been listed on an Alamos Christmas Bird Count. On 3 January 1981 one was found at Puerto Peñasco (D. Stejskal).

TOWNSEND'S WARBLER
Chipe Negroamarillo
Dendroica townsendi

Townsend's Warbler is an uncommon transient in Sonora, seen chiefly in the mountains and undetected in the thornscrub and desertscrub west of the Río Yaqui in the south and the Ríos Sonora and Magdalena in the north. It does, however, pass through the deserts of southwestern Arizona, especially in the spring. It would be expected on migration in northwestern Sonora, but there is only one record from the area (near El Doctor on 7 April 1994). There are but two coastal records of migrants: two specimens taken at Guaymas on 8 May (van Rossem 1945) and one seen at Punta Cirio on 10 April. Extreme spring migration dates are 12 March in Arroyo el Reparo and 31 May in the Sierra la Madera (van Rossem 1945) and the Sierra los Ajos (Marshall 1957). We have seen as many as 18 in spring migration (on 16 May) on the Yécora road at the divide below Mesa de Enmedio. A male collected (LSUMZ) in oaks in the Sierra Aconchi on 23 June 1954 by J. Marshall was extraordinary. Fall migration dates extend from 16 August (in pines southwest of Nacozari) to 19 September (in oak-juniper northwest of Cananea).

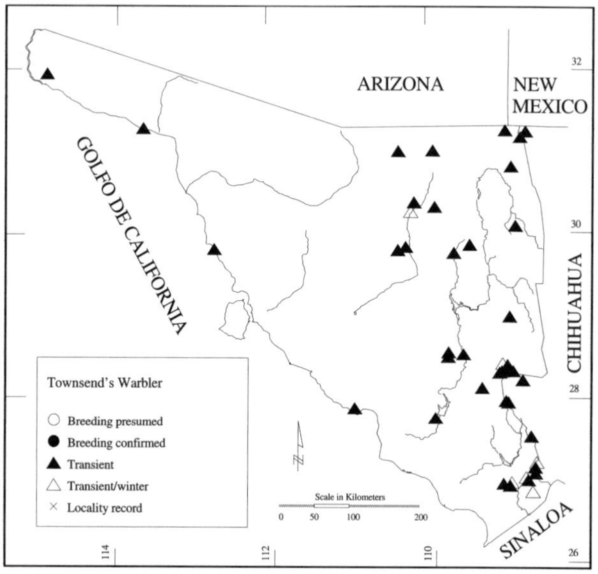

A few Townsend's Warblers have been recorded in winter in the southeast, chiefly from the Alamos area (where they are irregular on Christmas Bird Counts, but up to 27 were on one count). One was seen near Santa Rosa (1450 m) on 8 January. The species has been recorded three times on Christmas Bird Counts at Puerto Peñasco.

HERMIT WARBLER
Chipe Cabeza Amarilla

Dendroica occidentalis

Hermit Warblers are uncommon in Sonora, although many migrants presumably pass through the state in spring and fall. They have been found mostly in the mountainous eastern parts. There are but two records for the coast of El Golfo de California: one at Punta Cirio on 10 April 1970, and one near El Doctor on 9 April 1994. Although they are fairly common in spring and fall in the deserts of southwestern Arizona, there are no records for the desertscrub and thornscrub of Sonora. Aside from the above Punta Cirio and El Doctor birds, only four records exist for elevations below 1000 m.

Migration dates are, in spring, from 13 March (on Mesa de Enmedio) to 29 May (in the Sierra los Ajos, A. Phillips) and, in fall, from 28 July (in the Sierra San Luis, Marshall 1957) to 18 September (near Cananea and in the Sierra Oscura). The largest number of birds observed in one day has been 14 on Mesa de Enmedio on 16 May.

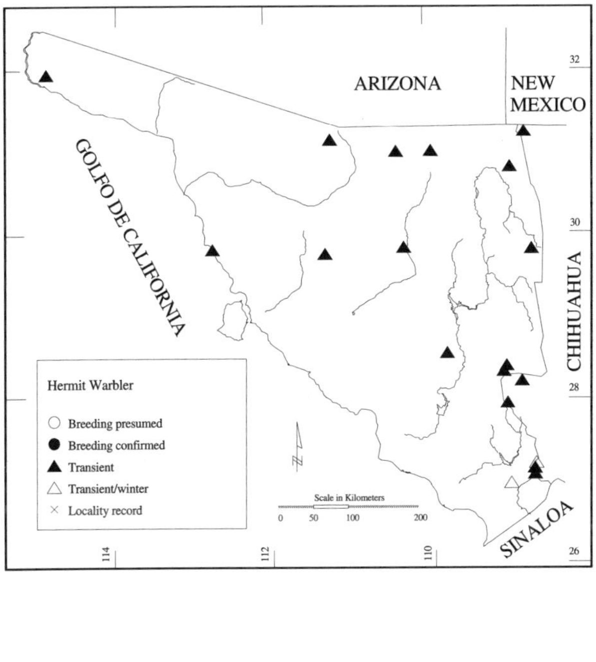

In winter, Hermit Warblers were seen by D. Stejskal on Alamos Christmas Bird Counts in 1983 (two) and 1989 (one). We saw one barely across the state boundary at Las Chinacas, Chihuahua (1500 m), northeast of Alamos, on 10 January 1981.

BLACK-THROATED GREEN WARBLER
Chipe Dorso Verde

Dendroica virens

This eastern species is casual in Sonora. One was collected at Huasabas (DVM) on 15 November 1953 (A. Phillips). A second specimen (UK) was obtained about 34 km south-southeast of Nogales on 20 December 1956 (Rising 1988).

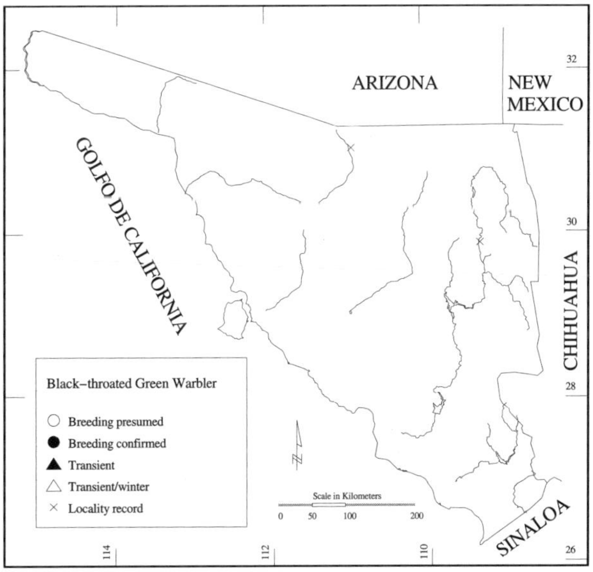

[BLACKBURNIAN WARBLER
Chipe Garganta Naranja
Dendroica fusca

There is only one record of this eastern wood-warbler in Sonora. An adult female was seen near Magdalena on 26 October 1980 (S. Terrill, A. Gast).]

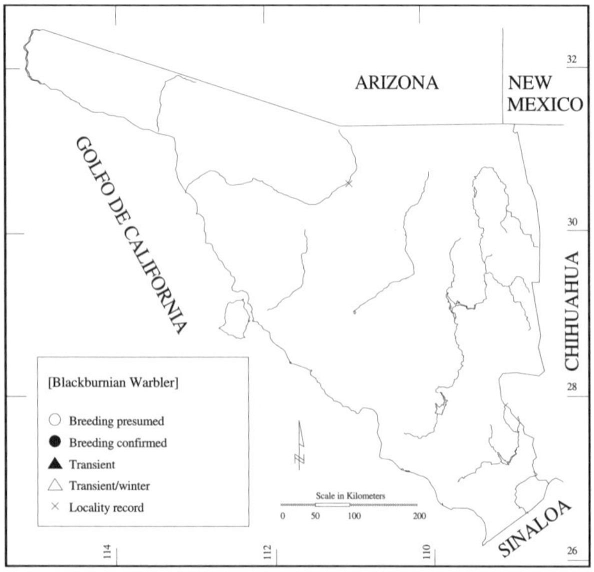

[YELLOW-THROATED WARBLER
Chipe Garganta Amarilla
Dendroica dominica

A. Rea observed a Yellow-throated Warbler feeding in an acacia with other wood-warblers at Quitovac, a village 40 km south-southeast of Sonoita, on 15 May 1982.]

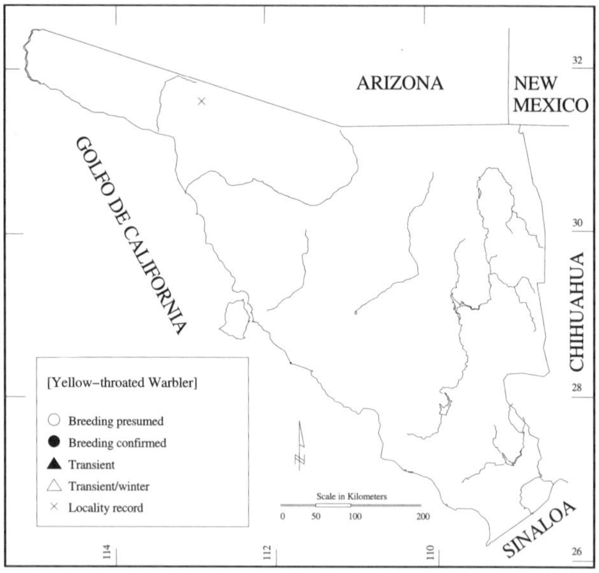

GRACE'S WARBLER
Chipe Ceja Amarilla
Dendroica graciae

Grace's Warbler is a common summer resident of pine forests and pine-oak woodlands of the higher mountains of Sonora. It forages chiefly in pines but is found now and then in oaks and other broad-leafed trees. The earliest spring records are of four on 13 March on Mesa Grande northwest of Yécora, and of six on 13 March on Mesa de Enmedio. The latest fall record is of one found on 17 September in the Sierra Oscura. Grace's Warblers

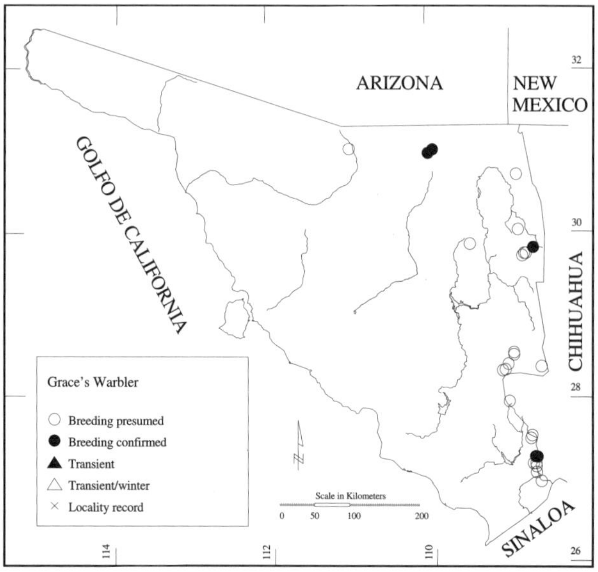

must migrate only through the mountains; there are no reports from habitats below pine-oak woodland. We have never noted large numbers of individuals indicative of transients. Our information indicates that nesting begins in May and extends into July, although territorial males have been noted in mid-March.

PALM WARBLER
Chipe Playero
Dendroica palmarum

P. Pyle and S. Howell found ten Palm Warblers, six of them in one flock, on 4 December 1983 on the edge of the Río Colorado delta about 15 km northwest of Golfo de Santa Clara. The observers substantiated the record with careful details. On 18 and 19 December 1993, three birds were found at Puerto Peñasco (S. Ganley).

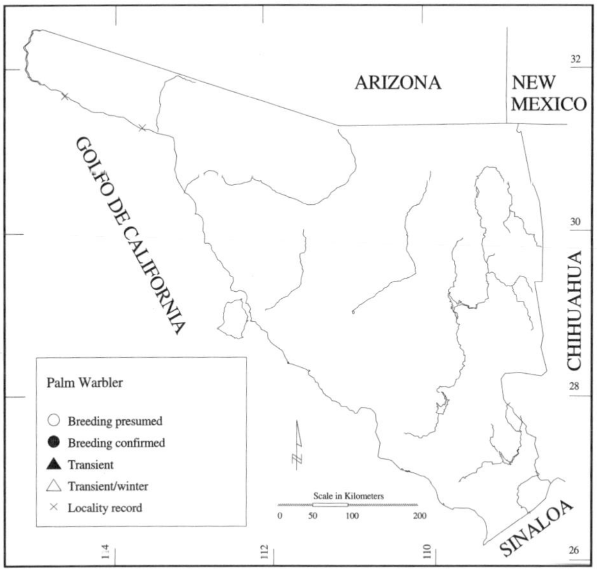

[BAY-BREASTED WARBLER
Chipe Castaño
Dendroica castanea

A male was observed on the Río Sonora west of Ures on 29 May 1980 by G. Rosenberg, B. Whitney, and R. Dummer.]

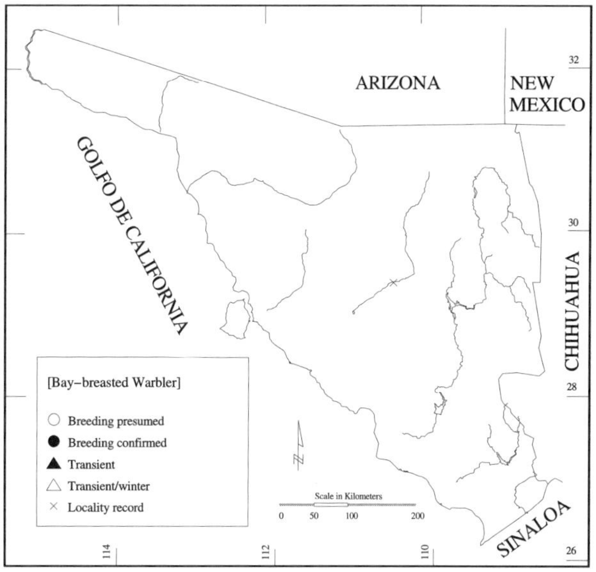

BLACK-AND-WHITE WARBLER
Chipe Trepador
Mniotilta varia

We know of ten records of the Black-and-white Warbler from Sonora, each involving one to four individuals on dates as early as 12 November and as late in spring as 15 March. Most observations have been of birds in tall riparian growth at elevations from 36 to 980 m, but one bird was in pine-oak woodland (1380 m, south of Milpillas).

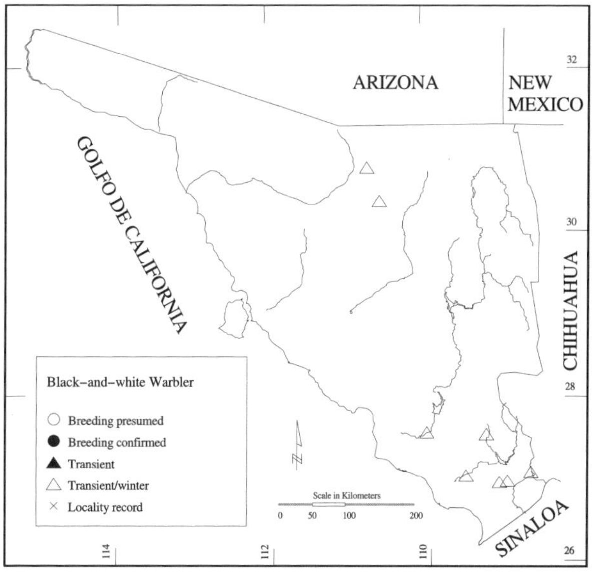

AMERICAN REDSTART
Chipe Flameante
Setophaga ruticilla

Van Rossem (1945) was the first person to note this warbler in Sonora. He saw one at Estero Tóbari on 26 April 1930 and collected another the next day. Subsequently there have been seven records: two immature males (one collected, LSUMZ) in thicket of silverleaf oak in the Sierra los Ajos on 16 July 1952 (J. Marshall, A. Phillips); one collected on the Río Sonora just south of Aconchi on 29 December 1952 (A. Phillips); one seen at Bácum near Cd. Obregon in December 1964 (Alden 1969); one at Sonoita on 11 November 1968 (R. Cunningham); a female observed in Arroyo Cajón Bonito on 9 September 1982; a male seen in Cajón la Calera in the Sierra Huachinera on 3 September 1983; and an immature female netted on the Río Bavispe about 7 km south of Bacerac on 4 September 1983.

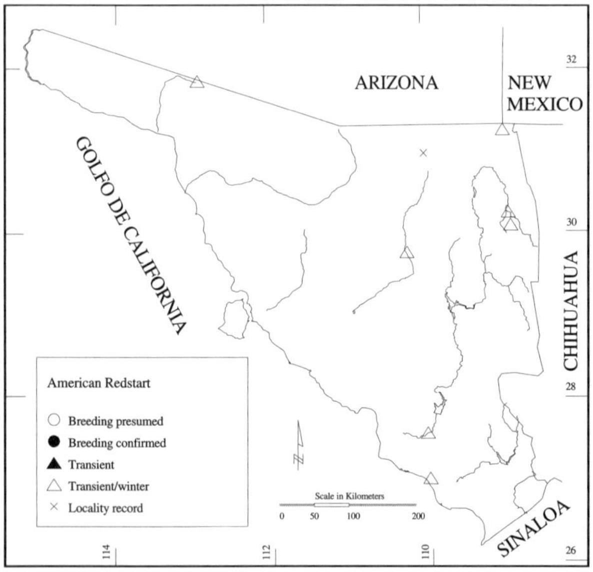

[PROTHONOTARY WARBLER
Chipe Dorado
Protonotaria citrea

Monson saw a male Prothonotary Warbler in batamotes along the Río Cuchujaqui south of Alamos on 26 April 1972.]

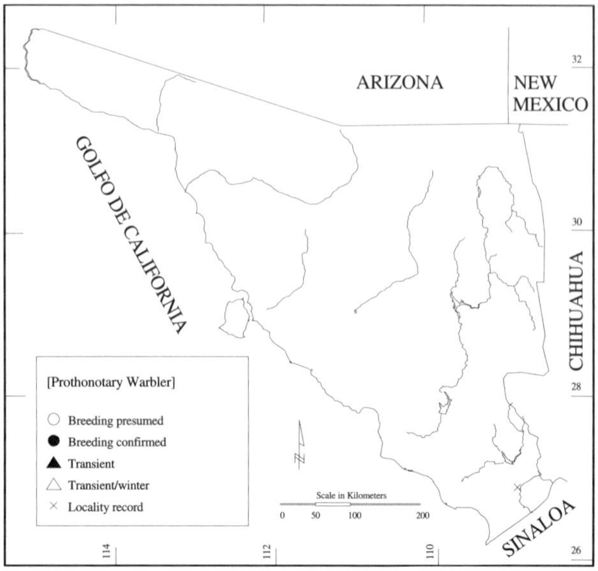

OVENBIRD
Chipe Suelero
Seiurus aurocapillus

Ovenbirds have been detected four times in Sonora: a singing male was collected (LSUMZ) at a spring in the Sierra Aconchi on 21 June 1954 (Marshall 1957); single birds were banded in mangroves near Punta Chueca on 21 June 1974 and 28 May 1975 (Russell and Lamm 1978); and one was seen in Arroyo Cajón Bonito on 10 May 1983 (R. Morse, R. Taylor).

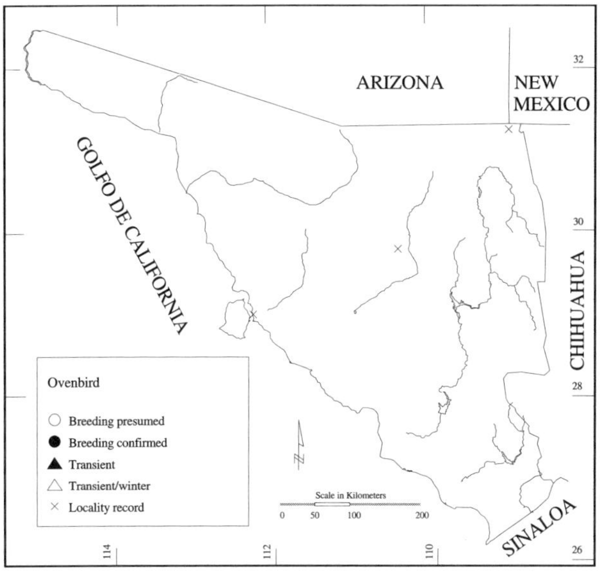

NORTHERN WATERTHRUSH

Chipe Charquero

Seiurus noveboracensis

The Northern Waterthrush is a fairly common fall transient and rare winter visitant and spring transient in mangroves along the coast of El Golfo de California. Away from the coast, it is a rare spring and fall transient in shaded aquatic habitats. There are two inland winter records (Lo de Campa on 8 January 1988 and east of Alamos on Arroyo el Palmarito on 5 January 1993).

Data from 27 birds banded in mangroves in the vicinity of Punta Chueca indicate that the Northern Waterthrush is a regular fall transient but is rare in spring. A few individuals winter in the mangroves. One wintering bird returned the next September and was still present on 29 November. On 18 September 1974, nine were netted in the same morning that seven Swainson's Thrushes were captured. An intense thunderstorm during the night just inland from the coast may have forced the birds from the air. Spring coastal migrants apparently move through in a short period in mid-April, as our only records are 12–15 April. Van Rossem (1931) reported several in Estero Tóbari mangroves on 26 April.

Dates of fall migration in the interior extend from 4 September (Mazatán and San Bernardino) to 13 September (Arroyo Santo Domingo). One was collected at Rancho la Arizona on 28 November (J. Marshall). For spring, we have only three records: one near Navojoa on 27 April (A. Phillips); one near Tepoca on 12 May; and one in Arroyo el Reparo on 14 May.

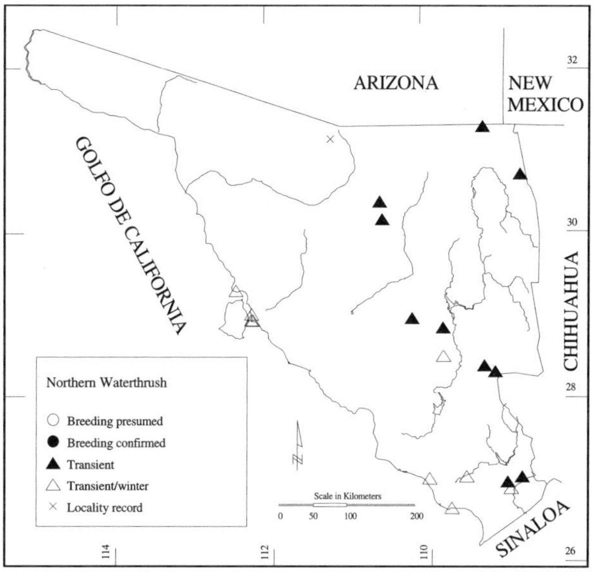

LOUISIANA WATERTHRUSH

Chipe Arroyero

Seiurus motacilla

The Louisiana Waterthrush is a rare transient and wintering bird in Sonora. The 12 records range seasonally from 12 October (1954, La Casita south of Nogales; A. Phillips) to 28 March (1888, Alamos; van Rossem 1945). All come from lowland streambeds.

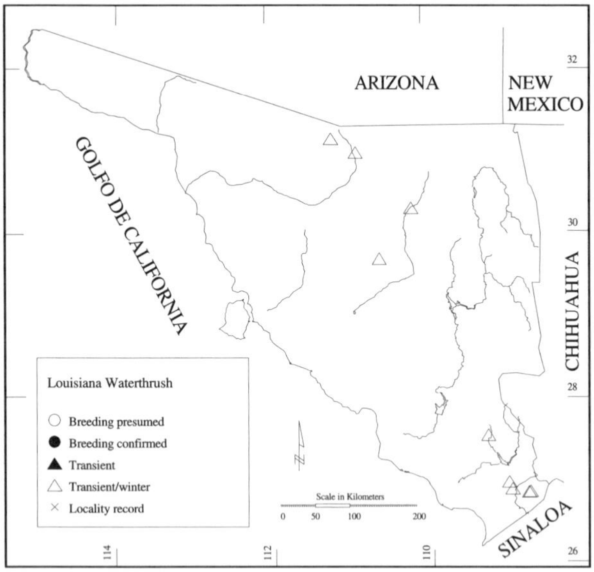

KENTUCKY WARBLER

Chipe Patilludo

Oporornis formosus

A Kentucky Warbler was taken at Rancho Santa Barbara about 9 km north of Milpillas on 12 June 1937 (van Rossem 1945), and a male was seen in a brushy area at the edge of the Río Yaqui below Presa Obregon on 15 May 1986 (A. Moorhouse).

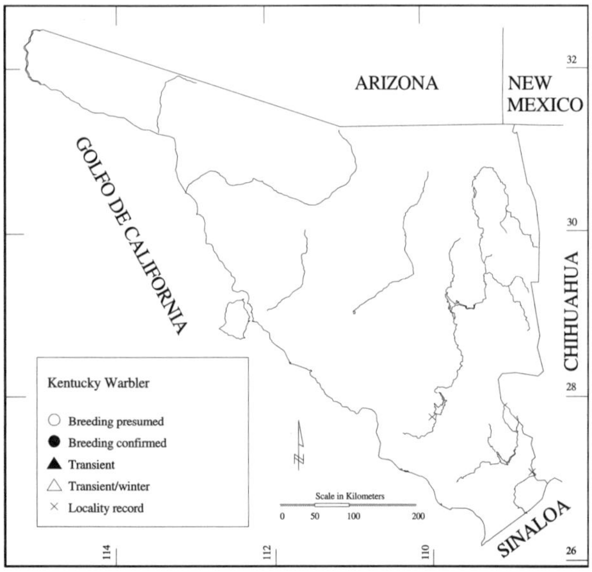

MACGILLIVRAY'S WARBLER

Chipe de Tolmie

Oporornis tolmiei

This wood-warbler forages in dense brush and the lower branches of trees. It is a common transient below 1200 m throughout all of Sonora, except along the coast where it is rare; in the higher mountains it is uncommon. It is fairly common in winter in underbrush of the tropical deciduous forest, especially along streams, but rare farther north. An unusual late fall record is that of one seen at Punta Sargento on 28 November.

Spring migration begins early in March and peaks rapidly (mid-March). Stragglers have been found until 21 May in far northeastern Sonora. The earliest fall migrants have been noted as early as 14 August, and most transients move through the state in the first half of September. In northern Sonora, the migrants have passed through by the end of September.

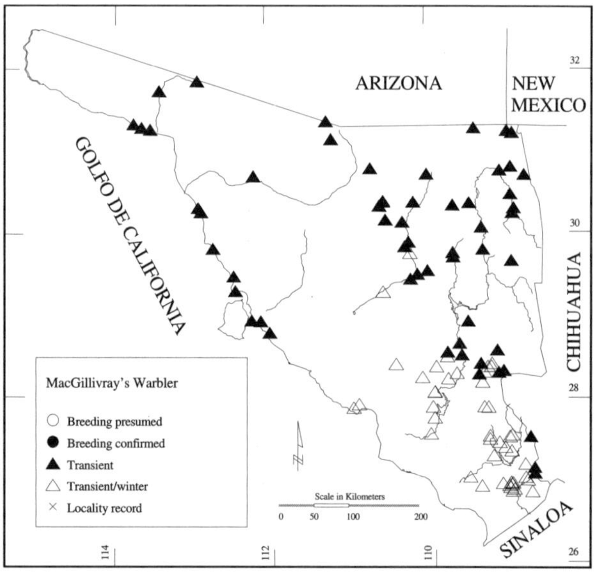

COMMON YELLOWTHROAT
Mascarita Común
Geothlypis trichas

The Common Yellowthroat is most frequently found in cattails, mangroves, or very dense shrubbery (such as batamote) along streams. Although the species is present throughout the year, Sonoran birds that have nested may migrate and be replaced by others from farther north. Migrants may move through habitats away from water and have been found in scrubby brush from the coast upward into the oaks. Migration dates are 20 March to early May and late August to October. At Puerto Lobos, where the species is a transient, all spring records are in the period of 20 March to 9 April.

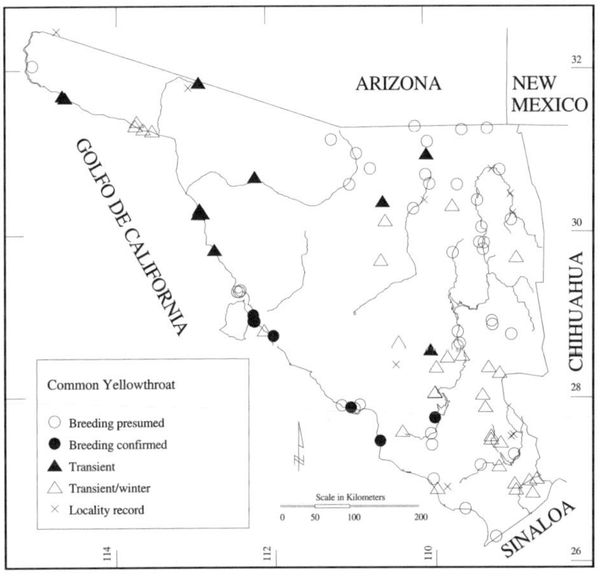

Yellowthroats are uncommon breeding birds of the coastal mangroves. Nests are typically constructed in the last two weeks of May in mangroves or salicornia within 1 m of the water or wet sand. The six clutches (WFVZ) examined each contained three eggs. Single broods may be normal, but we have found fledglings at Punta Chueca as late as 8 September. Our banding data from Punta Chueca suggest that the breeding birds migrate by late September and are replaced by northern birds, some of which spend the winter. The summer residents return in mid-March and April.

We have seen fledglings in the cattails of the lower Río Yaqui in early July. Yellowthroats observed singing in appropriate marshy habitats in April to June are likely to be nesting there. Our data are inadequate to verify nesting in most suitable places, much less establish where the population is resident or migratory. We think the breeding birds of the lower Ríos Yaqui and Mayo may be resident.

HOODED WARBLER
Chipe Encapuchado
Wilsonia citrina

An adult male Hooded Warbler was collected on the Río Cuchujaqui southeast of Alamos on 30 January 1966 (Russell and Lamm 1978). A. Moorhouse found one in Cañon Evans in the Sierra los Ajos on 6 July 1991.

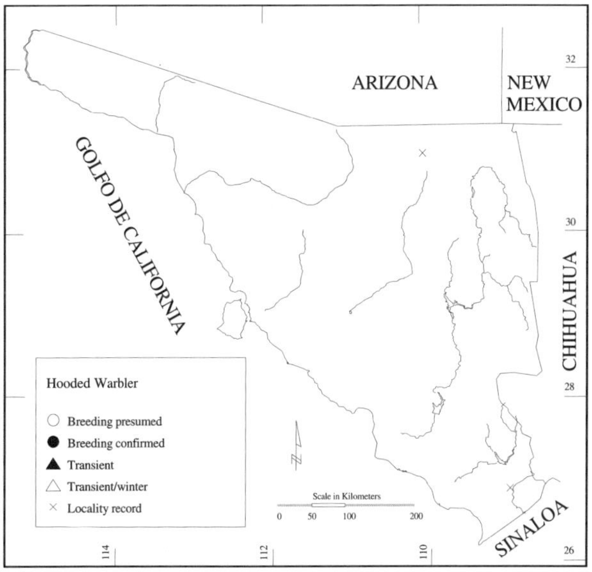

WILSON'S WARBLER

Chipe Corona Negra

Wilsonia pusilla

The Wilson's Warbler is one of the more numerous transient wood-warblers in Sonora. It is also fairly common in winter north to the lower Río Yaqui valley. On migration it is found in all nooks and corners of the state, so long as there are a few bushes and a tree or two. Ninety have been counted in one locality (Arroyo Santo Domingo, 11 September)—all singles, for this warbler does not travel in groups or flocks as do the Yellow-rumped Warbler and others. It is a lively, darting bird that makes short quick flights among trees and shrubs as it picks off its prey from the underside of leaves and twigs. Now and again it will descend to the ground to feed, which sometimes ends in disaster—one was removed from the belly of a Ridge-nosed Rattlesnake (*Crotalus willardi*) in the Sierra San Luis (Marshall 1957).

There are only a few records from winter north of the lower Río Yaqui. Exceptionally, one was banded at Punta Chueca on 17 December, and two were observed at Puerto Peñasco on 3 January. Another wintering bird was recorded near San Ignacio in the Magdalena area in December 1979–January 1980 (Terrill 1981).

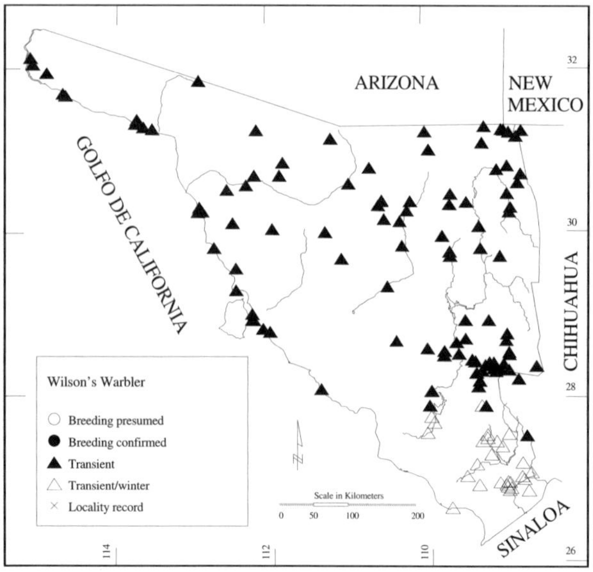

Migration takes place from mid-March through May and again from late August to early October. The species often forages in mangroves. Outside dates in spring are 11 March (El Carrizo) and 2 June (Sierra los Ajos; A. Phillips). In fall, they are 16 August (Rancho la Arizona; van Rossem 1945) and 25 October (Sierra San Jose; van Rossem 1945). One at Lo de Campa on 25 November could have been wintering. Along the coast this species is more common in spring than in fall.

RED-FACED WARBLER

Chipe Cara Roja

Cardellina rubrifrons

Red-faced Warblers are uncommon summer residents above 1700 m in some mountain ranges in eastern Sonora. The species is most frequently associated with deciduous trees (such as Gambel oak) in coniferous forest; its scarcity in Sonora is partially attributable to the scarcity of this habitat. We have never encountered more than four in an area. The earliest arrival date is 13 March and the latest fall date is 18 September, both in the Sierra Oscura. The species undoubtedly nests in Sonora, but we have no direct evidence that it does so.

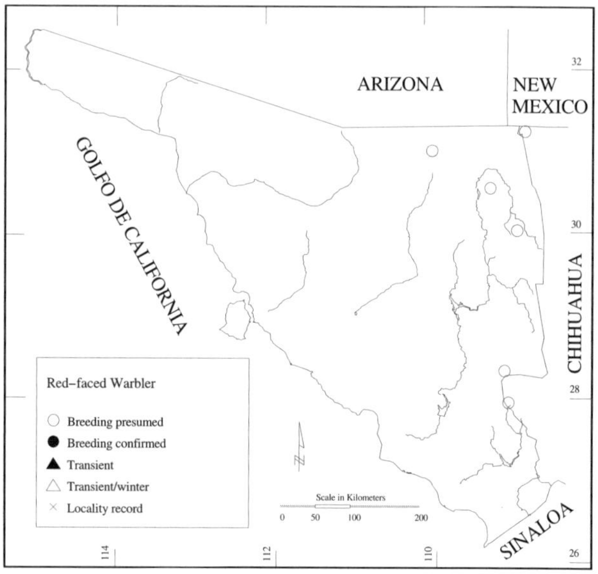

PAINTED REDSTART
Chipe Ala Blanca
Myioborus pictus

J. Marshall (1957) designated this species the most characteristic pine-oak bird. It breeds commonly in all pine-oak communities and is made conspicuous by its fluttery flights and often fanned wings and tail. It feeds by gleaning insects from vegetation at all heights but shows a preference for shaded places and north-facing slopes. It often is active in the near darkness of early morning and late evening.

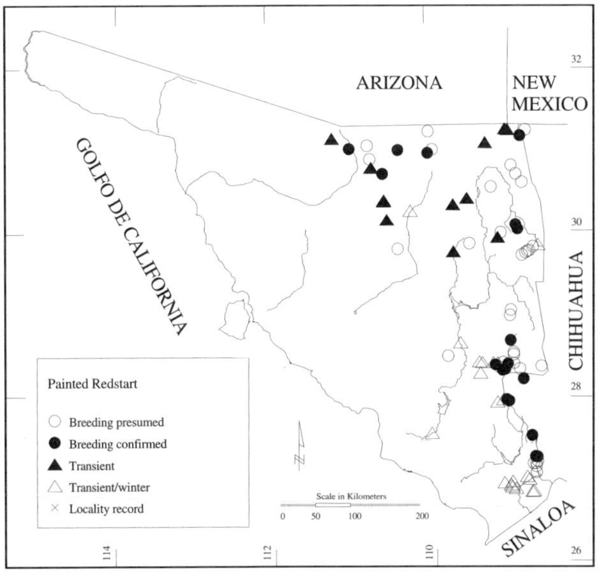

In the northern half of Sonora the species is migratory, although an occasional individual is noted in winter. It winters regularly in the lowlands near Alamos and not infrequently in the eastern valleys of southern Sonora, preferring lower oaks and riparian woodlands. Northward migration begins early; we found singing birds in the Sierra los Ajos on 3 March. In August some leave their breeding areas, but many remain into September.

Nests are placed on or near the ground, beginning in early April. Some pairs may be nesting as late as July; at the same time, juveniles from earlier broods are beginning to aggregate in loose flocks in canyon bottoms. Young birds lack red in their plumage but may attract attention through their singing, which may be well developed by the time the first red feathers appear in the prebasic molt.

SLATE-THROATED REDSTART
Chipe de Montaña
Myioborus miniatus

The Slate-throated Redstart is a rare summer resident in two mountain ranges along the Chihuahua border above 700 m. Two single birds were found near Sahuarivo on 15–16 July 1987, and a pair was on Mesa de Enmedio on 16 May 1988. Six or seven on Mesa de Enmedio on 13–14 March 1989, at least one of them singing, were thought to have just arrived. W. Leitner noted four on the slopes below the mesa on 26 May 1991. There are nine reports from elevations below 700 m, all within the period 29 November to 15 March.

A male at Mesa de Enmedio on 13 March foraged less than a meter above the ground in the bottom of a damp ravine, singing every 18–20 seconds. It moved frequently, appearing to glean an insect every 3–4 seconds. It caught insects by popping 0.3–1 m up from a low perch, often dropping back to the same perch. Later this bird foraged 6–8 m above the ground in oaks and madrones and 15 m in pines.

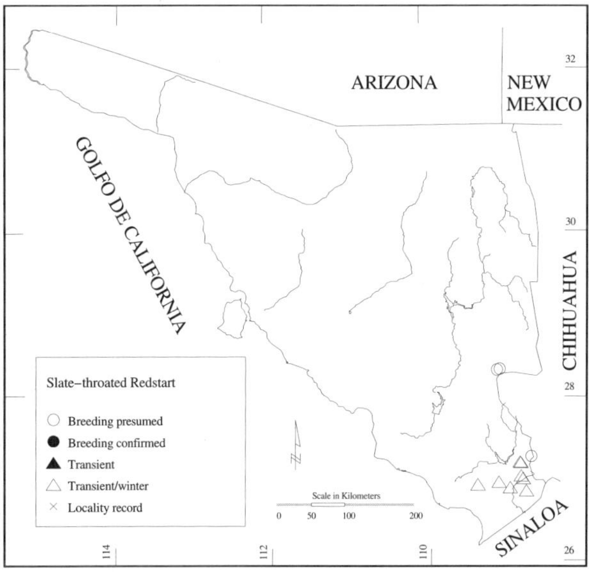

FAN-TAILED WARBLER
Chipe de Roca
Euthlypis lachrymosa

The Fan-tailed Warbler is local in Sonora, present from mid-March to late July. We found it to be common in only one place, the incomparable Arroyo el Reparo, on 13–15 May 1988. There, under the sycamores, we estimated ten territories per kilometer of riparian growth. Elsewhere we found these warblers in ones and twos at Sahuarivo on 16 July 1987; in an arroyo 4 km southwest of Guajaray on 16 May 1986; and at 2000 m in the Sierra Oscura on 13 March 1982. Winter records are lacking. A male found at the waterfall in Arroyo Cajón Bonito on 23 May 1982 was perhaps north of its usual range; six days later it had been joined by a female (D. Danforth, A. Moorhouse). Contrary to the usual ground-loving proclivities of the species, this bird was busily flying in a rough circle through the tops of tall sycamores, ashes, and walnuts, perching briefly and singing incessantly and seldom descending to the ground during the hour or so that it was under observation.

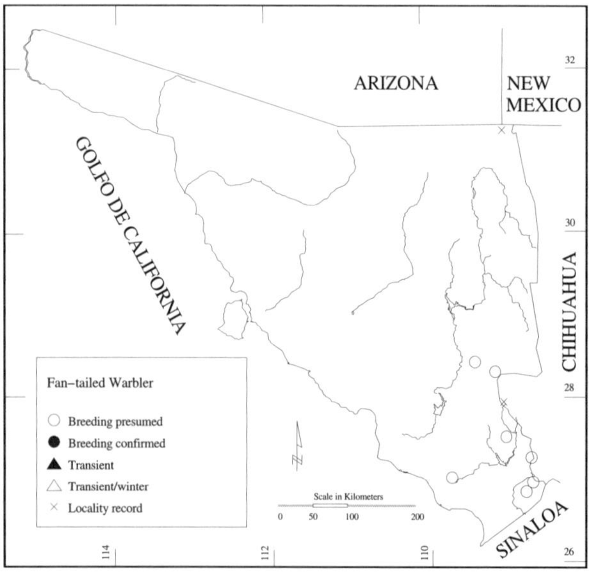

RUFOUS-CAPPED WARBLER
Chipe Gorra Rufa
Basileuterus rufifrons

The diminutive and very active Rufous-capped Warbler is a common to fairly common resident throughout the mountains and foothills of the eastern side of Sonora as far north as El Riito east of Nacori Chico, where it was found in June of 1953 and 1960 (J. Marshall). It occupies dry, rocky situations at the edges of canyon bottoms, where riparian plants give way to those of dry hillsides. There it builds its nest in open sites on the ground. Monson watched one in the Sierra el Tiznado east of Mazocahui on 10 June 1982 constructing a nest. Its mate was not seen. We have found this species up to 2000 m in the Sierra Oscura and as low as 205 m near San Antonio de la Huerta on the Río Yaqui.

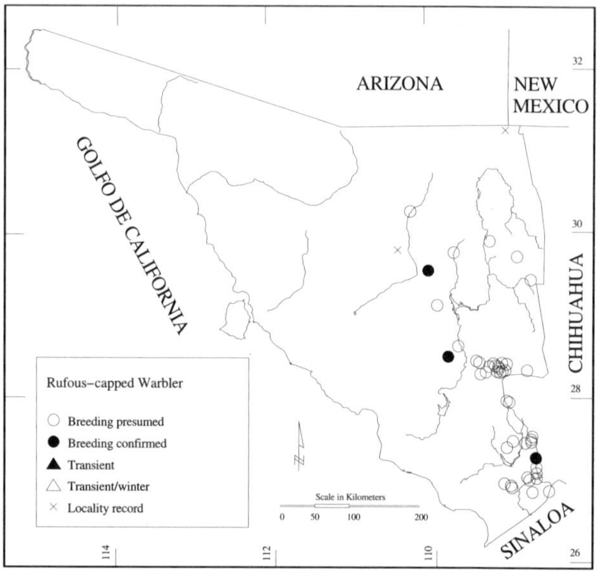

During the period 22–25 February 1957, J. Marshall found the Rufous-capped Warbler in the Sierra Aconchi. The most northern locality of record is Arroyo Cajón Bonito, where a singing male was found on 13 September 1975 (D. Danforth) and four singing birds were observed on 14 May 1976 (Russell and Lamm 1978); it has not been found there again. Pairs are regularly found throughout the winter, suggesting that male and female are permanently paired.

YELLOW-BREASTED CHAT

Buscabreña

Icteria virens

A perennial stream bordered by mesquites, willows, batamote, and arrowweed provides typical habitat for Yellow-breasted Chats in Sonora. They can be unbelievably abundant in this plant association—for instance, we counted 355 along 4 km of stream at El Carrizo north of Tonichi on 18 July. Summer days (and often nights) are filled with the loud jumbled whistles, catcalls, and harsh notes of these birds, which are the largest wood-warblers. If a large desertscrub wash is sufficiently bordered by dense, tall, tangled growth that it becomes lush with the summer rains, an occasional pair of chats will be found there. Chats range over much of Sonora from the southern lowlands to elevations of up to 1280 m (as at Arroyo Cajón Bonito, where willows and batamote are supplemented by

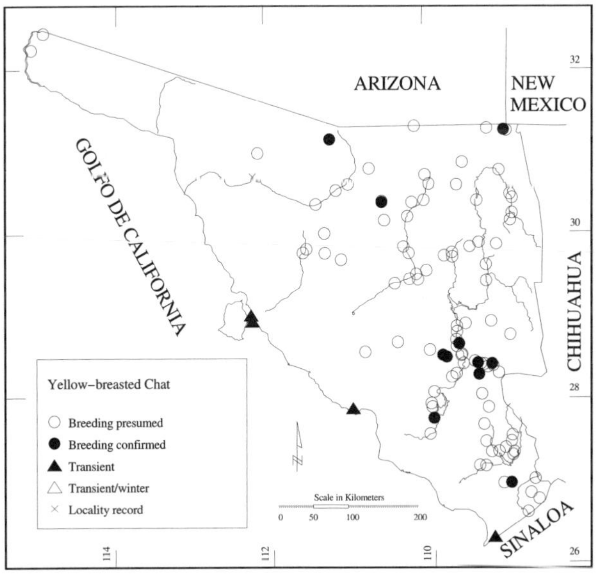

Senecio salignus). They are summer residents or migrants in Sonora, with but one winter record (near Alamos on 30 December 1984; D. Stejskal). The spring migration peaks in mid-May, the fall migration in mid-September. There are only three records in March and two in April. Two banded in mangroves near Punta Chueca on 30 May may have been migrants. Only two chats have been noted in October (at Lo de Campa on the 13th). Especially late was an immature male collected near Caborca on 2 November 1948 (Phillips and Amadon 1952).

OLIVE WARBLER

Ocotero Enmascarado

Peucedramus taeniatus

The Sonoran breeding population of the Olive Warbler is thinly scattered through the pine-oak woodland. These warblers are fairly common but hard to see among the thick pine needles where they feed; their presence is often revealed by faint calls reminiscent of those of Western Bluebirds. "They have a very erratic, swift, zigzag flight in entering a tree. They reverse position and twitch back and forth on a twig like a redstart" (J. Marshall, unpublished field notes). Individuals are paired and courtship is well under way as early as mid-March. Some birds are still nesting in late July. During the winter, they often join flocks of other bird species and may range down into oaks below the lowest pines (as near Santa Rosa, 1450 m, January).

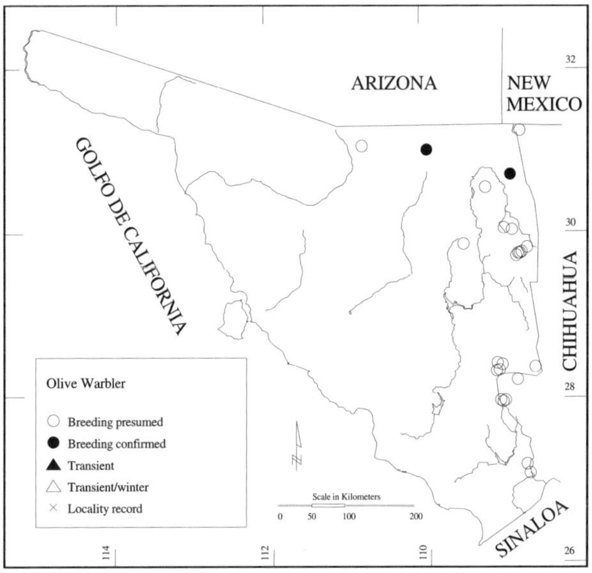

SUBFAMILY THRAUPINAE

Tanagers

SCRUB EUPHONIA

Eufonia Garganta Negra

Euphonia affinis

Euphonias are small finchlike tanagers, rare in Sonora. Van Rossem (1934) provided one record of the Scrub Euphonia: two birds collected by M. Frazar at Alamos on 16 and 21 March 1888. Two recent records are also from Alamos, of individuals seen on Christmas Bird Counts: three observed by D. Stejskal on 28 December 1982 and a single male identified by him on 28 December 1985. The birds were in a large north-facing canyon of the Sierra Alamos and had a very distinctive three-note whistled call.

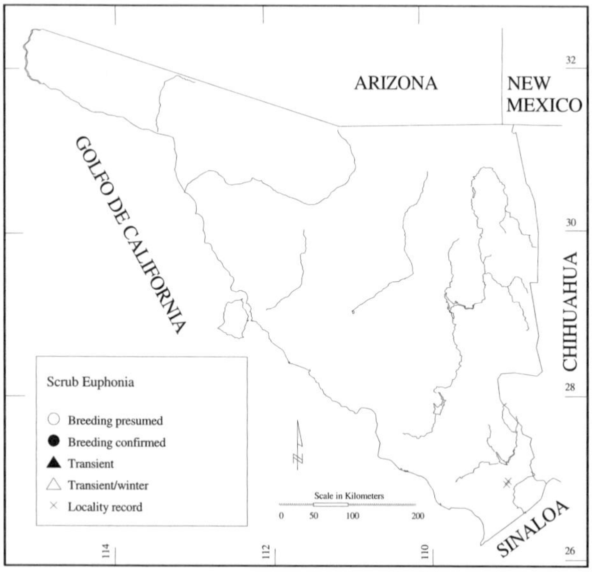

BLUE-HOODED EUPHONIA

Eufonia Capucha Azul

Euphonia elegantissima

This exceptionally beautiful euphonia is known from only eight Sonoran localities. Extreme dates of occurrence are 13 March (San Antonio, near Milpillas) and 15 July (Barranca las Colas near Sahuarivo), except for one early winter record. On 27 November 1993 on the highway east of Yécora and barely inside Sonora on the border with Chihuahua, seven were attracted to an imitation of pygmy-owl calls (M. Hage et al.).

All individuals have been observed in pine-oak woodland between 1250 and 1660 m, except a bird collected (MVZ, 24 April 1933) 2 km northwest of Alamos (390 m). The species feeds extensively upon the fruit of a large weeping mistletoe that we have seen only in oaks. There is no evidence that the Blue-hooded Euphonia has nested in Sonora, although the dates suggest it. The spotty distribution (in both location and dates) may reflect the perhaps unpredictable pattern of flowering and fruiting in its mistletoe food staple.

The common name of this species has been changed to Elegant Euphonia (AOU 1997).

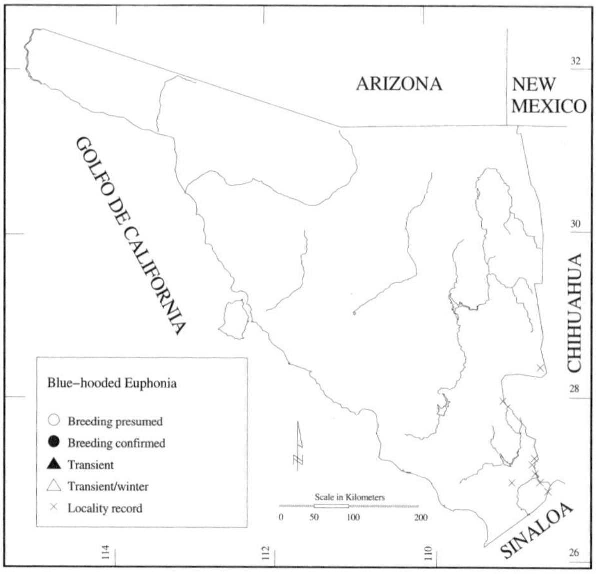

HEPATIC TANAGER
Tángara Encinera
Piranga flava

Hepatic Tanagers are common breeding birds in pine-oak woodland throughout the higher mountains of Sonora. Most arrive in mid-May and depart in September. Some individuals may be present in the breeding range of southern Sonora in winter, even at elevations up to 1800 m. Wintering Hepatic Tanagers will be found in small numbers from October and November through March at elevations below 1000 m in large riparian trees, including larger oaks, in northern Sonora. In riparian areas of the tropical thornscrub and in tropical deciduous forest they are fairly common in winter.

They nest from late April to early August, typically in the pine-oak association but sometimes in the lowest oaks in sycamore-cottonwood canyons, as in Arroyo Cajón Bonito. Marshall (1957:114) wrote: "Its song, the most conspicuous in the dawn chorus, begins while the caprimulgids are still singing. It continues for a half hour or so while the males are stationary in their respective pines."

Although a diet of insects is usual in summer (Marshall 1957), the Hepatic Tanager was among the several species of birds feeding on the drupelike fruits of *Bumelia persimilis* in Arroyo el Reparo in May 1988.

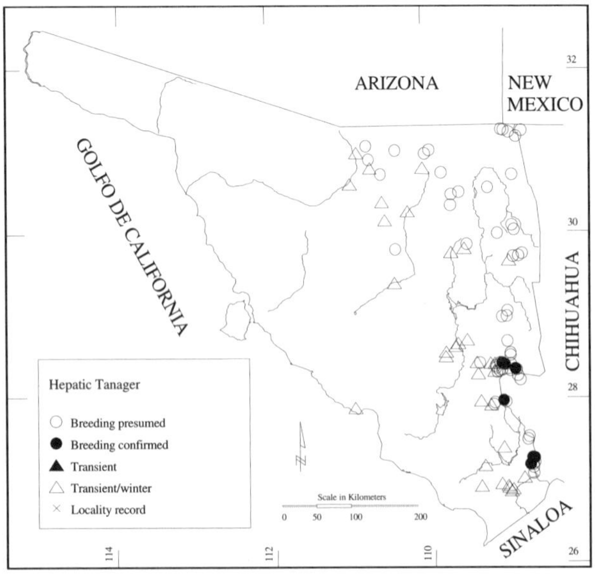

SUMMER TANAGER

Tángara Roja

Piranga rubra

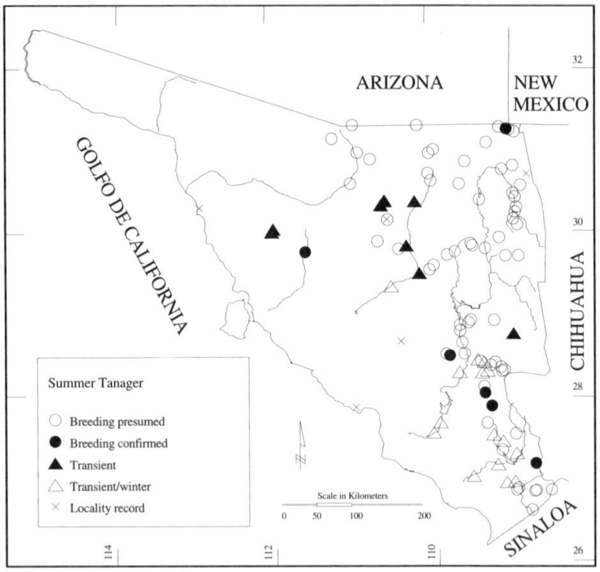

The Summer Tanager is a common summer resident in Sonora. It selects tall deciduous riparian trees such as willows, as well as chinos and feather trees growing along arroyos that may hold little or no water in the dry season. Its altitudinal range is approximately 50 m to as high as 1750 m, where it utilizes oaks and sycamores. Its abundance in summer is shown by 20 pairs counted in less than 2 km^2 at Lo de Campa on 17 July. Nesting is in June and July.

The migration period includes birds passing to and from nesting areas to the north. They are common in the lowlands at this time and stray upward into oaks. We believe that one on the Río Tepoca north of Nuri on 15 March provides an early spring record, although the main migration begins in mid-April. Fall migration takes place in September. At Lo de Campa, where they nest, a female banded on 6 May was still present on 27 September. We have only two records in October. The seasonal status of one at Puerto Lobos on 11 November 1986 (J. Bates) is unclear; Summer Tanagers are not otherwise reported from the coast of Sonora, except for a specimen mentioned by van Rossem (1945) as taken near Guaymas on 15 January 1933.

A few individuals winter in southern Sonora. Up to four have been listed on an Alamos Christmas Bird Count, and there are late December and January records from half a dozen other southern localities. Far to the north, one was found near Ures in December 1979–January 1980 (Terrill 1981).

SCARLET TANAGER
Tángara Escarlata
Piranga olivacea

An immature male Scarlet Tanager collected (DVM) near Huasabas on 26 November 1953 by A. Phillips is the only Sonoran record for this eastern bird.

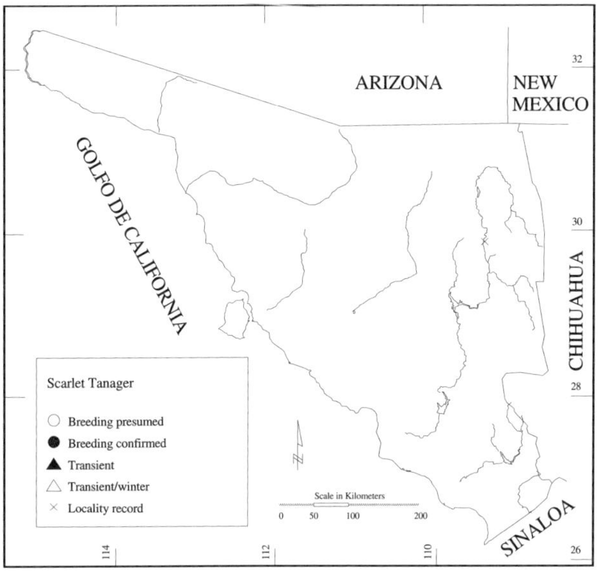

WESTERN TANAGER
Tángara Capucha Roja
Piranga ludoviciana

Although the species breeds in southern Arizona, a Sonoran nest of the Western Tanager has not been found. Marshall (1957) concluded that van Rossem (1945) must have erred in stating that the species nested, for their singing while in migration and the presence of migrants almost throughout the summer could provide misleading information. We have no evidence the species ever nested in Sonora.

Western Tanagers migrate through most of Sonora but are most common at middle elevations, where they are occasionally abundant. Northbound migrants arrive in mid-March (earliest date, 11 March) and pass in small numbers through April. Most migrants move through Sonora in the two central weeks of May, with lingerers as late as 10 June (singing at El Riito east of Nacori Chico on 10 June; J. Marshall, A. Phillips). They are rare on the coast at any season, but two were banded in the mangroves at Punta Chueca on 30 May.

Curiously, the earliest fall migrants were found in extreme southern Sonora on 12 July (Sejaqui) and on 13 July (La Tinaja, on the Río Fuerte). Although many move south from mid-July through August, most migrate through the state in the middle third of September. On 11 September, Russell counted 100 individuals in 2 km of riparian growth in Arroyo Santo Domingo (540 m) northeast of Cucurpe; 12 responded to a whistled imitation of the call of a Northern Pygmy-Owl. There are few October records, the latest being 16 October, when six were seen at Lo de Campa. One was collected near Caborca on 30 October (Phillips and Amadon 1952).

The only winter occurrences, with one exception, were at Alamos, where up to 20 have been listed on Christmas Bird Counts. Twenty observed there on 16 November (D. Stejskal) could have been late migrants, potential wintering birds, or some of each. One was noted in January 1980 on the Río Magdalena between Imuris and Magdalena (Terrill 1981).

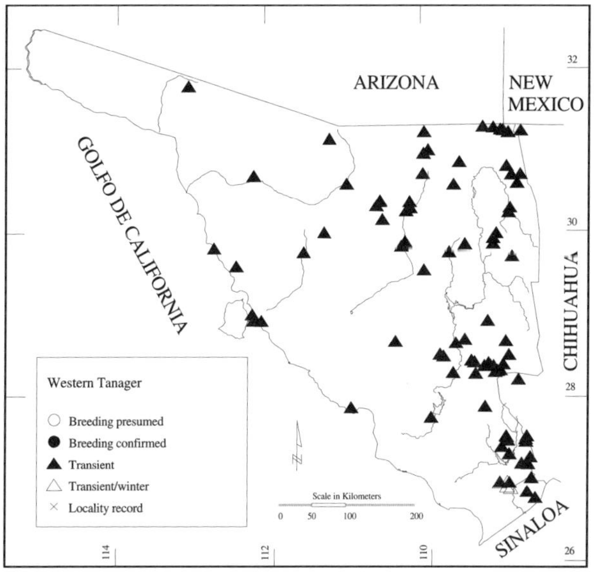

FLAME-COLORED TANAGER

Tángara Dorso Rayado

Piranga bidentata

This tanager has proven to be one of the more elusive birds of Sonora. Thus far, it has been found in only seven localities, all in the south. One was taken near Alamos on 30 March 1888 by M. Frazar (van Rossem 1945). Another was collected at Hacienda de San Rafael in the mountains east of Alamos on 1 March 1932 by J. Wright (OC). A pair was taken (both OC), and described as breeding (Miller et al. 1957), at Baromico (18 km northeast of Güirocoba) on 20 May 1933 by R. Moore and W. Burt. A male was seen 8 km north of Movas along a side canyon of the Río Chico on 4 May 1978 (Clark 1984). Several were seen, the males singing almost daylong from pines and Arizona cypresses, in and along Barranca las Colas near Sahuarivo on 16–18 July 1987. At least four pairs were observed in Arroyo el Reparo on 13–15 May 1988; some of these were feeding, with several other bird species, on *Bumelia persimilis* fruits, with no song or indication of breeding. The sole winter records are of one seen on 22 December 1989 and two on 22 December 1990 at about 1850 m in the Sierra Alamos (both by D. Stejskal) and one in Arroyo el Palmarito (670 m) on 5 January 1993.

Breeding in Sonora is unsubstantiated. There are several recent and well-documented records from southeastern Arizona, including breeding attempts.

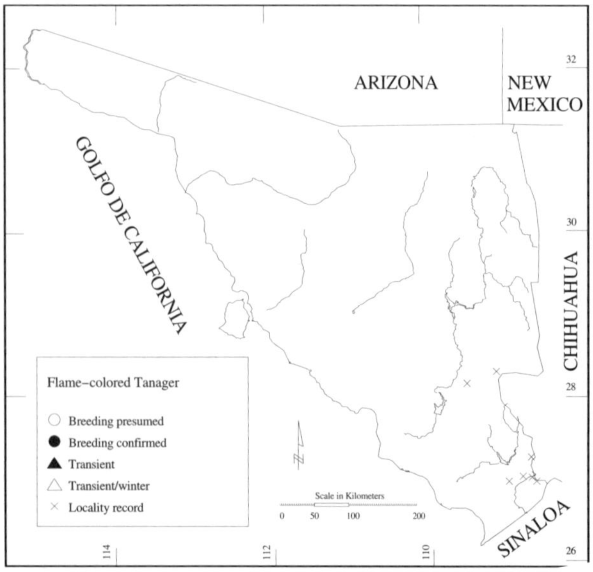

RED-HEADED TANAGER

Tángara Cabeza Roja

Piranga erythrocephala

The only Sonoran record of the Red-headed Tanager is that of four collected by M. Frazar near Hacienda de San Rafael in the mountains east of Alamos between 8 and 18 May 1888 (van Rossem 1945). The species (at its northern limit in Sonora) has been found in pine-oak woodlands elsewhere in the northern part of its range.

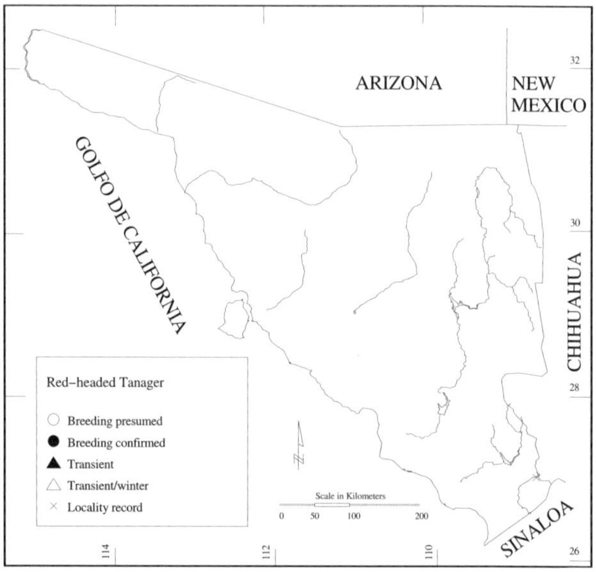

SUBFAMILY CARDINALINAE
Cardinals, Grosbeaks, and Allies

NORTHERN CARDINAL
Cardenal Rojo
Cardinalis cardinalis

The Northern Cardinal is a common, conspicuous bird in the lowlands of Sonora, where it is resident in thickets consisting mainly of leguminous shrubs (mesquites, feather trees, *Pithecellobium* spp., acacias, mimosas). It is also common in mangrove thickets along the coast of El Golfo de California. Its altitudinal range is from sea level to 1300 m, but most cardinals occur below 1000 m.

Cardinals are persistent singers through late winter, spring, and summer, their whistled song loud and rich. Both sexes sing, although the male is most often heard. Their singing and their bright red plumage make the males favorite cage birds among the people of Sonora, although keeping them is illegal.

The onset of nesting is modified by seasonal variables such as precipitation and temperature. In the mesic environment of the banks of the lower Río Yaqui in the south, nesting begins in early April, whereas pairs in drier situations not far away may not nest until May or June. Most of the many nests found in Sonora have been in July, reflecting the predictability of the summer rains that produce the insect food required by the young. In the coastal mangroves, nesting sometimes begins in March, but it may be delayed by cold weather. Pairs may nest more than once a season during favorable conditions. A nest with eggs was found as late as 20 August (east of Mazocahui, 1150 m). In the vicinity of Puerto Lobos, nesting usually is delayed until the summer rains, but there were fledgling young on 1 April after an exceptionally wet winter.

The highly sedentary nature of the species has been verified through our banding in the mangroves at Punta Chueca and Lo de Campa. Some young birds remain in the area where they were hatched, although others undoubtedly disperse. Northern Cardinals are rarely seen where they do not breed. D. Stejskal found one, possibly an escaped or released cage bird, on 3 September 1979 at Puerto Peñasco.

Vegetable matter, particularly seeds, form the bulk of the diet, but insects are readily taken. Cardinals nesting and roosting in the mangroves at Punta Chueca fly to adjacent areas of desertscrub, where they may forage in vegetation or on the ground; in late afternoon there may be a conspicuous flight of individuals returning to the mangroves. Cardinals live in some extremely arid regions of northwestern Sonora, selecting sandy arroyos with their somewhat more dense vegetation. In long periods of drought, the fruit of species of *Bursera* is regularly eaten and may be essential for the presence of the birds. We have seen them consume fruit of *B. hindsiana*, *microphylla*, or *laxiflora* when little else appeared to be available.

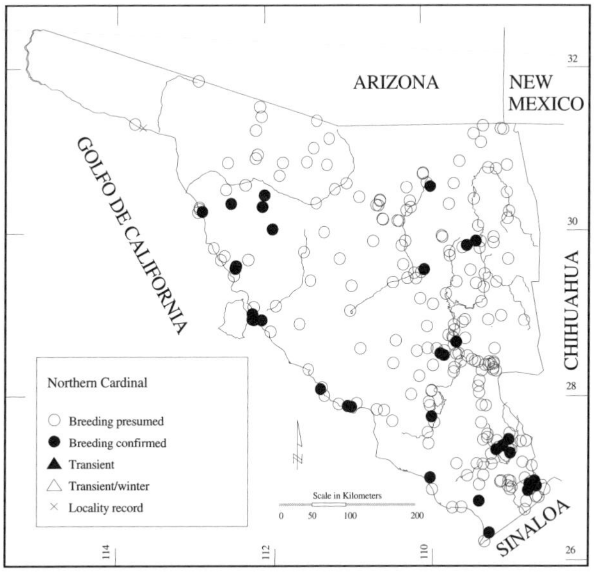

PYRRHULOXIA
Cardenal Pardo
Cardinalis sinuatus

The Pyrrhuloxia is a characteristic species of mesquite-dominated desertscrub at elevations from about 100 to 1000 m. Although it has been recorded from many localities, we do not think it is a permanent resident in any one locality in Sonora. At Lo de Campa, it tends to be more conspicuous in fall and winter, but few birds are present in summer. We saw two females with possible nest-building material on 17 March of one year and banded a juvenile on 27 September; this circumstantial evidence does not constitute proof of nesting. Of the 20 Pyrrhuloxias banded at Lo de Campa, only three were subsequently recaptured: two banded in October were taken again in December, and one banded in January was recaptured in March. The banding data do not suggest that the species is a permanent resident.

Short (1974) did not find Pyrrhuloxias near Alamos in July and August, nor did various collectors find the species nesting in or near Güirocoba. Yet

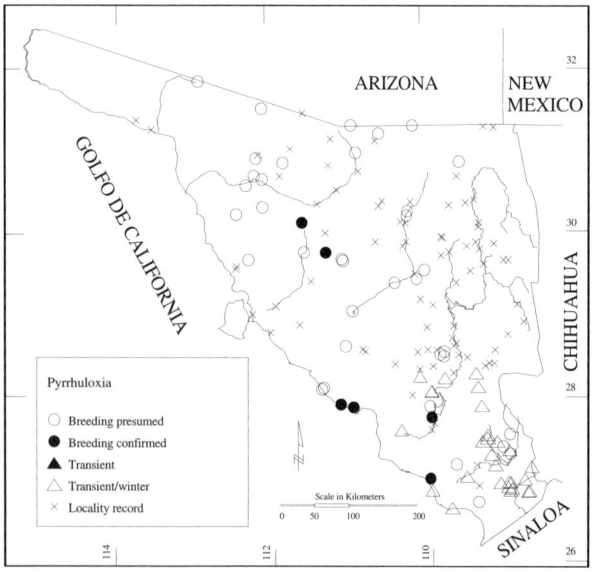

Pyrrhuloxias winter at Alamos in numbers; one Christmas Bird Count (1984) reported 80 individuals.

We have found most nests during the summer wet season in the rolling mesquite- and grass-dominated hills north and west of Hermosillo. On 6 August 1986, we found five nests containing eggs about 25 km south of Trincheras. In another year, two nests with eggs were noted west of El Oasis on 22 August. Only a long-term study of banded birds will reveal whether there is a resident population anywhere. Our interpretation of available information suggests that the Pyrrhuloxia is an opportunistic species, moving from an area at any time of the year when conditions are unfavorable. It does prefer nesting areas of mesquite grassland after summer rains. A nest containing three eggs (WFVZ) was found in a mesquite on an island in Estero Tóbari on 27 April 1930; we have few records at any season from mangroves or localities close to the coast.

In winter, Pyrrhuloxias may aggregate to share a favorable roost tree, probably gathering from a radius of 1 km or more. The birds may leave roosts in groups of 5 to 20.

YELLOW GROSBEAK
Picogordo Amarillo
Pheucticus chrysopeplus

Yellow Grosbeaks are common summer birds of the tropical deciduous forest, present wherever there are large trees but most numerous along well-wooded watercourses. The birds arrive in mid-May; there is an early record for 21 April. Their bright color and striking song make them conspicuous, seemingly everywhere in appropriate habitat, in late May, June, and July. Most

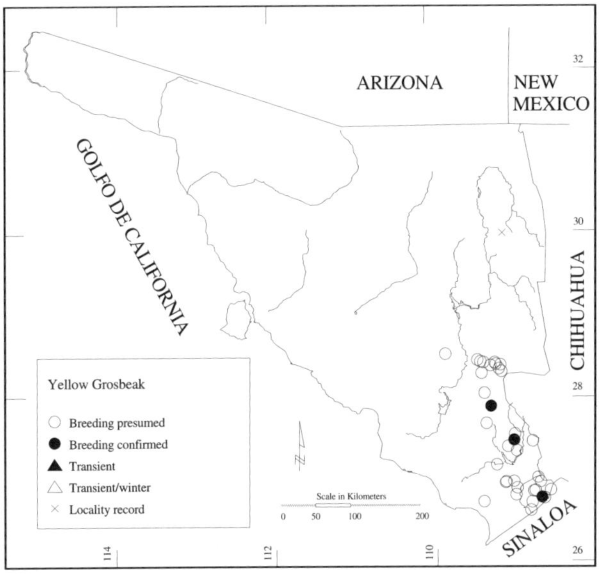

have departed for more southerly wintering areas by mid-September (latest date, 20 September). Occasional birds are seen in winter. Single birds have been found on only two Alamos Christmas Bird Counts.

Pairing takes place almost immediately upon arrival on the breeding grounds. Nest building commences in early June. Construction on one nest began on 13 July (the latest nest initiation we noted); we do not know if this was the initial attempt by the pair. One family group of six was still together on 16 September. Although individuals wander rarely to Arizona, we have only one record north of the breeding range, a bird in oaks near Aribabi (1300 m) on 19 July.

ROSE-BREASTED GROSBEAK

Picogordo Pecho Rosa

Pheucticus ludovicianus

There are three Sonoran records of this eastern grosbeak. A male was seen 8 km south-southeast of Puerto Libertad on 26 May 1975, and an immature male was collected near Carbo on 24 September 1975 (Russell and Lamm 1978). The third bird was observed near Alamos on 18 March 1982 by D. Stejskal.

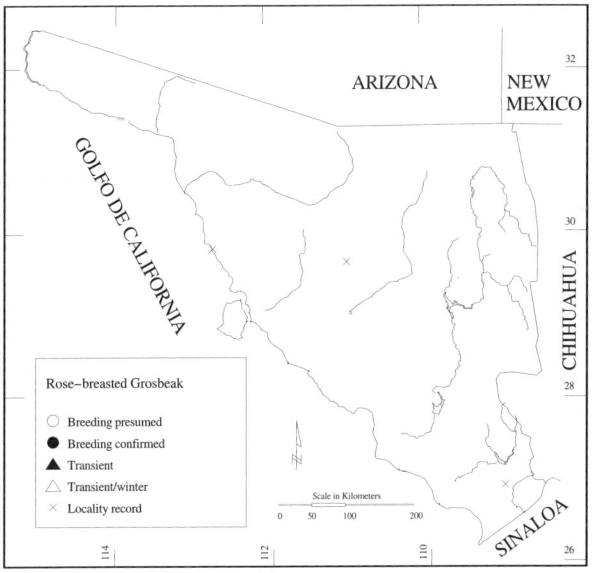

Yellow Grosbeak

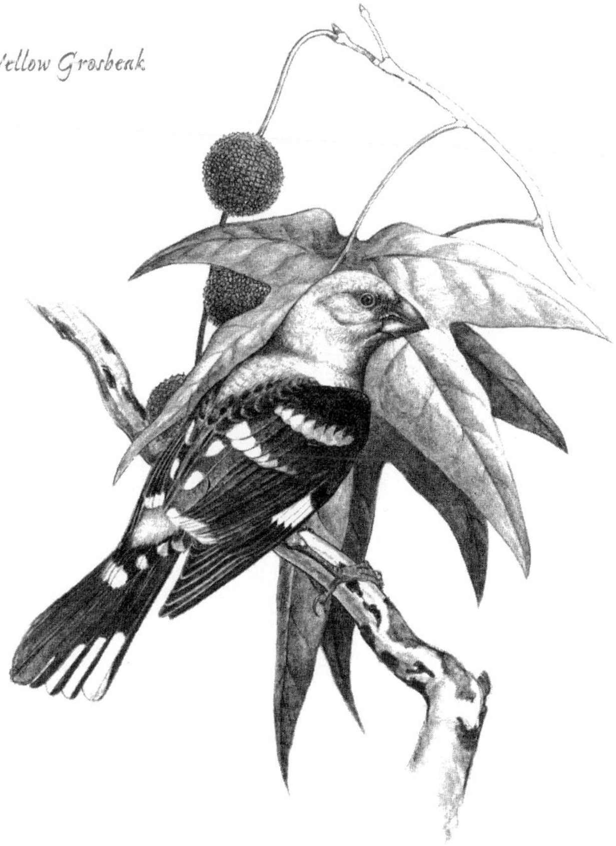

BLACK-HEADED GROSBEAK

Picogordo Tigrillo

Pheucticus melanocephalus

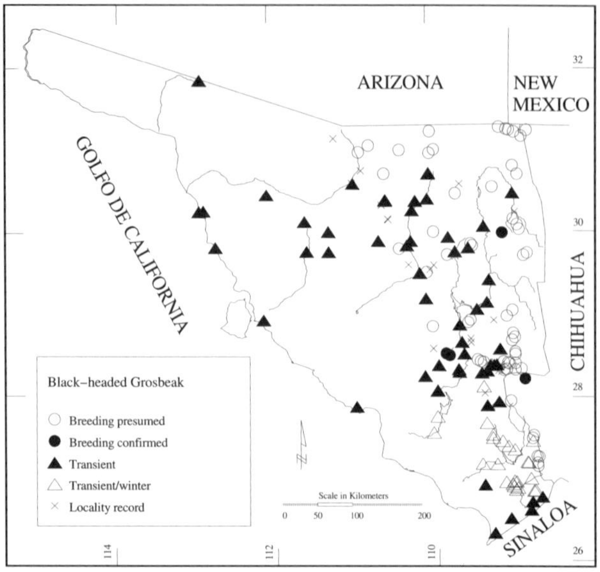

The Black-headed Grosbeak is generally presumed to be a breeding species in ponderosa pine forest and pine-oak woodland, less commonly so in riparian habitat within oaks but below pines. We also found the species breeding in tropical deciduous forest in the foothills. Marshall (1957:115) hinted at such a possibility when he reported a male with enlarged gonads collected "far down in the Chihuahua oak zone" of the Sierra Aconchi on 24 June 1954.

On 11 June 1982 Monson was in tropical deciduous forest in the Sierra el Tiznado (900 m) a few kilometers east of Mazocahui, where feather trees, kapok, tree ocotillo, mesquite, *Bursera*, and organ pipe cactus were growing in a canyon. Above this canyon on a relatively flat area were many summer-deciduous Chihuahuan oaks. He was surprised to find Black-headed Grosbeaks common and singing in the canyon well below the oaks, the males sometimes singing on the wing. He surmised that the birds were preparing to nest, and that they were not a grouping of late migrants.

Subsequently, in 1983 and following years they were found to be fairly common through the summer at Lo de Campa, on the canyon sides of Arroyo San Javier. Nests were not located, but females with functional brood patches were banded, as well as a bobtailed young bird (on 19 August). Females netted at Lo de Campa in early June did not yet have brood patches, but males had fully developed cloacal protuberances. Here there are no oaks, but tropical deciduous forest covers the canyon sides, feather trees being a common component.

East of the Río Yaqui, Black-headed Grosbeaks are common in rugged foothills. In most situations, Chihuahuan oak is either present or not far away, but the vegetation is nevertheless predominantly tropical deciduous forest. Summering birds were found south to near Nuri, where two birds were observed on 7 June 1984, and Rancho Jacalón northeast of Rosario, where three were seen on 11 June 1984. To the northwest, at least four observed at Rancho los Pozos 21 km east of Mazatán on 25 June 1983 (S. and J. Levy) suggested that breeding was taking place there; the locality is at the edge of tropical thornscrub, and there are no oaks for many kilometers. We conclude that Black-headed Grosbeaks do not nest in the northern portion of the tropical deciduous forest until summer rains have begun falling.

They are common to uncommon in pines and pine-oak woodland in summer, and we think they nest at elevations of about 1900 m without waiting for the rains to begin; we found a nest southeast of Yécora on 10 May.

Some Black-headed Grosbeaks winter in riparian woodlands in north-central and eastern Sonora, but more commonly they are found from Quiriego southward. Up to 43 have been seen on an Alamos

Christmas Bird Count. Among the more northern wintering sites are the Río Magdalena in the Imuris-Magdalena district, Rancho Saracachi, and Moctezuma.

In southern Sonora, northbound migrants appear in mid-March, but migrants (as well as breeding birds) do not reach the northern border until late March or early April. Migrants are noted only infrequently in arid regions of the northwest, but elsewhere they are quite common. Fall migration is well under way in early July. Migrants have been noted near the Sonora-Sinaloa border (Río Fuerte and Agiabampo) by 11 July. Peak numbers are present in September. Because northbound migrants may be present into June, it is difficult to determine if individuals seen in summer are transients or local breeders unless nesting behavior is noted.

BLUE GROSBEAK

Picogordo Azul

Guiraca caerulea

The Blue Grosbeak is a common summer resident of open areas along streams, of washes where some grass is present, of farmlands, and of oak savannas up to 1600 m. It is most common where willows, cottonwoods, and mesquites are all present. Nowhere does it come near the coast; a stray was found at Guaymas on 9 March 1905 (van Rossem 1945), and another at San Miguel on 22 May 1993. It is a very persistent songster during the summer rainy season. Nesting begins in mid-July and continues to mid-September. In some nonfarming areas this grosbeak may not show up until after the rains arrive and will then breed, as in west-central Sonora north of Hermosillo.

Spring migration begins about mid-March (two on the Río Sahuaripa above Guisamopa, 14 March), but individuals do not typically arrive on their breeding grounds until June. Fall migration ends about the last of September, but three were seen on 3 November at Caborca (Phillips and Amadon 1952). We have a rather small number of records from winter months (but up to 27 were seen on an Alamos Christmas Bird Count) in southern Sonora, and it is not found there every winter.

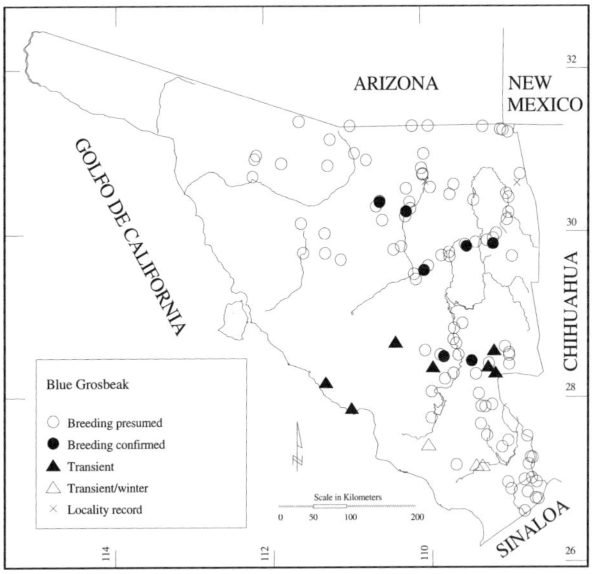

LAZULI BUNTING

Colorín Lázuli

Passerina amoena

The Lazuli Bunting is a transient and winter resident in Sonora. As a transient, it is quite common in eastern Sonora, mainly below oaks. There are only two coastal records, both in April, one at Puerto Lobos and one at Puerto Peñasco. The first autumn birds arrive in August (earliest date, 9 August, at Rancho San Bernardino; van Rossem 1945), and most have departed northern Sonora by mid-October. Spring migration takes place from mid-March to about 22 April. Some birds may linger into June: one seen at Huasabas on 8 June (A. Phillips); one at El Saucito northwest of Nacori Chico on 13 June (J. Marshall).

The species winters north to Lo de Campa, the birds being most common in brush associated with weedy Johnson grass farmlands. A number were recorded from near Ures in November, December, and January of 1979 (Terrill 1981). At Alamos the species is regularly recorded on Christmas Bird Counts; up to 64 have been noted on a single count day.

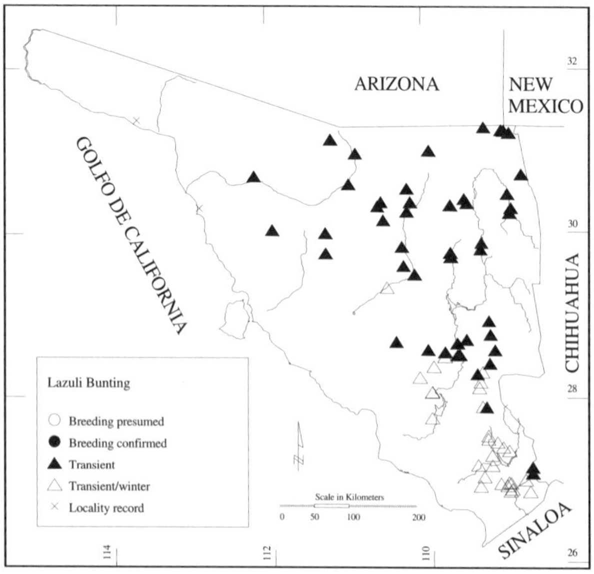

INDIGO BUNTING
Colorín Azul
Passerina cyanea

We have 15 records of the Indigo Bunting from Sonora: one in March, two in April, four in May, two in July, two in August, two in September, one in November–December, and one in December. The localities represented span elevations from sea level to 1500 m. Among the more interesting observations are a singing territorial male 65 km south of El Sasabe on 25 May 1975 (Russell and Lamm 1978) and a singing male on the Río Cuchujaqui southeast of Alamos on 25 July 1993 (R. Bailowitz). Two were reported from Ures on 30 November–2 December 1979 (Terrill 1981) but without any supporting details.

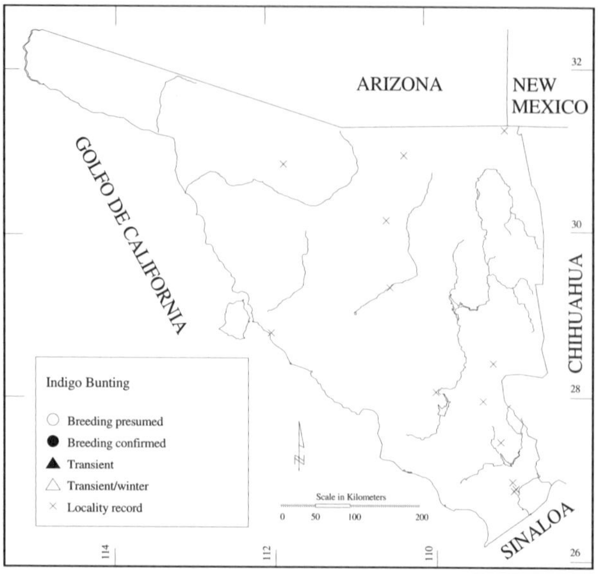

VARIED BUNTING
Colorín Morado
Passerina versicolor

Varied Buntings are common summer residents of lowland Sonora, sometimes present as high as the lower limit of oaks. They nest in dense desertscrub, thornscrub, and tropical deciduous forest, being especially numerous in scrub adjacent to riparian habitat. Their nesting density in such places is as high as that of any other Sonoran breeding bird. They are uncommon and local in more arid western areas.

Nesting usually begins in May, although it may be delayed until the onset of summer rains in drought years. We netted many young birds at Lo de Campa in early June in a normally dry season, however. Nesting continues at least to mid-August. Postbreeding adults and juveniles may move from their nesting areas into riparian thickets and weed patches by mid-August.

Varied Buntings winter in small numbers in the extreme south. Whether these are resident birds or individuals from farther north, we do not know. By mid-March, a few birds are migrating northward, but most birds arrive in late May and early June. Indicative of this pattern was an observation of at least 250, mostly males, seen in a single field west of Ures on 29 May 1980 (G. Rosenberg).

The birds remain in the vicinity of the nesting grounds until early September in the north. Farther south, they stay later in the season. We have counted up to 140 in weedy tall-grass fields at Lo de Campa in the last days of September, and many remain there until mid-October. Only two of the nearly 250 banded at Lo de Campa were recaptured a year later.

A nonbreeding female was netted in mangroves near Punta Chueca on 22 June 1974 (Russell and Lamm 1978). Its partially ossified skull was indicative of a young bird. There had been little winter rain in the area, and it is unlikely that Varied Buntings had nested in the vicinity. There is no evidence that the species nests in mangroves.

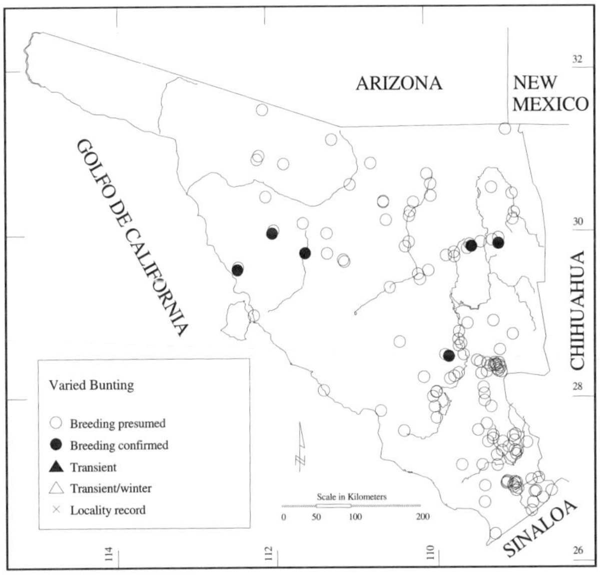

PAINTED BUNTING

Colorín Sietecolores

Passerina ciris

Painted Buntings are uncommon fall migrants in Sonora; they may be absent in some years. They are most often found in areas with much Johnson grass. The first individuals appear in mid-July, and numbers peak from mid-August to mid-October as these buntings pause in their migration in order to molt. These may be birds that nested in Texas or Chihuahua and then flew over the Sierra Madre into Sonora, as suggested by Thompson (1991). The variability in their pattern of occurrence is demonstrated by A. Phillips, who wrote in his journal: "Seem to be all over Sonora [in 1954]—probably an exceptional flight?"

On 19 August 1984 we banded eight Painted Buntings at Lo de Campa. Three of the birds had incompletely ossified skulls and were in their first prebasic molt. There is a possibility that these birds were hatched at Lo de Campa, but in the absence of other evidence of nesting (lack of song, parental behavior, or nests) it is more likely that they migrated soon after fledging, as reported by Thompson (1991). A second-year male in heavy prebasic molt had lost all of his tail feathers prior to replacing them. He was capable of only short flights. We netted ten birds at the same location on 13 October 1983. Some of these birds had finished their adult prebasic molt. Others had completed molt of flight feathers but not all body feathers. One juvenile was in heavy presupplemental molt.

There are only two reports of this species in other seasons. We saw a female below Presa Obregon on 15 May 1986. An adult male was collected (UA) at Bahía la Choya near Puerto Peñasco on 4 December 1965.

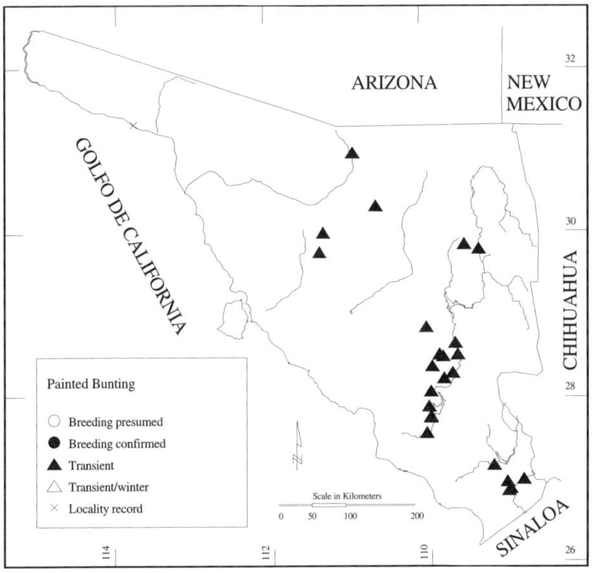

DICKCISSEL

Arrocero Americano

Spiza americana

The Dickcissel is a casual fall visitant in Sonora. Van Rossem (1945) told of one being carried by a Merlin when the latter was shot at the United States–Mexico boundary on the Río San Pedro on 10 October 1892 by E. Mearns; the Dickcissel victim was not preserved, and it is not known on which side of the boundary the action took place. Three other Dickcissels were collected the same year by Mearns but on the Arizona side of the line at the San Bernardino Ranch. A. Phillips noted one on the Río Yaqui near Esperanza in late September 1954. Russell saw 12 in a flock at Punta Chueca on 24 September 1977, flying north! We saw two females (or juveniles) at Lo de Campa on 17 October 1983. Six were found by G. Homel at San Carlos on 25 October 1992.

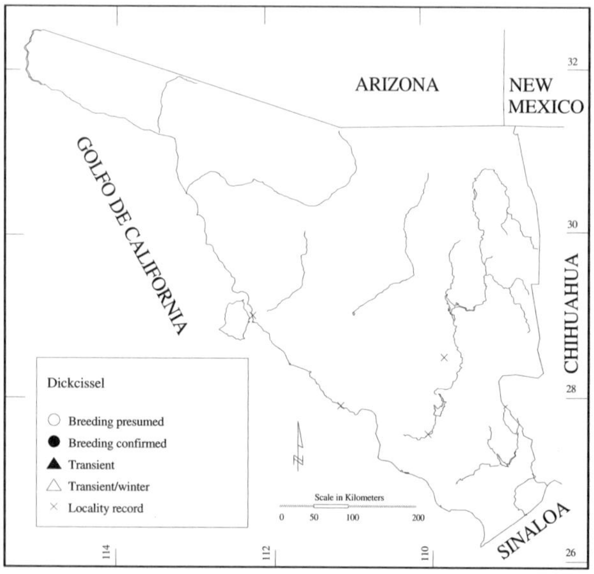

SUBFAMILY EMBERIZINAE

Emberizines

[RUFOUS-CAPPED BRUSH-FINCH

Atlapetes Gorra Rufa

Atlapetes pileatus

A. Moorhouse and C. Green found a single individual on 27 November 1993. The bird was in a deep canyon (1600 m) in pine-oak along the Hermosillo-Yécora highway, on the west side of the divide and below Mesa de Enmedio.]

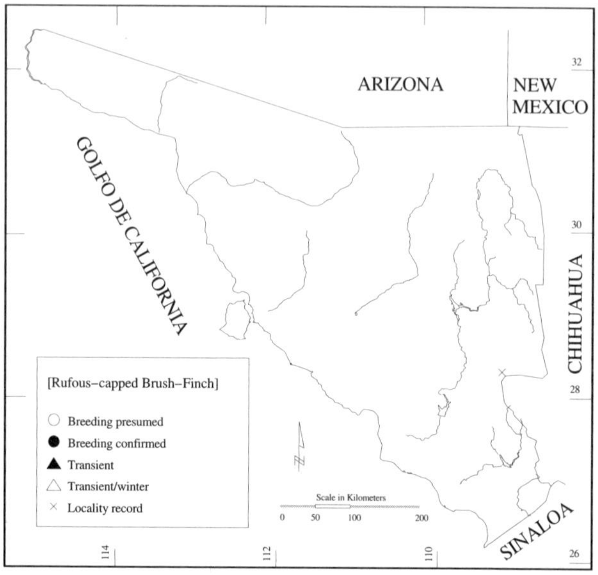

RUSTY-CROWNED GROUND-SPARROW

Rascador Nuca Rufa

Melozone kieneri

This species is known from only five localities (all between 400 and 1000 m) in Sonora. According to van Rossem (1945), 28 specimens were taken between 3 and 19 May 1888 at Hacienda de San Rafael east of Alamos; at least one was collected in Cañon de San Francisco on 30 May 1937; and one was seen at Güirocoba on 3 June 1937. Subsequently, three were collected at El Mirasol (northeast of Güirocoba) on 25

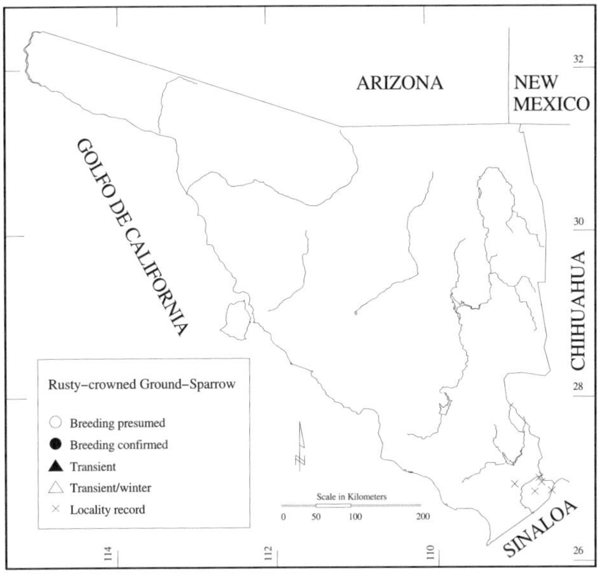

May 1945 and 1 June 1947. The label on a male taken on the latter date was marked "breeding" by W. J. Sheffler, its collector, but we do not accept this designation as evidence of nesting. Recently, one was seen on each of two Alamos Christmas Bird Counts, in 1982 (by D. Stejskal) and 1983 (by J. Bock), and two were noted 22–25 November 1984 (D. Stejskal), all in the same area on the northeast slope of the Sierra Alamos.

GREEN-TAILED TOWHEE

Toquí Cola Verde

Pipilo chlorurus

This terrestrial finch is an abundant transient and winter resident of brushy thickets, primarily at elevations between 200 and 700 m (rarely to 1500 m). The Green-tailed Towhee is rather reclusive; it does not flock, and because it hides in thickets, the observer finds far fewer than are actually present. The call note is distinctive and given often when the bird is disturbed.

Southbound migrants appear in northern Sonora in the first half of September (earliest date, 4 September). The migration tends to be steady and progressive with birds well distributed throughout the southern part of the state by the end of September. Many of them spend the winter in Sonora. Counts of more than 50 birds by one person are not unusual in areas affording dense cover and abundant small seeds.

The spring migration is not conspicuous. There is never a swelling of numbers, as might be expected from a surge of migrants. Instead, winter residents depart gradually in late March and early April, as migrants drift northward (latest date, 6 June). Migrants are likely to appear at higher or lower elevations than those used by wintering birds and in places with little food or cover. A few birds migrate along the coast, and in good years they will winter there: for example, nine were noted at Bahía Kino on 20 December 1980. At Punta Chueca we netted a few towhees for banding, capturing them as they flew into the mangroves for cover at dusk, or departed from the mangroves at daybreak.

At Lo de Campa, Green-tailed Towhees may be numerous in winter. We banded 114 there, of which 25% were recaptured at least two months later. Six birds returned in subsequent winters. The quantity of food available for wintering sparrows varied from year to year, depending upon the summer precipitation. The populations of seed-eating birds, including wintering towhees, varied accordingly. Although the recapture of banded towhees indicates a high level of winter site fidelity, it must have been necessary for returning birds

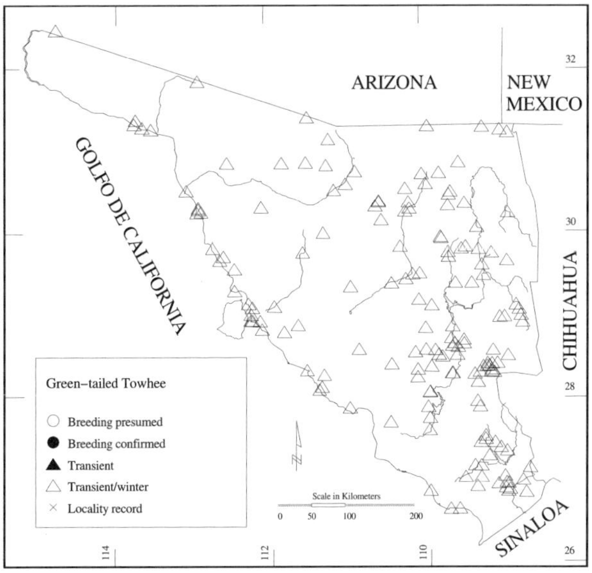

in some years to search elsewhere for food. In some summers at Lo de Campa, intensive livestock grazing destroyed the potential winter food supply; subsequently, bird densities were low.

SPOTTED TOWHEE

Toquí Pinto

Pipilo maculatus

Spotted Towhees breed commonly in chaparral and oak brush within pine-oak woodland, where they are resident, in north-central and northeastern Sonora. Marshall (1957:117) indicated that the main nesting season is in July and August. He noted, "Singing increases greatly in July and August as contrasted with April through June. Spotted Towhees are sedentary, and their territories, battled over in spring, are small." Because they like brush or shrubbery (especially *Quercus hypoleucoides*), they do well in sites that have recently been logged.

Towhees from the western United States migrate into various brushy habitats in Sonora in the fall. One strayed south to the Sierra Alamos (29 December 1987;

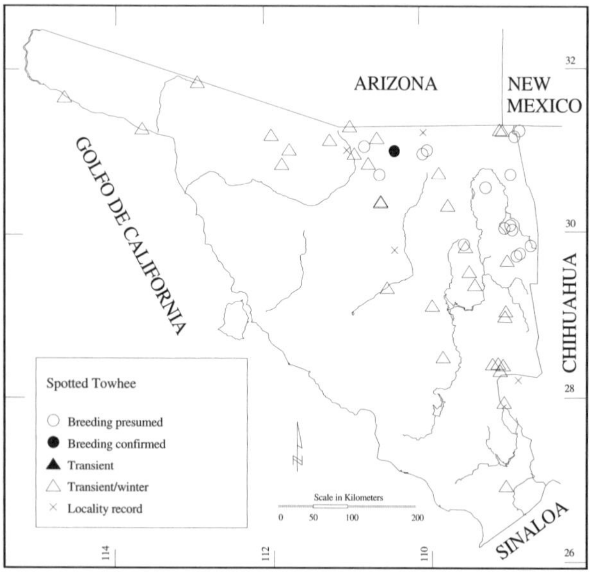

N. Krug and D. Stejskal). Visitants from the north are never numerous, the usual number observed being one to five in a day. They are more common in some years than in others. Along the upper portion of El Golfo de California they have been observed three times (two at Puerto Peñasco on 3 January 1965, one at Golfo de Santa Clara on 29 February 1968, and one at Puerto Peñasco on 16 December 1989 [the latter noted by D. Stejskal]). Outside migration dates are 13 October (1954, La Casita south of Nogales; A. Phillips) in fall and 31 March (1983, north of Altar). A single bird in the Sierra Metate southeast of Yécora on 10 May 1983 may have been a late migrant, or it may have been part of a breeding population.

CANYON TOWHEE

Toquí Pardo

Pipilo fuscus

The Canyon Towhee is one of the most common and ubiquitous resident birds of Sonora, living from coastal scrub to clearings in the highest mountains. It scratches for food on open ground, which must be adjacent to cover. Although widespread, its distribution may be spotty; it is almost abundant in some localities while in other seemingly favorable places (especially in southern Sonora) it is scarce or absent. Its distribution probably reflects the availability of the small seeds upon which it feeds. It shows an affinity for the works of man, such as corrals and outbuildings of ranches and the outskirts of pueblos, and the various structures of abandoned mines and sawmills.

The Canyon Towhee has a long nesting season, extending from March to September. Pairs may be seen in any month but most frequently during the nesting

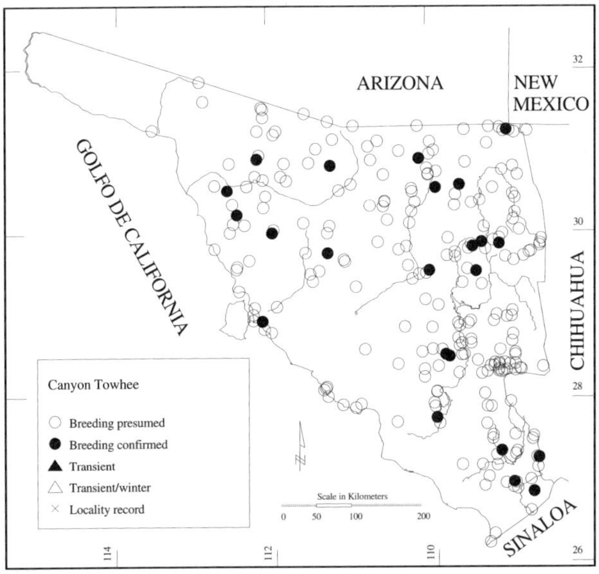

season, when the two birds associate closely in all their activities, maintaining contact through call notes. Singing begins in March and intensifies in July. By September when family groups are widespread, they no longer sing. Early nesting is dependent upon mild temperatures and unusually heavy winter rain. Most pairs nest in July and August (74% of the 27 nests we have noted), coincident with the time of summer precipitation.

ABERT'S TOWHEE
Toquí Enmascarado
Pipilo aberti

Prior to the construction of Hoover Dam and the subsequent draining and development of the Río Colorado delta, the Abert's Towhee was a common resident of that part of Sonora. It persists in riparian areas where irrigation water and cover are available, especially at the edge of the Ciénega de Santa Clara. The species continues to be common in adjacent portions of the Río Colorado in the United States (Rosenberg et al. 1991) and Baja California.

Recently, the Abert's Towhee has spread upstream in the valleys of the Santa Cruz and San Pedro Rivers in Arizona, and into Sonora along the same rivers. These towhees are regularly seen in Sonora where the San Pedro crosses the international boundary (D. Danforth, T. Huels, A. Moorhouse), and one was seen 3 km into Sonora along the Santa Cruz on 31 January 1987 (A. Moorhouse).

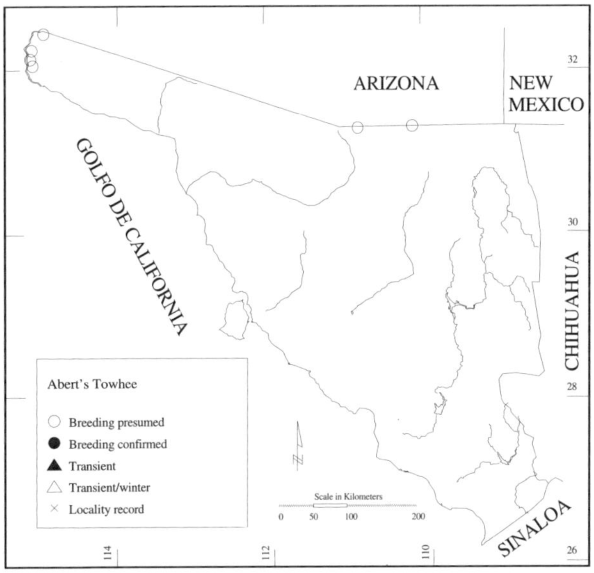

BLUE-BLACK GRASSQUIT
Semillero Brincador
Volatinia jacarina

The tiny Blue-black Grassquit is locally common in summer from the Alamos region south; it has been found farther north only once, when nine were observed a few kilometers north of Quiriego on 11 July 1986. There are fewer than a dozen records from Sonora. We found it in foothills close to the Sonora-Sinaloa border on 12–13 July 1987. Short (1974) said it was common locally in bushes and fields along arroyos near Alamos in the summer of 1971. Van Rossem (1945) mentioned specimens from Güirocoba on 2–3 August 1933 and a sight record for Baromico on 17 May 1933. There is no definite nesting record for Sonora. According to Short (1974:30), the species "occurs in the Alamos region only in mid- and late summer as a rainy season breeder."

The male has concealed patches of white beneath the forewings that show in flight; when singing, the male repeatedly flies up from a perch for a meter or less, delivers a very brief song, then falls back to the perch. These grassquits usually are found in patches of tall grass and annuals (such as *Amaranthus*) that spring up with the summer rains.

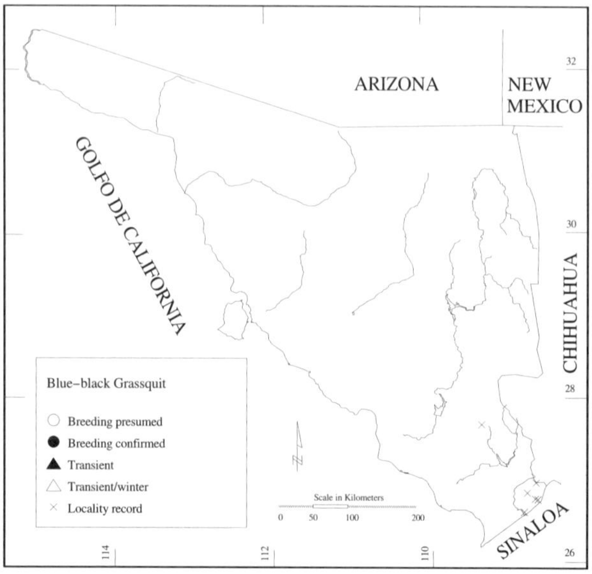

BOTTERI'S SPARROW

Zacatonero de Botteri

Aimophila botterii

Botteri's Sparrow is a fairly common summer resident, chiefly in oak-mesquite woodland with considerable grass in northeastern Sonora, where it is found from 900 to 1360 m. It prefers tall coarse grass with associated scattered shrubs where it can perch to sing. It nests on or close to the ground in the grass. Most individuals

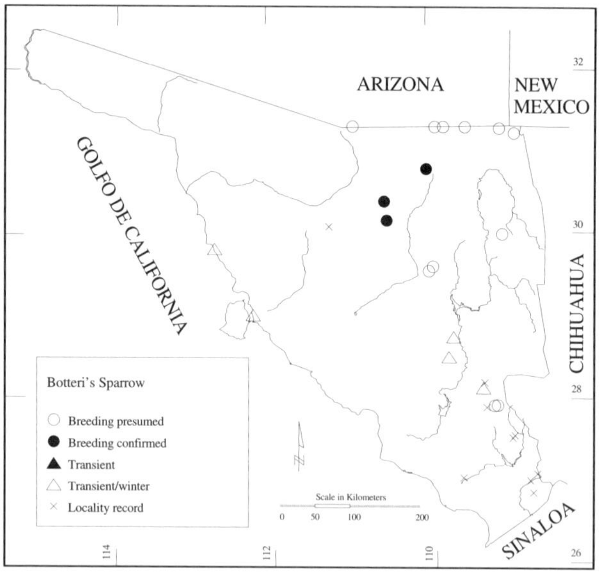

arrive on the breeding grounds in the latter part of May, and this species' numbers increase rapidly. It sings on arrival but reserves full song until the summer rains begin and nesting commences. It continues to sing into mid-September, only to disappear by early October (latest northern date, 7 October near Agua Prieta; van Rossem 1945).

Some Botteri's Sparrows winter in southern Sonora in grassy, weedy fields, perhaps as far north as Soyopa on the Río Yaqui, where one was seen on 15 March. Not much farther south, we found one at Lo de Campa on 6 January (we have seen two there in October). We also have early May records from Nuri and a January record from 14 km northeast of Rosario. Because these sparrows are skulkers and not singers during the nonbreeding season, they are extremely difficult to observe and are easily overlooked.

Two singing birds were noted on Rancho el Carrizo south of Benjamin Hill on 2–3 August (J. and S. Levy), indicating that some birds may nest in the central plains, at least in extra-wet summers, far indeed from the nearest oaks. Russell identified four in dense vegetation along a sandy desert wash at Punta Chueca on 26 November 1977 and a single bird in similar heavy vegetation at Punta Cirio two days earlier. These birds were seen after an unusually wet summer had produced many weeds and much grass; only under such conditions could this species be expected so far west. One was observed drinking on 17 May 1986 from the Río Guajaray, where the birds could possibly breed. We are uncertain of the status of birds in extreme southeastern Sonora. None of the records from the Güirocoba, Alamos, Milpillas, or Guajaray areas involve summer birds or even singing individuals.

CASSIN'S SPARROW
Zacatonero de Cassin
Aimophila cassinii

The thin, haunting song of the Cassin's Sparrow, delivered on the wing as the singer flutters from perch to perch, is heard through mesquite savannas in northern Sonora, mostly after the summer rains have begun. The singing persists occasionally into early September.

We found Cassin's Sparrows wintering rather commonly through much of Sonora, especially south of the latitude of Hermosillo. Winter numbers are dependent upon the availability of grassy, weedy places bordered or flecked with small mesquites or other leguminous shrubs. The dense ground cover depends upon favorable summer rain and little or no cattle grazing. Under excellent conditions the sparrows become abundant. More than 50 have been listed on an Alamos Christmas Bird Count. At least 100 were estimated to be present in the vicinity of Lo de Campa on 6 January in an ungrazed meadow after abundant precipitation during the previous growing season. In years without dense grass no Cassin's Sparrows were present. In the nonbreeding season, there is no singing and the birds are extremely reclusive. The large number of recent distributional records is attributable to the widespread use of tape recorder playbacks of the song of the Cassin's Sparrow. Although individuals do not often respond by singing, they may utter chip-notes or move to a point where they may be observed for a moment.

In years of above-normal rainfall, Cassin's Sparrows appear in unexpected places. For instance, in late March and early April 1984 a dozen were singing among the dunes back of the beach north of Puerto Peñasco. This is the only time they have been detected along the coast (aside from one at San Carlos on 25 October 1974 and one collected at Puerto Libertad on 2 February 1946, UCLA). Similarly, a few birds were singing near Caborca in late March and April of 1983. In wet winters an occasional individual may sing as early as mid-March anywhere in the wintering range.

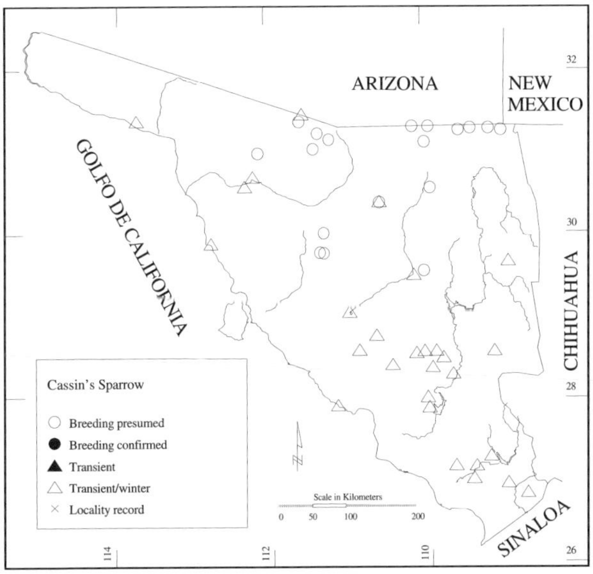

RUFOUS-WINGED SPARROW
Zacatonero Ala Rufa
Aimophila carpalis

The Rufous-winged Sparrow's entire range lies mostly in Sonora. Elsewhere, it occurs only in south-central Arizona and northern and central Sinaloa. Its habitat is desertscrub and tropical thornscrub with grass, the grasses ranging from the tall, relatively coarse tobosas to the short, ground-hugging false grama. Its occurrence in the tropical deciduous forest is limited to open farmland. North from the Río Yaqui delta, it is found casually on the coast of El Golfo de California; there are records for San Carlos (van Rossem 1945) and Estero Tastiota (A. Phillips *in* Bent 1968). Over its entire range it is a permanent resident, quite numerous in places. Altitudinally, it nests from near sea level at Agiabampo to above 1200 m in the northeast.

Singing may be heard in any month, although most frequently from March through September. The species nests through the rainy season of July, August, and September (and at least once in late October; Pitelka 1951). Spring nesting is rare (one building 19 km north of Altar, 31 March 1983) in Sonora because the essential rains are typically in the summer.

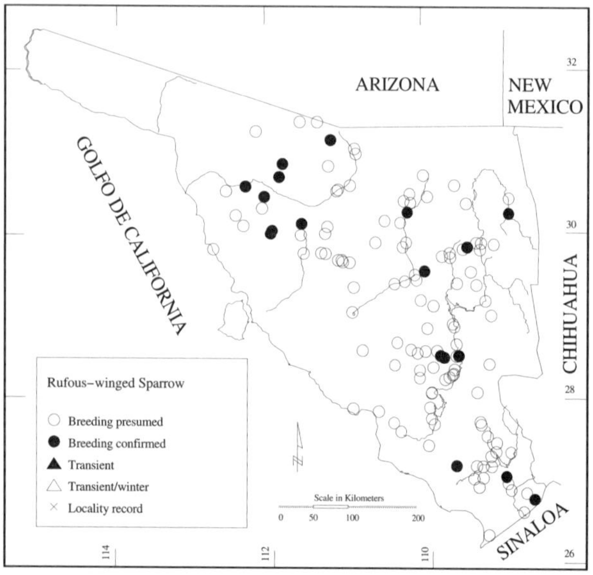

The species is most abundant in a broad band from the plains northwest of Hermosillo extending southeast to the Navojoa region. In optimal grass and shrub habitats, densities may be a pair per hectare. Although the species typically waits for summer precipitation, we found nests in July of the drought year of 1987. These birds, in the tropical thornscrub northwest of Navojoa, had nested even without rain, rather than skip a breeding season.

RUFOUS-CROWNED SPARROW
Zacatonera Corona Rufa

Aimophila ruficeps

This Aimophila is a common resident of rocky mountainsides and canyon walls all through northern and eastern Sonora. Its habitat consists of grass with various shrubs, sotol, agave, yucca, and cacti, from below the oak zone up into high pine-oak woodlands. It may reside in relatively flat rocky places if the grass is thick and the requisite shrubs occur. The elevational range is from 700 m (near San Nicolás) to 2100 m.

The nesting season varies with the yearly rainfall pattern. It may begin as early as mid-May (fledgling at El Terreito on 19 May 1986) and may last into September, but generally the peak of breeding activity is in July and August. The nest is placed on the ground beside or beneath shelter, such as a clump of grass or overhanging small bank. We have little evidence of wandering out of the breeding range. One, probably the same individual each date, was seen on 11 and 13 October 1973 on Cerro Prieto just west of Bahía Kino and only 5 m above the high-water mark of El Golfo de California. It foraged in a tangle of weeds, grass, and shrubs made green by a leaking water reservoir.

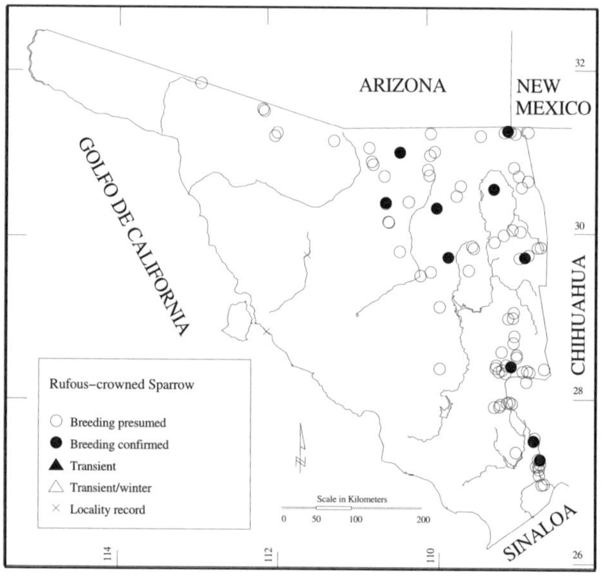

Two counter-sang, one on rocky slopes on each side of the Río Bacoachito (southwest of Félix Gómez at 605 m) on 10 March 1989. This was a warm spring following a wetter than average winter. Although the species may nest there, we had not found it on previous visits.

RUSTY SPARROW
Zacatonero Rojizo
Aimophila rufescens

This is a large sparrow superficially resembling a Rufous-crowned Sparrow or even a Canyon Towhee. It has a considerable resident range in mountainous eastern Sonora between 800 and 2000 m but is uncommon and local in its distribution.

Marshall (1957:119) stated, "The species occurs in 'pockets' or colonies . . . , but apparently with no regard for terrain or vegetation." We have found it chiefly in pine-oak woodland, where it often is associated with patches of *Rubus*. It was seen in lower, drier habitat in Arroyo el Reparo (920 m), and Marshall (1957) found it in similar habitat on El Rijito (915 m) east of Nacori Chico. Near El Trigo, it was present on a rocky hillside above a riparian area dominated by Arizona cypress. Only at Sahuarivo (1500 m) do we have evidence of nesting; an adult fed a heavily streaked juvenile on 16 July in an open oak area with large boulders.

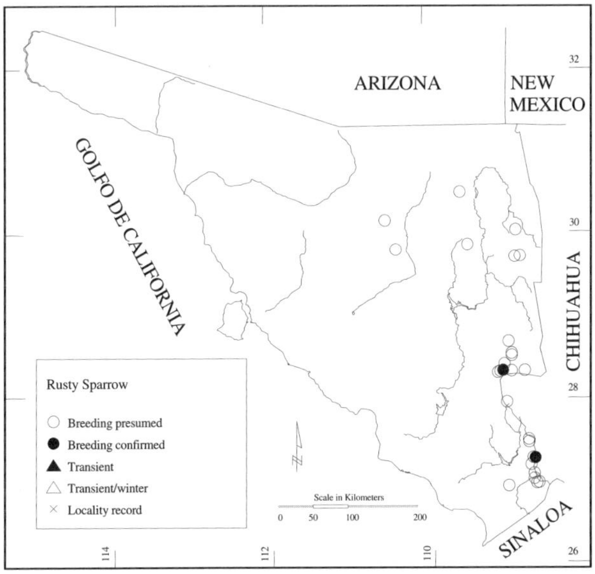

There have been no reports of the Rusty Sparrow from the most westerly population (in the Sierra de San Antonio) since specimens were collected by W. Brown, Jr., at La Chumata Mine in 1905 (Thayer and Bangs 1906). (The exact location of La Chumata Mine is vaguely known; see gazetteer appendix.) But we have been in that area only once and are unaware that anyone else has looked for them. Marshall (1957) reported them from the Sierra Aconchi in June 1954.

[STRIPED SPARROW
Zacatonero Rayado
Oriturus superciliosus

Ascription of the Striped Sparrow to the list of Sonoran birds rests on specimens taken by F. Robinette of the Lumholtz expedition on the "upper Río Bavispe" on 21 and 26 December 1890 (van Rossem 1945:267). It is probable that these specimens were taken in Chihuahua and not in Sonora, but the dates they were taken argue in favor of Sonora. Marshall (1957) found the species in Chihuahua in mountain meadows of tall grass in the headwaters of the Río Gavilan (2130 m), a tributary of the upper Río Bavispe. We have not found this habitat in Sonora, where the highest point on the Río Bavispe is 1400 m.]

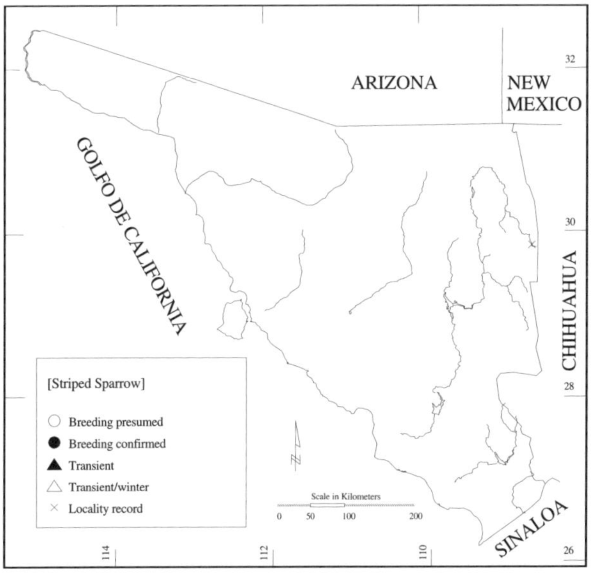

CHIPPING SPARROW
Gorrión Ceja Blanca
Spizella passerina

The Chipping Sparrow is local and uncommon in Sonora as a breeding species. While nesting it is confined to open spaces in pine-oak woodland. It has been reported as nesting at only two localities. Van Rossem (1945:283) noted it at Santa Barbara as "common in Transition Zone where breeding in June 1937." At Sahuarivo on 15 and 16 July 1987, we saw pairs copulating, adults feeding nestlings, and several adults feeding fledglings. We indicate several localities on the map where we think they nest, based on the presence of singing, paired birds.

The distribution of the Chipping Sparrow as a migrant and wintering bird is quite another matter; in the nonbreeding season it is common, often abundant in large flocks, in the eastern two-thirds of the state. It utilizes open pine-oak woodland, open woods along streams, and trees and tall brush in irrigated farmland. Water is often present. This species is not often found in desertscrub except in winters of good rains. It is occasionally seen along the coast as a migrant. The first fall birds are seen in mid-August in the north, mid-October in the south. Flocks in August and early September may contain heavily streaked juveniles; these birds have already migrated a considerable distance from their breeding grounds. In the spring, they are migrating northward in early April in the central and southern regions but may be seen into late May in the north.

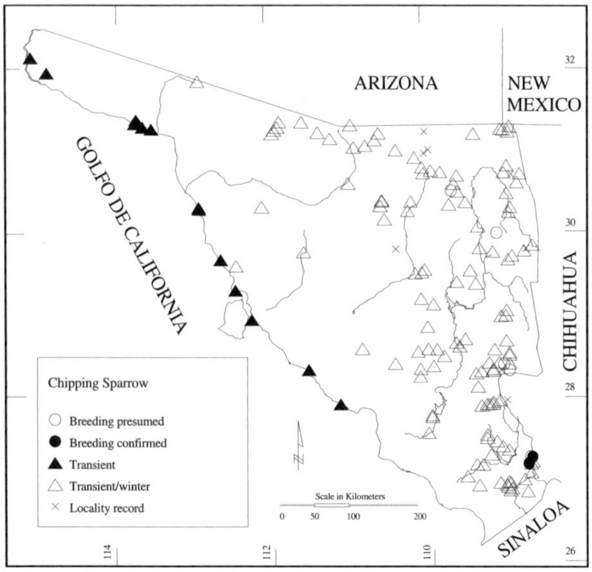

CLAY-COLORED SPARROW
Gorrión Pálido
Spizella pallida

The Clay-colored Sparrow is locally common in migration and is quite widespread in winter in southern Sonora, though in lower numbers than during migration. It is fairly common in winter along the coast of El Golfo de California from San Carlos southeastward. Inland, it is fairly common to common as a migrant and winter resident south of the latitude of Hermosillo and east of Mexico Highway 15, exclusive of the higher pine-oak woodland. We have seen as many as 80 in a day. North of the latitude of Hermosillo, the species is found only in some years and almost exclusively as a fall migrant and as single birds.

The earliest fall migration date is 1 September (San Miguelito on the upper Río Bavispe), while the latest spring departure date is 3 May (about 8 km northeast of Nuri). The most pronounced migration is in April and September.

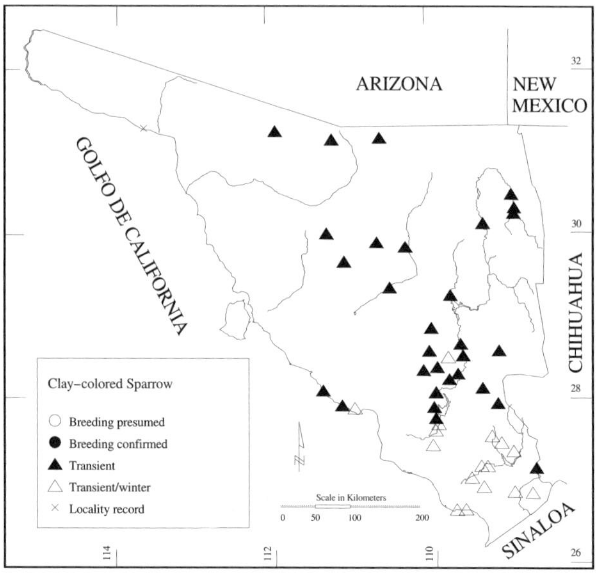

BREWER'S SPARROW
Gorrión de Brewer
Spizella breweri

This species, the plainest of all the sparrows, is a common, often abundant, transient and winter visitant in Sonora, mostly below 1200 m in the north and 500 m in the south. Its preferred habitat is brush in and adjacent to grass and summer annual growth in desertscrub. Here it is to be seen in various-sized flocks, where, when not feeding, the birds perch in low bushes and sing as a chorus, as do the White-crowned Sparrows with which they commonly associate. It is fairly common in openings in tropical deciduous forest. Among the high counts of this sparrow are 500 at Puerto Peñasco on 31 March and 400 on 2 January; 200 at Altar on 31 March; 200 north of Rancho Saracachi on 14 November; and 250 at El Represito de Luján about 50 km south-southeast of Hermosillo on 4 November. It is not common in areas that have received little summer rainfall; it moves elsewhere in search of seeds.

The peak of migration is in mid-September and early April. The earliest fall date is 1 September at San Miguelito on the upper Río Bavispe, while the latest spring date is 6 May at Granados (van Rossem 1945).

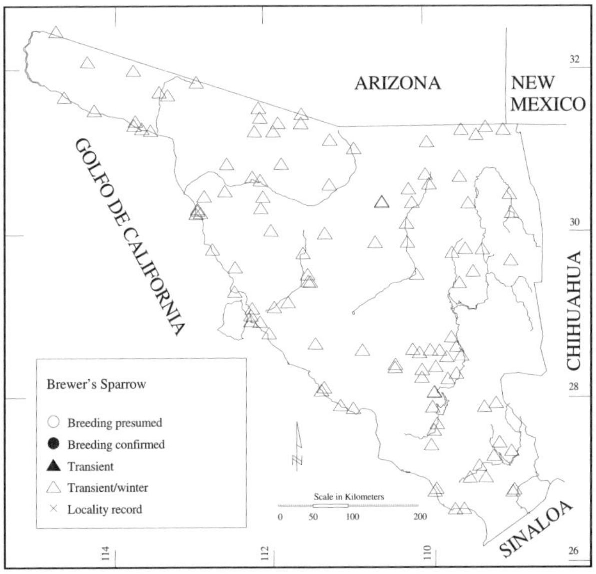

BLACK-CHINNED SPARROW
Gorrión Barba Negra
Spizella atrogularis

The Black-chinned Sparrow is an irregular to uncommon winter visitant to northern and eastern Sonora. All records are of fewer than four individuals, except for two flocks of about 25 birds (Sierra el Humo on 12 November and near Rancho Saracachi on 24 November).

Breeding in Sonora has not been documented. Van Rossem (1934) reported a specimen taken in the Sierra la Madera on 10 May 1887. Although the date suggests nesting, no other individuals have been found there, although J. Marshall and others have visited the mountains. Two presumed pairs, with the males singing, were noted in Arroyo Cajón Bonito on 16 and 17 March 1976. Habitat apparently suitable for nesting (much grass, agaves, sotol, shrubs, boulders) does occur on the canyon slopes. The species nests in nearby areas in Arizona.

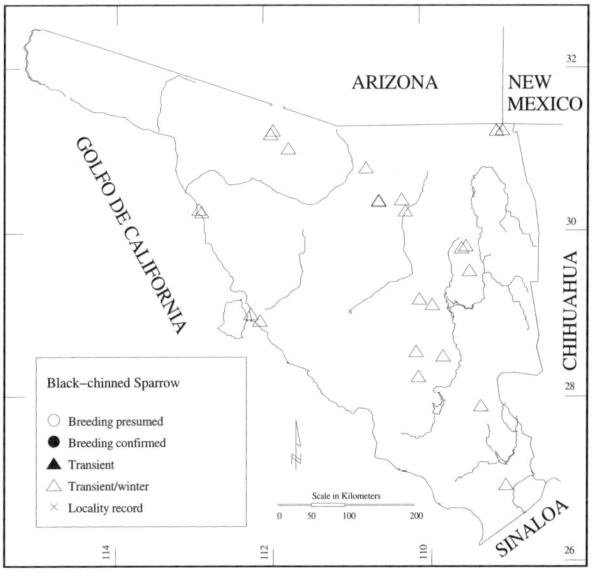

VESPER SPARROW
Gorrión Cola Blanca
Pooecetes gramineus

The Vesper Sparrow is a medium-sized sparrow with outer tail feathers that flash white when it flies from the ground where it feeds. It is a common transient and winter resident in Sonora wherever there is open ground with low annual plants and especially grasses with nearby shrubbery for escape cover. It often associates with other sparrows, especially White-crowned, Lark, and Brewer's Sparrows. Although most common below 1200 m, it is also occasional in openings and farm clearings in pine-oak woodland. Sometimes quite large gatherings can be found, illustrated by 420 on overgrazed flat ground 55 km northwest of Hermosillo on 11 November and 200 at the Saracachi ciénega on 12 November.

Outside migration dates are 9 September (Arroyo Cajón Bonito) and 16 April (Güirocoba; van Rossem 1945). The principal migration periods are in November and March.

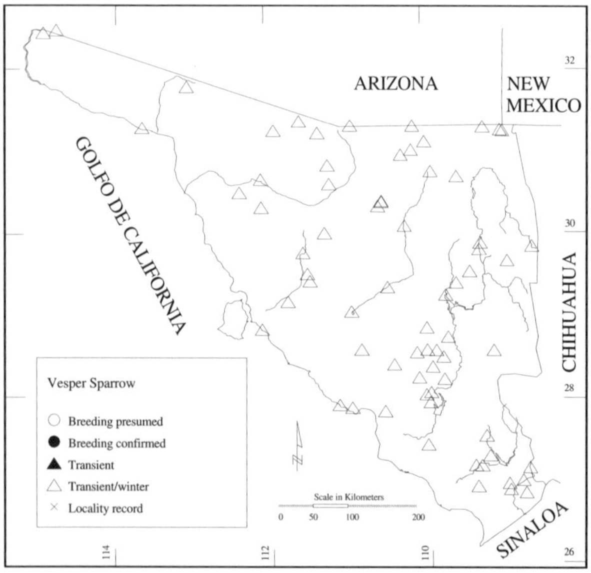

LARK SPARROW
Gorrión Arlequín
Chondestes grammacus

This is a medium-sized sparrow, boldly and handsomely marked on the head, with broad white areas in the tail. It is a common migrant and winter resident over much of Sonora, especially in weeds and grass. Migration begins very early in the fall; birds have been seen in Navojoa on 19 July and at Alamos on 26 July (Short 1974). Flocks of migrant birds are most noticeable in mid-September. We saw hundreds several kilometers north of Cumuripa on 20 September in tall grass and dense weeds in rocky terrain. The species is most common (even abundant) in the far south in winter. Northbound birds are evident from mid-March into April (latest date, 13 May, below Presa Novillo). Lark Sparrows are seen along the coast infrequently and only in small numbers.

No nests of Lark Sparrows have been found in Sonora. J. Marshall collected (WFVZ) a female in breeding condition near Naco on 6 June 1958. Other locations where we presume they may breed (singing birds, appropriate dates) are indicated on the map. Five locations on the map, identified only as locality records, may be within the breeding range in some years. Acceptable nesting habitat is provided by mesquite grassland at elevations of about 1100 to 1500 m.

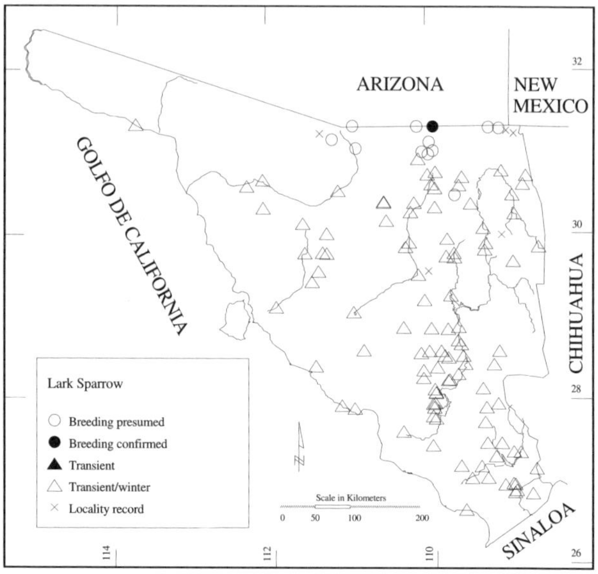

BLACK-THROATED SPARROW
Zacatonero Garganta Negra
Amphispiza bilineata

Once called the Desert Sparrow, this strikingly marked small sparrow may be found commonly throughout the year in desertscrub, semidesert grassland, and tropical thornscrub. It is most common where mesquite, palo verde, creosote bush, various low shrubs, and cacti grow in abundance. Although some of these sparrows may be found year-round in these communities, we do not believe that the species is sedentary and a permanent resident. There are fewer of them where the vegetation is less dense, as in the extreme northwest; this sparrow is absent in the Río Colorado delta.

The Black-throated Sparrow has a pleasant tinkling song that is heard through the spring and summer and occasionally in fall and winter. The nesting season is protracted, beginning in March and extending into September. Seasonal precipitation greatly influences nesting. Nesting may be deferred in the absence of rain, or the birds may leave an area during drought in search of more favorable conditions. The presence of birds in streaked juvenal plumage does not signify nesting where seen. Young birds often disperse with their parents well before their first prebasic molt; juveniles may be seen from April until at least mid-October. Small flocks of adults and their young form late in the breeding season. The flocks may wander and be augmented by birds from other parts of the breeding range. Flocks of 20 to 30 birds are not unusual in winter, and many such flocks may be in an area with abundant small seeds.

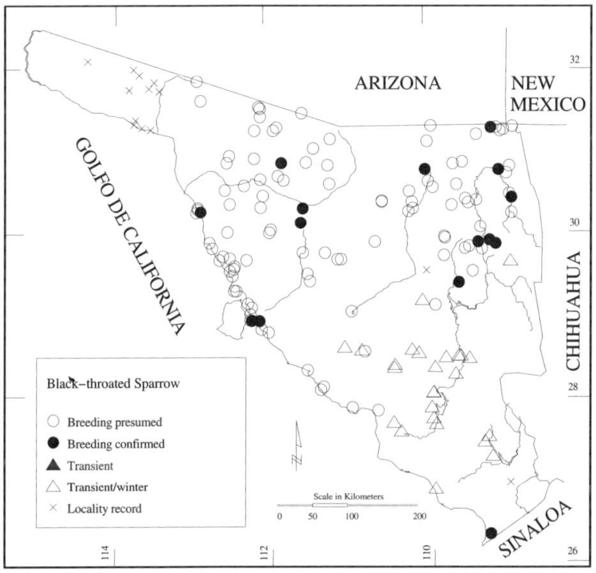

Although widespread, conspicuous, and common, the Black-throated Sparrow is one of the most poorly known Sonoran birds. We do not know if there are any resident populations, nor do we know the status of the species in the more arid deserts of the northwest.

SAGE SPARROW
Zacatonero de Artemisa
Amphispiza belli

The Sage Sparrow is an uncommon winter visitant in coastal dunes, in open creosote bush with bur sage, in salt bush communities, and particularly in growths of iodine bush and *Frankenia*. It is highly terrestrial, running rapidly from bush to bush as the observer approaches, with its tail cocked at a 45° angle. The species should be expected in high desertscrub, especially in the vicinity of Agua Prieta.

Outside migration dates for Sage Sparrow in Sonora are from the Puerto Peñasco region: fall arrival, 23 September; spring departure, 31 March.

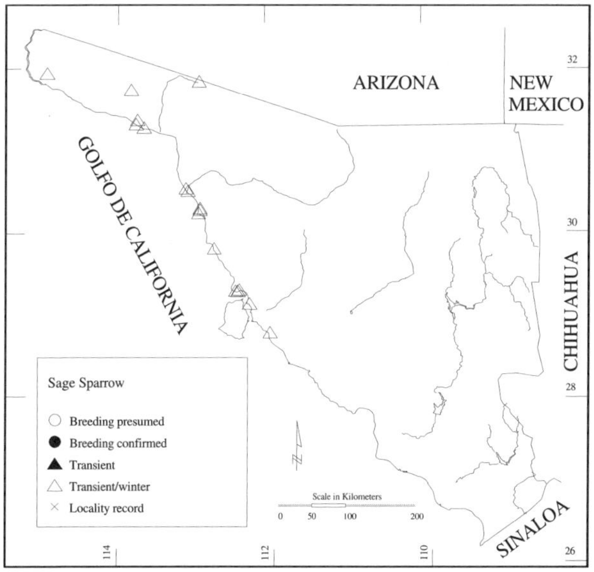

FIVE-STRIPED SPARROW

Zacatonero Cinco Rayas

Amphispiza quinquestriata

The Five-striped Sparrow is fairly common on rocky, well-vegetated hillsides and canyon slopes throughout much of Sonora. It sings little until the onset of the summer rains, when it commences nesting. It then becomes relatively conspicuous. In July of a wet summer it sings from seemingly every slope along the highway from Tonichi to Tepoca. There the habitat is thornscrub and tropical deciduous forest. Aside from the breeding period, it is silent, foraging inconspicuously on the ground. We have no winter records in extreme northern Sonora. It may be migratory there, but it can also be easily overlooked. We have never seen more than six birds together, and we think those small flocks may be family groups.

Birds have been found at elevations ranging from 50 m on a rocky hill covered with desertscrub-thornscrub near San José Masiaca (paired and singing in July) to 1370 m at the southwest base of the Sierra Oscura.

The Five-striped Sparrow has been moved from *Amphispiza* to *Aimophila* and is now *Aimophila quinquestriata* (AOU 1997).

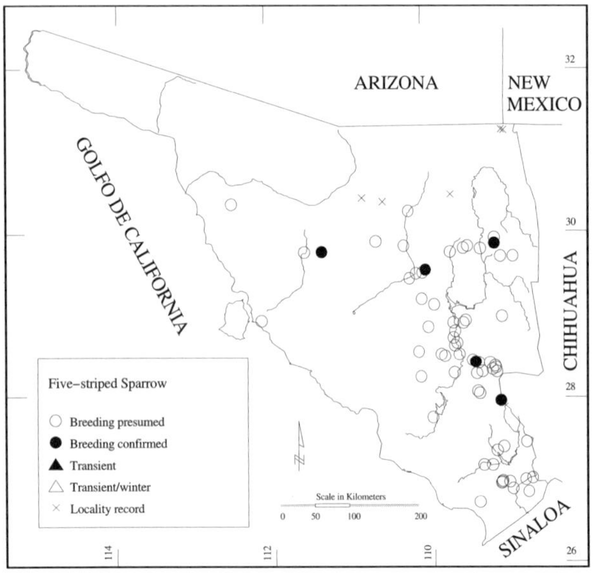

LARK BUNTING

Gorrión Ala Blanca

Calamospiza melanocorys

Lark Buntings are common (in some years, abundant or rare) wintering birds in open country, feeding on the ground, perching on fence wires and bush tops, and flying up from roadsides, in closely packed flocks. They may spend more than half the year in Sonora, where they arrive from the Great Plains as early as 6 August; by mid-August they may be abundant (at least 420 in one flock northwest of El Oasis on 17 August 1976). Many remain into mid-April. The latest date is of one in grassland above Arroyo Cajón Bonito on 10 May 1983.

Their occurrence in far western Sonora varies greatly from year to year depending on the food supply. They have been seen on four of ten Christmas Bird Counts at Puerto Peñasco (as many as 800 in 1984, a year of unusual summer rainfall). Three hundred were observed at Estero Tastiota on 16 March 1989. Most are in agricultural areas, where one of their main foods is waste grain. Flock after flock have been seen along Mexico Highway 15 from Hermosillo to the Sonora-

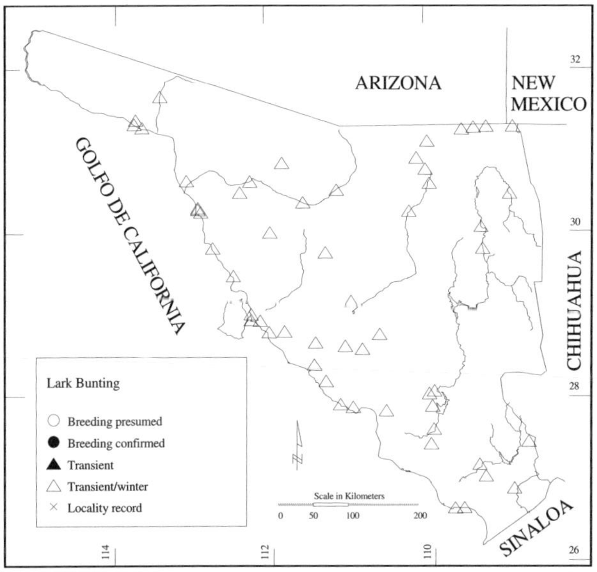

Sinaloa border, the attraction of the highway being not only the weed seeds on the berms but grain lost from trucks. Lark Buntings are not found in mountains.

SAVANNAH SPARROW
Gorrión Sabanero
Passerculus sandwichensis

The Savannah Sparrows of Sonora are of two distinct groups: one a transient and winter visitant, the other a permanent resident of saltwater marshes along El Golfo de California and formerly called the Large-billed Sparrow (*P. s. rostratus*).

The nonresidents are apt to be found any place that is damp and grassy, such as freshwater shorelines, pastures, dense grassland, and alfalfa fields. They have been found most commonly in the northern half of the state and occur only sparingly in the southern half; they are rare at Alamos. Outside migration dates are 20 September (1974, at least 20 at Pozo Coyote about

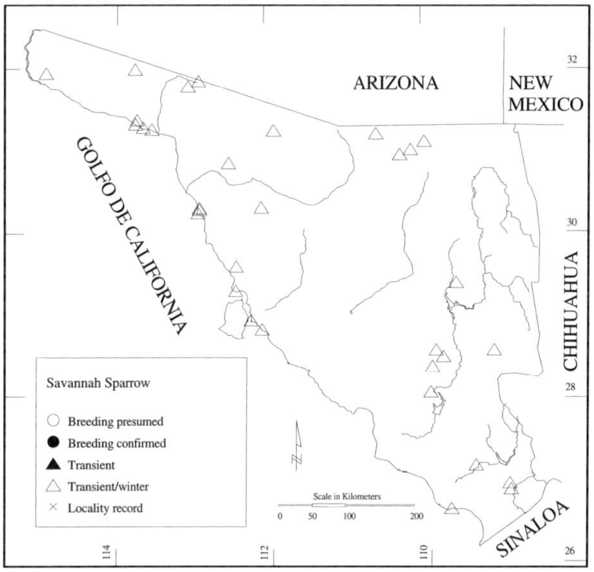

13 km north of El Desemboque de San Ignacio) and 9 April (1986, one at Piojillo at the head of Presa Obregon).

"Large-billed Sparrows" are resident in open saltwater marshes from the Río Colorado delta all the way to the Sonora-Sinaloa border. They begin singing by early February and continue to do so into the summer. Nests are placed in salicornia, iodine bush, and beach grass, and occasionally in the *Frankenia*-dominated areas back of the beach. Most nests have been noted in March and April, but one bird was building a nest on 20 June at Punta Santa Rosa. We have seen these sparrows foraging in eelgrass and seaweed at low tide more than 200 m from their territories.

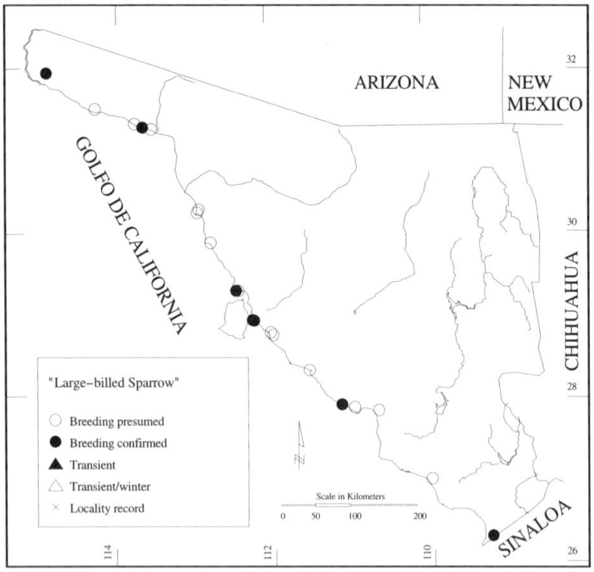

BAIRD'S SPARROW

Gorrión de Baird

Ammodramus bairdii

Baird's Sparrow is a rare winter resident of grassy plains and mesas in extreme northern Sonora. It spends all of its time on the ground in dense grass that is not tall or coarse. When flushed, it does so unexpectedly, flies off for some distance, and abruptly drops into the grass and disappears.

Because of the conversion of almost all of its historical habitat in the northern Great Plains to farmland, the Baird's Sparrow is much less common than formerly. Nevertheless, it continues to be found on its winter range in grassland that is not heavily grazed by domestic livestock, a type of grassland not easily found on either side of the border. The amount of cover available to the Baird's Sparrow is also governed by the amount and frequency of summer rains.

We know of only four records from Sonora. E. Mearns took two specimens on the border in El Valle de Animas on 2 October 1893 (van Rossem 1945). An immature female was collected in Sonora in the same valley on 23 or 24 November 1951 (J. Marshall). An adult male was taken 16 km south-southwest of El Sasabe on 23 October 1948 (Phillips and Amadon 1952). We saw five or six on Mesa Green 22 km east-southeast of Cananea on 4 March 1984.

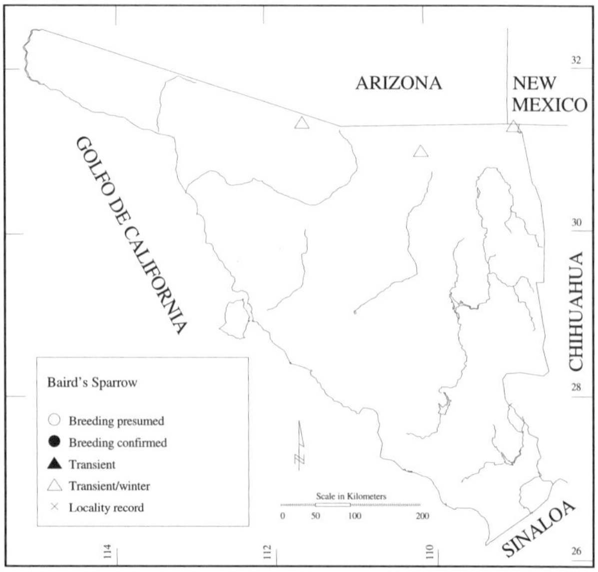

GRASSHOPPER SPARROW

Gorrión Chapulín

Ammodramus savannarum

This sparrow, well known as a breeding bird in parts of southeastern Arizona (Phillips et al. 1964), probably nests across the line in adjacent Sonora, where singing birds are regularly present locally in summer. The breeding habitat is grassland, where the grass is not very tall and may be grazed rather heavily. An occasional small mesquite or mimosa or a fence wire may provide a singing perch. We assume that any breeding birds in Sonora are permanent residents, but it is possible that some move south in the fall, especially if there is snowfall or if food conditions are not favorable. The probable breeding range also may be occupied by wintering birds from farther north.

Grasshopper Sparrows are locally common winter residents of eastern Sonora, being found in mesquite savannas, in grassy pastures, and on grassy foothill slopes having various deciduous shrubs. As many as 27 have been found on an Alamos Christmas Bird Count. We flushed at least 30 with Vesper and Clay-colored Sparrows in mesquite and grass on a bench above the Río Sahuaripa south of Guisamopa on 13–14 March 1984. Unexpected were three in a pasture on Rancho Tres Ríos along the Río Bavispe (1380 m) on 21 March 1985; one on Mesa Grande (1820 m) northwest of Yécora on 8 January 1983; and one in a grassy patch in a canyon bottom (1480 m) near El Ocote in mountains east of Sahuaripa on 16 March 1984.

Satisfactory arrival and departure dates for transients are lacking, but one banded near Rancho Saracachi on 16 April may be representative of a spring departure date.

The only record of a Grasshopper Sparrow from western Sonora is that of a single bird singing at Bahía la Choya near Puerto Peñasco on 12 March 1983 (T. Clark).

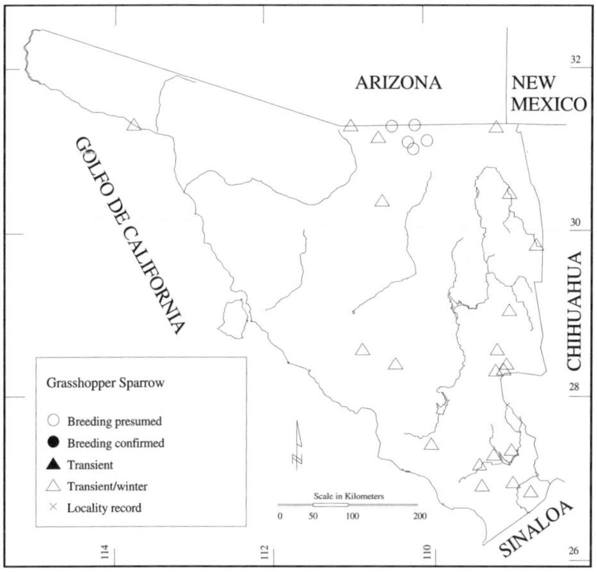

FOX SPARROW
Gorrión Rascador
Passerella iliaca

The Fox Sparrow has been reported four times in Sonora. One was collected on 22 December 1954 near Santa Cruz about 30 km east of Nogales (A. Phillips); one was observed on the Río Sonoita near Sonoita on 23 December 1968 (R. Cunningham); one was seen a few kilometers west of Ures on 14 January 1980 (Terrill 1981); and one was banded on the Río Cedros about 8 km above Quiriego on 5 January 1986. The Fox Sparrow may occur more often in northern Sonora than the few records indicate. It is to be expected in dense riparian thickets.

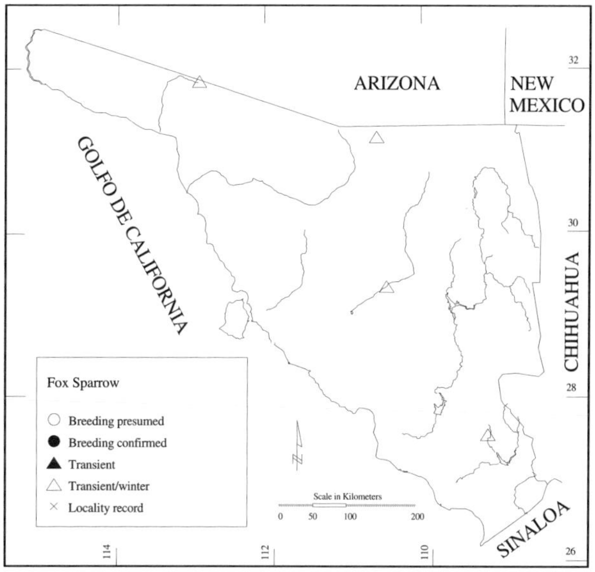

SONG SPARROW
Gorrión Cantor
Melospiza melodia

Song Sparrows are fairly common but local residents along permanent streams and freshwater marshes. Habitat requirements include a flowing stream or ciénega with thickets of *Senecio salignus*, batamote, and willow. The species is a common breeder in the

Ciénega de Santa Clara, chiefly in cattail associations. One collected on the Río Yaqui north of Esperanza on 28 June 1959 (A. Phillips) hints that the species formerly nested there, although we have not found it. Nesting takes place from mid-April to mid-June.

Some Song Sparrows migrate into Sonora from the north, arriving as early as 10 September. Many spend the winter and depart in March and April (latest date, 12 April). Migrants and wintering birds are uncommon and are typically darker than the resident form. At Lo de Campa, where we have found many emberizids, the Song Sparrow has been noted only once (on 6 January 1983).

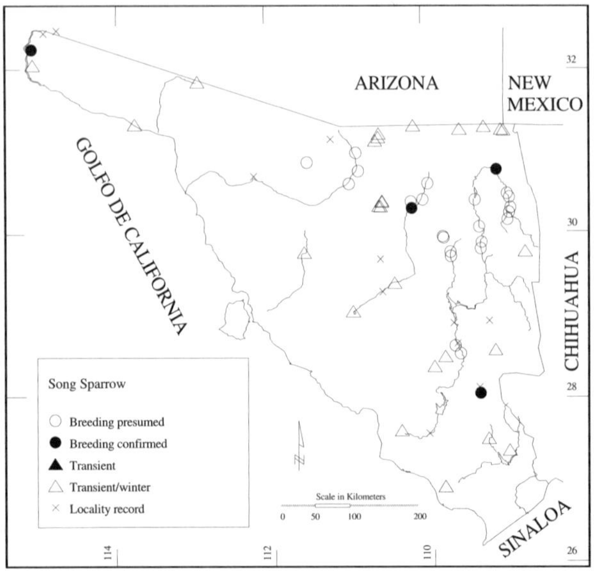

LINCOLN'S SPARROW

Gorrión de Lincoln

Melospiza lincolnii

Lincoln's Sparrows are common winter visitants in Sonora wherever dense weeds and grass are intermixed with thick shrubbery and tangled vines. They are scarce in the mountains above 1000 m but range virtually to the seashore in the lowlands. Fall migration consists of a gradual and prolonged southward movement (earliest date, 10 September); many birds do not arrive until November. Their secretiveness belies their abundance in winter; they do not flock and tend to remain hidden in thickets. We counted more than 250 at Lo de Campa on a January morning in a year with much food and cover. Our banding at Lo de Campa indicates a sedentary winter population. Some individuals returned in subsequent years. Northbound migrants are most numerous in March, particularly in the latter half, and may be seen almost anywhere there is cover. By mid-April, almost all Lincoln's Sparrows have left Sonora (latest date, 23 May, in Arroyo Cajón Bonito).

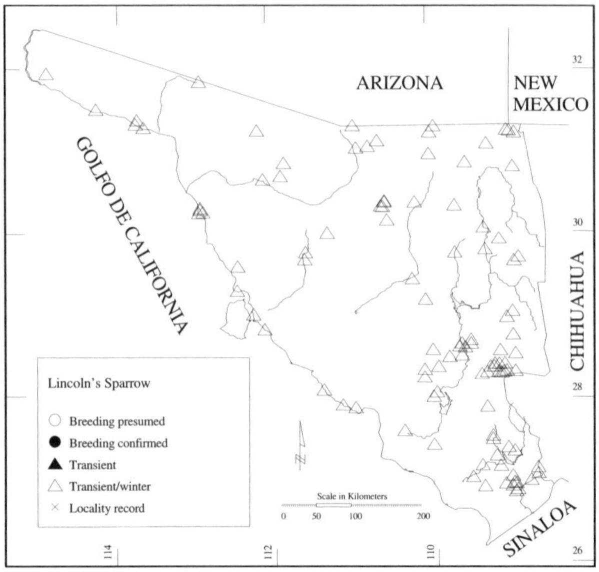

SWAMP SPARROW

Gorrión Pantanero

Melospiza georgiana

The Swamp Sparrow is a rare winter visitant, recorded from only a few localities in Sonora. It has been found typically in dense vegetation along and in freshwater, particularly where cattails are present. Most reports are of single birds, but Terrill (1981) found a total of 13 west of Ures from November 1979 through January 1980. Outside dates are 31 October and 1 May.

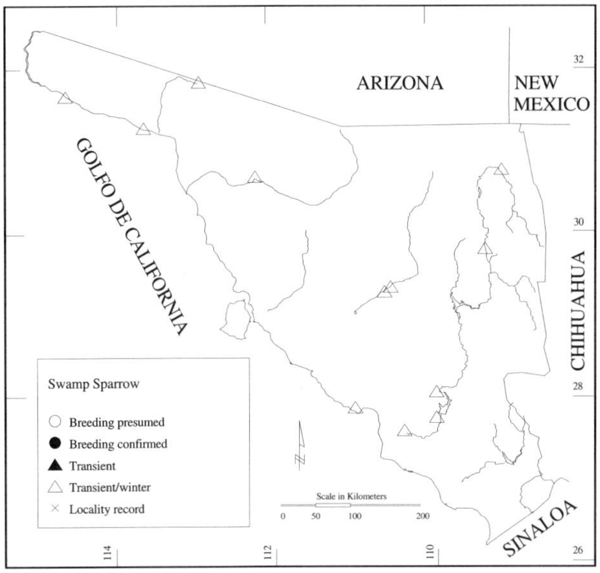

WHITE-THROATED SPARROW

Gorrión Garganta Blanca

Zonotrichia albicollis

This sparrow from eastern North America is a rare winter visitant in Sonora. Eleven records (including two specimens; A. Phillips) between 1 November and 1 April are of birds seen in or near thickets, often in the company of other sparrows or juncos. One of these records was a total of 15 seen along the Río Magdalena (13 at Cibuta and 2 near San Ignacio) on 22 January 1977 (K. Kaufman, T. Parker, and associates); this was a winter when unusual numbers reached southern Arizona. The elevational range is from sea level to about 1300 m (along the Río Santa Cruz and Arroyo Cajón Bonito).

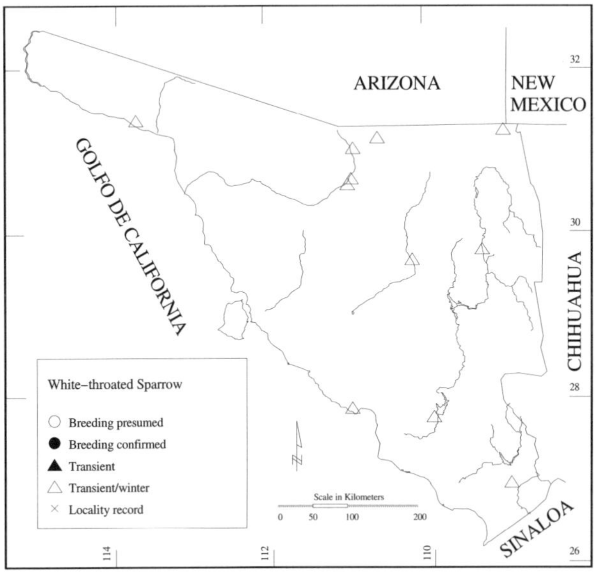

GOLDEN-CROWNED SPARROW

Gorrión Corona Dorada

Zonotrichia atricapilla

There are four sight records of this sparrow in Sonora: one near Caborca on 2 November 1948 (Phillips and Amadon 1952); three near Puerto Peñasco on 3 January 1965 (Russell); one in Arroyo Cajón Bonito on 21 May 1982 (R. Russell); and one on an Alamos Christmas Bird Count on 28 December 1982 (R. Bowers).

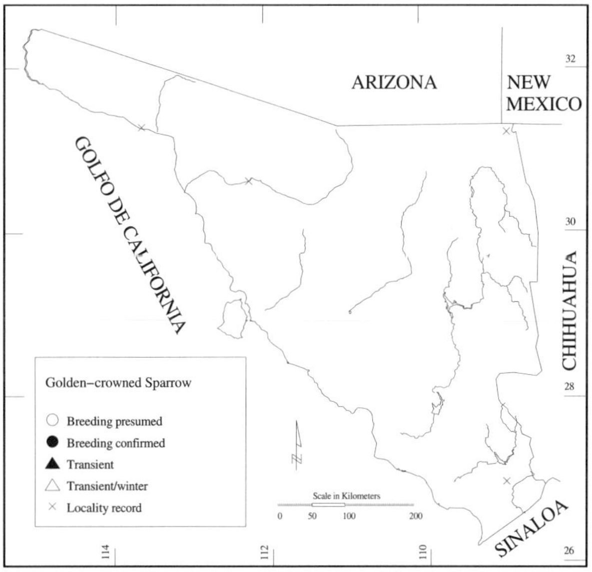

WHITE-CROWNED SPARROW

Gorrión Corona Blanca

Zonotrichia leucophrys

The bold, contrasty head stripes impart a regal mien to this common, often abundant, migrant and wintering species. This sparrow generally is seen in conspecific or mixed-species flocks of ten to more than a hundred

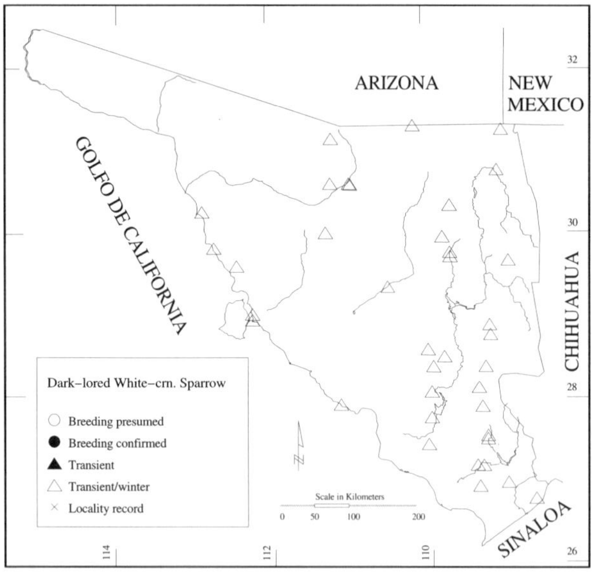

emberizids, feeding on the ground near bushes that provide cover when the birds are disturbed. As is the case with Brewer's and Lark Sparrows, the White-crowned Sparrow occasionally sings through the winter, the singing increasing as the season progresses. Its presence depends upon adequate quantities of small weed and grass seeds in open areas near cover.

The White-crowned Sparrow is most abundant in desertscrub, especially where mesquite is a dominant tree, in years in which rain has produced an abundant seed crop. It may be plentiful along the coast of El Golfo de California north of Bahía Kino, but not every winter. It is not numerous in tropical thornscrub and

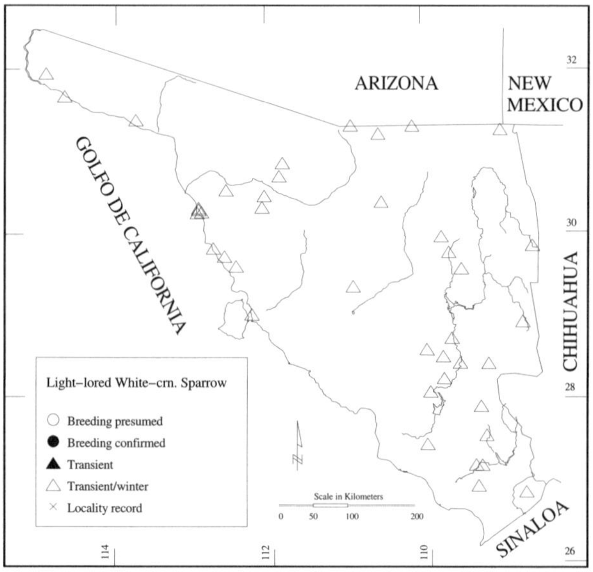

tropical deciduous forest. It is rather scarce in the mountains, and there are few records above 1300 m.

The White-crowned Sparrow migrates into and through Sonora mainly in late September and October and departs in late March and April. Most have left by 20 April. Extreme seasonal records are of two or three on the Río Babasac northeast of Imuris on 14 September 1952 (R. Dickerman) and noted about 20 km north of Bavispe on 2 June 1956 (A. Phillips, W. Lanyon). One seen on 13 July 1987 (A. Moorhouse) at Rancho la Tinaja (close to the Río Fuerte in the section where the river flows through Sonora) failed to migrate for some reason.

Two types of White-crowned Sparrows are usually distinguishable in the field in both adult and immature plumages: one with dark lores, the other with light lores. The former type is less common than the latter, especially in winter; it is the first to arrive in the fall and the last to depart in the spring, when for short periods it outnumbers the light-lored birds. The individual seen on 13 July was a dark-lored bird.

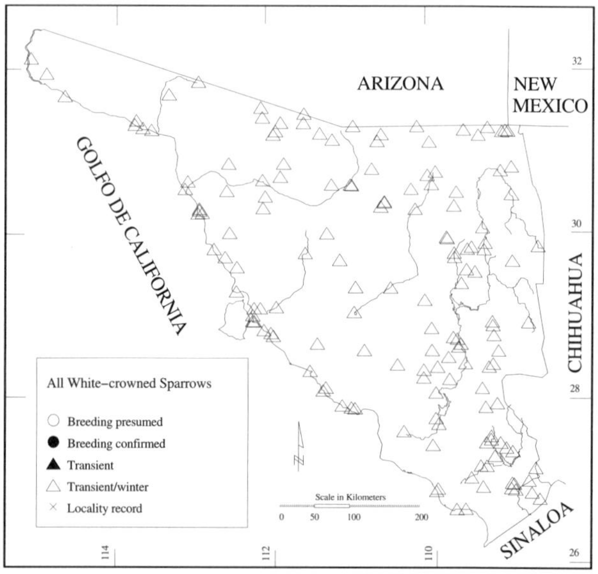

DARK-EYED JUNCO

Junco Ojo Oscuro

Junco hyemalis

The Dark-eyed Junco is a common winter visitant to Sonora, where it finds its southern limits on the Pacific slope of Mexico. It is most abundant in open patches in woodlands and forests, but a few birds will be found in some winters in desertscrub. It feeds on the ground, often in company with Chipping Sparrows. The Dark-eyed Juncos of Sonora consist of three groups, each embracing one or more subspecies and often recognizable in the field. Our records indicate different distributions and abundance of the "Slate-colored Junco," the "Gray-headed Junco," and the "Oregon Junco."

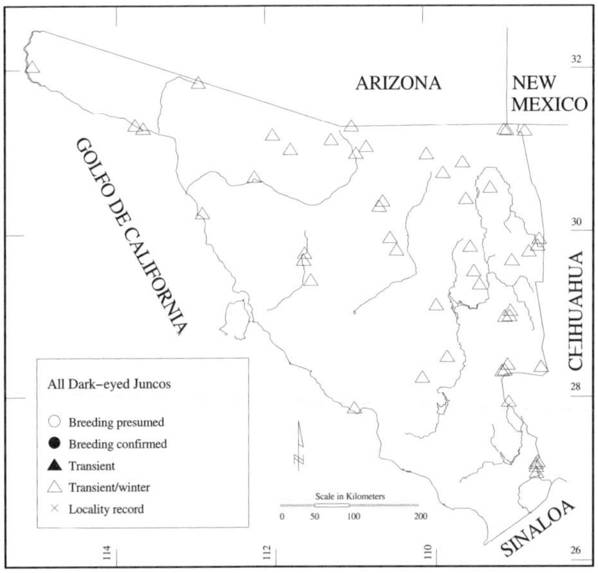

Only four records exist for the Slate-colored Junco. One was collected near Sonoita on 6 November 1948 (Phillips and Amadon 1952). Single birds were identified near Bahía la Choya on 1 November 1975 (R. Witzeman) and on Puerto Peñasco Christmas Bird Counts on 3 January 1981 and 17 December 1983.

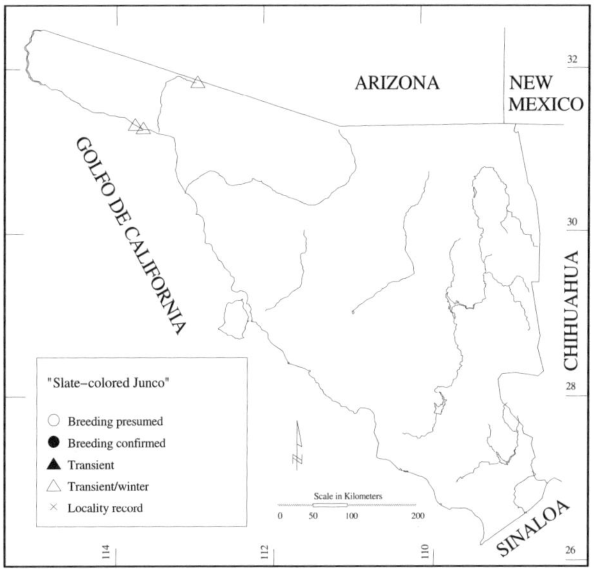

Gray-headed forms are generally common to abundant in the mountains but scarce or rare in the lowlands. They have been noted from 26 October to 3 May.

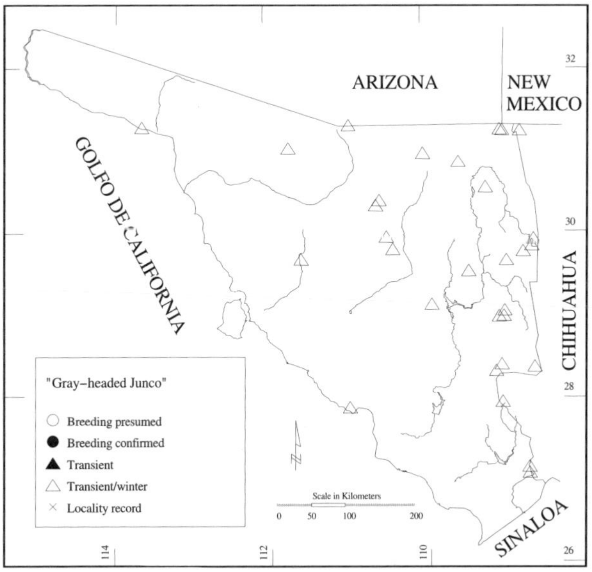

The Oregon Junco is often common in the north, becoming scarcer to the south. In mountains it is sometimes outnumbered by the Gray-headed Junco. Dates when present in Sonora extend from 14 October to 12 May.

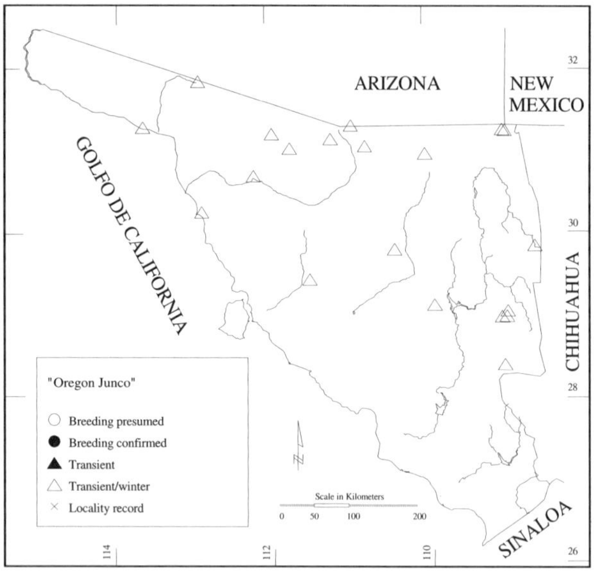

YELLOW-EYED JUNCO
Junco Ojo de Lumbre
Junco phaeonotus

The Yellow-eyed Junco is similar in appearance to the Gray-headed form of the Dark-eyed Junco except that the adults possess brilliant yellow eyes. It is resident in the highest mountains of Sonora. The lowest point we have found it is Rancho Tres Ríos on the Río Bavispe at 1380 m, where it was present on 22 March 1985. Possibly there is downward movement in winter. A. Phillips collected it in the Sierra Aconchi in February 1957, but J. Marshall was unable to locate it there in June 1954.

The Yellow-eyed Junco is rather inconspicuous and joins flocks infrequently. It is eminently a denizen of the pine-oak woodland and tall coniferous forest, where it prefers to feed under and in pines. It is almost sluggish in behavior, and when on the ground it walks, unlike the Dark-eyed Junco, which hops. It is heavier than other juncos and dominates them when in mixed-species flocks. Its song is heard from late winter through the summer; Marshall (1957) noted an increase in the amount of singing in July and early August. The nesting season produces the first young in June and continues into August.

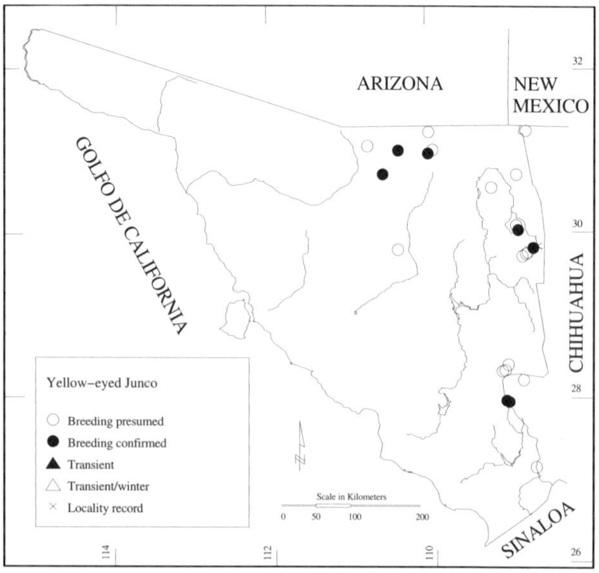

MCCOWN'S LONGSPUR
Escribano de McCown
Calcarius mccownii

Now one of the most uncommon birds in North America, the McCown's Longspur flies southwest from its breeding grounds in the western Great Plains to winter, sometimes as far as grasslands in northeastern Sonora. How regularly it does this is unknown. A specimen (USNM) was taken along the Río Santa Cruz on 20 October 1893, and it was observed by E. Mearns about 53 km west-northwest of El Sasabe, between 29

December 1893 and 8 January 1894 (van Rossem 1945). The latter locality is west of any grasslands existing today. Since these 19th-century records, McCown's Longspur has been found only once in Sonora; on 23 January 1977, a flock of about 250 feeding in heavily overgrazed grassland was discovered along Mexico Highway 2 east of Cananea (K. Kaufman, T. Parker, D. Stotz, and J. and R. Witzeman).

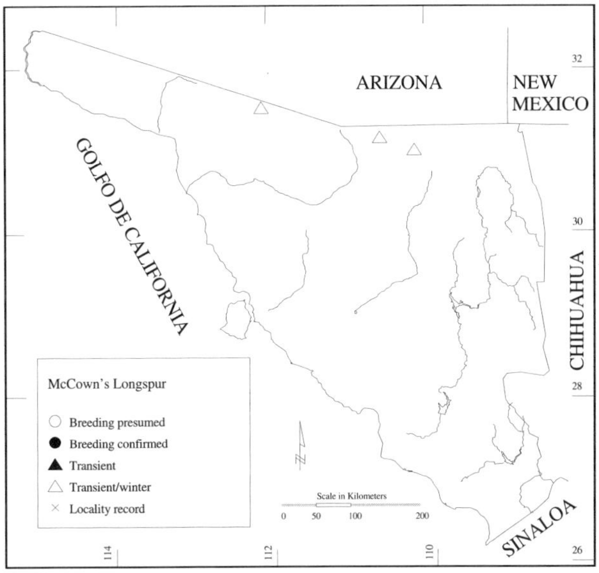

CHESTNUT-COLLARED LONGSPUR
Escribano Collar Castaño
Calcarius ornatus

The Chestnut-collared Longspur is a regular winter visitant to grasslands across northern Sonora west at least to the El Sasabe area. In grassy rangeland from the Río Santa Cruz east of Nogales to the northeast corner of Sonora it may be expected in most winters, sometimes flushing in flocks that may be of considerable size, then flying about en masse apparently aimlessly for a time before alighting again. Elsewhere, 24 were found on 24 October 1976 at Puerto Peñasco (D. Stejskal), and three were seen at San Carlos just in from the beach on 21 October 1975. Sonoran records are from 21 October to 4 March.

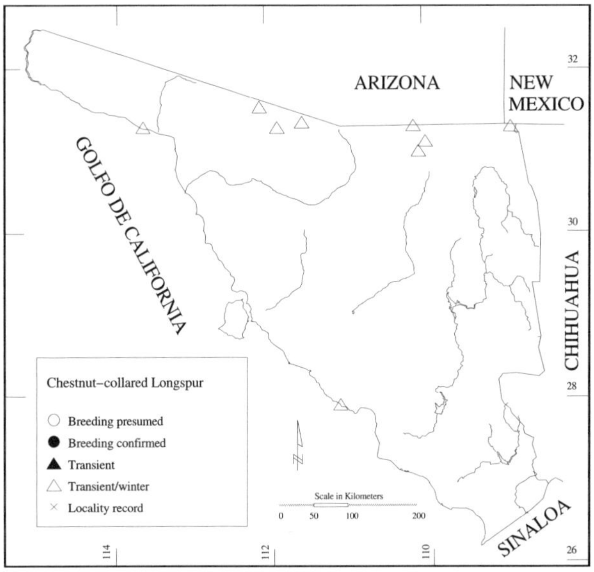

SUBFAMILY ICTERINAE
Icterines

RED-WINGED BLACKBIRD
Tordo Sargento
Agelaius phoeniceus

One of the most widespread of North American birds, the Red-winged Blackbird is distributed throughout Sonora wherever there are cattails growing in freshwater or brackish marshes. In irrigated areas, Red-wings frequent alfalfa fields that remain unmowed sufficiently long to allow the completion of a nesting cycle, and we believe that at times they may nest in standing fields of wheat and other small grains. When the breeding season is over, probably by August, the birds gather into flocks that feed mainly in grain fields but roost in cattail marshes.

By early September, and through October, Red-winged Blackbirds from north of Mexico arrive in Sonora to spend the winter. Ten seen at Puerto Peñasco on 3 September (D. Stejskal) were early migrants. They become numerous at cattle pens, where they feed with other blackbirds and cowbirds on grain put out for the stock. Large preroosting concentrations of several thousand birds have been observed on transmission wires, fences, and other elevated perches in the irrigated farmlands along the Río Asunción west and southwest of Caborca. The spring departure time of wintering birds is uncertain; it may be spread over a considerable period. A flock of more than 40 singing birds was seen in farmland about 6 km east of Magdalena on 26 April.

The increase in agricultural lands resulting from the expansion of irrigation systems has contributed to an increase in numbers of this blackbird. Our knowledge of its actual breeding range is inadequate. No nests have ever been reported, although displaying territorial males have been seen in several places.

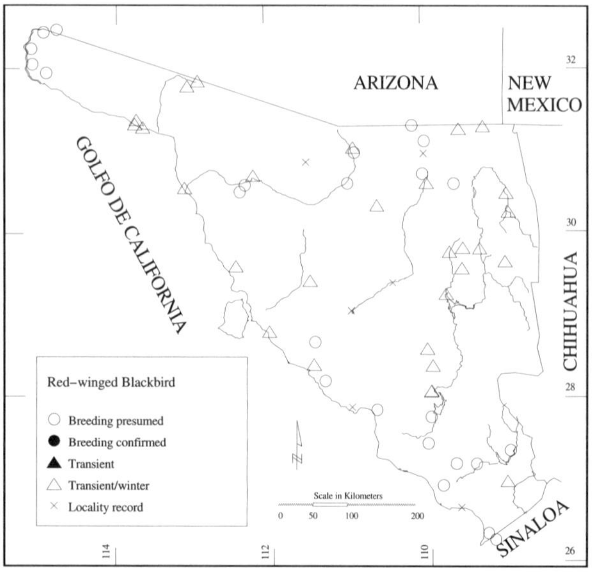

EASTERN MEADOWLARK

Pradero Tortilla-con-chile

Sturnella magna

The Eastern Meadowlark is resident in the grasslands of northeastern Sonora. Singing is heard from January through the summer and occasionally in the rest of the year.

There are widespread winter records, probably of migrants from the United States. Specimens have been taken about 20 km south of Santa Ana on 2 December 1932 and at Camoa on the Río Mayo northeast of Navojoa on 20 February 1931 (both van Rossem 1945).

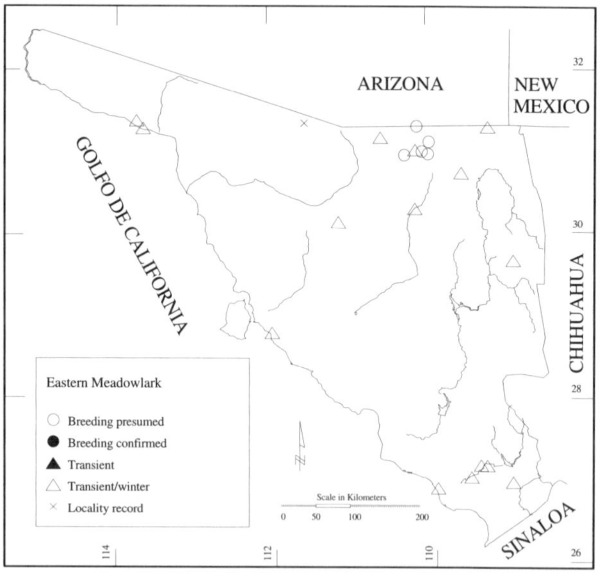

There are sight records, with identity based on song, from near the coast of El Golfo de California northwest of Huatabampo on 11 December 1983 and from Puerto Peñasco on 2 January 1984 (H. Hobart). Sight records, with identity based on call notes, come from Puerto Peñasco on 6 October 1979 (D. Stejskal) and 22 February 1979.

Christmas Bird Count records of the species from Alamos in 1980 and 1982 are not verifiable. A female (of unknown age) was taken at Bahía Kino on 1 July 1963 (UK). Unless the observer is very familiar with the call notes and songs of both species, Eastern and Western Meadowlarks are not reliably separable in the field in Sonora.

WESTERN MEADOWLARK

Pradero Occidental

Sturnella neglecta

The Western Meadowlark is a common transient and winter visitant in Sonora, sometimes in large flocks, in irrigated fields, near reservoirs, and in open desert where annual forbs and grasses have grown after heavy summer rainfall. It is often seen in grassy, weedy growth along the coast of El Golfo de California at least as far south as San Carlos, and it has been collected at Alamos (van Rossem 1934), where it was also reported on the 1984 Christmas Bird Count. We have seen one bird in mountains (east of Sahuaripa [1475 m] on 15 March, undoubtedly a migrant). The earliest fall record is 3 September 1979, when D. Stejskal counted 33 at Puerto Peñasco.

A nest of the Western Meadowlark has not been found in Sonora, nor are there observations of the species between 3 May and 3 September. We think these meadowlarks may nest as far south as Puerto Lobos following very wet winters, when courting pairs have been seen. Males become territorial in February, and pairs are noted into April. Van Rossem (1945) reported a breeding record from San Luis Río Colorado on 3 May. Song alone is not suggestive of breeding, as spontaneous singing may be heard in any month Western Meadowlarks are present.

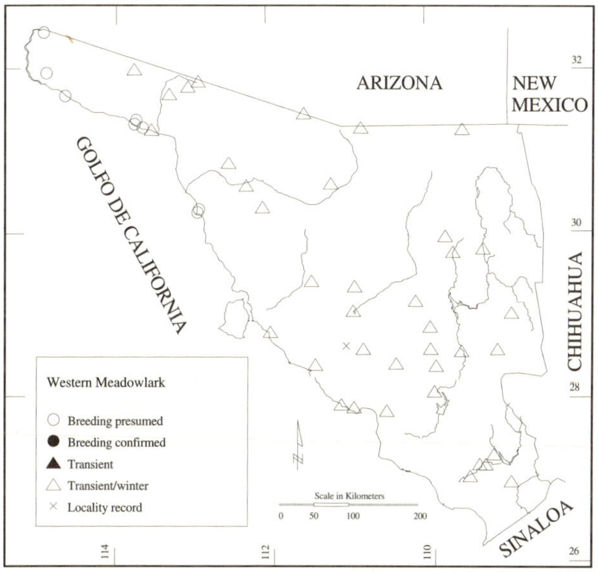

YELLOW-HEADED BLACKBIRD

Tordo Cabeza Amarilla

Xanthocephalus xanthocephalus

The first recorded sighting of the Yellow-headed Blackbird in Sonora was by J. W. Audubon (1906:141–142) on 30 August 1849, when he wrote of "large flocks of the yellow troupial in noisy bustle settling themselves in the rushes and willows bordering the little stream we are now fording" at Ures.

This blackbird is a common transient and winter visitant in irrigated fields near streams, along major canals and reservoirs, and in and near cattle pens on the outskirts of cities and corrals on ranches. In such places it often associates with other blackbirds, cowbirds, and grackles. It usually roosts in marshes, as noted by Audubon.

In 1987 Yellow-headed Blackbirds were seen at several southern Sonoran localities early in July. These were probably early fall migrants, although they may have been summering nonbreeding birds. Fall migration lasts sporadically into September and early October. Northbound migration occurs mostly in April. The latest spring record was that of a bird collected on or about 31 May 1952 on the Río Babasac northeast of Imuris (A. Phillips). One was collected at Güirocoba on 10 May 1931 (van Rossem 1945). Of interest is an observation of more than 3000, most if not all males, flying northwesterly over north-central Hermosillo from 1825 to 1900 hours on 26 April 1977. Eight evenings later in the same time period, only about 50 were counted, so evidently northward migration had occurred in the interim.

Yucateco, or shining fig, lines Mexico Highway 15 as it passes through the city of Hermosillo. Some very large examples of this tree stand before the post office, and others grow in the *zocalo*, or public square, before the cathedral. Both places are well lit throughout the night. On the evenings of 4 and 5 April 1990, from 1815 until 1840 hours, Monson watched many swirling flocks of Yellow-headed Blackbirds flying into these trees to roost. They arrived from the east, apparently from feedlots and farmlands beyond Presa Rodriguez. It was difficult to estimate their numbers, which were in the thousands. The din as they settled almost drowned out the noise of city traffic. Monson estimated that about 75% were males. Other blackbirds also roosted in the figs with the Yellow-heads. Yellow-heads were estimated to constitute 85% of the total birds, Brown-headed Cowbirds 10%, and Great-tailed Grackles and Bronzed Cowbirds 5%. For how long a period the Yellow-heads use the roosts is not known.

As van Rossem (1945) mentioned, the Yellow-headed Blackbird may once have nested in the Río Colorado delta, and it may still do so in the bulrushes of the Ciénega de Santa Clara.

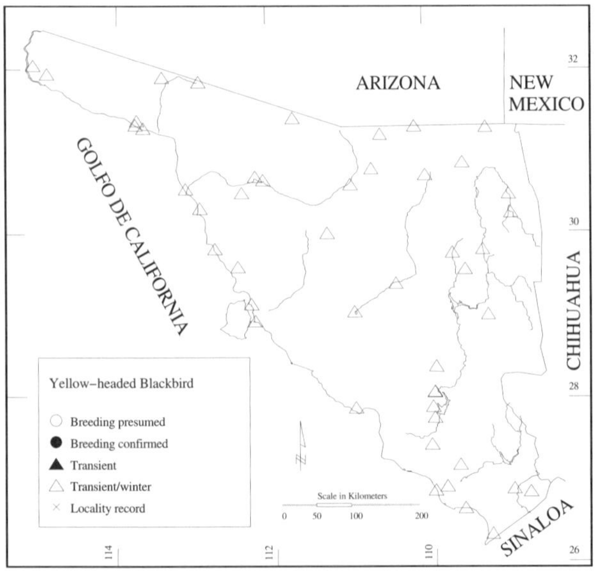

[RUSTY BLACKBIRD
Tordo Canadiense
Euphagus carolinus

A Rusty Blackbird was observed at Puerto Peñasco on 16 December 1989, and good details were provided (D. Stejskal, S. Ganley). What was ostensibly the same bird was seen in the same locality on 12 January 1990 (T. Wurster).]

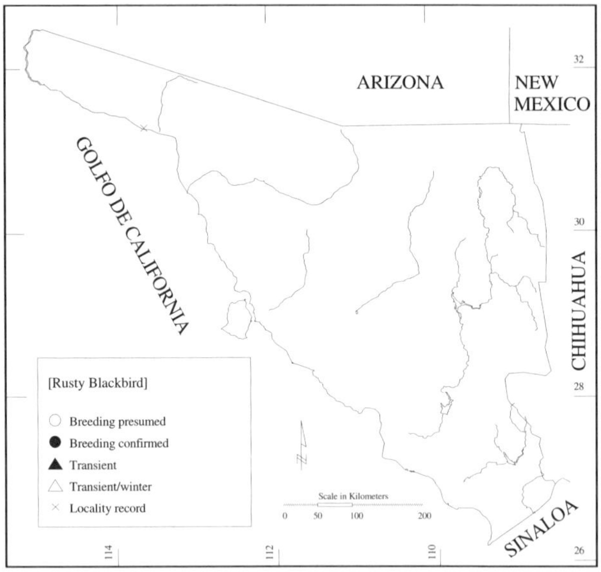

BREWER'S BLACKBIRD
Tordo Ojo Amarillo
Euphagus cyanocephalus

Brewer's Blackbird is a common winter visitant to Sonora, usually present in flocks in and near cattle feeding pens and corrals and in irrigated farm fields. It is often seen following earth-turning farm machinery, and occasionally it is observed in villages and towns. It migrates in March and early April and again in late September and October.

The earliest fall arrival date is 15 September, on the Río San Pedro at the United States—Mexico boundary (van Rossem 1945) and the Río Bacanuchi north of Arizpe. The latest spring departure date is 18 April, at Mututicache southeast of Cananea (A. Moorhouse).

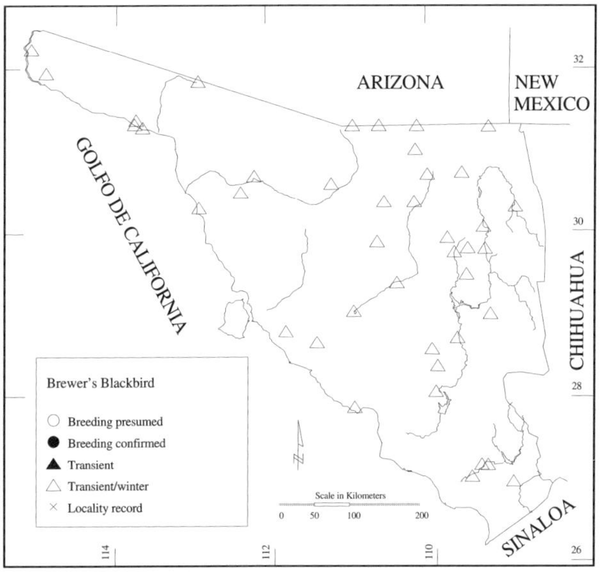

GREAT-TAILED GRACKLE

Zanate Mexicano

Quiscalus mexicanus

The Great-tailed Grackle is a familiar dooryard bird of Sonora. It is a common resident of city and pueblo and associated farmland, where it seems as though one or two of the large-tailed birds are constantly in flight all day long. It nests in colonies in spring and summer, usually in a small group in a deciduous tree where the bunchy, surprisingly small nests are exposed when the leaves have fallen; often the tree's branches are over someone's house. The grackles are noisy birds, with a variety of "songs" and notes, one of the most common being a loud, liquid gurgling that is quite arresting when one hears it for the first time. Grackles sometimes gather into large postbreeding foraging bands in fields, particularly irrigated pastures. They are omnivorous; in addition to grain and small insects, we have seen them capture and eat a large June beetle, a mouse, and minnows in a shallow creek. Along the coast of El Golfo de California, grackles comb the beaches for edible marine life. In southern Sonora, they even nest in mangroves. They were seen as high as 1880 m, at Mesa Tres Ríos on 21 March 1985, although we do not know if they were there all through the winter.

Van Rossem (1945) considered them to be resident in the coastal plain but no farther north than Rancho Costa Rica east of Bahía Kino, and he did not cite any records from Hermosillo. We have few records documenting the northward spread of the Great-tailed Grackle in Sonora. J. Marshall noted the species at Arizpe in July 1952 and found it on the Río Bavispe below the present Presa Angostura and at Huachinera in 1953. A. Phillips collected specimens at Bacoachi and Morelos in June 1953. It appears that the Great-tailed Grackle expanded its range northward rapidly between 1945 and 1953.

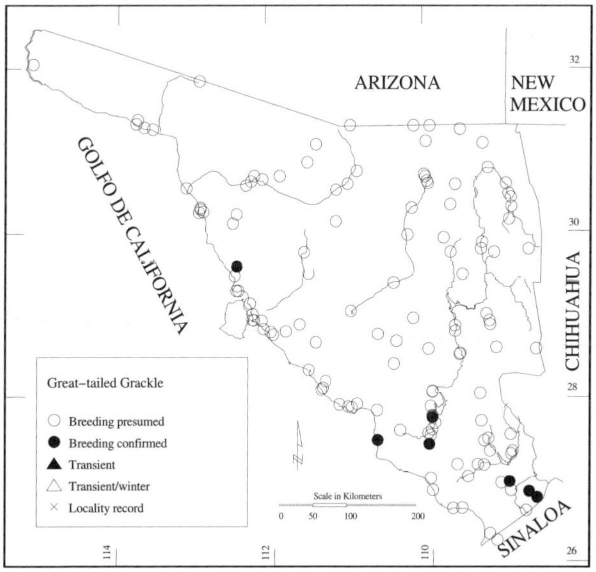

BRONZED COWBIRD
Tordo Ojo Rojo

Molothrus aeneus

The red-eyed Bronzed Cowbird is chiefly a summer resident, found commonly in all well-vegetated parts of the state and at all elevations, and usually associated with cows or horses, in whose shadows it often feeds. It arrives from the south in early May, breeds from mid-May to early August, and departs by the end of September. There are winter records from a number of localities, most from the far south, where it seems to be an uncommon winter resident in and near active corrals and feedlots.

In Sonora we have found eggs or young of the Bronzed Cowbird in nests of the Black-tailed Gnatcatcher, Sinaloa Wren, Bell's Vireo, and Hooded, Scott's, and Streak-backed Orioles. Many other bird species are known to be hosts to Bronzed Cowbird eggs elsewhere.

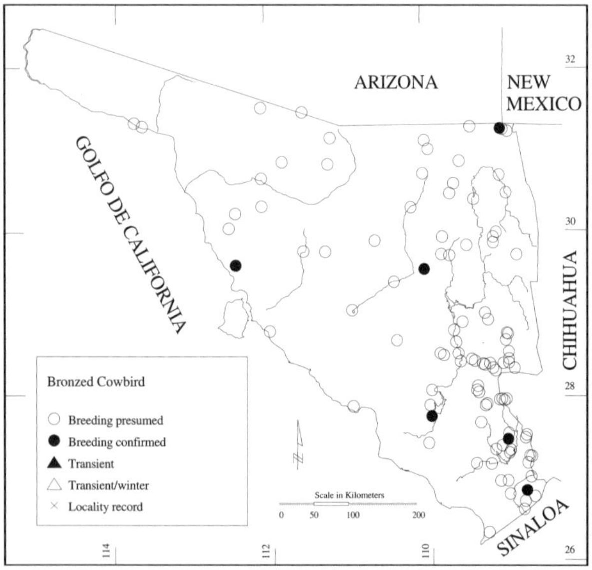

BROWN-HEADED COWBIRD
Tordo Cabeza Café

Molothrus ater

A common summer resident at all elevations in Sonora, the Brown-headed Cowbird gathers in large flocks in the fall and winter to feed in irrigated fields and at feeding pens (e.g., at Hermosillo and Caborca), spending the night in large roosts, frequently in company with other "blackbirds," especially Red-winged and Brewer's Blackbirds. These winter flocks, often numbering hundreds or thousands of individuals, undoubtedly include many migrants from north of Sonora. Smaller postbreeding flocks are found in villages and on small farms, usually associated with livestock. During the fall and winter, the birds are absent from desertscrub, thornscrub, and woodlands. We believe it is likely that a good many of the birds that breed in Sonora go farther south for the winter.

In March and April, the winter flocks disintegrate and cowbirds scatter everywhere for the long breeding season that extends to late September. They are most conspicuous near agricultural areas, ranches, and pastures where grain may supplement other foods. They have a detrimental impact on songbirds in the same habitat, whose nests they parasitize with their eggs. Many species are victimized by the cowbirds—chief among them gnatcatchers, vireos, warblers (especially chats), grosbeaks, buntings, and sparrows.

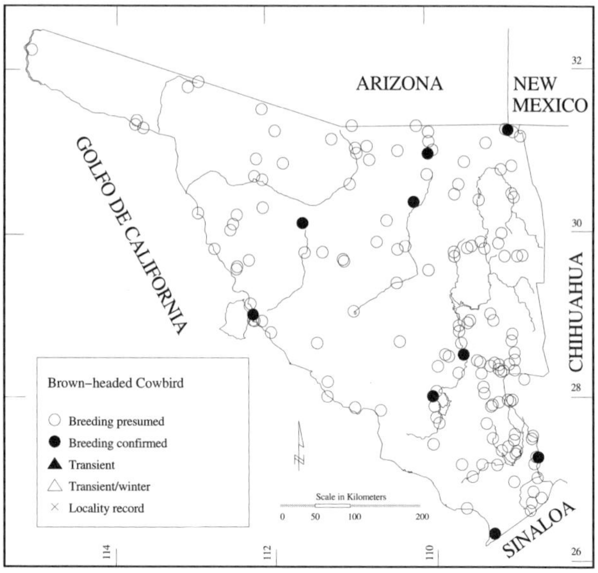

BLACK-VENTED ORIOLE
Bolsero de Wagler
Icterus wagleri

The Black-vented Oriole is a moderately common resident of the foothills and lower mountains of eastern Sonora, where it is local and usually on canyon slopes in tropical deciduous forest. Representative habitats typically have streams with vegetation such as *Ipomoea* and palms (*Sabal*), the latter regularly used for nesting (June, July). Groups consisting of two adults and up to four juveniles, presumably family units, are seen into November. In winter up to 40 Black-vented Orioles may roost communally in as few as two palms.

Our experience with the Black-vented Oriole has been limited to a few localities between 200 and 1500 m in tropical deciduous forest. Van Rossem (1945) reported the presence of nesting birds at Agiabampo; we have not found the species near the coast. J. Marshall collected an immature male at 2100 m in pine-oak woodland in the Sierra Nacori. We have seen these orioles in oak woodland only at low elevations at the edge of tropical deciduous forest. Their presence in pine-oak woodland is unusual.

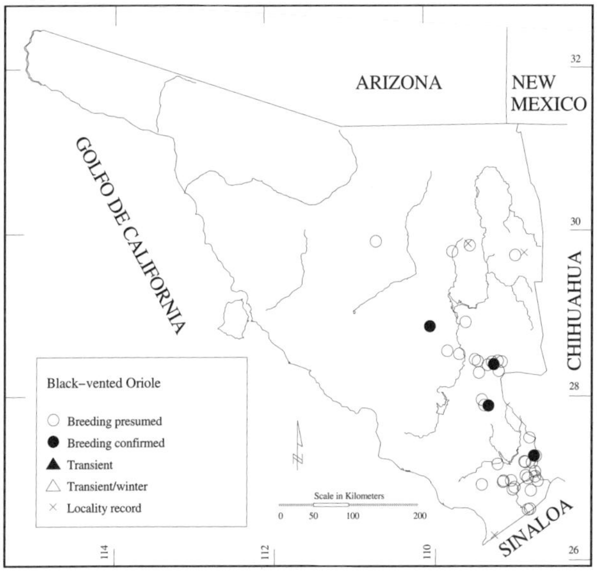

Black-vented Oriole

Black-vented, Streak-backed, Hooded, and Scott's Orioles have been observed in close proximity. The Black-vented Oriole is the most local and least abundant of the four orioles. We know little of the factors that limit its distribution or of its interactions with other orioles.

ORCHARD ORIOLE
Bolsero Castaño
Icterus spurius

Orchard Orioles are moderately common summer residents of tall cottonwoods along the lower Ríos Yaqui and Mayo. The species was first recorded from Sonora by A. Phillips, who collected several specimens on the Río Yaqui near Esperanza in September 1954. Although the Orchard Oriole breeds on the Mexican Plateau from Durango southward, it is present on the coastal plain only in Sonora. We do not know if this is a recent range extension or if Sonoran birds had been overlooked. Although we have found no nests, we have seen fledglings in early July. Inclusive dates extend from 7 April to 13 October; some spring and fall records may be of transients en route to and from breeding areas in the United States and Canada.

On 28 May 1980, G. Rosenberg and others found eight pairs in cottonwoods on the Río Sonora west of Ures. One pair copulated, suggesting a recent extension of the Sonoran nesting region.

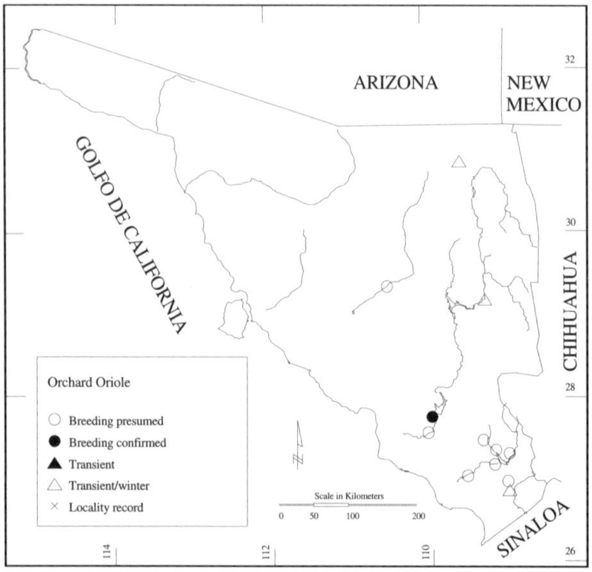

HOODED ORIOLE

Bolsero Encapuchado

Icterus cucullatus

The male Hooded Oriole is one of the most brightly colored and conspicuous birds of Sonora. The species is present from mid-March to the end of September. It is a common resident of much of the state below 1200 m, often nesting in village plazas, yards, and ranches. It is an uncommon breeding bird in the extreme south, especially where the larger Streak-backed Oriole is common. Migrants augment the number of local birds from March into May and in August and September. The species winters regularly in small numbers in the extreme south and occasionally in the north. Fourteen birds were netted and banded as they entered a mangrove swamp near Punta Chueca at dusk to roost on 27 November 1977. Eleven of the birds were females, and only two were young of the year. One had been banded there on 23 September. Although Hooded Orioles were not in the mangroves later in winter, the habitat may be important for southbound migrants.

Palms characterize many of the places in which this and other orioles nest. Palms provide not only shelter and structural support for the nests but also fibers, woven into the birds' intricate, often hanging nests. In the absence of palms, this oriole may utilize other trees, including mesquites, willows, cottonwoods, and sycamores. Twice we have seen nests attached to the underside of inactive nests of the Rose-throated Becard. Nesting may begin in early April and continue into August; a pair may have more than one brood. Nesting pairs of Hooded Orioles are usually spaced some distance apart, but they do sometimes form small colonies. In tropical deciduous forest just west of the divide (960 m) on the highway between Tonichi and Tepoca, we saw three nests on 30 July 1988 and noted at least eight pairs, all within 250 m of each other. At least 25 individuals, including some juveniles, were present. The locality is distinctive in being more mesic than other areas in these hills, and many palms grow there.

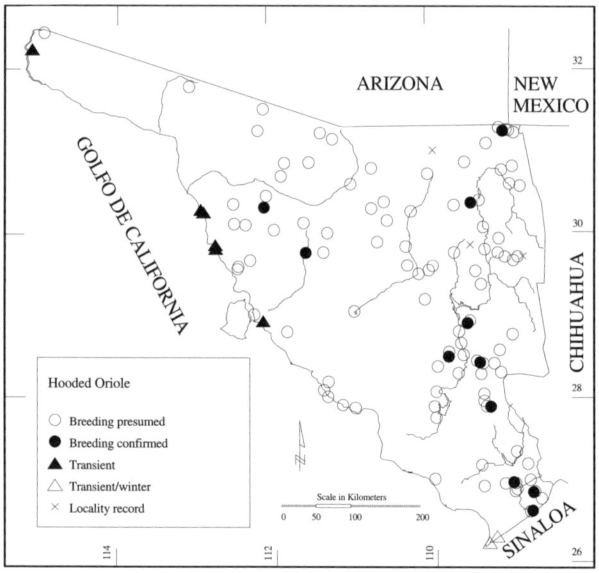

STREAK-BACKED ORIOLE
Bolsero Dorso Rayado
Icterus pustulatus

The Streak-backed Oriole is a common, sometimes abundant, resident of southern towns and cities, as well as riparian vegetation, tall trees, and brush of the tropical thornscrub and tropical deciduous forest. It is also common in some tall riparian habitats in desertscrub. The upper elevational limit is about 1000 m, but we have seen them higher a few times, most notably four feeding in a flowering oak at 2000 m on Mesa de Enmedio on 16 May. Many but not all individuals from the northern portion of the range are migratory, departing in October and returning in March. The species is basically insectivorous, but we have seen birds eating figs and guamachils.

The nest resembles a gourd with a very long neck. It is hung with little regard to concealment, often over a stream, and is so well built that it endures through the winter. Nesting takes place from May into August; July is the peak of nesting activity. Nesting regularly begins before the onset of summer rains;

Streak-backed Oriole

young fledge coincident with summer storms. Some nests contain young that do not fledge until September, probably the result of a second brood. Cowbird parasitism, particularly by the Bronzed Cowbird, is frequent. Streak-backed Oriole pairs may build within sight of one another. Although the species is presumably monogamous, Russell watched two females building about 20 m apart near Nuri on 7 June 1984. A second less colorful but adult female accompanied each builder to her nest. A single very vivid male remained close to the two nests. These observations, though short in duration, suggest the possibility of polygyny and helpers at the nest.

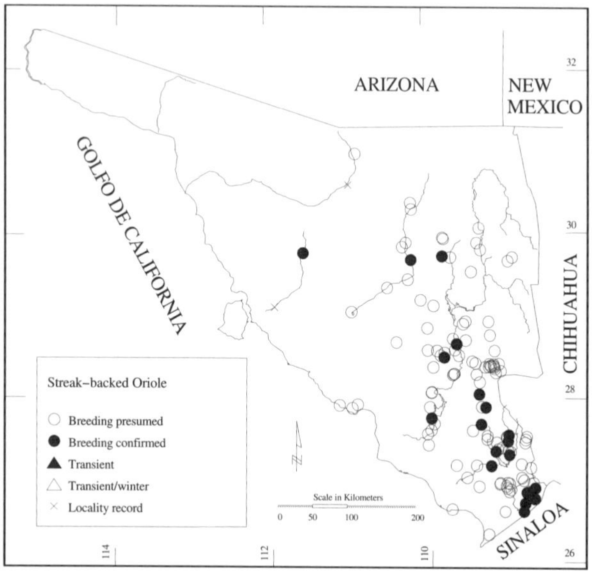

BALTIMORE ORIOLE
Bolsero de Baltimore
Icterus galbula

The Baltimore Oriole has been reported from Sonora four times (all single adult males): collected near La Casita on 12 October 1954 (A. Phillips); seen 24 km southeast of Magdalena on 27 April 1981 (Monson); seen at San Ignacio near Magdalena on 27 January 1980 (Terrill 1981); and seen near Alamos on 19 March 1982 (D. Stejskal).

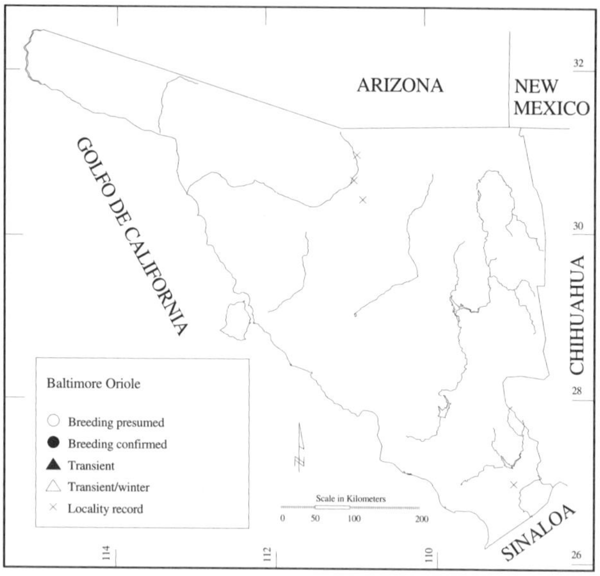

BULLOCK'S ORIOLE
Bolsero Calandria
Icterus bullockii

Bullock's Oriole has nested across the northern edge of Sonora (van Rossem 1931 and 1945; Phillips et al. 1964), but we are not aware of nesting records during the past 50 years. Even in suitable habitat in southern Arizona, where it selects cottonwoods and willows as nest sites, nesting is irregular.

It is a common transient almost statewide. Like some other land birds, such as the Western Kingbird, it is a daytime migrant. We saw 30 birds fly by us on their southward journey in an hour in the morning of 11 July near Agiabampo. On 23 September 1977 we netted and banded 12 birds that were flying into the mangroves near Punta Chueca to roost. Most fall migrants have left the state by late September. The latest fall date is 16 October (La Casita; A. Phillips).

There are half a dozen observations in the period of November through February, plus single birds reported on the Alamos Christmas Bird Counts. Although we question some of the reports, it does appear that Bullock's Orioles do occasionally winter. One was collected south of Fronteras on 12 November 1953 (A. Phillips).

The spring migration lasts from mid-March to mid-May, with a few birds lingering to late May. The earliest record is of one bird at Puerto Lobos on 2 March 1987 (J. Bates). Most spring migrants move through from mid-March through April, while the latest is of a female in the Sierra Huachinera on 1 June 1956 (J. Marshall). The birds are found in a variety of habitats while migrating, including the coast of El Golfo de California, desert situations, and mountains up to 2000 m.

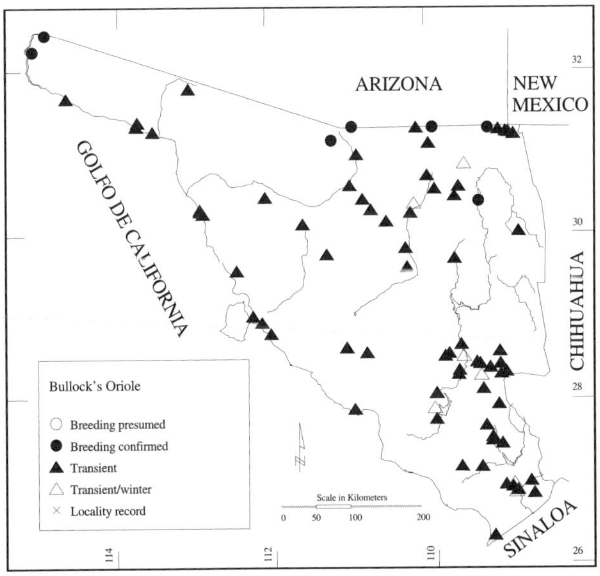

SCOTT'S ORIOLE

Bolsero Tunero

Icterus parisorum

Scott's Oriole is a moderately common summer resident in many mountains of Sonora and in parts of the low desert. It is most numerous in the oak zone, where it uses such plants as sotol, agave, and ocotillo for foraging and nesting. It ranges into the pine-oak woodland where these plants are present, but even in these optimal habitats and during the season of greatest song, it would be unusual to find as many as ten individuals in a morning. Its flutelike notes make a repetitious but pleasant sound, considered by Marshall (1957:113) as "the finest to be heard" in encinal and pine-oak woodlands. It nests from May into July, placing the nest in sotol or various vines and shrubs.

Migration is mainly in March and April and again in August and September. Extreme dates are 4 March (Cañon Evans, Sierra los Ajos) and 28 September (Lo de Campa). It occurs occasionally in valleys and on plains.

Scott's Oriole winters sparingly in the mountains of the southeast. It was seen in pines on Mesa Grande in January 1983 and near the San Antonio sawmill northeast of Alamos on 14 November 1973, and it is sometimes observed on Alamos Christmas Bird Counts. Other winter records are one near Moctezuma on 4 February 1983 and two in the foothills southeast of Tonichi on 7 January 1983.

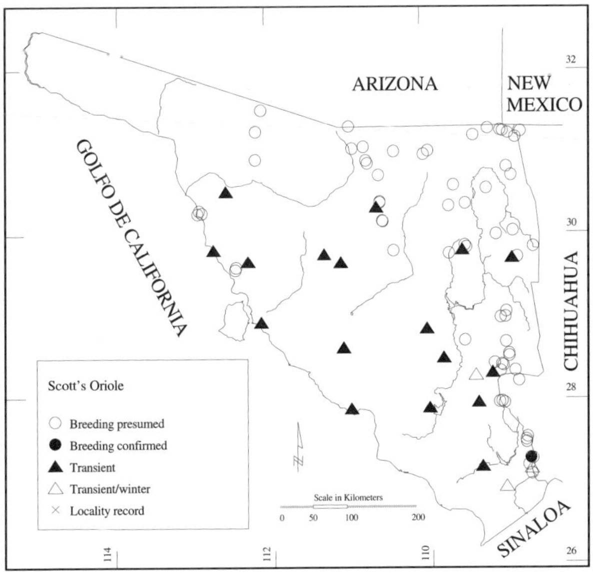

YELLOW-WINGED CACIQUE
Cacique Mexicano
Cacicus melanicterus

Van Rossem (1945:234–235) wrote of the Yellow-winged Cacique in Sonora, "Known to have nested near Alamos in the summer of 1931, but not found in the State before or since that time. In early September of that year W. J. Sheffler photographed this (then abandoned) colony which contained over thirty nests in a single tree. With native 'sling-shot' assistance he secured one bird, apparently a fully grown male of the year, which is now in his collection."

Since van Rossem's book was published, only one Yellow-winged Cacique has been detected in Sonora. That was a single bird along the Río Cuchujaqui south of Alamos on 13 March 1982 (D. Stejskal, R. Bowers). As far as is known, the species does not presently nest closer to Sonora than central Sinaloa.

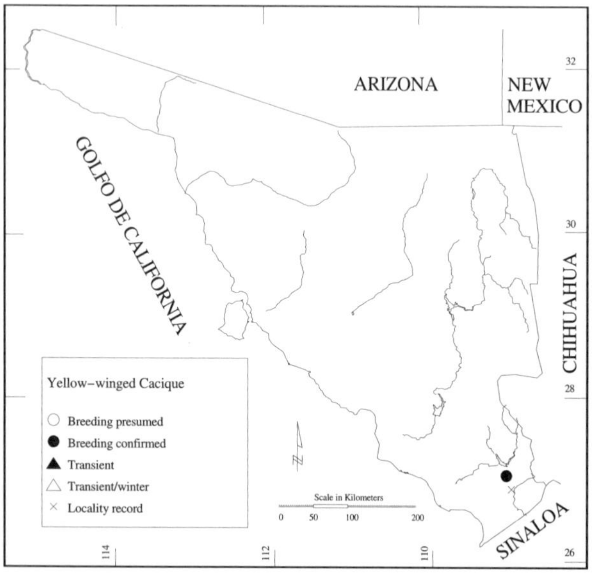

FAMILY FRINGILLIDAE

Fringilline and Cardueline Finches and Allies

PURPLE FINCH
Pinzón Purpúreo
Carpodacus purpureus

The Purple Finch is a casual winter visitant to Sonora. There are two records, both from elevations below 260 m: J. Bates, G. Monson, and A. Moorhouse saw two males and a female at Rancho el Puerto (15 km northeast of Punta Chueca) on 10 November 1984; and an adult male was found in a flock of House Finches at a roadside pond about 6 km south of Alamos on 27 December 1985 (D. Stejskal).

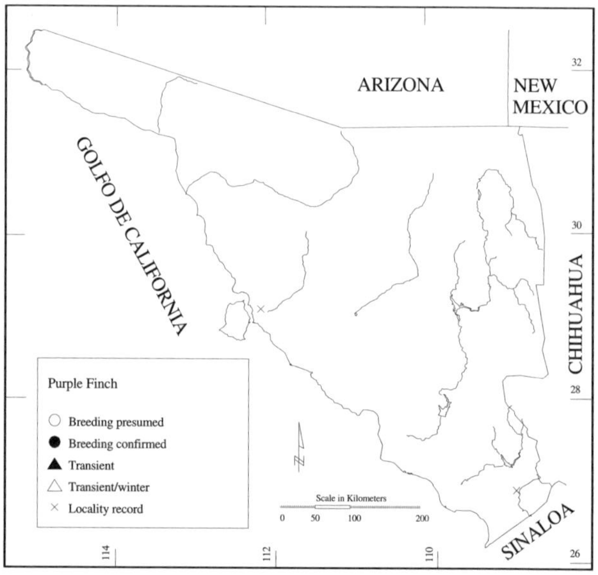

CASSIN'S FINCH
Pinzón de Cassin
Carpodacus cassinii

Cassin's Finches are probably more frequent winter visitants to Sonora than these four records indicate: an immature female and an immature male collected in pine forest at the summit of the Sierra el Pinito (2070 m) on 27 November 1954 (J. Marshall); two (an adult female and one of unknown sex) collected in pine-oak woodland of the Sierra los Ajos on 28–29 December 1963 (UA; R. Ohmart); two found in pine-oak woodland on Mesa Grande (1820 m) on 9 January 1983 (B. and R. Chapin); and at least one female observed in oaks near Tres Ríos (1380 m) on 22 March 1985 (J. Bates, A. Moorhouse).

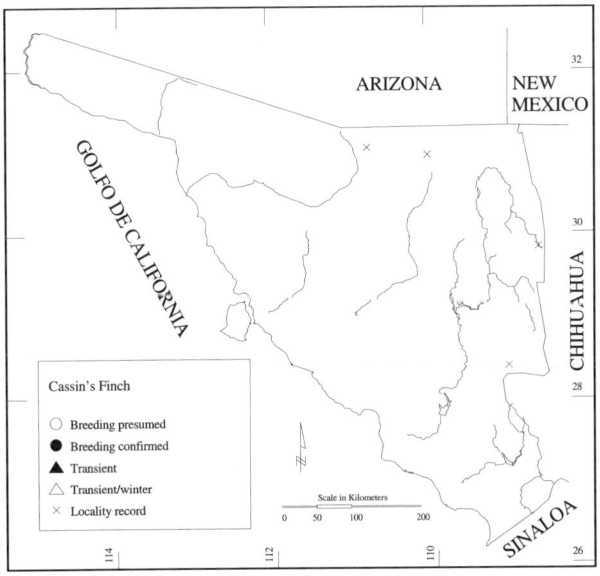

HOUSE FINCH
Pinzón Mexicano
Carpodacus mexicanus

The House Finch is the most widespread bird species in Sonora; except in woodlands and forests, its song can be heard from almost any spot. It is an adaptable species, resident from arid coastal desert to clearings in the mountains up to 1900 m. It would appear to occupy every dooryard in the state, usually in pairs or small groups. In fall and winter it may aggregate by the hundreds where seeds and grains are locally plentiful; it is then conspicuous on fences and bush tops.

In the warmer lowlands, song frequency and intensity increase in February just before nesting begins in March. In cooler regions, nesting commences in April; in all places nesting may extend into July. Nests are constructed in a variety of trees and shrubs, but House Finches select chollas, when available, over other plants. Occasionally they build in a shallow cavity in a cactus or in an abandoned oriole nest or on a convenient flat sheltered place on a house. Chollas are also

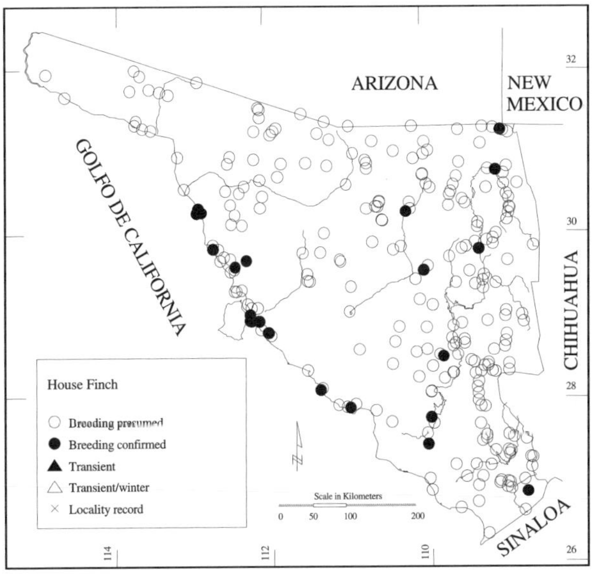

used as roosts through much of the year; a sizable mound of excrement indicates extended nocturnal occupancy. Near Punta Chueca, large numbers of the finches roosted in mangroves throughout the year. Arriving birds flew high over the 4 m tall mangroves and dropped almost vertically into their roost trees. During the breeding season, males utilized the mangrove roosts while females remained at their nests elsewhere.

House Finches eat fruit and seeds ranging from tiny weed seeds to grain and cactus seeds. Normally brown females may be tinged with pink or purple when they are consuming cactus fruit. The pungent and acrid fruit of several species of *Bursera* is regularly eaten and undoubtedly is an important food source during drought periods. House Finches must drink water; their presence in deserts is dependent upon a water source within a few kilometers, even though their food may contain much moisture.

RED CROSSBILL
Picotuerto Rojo
Loxia curvirostra

Red Crossbills are erratic in habits and distribution, being dependent on the patchy seed crop of coniferous trees. They nest at almost any season, often in winter. In Sonora they have seldom been seen, and they have not been found breeding. Marshall (1957) noted a few in the summers of 1952 and 1953 in the Sierra Azul, Sierra Cananea, Sierra los Ajos, Sierra los Ojos Azules, and Sierra Nacori. Our only records are of ten in the Sierra Oscura on 15 March 1982 and two near Sahuarivo on 15 July 1987. The species was heard at Fronteras in November 1953 (A. Phillips). It was not mentioned by van Rossem (1945).

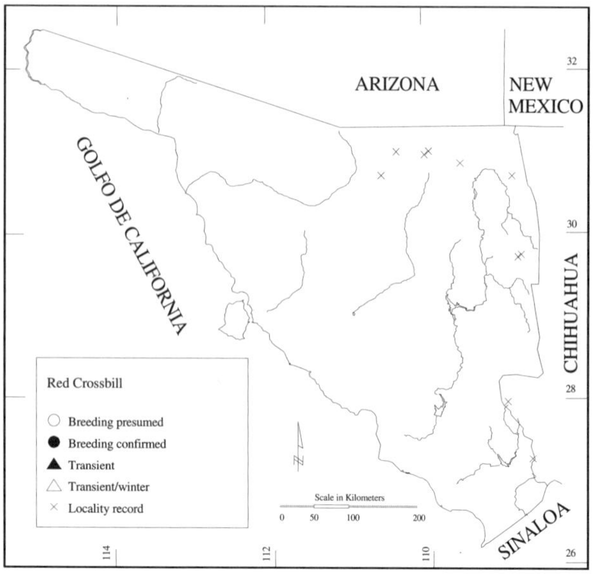

PINE SISKIN
Jilguero Pinero
Carduelis pinus

Small flocks of Pine Siskins are winter visitants to Sonora, found usually in pine-oak woodland but also in deciduous riparian woodland and rarely in favorable spots in the western desert. They are more common in some years than in others. Pine Siskins are bud-eaters in such trees as conifers, alders, willows, and cottonwoods, as well as consumers of seeds of composites, often in the company of goldfinches.

Although they do not appear in the fall until mid- or late October, flocks linger into May and even the first days of June. The presence of this species late in May or in June does not normally indicate nesting. But J. Marshall and A. Phillips found them singing in the Sierra los Ajos on 29 May to 1 June 1953. The well-developed condition of the gonads of three males they collected indicated that the birds could have been nesting.

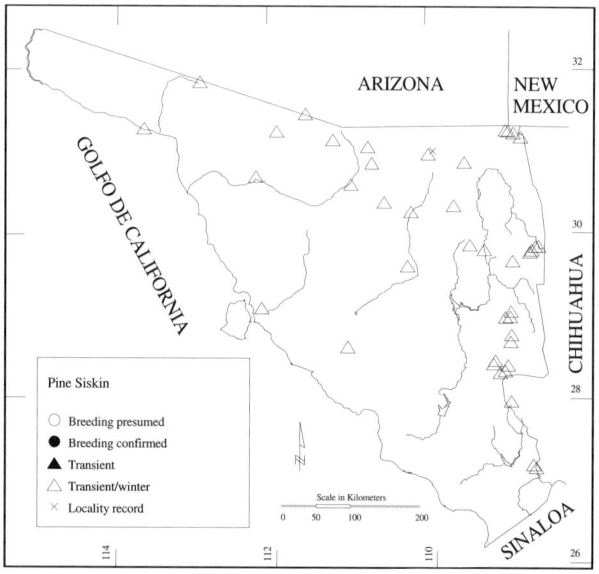

BLACK-HEADED SISKIN
Jilguero Encapuchado
Carduelis notata

Black-headed Siskins are uncommon and local summer residents above 1400 m in pine and pine-oak forests in eastern Sonora. Nesting is in June and July, as we have seen courtship in May and parents feeding their fledglings in mid-July. We have no records for the months August through February, and we assume the species is migratory. There are only two records from lower elevations, both in May: Baromico (880 m) and Arroyo el Reparo (920 m).

Black-headed Siskin

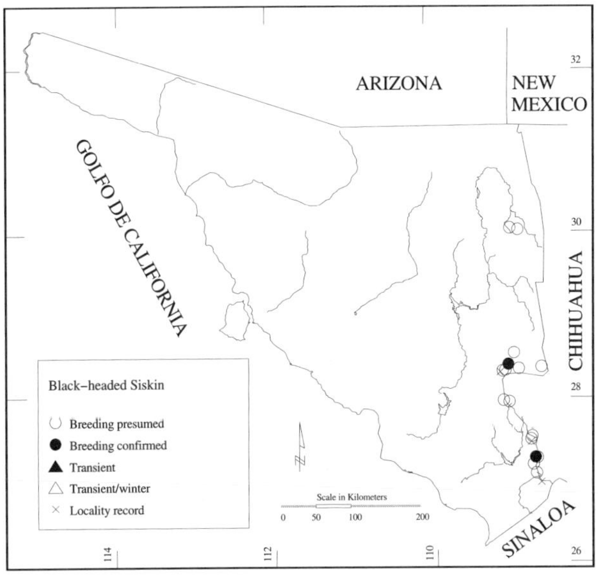

LESSER GOLDFINCH
Jilguero Dominico
Carduelis psaltria

This goldfinch is common throughout much of Sonora, breeding from the coastal plain into the pine-oak woodland of the mountains. It is a species typical of riparian areas with willows or cottonwoods and an abundance of composites, the seeds of which are extensively used as food. Although Lesser Goldfinches may be resident in some localities, it is more likely that they move considerably in search of food. During late fall and winter, flocks of up to 100 birds may forage in extensive tracts of wild sunflower, although most groups are of fewer than 20 individuals.

The nesting season is very long and may begin early in February; some pairs are nesting in September (if not later). Food availability and temperature probably have much to do with the timing of reproduction. During the breeding season, Lesser Goldfinches sing elaborate songs that often incorporate phrases from other species that they may have heard either in the area or on their wanderings.

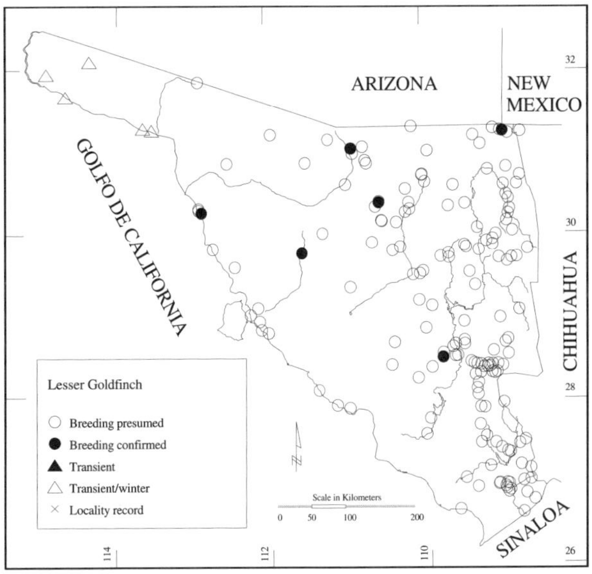

LAWRENCE'S GOLDFINCH

Jilguero Gris

Carduelis lawrencei

Lawrence's Goldfinch is an erratic winter visitant in small numbers to northern Sonora, casually to southern Sonora. It is normally resident in central and southern California south into northwestern Baja California, but every two to five years it moves eastward for the winter in some numbers into southern Arizona and northern Sonora. It has been found south to Tecoripa, where eight were collected on 12–20 March 1929 (van Rossem 1931). The earliest fall record is of eight at Félix Gómez on 12 November 1984, while the latest spring record is of three at Puerto Peñasco on 21 or 22 March 1978 (D. Stejskal). Lawrence's Goldfinch is most apt to be seen in and around weedy fields on farms and ranches, especially where water is available. The largest number observed at one time in Sonora was 40 at Arroyo Santo Domingo on 21 February 1982.

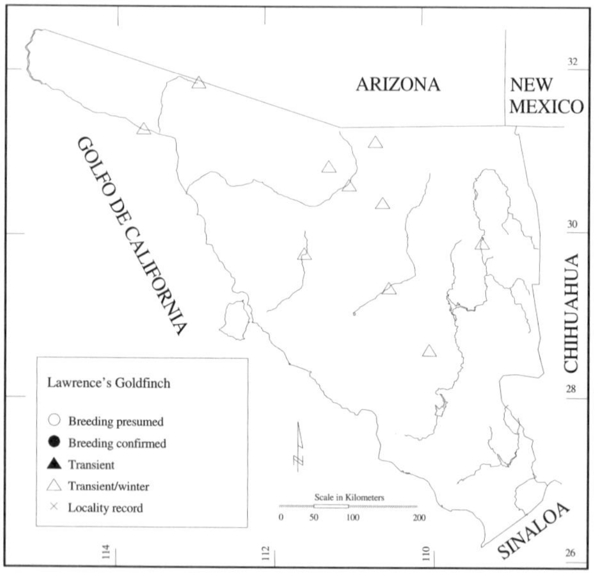

AMERICAN GOLDFINCH

Jilguero Canario

Carduelis tristis

The American Goldfinch has been recorded ten times in Sonora but is to be expected regularly along the northern border in patches of wild sunflowers and in budding, flowering, and fruiting cottonwoods and willows. A most unusual record was that of a singing male collected at Rancho Saucito in oaks northwest of Nacori Chico on 10 June 1960 (J. Marshall). Our records fall from late October to early April.

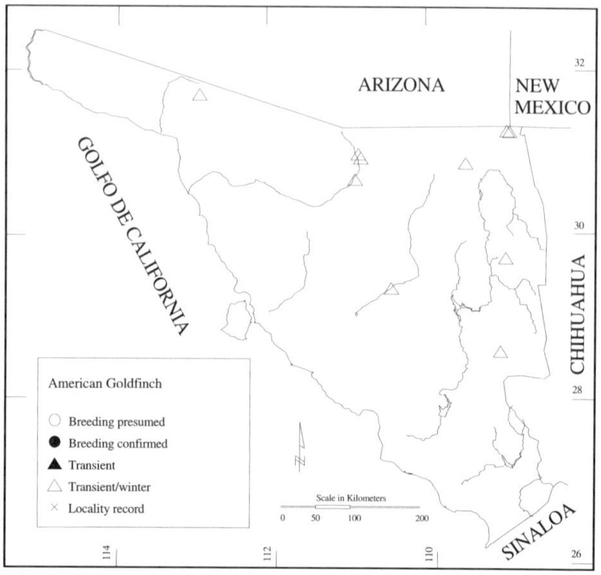

HOODED GROSBEAK

Picogrueso Encapuchado

Coccothraustes abeillei

The first known Hooded Grosbeaks in Sonora were discovered on 15 July 1987 by J. Bates, M. Kot, and A. Moorhouse. Two males and two females were found at the bottom of the Barranca las Colas at Sahuarivo; these or others were seen in the same vicinity on the next two days by several observers. The birds fed upon the fruit of *Prunus* just below the level of pines.

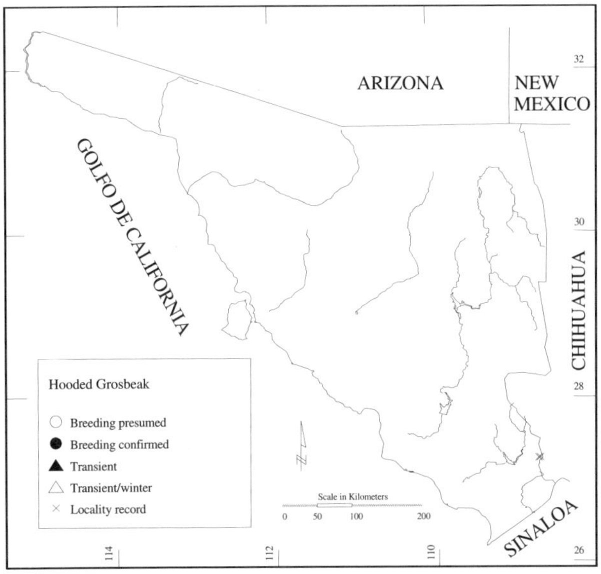

EVENING GROSBEAK
Picogrueso Norteño
Coccothraustes vespertinus

Evening Grosbeaks have been recorded in Sonora only six times: J. Marshall considered them abundant southwest of Mesa Tres Ríos on 10 June 1953, and he collected a male in pine forest in the Sierra Huachinera on 5 July 1954; four were seen near Milpillas on 17 March 1982 (D. Stejskal); we observed two at Metate southeast of Yécora on 10 May 1983; A. Moorhouse found ten on Mesa Grande on 13 March 1986; and we saw six on Mesa de Enmedio on 13 March 1989. All records are from pine forest. The species probably breeds in Sonora in some years.

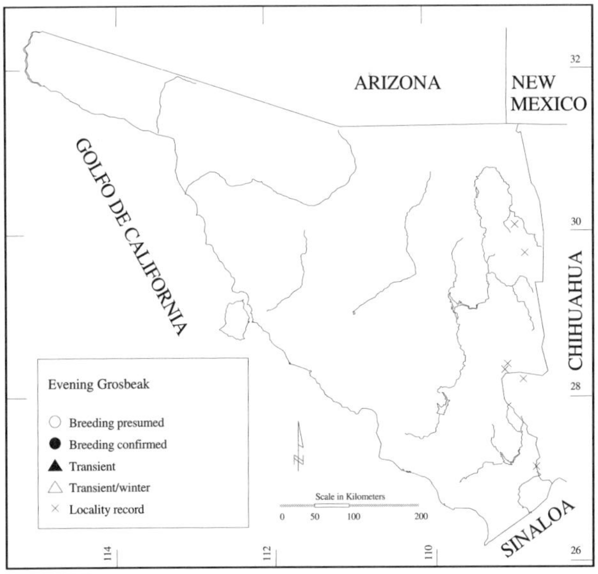

FAMILY PASSERIDAE
Old World Sparrows

HOUSE SPARROW
Gorrión Casero
Passer domesticus

This species introduced from Europe is a common dooryard resident of all cities, towns, and ranch buildings throughout Sonora. Its arrival in the state went unrecorded, but if we employ comparative data from Arizona (Phillips et al. 1964) we can postulate that it first appeared in the early 1900s and was securely established and widespread by 1920. Many field observers do not record the presence of House Sparrows, and our distributional map does not indicate all localities where the species is present.

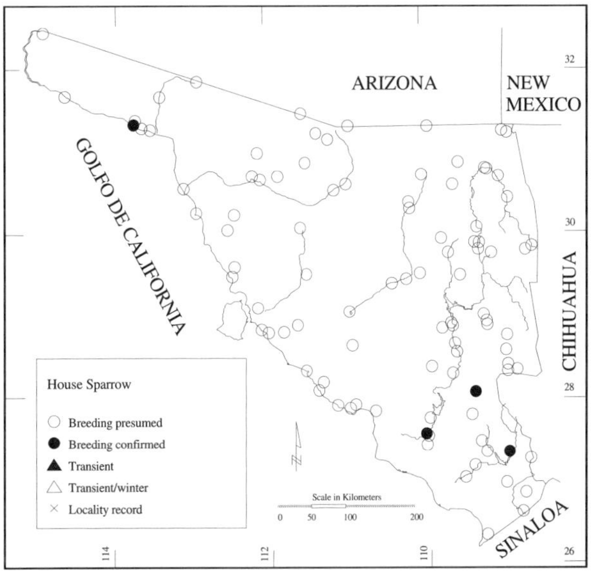

Appendix A

PLANTS NAMED IN THE TEXT

acacia. Mimosoideae. Sonora species include whiteball acacia *(Acacia angustissima),* boatspine acacia *(A. cymbispina),* and *A. willardiana.*

Acacia angustissima. See acacia.

Acacia willardiana. See acacia.

agave. Agavaceae. *Agave* spp.; many species in Sonora.

alder (Arizona). Betulaceae. *Alnus oblongifolia.*

alligator bark juniper. Cupressaceae. *Juniperus deppeana.*

amapa. Bignoniaceae. *Tabebuia* spp.

Amaranthus sp. Amaranthaceae. Refers mainly to *A. palmeri.*

Apache pine. See pine.

Arizona cypress. Cupressaceae. *Cupressus arizonica.*

Arizona oak. See oak.

arrowweed. Compositae. *Pluchea sericea.*

ash. Oleaceae. *Fraxinus velutina.*

barnyard grass. Gramineae. *Echinochloa crusgalli.*

barrel cactus. Cactaceae. *Ferocactus* spp.

batamote. Compositae. *Baccharis salicifolia.*

beach grass. Gramineae. *Monanthochloe littoralis.*

black mangrove. See mangrove.

blue palo verde. See palo verde.

boatspine acacia. See acacia.

bracken fern. Polypodiaceae. *Pteridium aquilinum.*

buffel grass. Gramineae. *Pennisetum ciliare.*

Bumelia persimilis. Sapotaceae.

bur sage. Compositae. *Ambrosia deltoidea* or *A. dumosa.*

Bursera spp. Burseraceae.

Bursera hindsiana. Burseraceae.

Bursera laxiflora. Burseraceae.

Bursera microphylla. See elephant tree.

canyon ragweed. Compositae. *Ambrosia ambrosioides.*

cardón (cactus). Cactaceae. *Pachycereus pringlei.*

Cathestecum brevifolium. See false grama.

cattail. Typhaceae. *Typha* spp.

Ceanothus sp. Rhamnaceae.

Celtis sp. See hackberry.

Cercidium sp. See palo verde.

cherry (wild cherry). Rosaceae. Any one of two or more species of *Prunus.*

Chihuahua pine. See pine.

Chihuahuan oak. See oak.

chino. See *Pithecellobium* sp.

cholla (cactus). Cactaceae. Round-stemmed cacti of the genus *Opuntia.*

chuparosa. Acanthaceae. *Beloperone californica.*

coral-bean. Papilionoideae. *Erythrina flabelliformis.*

coral-bell. Saxifragaceae. *Heuchera sanguinea.*

cottonwood. Salicaceae. Three species in Sonora: Fremont cottonwood *(Populus fremontii)* of the north, white cottonwood *(P. monticola)* of the mountains, and Mexican cottonwood *(P. mexicana)* of the south.

creosote bush. Zygophyllaceae. *Larrea tridentata.*

Croton fragilis (vara blanca). Euphorbiaceae.

desert hackberry. See hackberry.

desert willow. Bignoniaceae. *Chilopsis linearis.*

Douglas-fir. Pinaceae. *Pseudotsuga menziesii.*

eelgrass. Potamogetonaceae. *Zostera marina.*

elephant tree. Burseraceae. *Bursera microphylla.*

elderberry (Mexican). Caprifoliaceae. *Sambucus mexicana.*

Emory oak. See oak.

Epipactis sp. (orchid). Orchidaceae.

eucalyptus. Myrtaceae. Any one of the introduced trees of the genus *Eucalyptus.*

euphorb. Any plant of the family Euphorbiaceae.

false grama. Gramineae. *Cathestecum brevifolium.*

feather tree. Mimosoideae. Two common species: *Lysiloma divaricata* and *L. microphylla.*

fig. Moraceae. Several species of the genus *Ficus,* including the rock fig *(F. petiolaris).*

fir. Pinaceae. The only Sonoran fir is the white fir, *Abies concolor.*

Forchammeria watsoni. Capparidaceae.

Frankenia sp. Frankeniaceae. *Frankenia palmeri.*

Gambel oak. See oak.

gramas. Gramineae. *Bouteloua* spp.

guamachil. See *Pithecellobium* sp.

guácima (prickle-nut tree). Sterculiaceae. *Guazuma ulmifolia.*

guayacán. Zygophyllaceae. *Guaiacum coulteri.*

hackberry. Ulmaceae. Any one of a species of *Celtis,* including netleaf hackberry *(C. reticulata)* and desert hackberry *(C. pallida).*

Haplopappus sp. Compositae.

hecho (cactus). Cactaceae. *Pachycereus pectin-aboriginum.*

iodine bush. Chenopodiaceae. *Allenrolfea occidentalis.*

Ipomoea sp. Convolvulaceae. Several species in this genus, including tree morningglory.

ironwood. Papilionoideae. *Olneya tesota.*

Johnson grass. Gramineae. *Sorghum halapense.*

juniper. Cupressaceae. *Juniperus* spp.

kapok. Bombacaceae. *Ceiba acuminata.*

Lobelia laxiflora. Campanulaceae.

Lycium sp. Solanaceae. Any one of several species.

madrone. Ericaceae. *Arbutus arizonica* or *A. xalapensis.*

maidenhair fern. Polypodiaceae. *Adiantum capillus-veneris.*

Mandevilla foliosa. Apocynaceae.

mangrove. Rhizophoraceae. Two species, often growing together: black mangrove *(Avicennia germinans)* and red mangrove *(Rhizophora mangle).*

manzanita. Ericaceae. *Arctostaphylos pungens.*

maple. Aceraceae. *Acer grandidentatum.*

Maytenus phyllanthodies. Celastraceae.

mesquite. Mimosoideae. Most mesquites in Sonora are *Prosopis glandulosa* or *P. velutina.*

Mexican white pine. See pine.

mimosa. Mimosoideae. *Mimosa* spp. Several occur in Sonora.

mistletoe. Viscaceae. Any one of several species.

Montezuma baldcypress (sabino). Taxodiaceae. *Taxodium mucronatum.*

netleaf hackberry. See hackberry.

oak. Fagaceae. Trees of the genus *Quercus.* Sonora has numerous species, mostly spring-deciduous. They include Emory oak *(Q. emoryi),* Gambel oak *(Q. gambelii),* Chihuahuan oak *(Q. chihuahuensis),* Arizona oak *(Q. arizonica), Q. tuberculata, Q. albocincta, Q. epileuca,* silver-leaf oak *(Q. hypoleucoides),* willow-leaved oak *(Q. viminea),* and hand basin oak *(Q. pennivenia).*

ocotillo. Fouquieriaceae. *Fouquieria splendens.*
organ pipe cactus (pitahaya). Cactaceae. *Stenocereus thurberi.*
Oxalis sp. Oxalidaceae.
palm. Palmae. One species is common, the Ures palm *(Sabal uresana).*
palo blanco. Mimosoideae. *Piscidia mollis.*
palo de asta. Boraginaceae. *Cordia sonorae.*
palo verde. Caesalpinioideae. Genus *Cercidium.* May include blue palo verde *(C. floridum),* foothill palo verde *(C. microphyllum),* or *C. praecox.*
papache. Rubiaceae. *Randia echinocarpa.*
pine. Pinaceae. Trees of the genus *Pinus.* Sonora has several species, including ponderosa pine *(P. ponderosa),* Apache pine *(P. engelmannii),* Mexican pinyon *(P. cembroides),* Chihuahua pine *(P. leiophylla),* egg cone pine *(P. oocarpa),* and Mexican white pine *(P. ayacahuite).*
pinyon pine. See pine (Mexican pinyon).
pitahaya (cactus). See organ pipe cactus.
Pithecellobium sp. Mimosoideae. This genus includes chino (or palo chino) *(P. mexicanum),* and guamuchil *(P. dulce).*
ponderosa pine. See pine.
prickle-nut tree. See guácima.
prickly pear (cactus). Cactaceae. Flat-stemmed cacti of the genus *Opuntia.*
Prosopis sp. See mesquite.
Prunus sp. See cherry (wild cherry).
Quercus albocincta. See oak.
Quercus chihuahuensis. See oak.
Quercus epileuca. See oak.
Quercus hypoleucoides. See oak.
Quercus tuberculata. See oak.
rock fig. See fig.
rothrock grama. Gramineae. *Bouteloua rothrockii.*
Rubus sp. Rosaceae.
sabal palm. See palm.
sabino. See Montezuma baldcypress.
saguaro (cactus). Cactaceae. *Carnegiea gigantea.*
salicornia. Chenopodiaceae. *Salicornia virginica.*
salt bush. Chenopodiaceae. *Atriplex* spp.
saltgrass. Gramineae. *Distichlis spicata.*
scrub mesquite. Mimosoideae. *Prosopis glandulosa* var. *torreyana.*
sea blite. Chenopodiaceae. *Suaeda* spp.
seaweed. Any one of a number of nautical plants.
Senecio salignus. Compositae.
silver-leaf oak. See oak.
sitavaro. Apocynaceae. *Vallesia glabra.*
sotol. Agavaceae. *Dasylirion wheeleri.*
sunflower (wild). Compositae. *Helianthus annuus.*
sycamore. Platanaceae. *Platanus wrightii.*
tamarisk. Tamaricaceae. *Tamarix pentandra* or *T. aphylla.*
tarbush. Compositae. *Flourensia cernua.*
tesajo (cactus). Cactaceae. *Stenocereus alamosensis.*
tescalama (rock fig). See fig.
Tillandsia sp. Bromeliaceae. *Tillandsia recurvata.*
tobosa. Gramineae. *Hilaria* spp., usually *H. jamesii, H. mutica,* or *H. rigida.*
tree morningglory. Convolvulaceae. *Ipomoea arborescens.*
tree ocotillo. Fouquieriaceae. *Fouquieria macdougalii.*
tree tobacco. Solanaceae. *Nicotiana glauca.*
twin-flower cassia. Caesalpinioideae. *Cassia biflora.*
Usnea. Usneaceae. A genus of lichens.
uvulama. Verbenaceae. *Vitex mollis.*
vara blanca. See *Croton fragilis.*
walnut. Juglandaceae. *Juglans major.*
weeping mistletoe. Viscaceae. *Phoradendron scaberrimum.*
whitethorn acacia. Mimosoideae. *Acacia neovernicosa.*
wild cherry. See cherry.
willow. Salicaceae. *Salix bonplandiana, S. exigua, S. gooddingii, S. lasiolepis,* or *S. taxifolia.*
yellow columbine. Ranunculaceae. *Aquilegia chrysantha.*
yellow monkey-flower. Scrophulariaceae. *Mimulus guttatus.*
yucateco (shining fig). Moraceae. *Ficus nitida.*
yucca. Agavaceae. *Yucca* spp., several in Sonora.

Appendix B

GAZETTEER

This gazetteer contains entries for all Sonoran localities we mention in the text. In addition, it includes several names cited by van Rossem (1945) and Marshall (1957) because the original descriptions were inaccurate or vague. Other entries represent locality records provided to us by many individuals who have observed birds in Sonora.

Place names, coordinates, and elevations are based upon the 1:50000 maps published by the Comisión de Estudios del Territorio Nacional (Mexico, 1975–1982). Spelling follows the Spanish form used on the maps, including accents and diacritic marks. Some usually accented names are printed unaccented on the 1:50000 maps (e.g., Sierra San Jose); we adhere to the unaccented format to facilitate finding localities. Where local use of a name differs from that on the maps, we provide alternatives within parentheses in this gazetteer. Localities not appearing on the 1:50000 maps have been identified using various older maps, and their coordinates (or modern equivalents) are provided. Natural geographic features are indexed on the second part of the name, as are features following el, la, los, las, presa, puerto, and rancho. Locations are given as latitude north, longitude west, and the elevation (in meters). Coordinates are provided to the second where the maps provide adequate detail.

Many published records, especially those in van Rossem 1945, refer to specimens taken long before detailed maps were available. The localities associated with the records are often vague. For example, a specimen labeled by Frazar in 1888 as from Alamos may have been taken anywhere within several kilometers of Alamos. The 1:50000 maps used for this gazetteer, which were published between 1975 and 1982, are the first reliable and detailed maps available for Sonora. Using the very detailed sketch maps in Marshall's field journals, we have been able to determine the exact location of his observations. In many cases we have consulted with observers who have contributed records, in order to plot localities they cite.

The border between Sonora and Chihuahua is only vaguely established on maps. We have been in a number of places where some residents claim they live in Sonora and neighbors claim Chihuahua. These places, mostly in the Sierra Oscura, Sahuarivo, Rancho Quemado, and Sierra San Luis areas, are identified by precise coordinates in the gazetteer. All such debatable sites are certainly within 2 to 3 km of the border, although symbols on some of the range maps are situated on or slightly outside the sometimes poorly defined Sonora border.

Abundancia, Mina. See San Antonio.

Aconchi. Village on Río Sonora. 29°49′30″, 110°13′45″, 600 m. Phillips collected at Estancia la Aconchi, on the Río Sonora about 4 km ESE of the village: 29°47′30″, 110°12′45″, 570 m.

Aconchi, Sierra. Mountains between the Río San Miguel and the Río Sonora. The highest point is Picacho Alto la Bonancita. 29°49′10″, 110°21′00″, 2190 m. Coordinates of Marshall's pine-oak woodland site: 29°50′00″, 110°22′30″, 1600 m.

Aduana. Road junction 6 km SSW of El Sasabe. 31°25′20″, 111°33′40″, 1020 m.

Aduana, La. Village; a former mining center W of Alamos. 27°03′, 109°21′, 520 m.

Agiabampo. Village on bay called Estero Agiabampo. 26°22′00″, 109°08′30″, 1 m.

Agua, Punta de. Village 1 km below Presa Francisco L. Alatorre, on the lower Río Matape. 28°25′40″, 110°24′20″, 220 m.

Agua Blanca. Ranch in oaks on road to Sierra Oscura. 27°59′00″, 109°05′20″, 1000 m.

Agua Blanco, Puerto. Pass in pine-oak and juniper woodland, about 6 km SSE of El Trigo de Cororepe. 28°33′35″, 108°55′35″, 1500 m.

Agua Caliente. A holding reservoir 18 km N of Esperanza on irrigation canal from Presa Obregon. Coordinates at access on SE shore: 27°43′50″, 109°51′40″, 70 m.

Agua Caliente. Large ranch in area of extensive clearing for buffel grass. 28°04′05″, 110°09′10″, 190 m. The usually dry Río Agua Caliente arises in this area and drains SE into Presa Obregon.

Agua Fria. Ranch. 28°24′30″, 110°13′50″, 360 m.

Aguaitos, Los. Spring (unnamed on map) in oaks on steep-sloped road between Santa Rosa and Yécora. 28°27′20″, 109°00′55″, 1700 m.

Aguajito, El. Ranch on small arroyo in deciduous forest on road between San Nicolás and Santa Rosa. 28°27′30″, 109°07′30″, 900 m.

Agua Marin. See Higuera, La.

Agua Prieta. City on border, opposite Douglas, Arizona. 31°20′00″, 109°33′30″, 1200 m. Water birds are often noted at the sewage ponds at 31°17′30″, 109°34′20″, 1175 m.

Ajo Canyon. See Evans, Cañon.

Ajos, Sierra los. Mountains E of Cananea. The highest point is Cerro las Flores. 30°55′40″, 109°57′20″, 2630 m. Also see Evans, Cañon; Claro, Río.

Alamo, El. A van Rossem locality NW of Magdalena de Kino. 30°51′00″, 111°13′20″, 800 m.

Alamos. Historic mining town in tropical deciduous forest area; center for earlier collectors and destination of contemporary birders. 27°01′30″, 108°56′00″, 390 m. The "Alamos Wash" or Río Alamos flows through the town and SE into the Río Cuchujaqui.

Alamos, Pueblo de. Village NW of Mazatán in desert–thornscrub. 29°12′20″, 110°08′25″, 580 m.

Alamos, Sierra. Small rugged mountain mass SW of Alamos; covered mostly with tropical deciduous forest but having some pines and oaks on top. 26°58′21″, 108°59′22″, 1765 m.

Alatorre, Presa Francisco L. See Agua, Punta de.

Alcantar, Potrero (Potrero de Alcántar). Village in thornscrub–deciduous forest. 26°48′40″, 108°45′30″, 280 m.

Alcatraz, Isla. Small rocky island about 2 km offshore from the village of Bahía Kino. 28°48′30″, 111°58′00″, 100 m.

Algodones, Estero los. Estero at mouth of Río Yaqui; a van Rossem locality. 27°40′, 110°35′.

Algodones, Los. Village N of San Bernardo in deciduous forest. 27°30´30″, 108°44´00″, 850 m.

Altar. Town on Río Altar (which drains into the Río Asunción about 14 km SW). Extensive mesquite bosques persisted here until they were cut in the 1940s. 30°43´00″, 111°50´00″, 420 m.

Altar, Desierto de. Vast desert area in NW Sonora, between San Luis Río Colorado and Puerto Peñasco. Sometimes called the Gran Desierto.

Altar, Río. Tributary of the Río Magdalena. See Saric; Arizona, La.

Amador, Cajón de. Stream in thornscrub. Coordinates 6 km N of Nuri where Nuri–San Nicolás road crosses: 28°10´00″, 109°19´00″, 340 m.

Angostura, Presa la. Impoundment on Río Bavispe NE of Nacozari. Also called Presa Lázaro Cárdenas. 30°26´20″, 109°22´40″, 760 m.

Animas, El Valle de. Drainage N into New Mexico from extreme NW portion of Sierra San Luis. Coordinates at El Valle: 31°19´55″, 108°49´19″, 1575 m.

Apache, El. One of Marshall's sites in the Sierra Huachinera, in pine-oak woodland. 30°04´40″, 108°53´00″, 1760 m.

Arenas, Las. A van Rossem locality NE of Guaymas in desertscrub in the upper basin of the Río Agua Caliente. 28°15´20″, 110°18´30″, 265 m.

Arenas, Punta. See Viboras, Punta.

Aribabi. Village in oaks W of Sierra Huachinera. 30°04´10″, 109°05´10″, 1450 m.

Arivechi. Town on Río Sahuaripa. 28°55´45″, 109°11´10″, 490 m.

Arizona, La. Important van Rossem (1945) locality called "Rancho la Arizona" by him; on tributary of the Río Altar. 31°10´00″, 111°11´40″, 970 m.

Arizpe. Town on Río Sonora. 30°20´00″, 110°10´00″, 820 m.

Aros, Río. A major tributary of the Río Bavispe; its confluence with the Río Bavispe forms the Río Yaqui. Coordinates at confluence with the Río Bavispe: 29°31´45″, 109°14´20″, 430 m. The Río Aros is formed by the confluence of the Ríos Mulatos and Papigochi (but the 1:50000 maps refer to the lower portion of the Río Papigochi as the Río Aros; Papigochi is often spelled "Papagóchic"). The Río Mulatos is formed by Arroyo (Río) Yécora and Arroyo (Río) Maycoba.

Arpa, El. Ranch in "plains" about 29 km WNW of El Oasis. 29°48´50″, 111°18´10″, 790 m.

Aserraderos, Los. Once a sawmill, now a small village at W base of Sierra Oscura in pine-oak woodland. 28°00´40″, 109°02´15″, 1250 m.

Asunción, Río. The middle portion of the Río Magdalena, including that part in the Caborca-Altar area. The river is called Río Magdalena above the Altar vicinity.

Azul, Sierra. Mountains NE of Magdalena de Kino. The highest point is Cerro Azul. 30°44´20″, 110°34´25″, 2450 m. Marshall had a study site in pine-oak woodland about 1 km NE of Cerro Azul.

Babasac, Río. Tributary of the Río Magdalena, E of Imuris. Coordinates of a Phillips and Marshall site: 30°49´00″, 110°42´30″, 940 m.

Baboyahui. Village 13 km ENE of Güirocoba. 26°57´30″, 108°35´20″, 380 m.

Bacadehuachi (Bacadéhuachi). Town between Huasabas and Nacori Chico. 29°48´30″, 109°08´30″, 680 m. The Arroyo Bacadehuachi drains to the SW and into the Río Bavispe, about 20 km from the town.

Bacanuchi. Village NW of Arizpe. 30°36´30″, 110°14´10″, 1060 m.

Bacanuchi, Río. Tributary of the Río Sonora; Cananea-Arizpe highway crosses it about 7 km ESE of Bacanuchi. Coordinates at crossing (willows, cottonwoods): 30°34´30″, 110°13´15″, 1040 m.

Bacerac. Town on upper Río Bavispe. 30°21´20″, 108°55´50″, 1040 m.

Bacoachi. Town SE of Cananea on Río Sonora, has small permanent pond. 30°38´00″, 109°58´00″, 1060 m.

Bacoachi, Río. Drains extensive basin (including the Río Bacoachito) NW of Hermosillo. Water occasionally collects in Playa San Bartolo NE of Bahía Kino but rarely reaches El Golfo de California.

Bacoachito, Río. Usually dry river NW of Hermosillo, draining into the Río Bacoachi. See Vinatería, La.

Bacobampo. Town SW of Navojoa in agricultural area. 26°59´00″, 109°39´00″, 8 m.

Bacorehuis. Village, barely in Sinaloa, in vast agricultural region. 26°18´00″, 109°05´00″, 5 m.

Bácum. Town on S bank of Río Yaqui near Ciudad Obregon. 27°32´30″, 110°05´00″, 25 m.

Badesi. Village on Río Yaqui. 29°22´15˝, 109°15´00˝, 400 m.

Bahía Kino. See Kino, Bahía. The name is applied to both the town and the bay.

Bamuri (Bámuri). Ranch S of Pitiquito with occasionally flooded large pond. 30°20´20˝, 112°02´40˝, 390 m.

Banámichi. Town on Río Sonora. 30°00´30˝, 110°13´00˝, 680 m.

Banderita, La. Russell site in deciduous forest in upper Arroyo la Banderita, E of Tepahui. 27°26´50˝, 109°00´35˝, 480 m.

Barajitas, Las. Ranch WNW of Hermosillo at edge of "Plains of Sonora." 29°29´00˝, 111°22´30˝, 390 m.

Baromico. Ranch in Río Chinipas drainage, about 18 km NE of Güirocoba; R. Moore locality near Sonora-Chihuahua border. 27°01´30˝, 108°34´50˝, 880 m. "Maromicon" has been used as an alternate name by Moore.

Barrera, Juan de la. Village in area of agriculture and scrub. 26°24´30˝, 109°06´, 30 m.

Batacosa. Village in thornscrub W of Quiriego. 27°31´45˝, 109°24´00˝, 230 m.

Baviácora. Village on Río Sonora. 29°43´00˝, 110°10´00˝, 560 m.

Bavispe. Town on upper Río Bavispe. 30°28´45˝, 108°56´20˝, 1000 m.

Bavispe, Río. The major river in NE Sonora; drains W slope of Sierra Madre, flowing N to within 55 km of the U.S. border, then S to become the Río Yaqui at its confluence with the Río Aros.

Benjamin Hill. Town S of Santa Ana; important rail junction. 30°10´00˝, 111°07´00˝, 710 m.

Bird Island. See Pajaros, Isla.

Blanco, Cerro. Volcanic area 5 km W of Divisadero, cited by van Rossem. Listed as "Volcan Cerro Blanco" on 1:50000 map. 29°36´10˝, 109°31´30˝, 840 m.

Bonito, Cajón. A major drainage of the NW slope of the Sierra San Luis, near the U.S. border. Cited as Arroyo Cajón Bonito throughout. Coordinates for Rancho Nuevo in the floor of the canyon: 30°15´35˝, 108°57´00˝, 1360 m. Small falls are present where a stream converges from the S, 7 km W of Rancho Nuevo. 31°17´00˝, 109°00´50˝, 1280 m.

Brisca, La. Ranch on Arroyo Santo Domingo, 6 km N of Saracachi. Coordinates in riparian community at confluence of Arroyos Salsipuedes and Santo Domingo: 30°24´40˝, 110°33´00˝, 980 m.

Caborca. Major city in vast agricultural area of the lower Río Magdalena Valley. The river is called the Río Asunción in this area. 30°43´00˝, 112°09´00˝, 280 m.

Caborqueños, Los. Agricultural area 34 km by road N of Caborca. 31°00´00˝, 112°05´15˝, 530 m.

Cachuta. See Cuchuta.

Cajeme. See Ciudad Obregon.

Calabazas, Las. Russell site in hills 6 km SE of Bacoachi. 30°35´20˝, 109°55´20˝, 1060 m.

Calera, Cajón la. Canyon in pine-oak woodland on NW slope of the Sierra Huachinera. 30°08´00˝, 108°54´30˝, 1750 m.

Calles, Presa Plutarco Elias. See Novillo, Presa el.

Camoa. Village on S bank of Río Mayo; it was an important early collecting site, now in an agricultural area. 27°13´30˝, 109°15´30˝, 75 m.

Campa, Lo de. See Lo de Campa.

Campanero, Mesa. See Enmedio, Mesa de.

Cananea. Mining city in NE. 30°59´00˝, 110°18´00˝, 1540 m.

Cananea, Sierra. See Mariquita, Sierra.

Capulin, El (El Capulín). Ranch on Arroyo el Carrizo, 12 km N of Tonichi. 28°42´10˝, 109°35´30˝, 260 m.

Capulin, El (El Capulín). Ranch on Río Tepoca. 28°21´20˝, 109°16´25˝, 460 m.

Caracol, El. Ranch in desertscrub between Caborca and Libertad. 30°02´00˝, 112°27´20˝, 400 m.

Carbo (Carbó). Town on railroad about 65 km N of Hermosillo. 29°41´00˝, 110°57´10˝, 470 m.

Carbon, El (El Carbón). Ranch between Caborca and Libertad. 30°08´20˝, 112°25´00˝, 500 m.

Cárdenas, Presa Lázaro. See Angostura, Presa la.

Carnero, Cerro. Small hill about 10 km SE of Altar in desertscrub. 30°39´30˝, 111°46´40˝, 1000 m.

Carretas, Arroyo las. Large arroyo arising in a broad plain but situated among oak-dominated hillsides in southern portion. Coordinates about 19 km SSW of Monte Verde, Chihuahua, on road from San Miguelito to Monte Verde: 30°45´00˝, 108°47´50˝, 1520 m.

Carrizal, El. Agricultural area 27 km WSW of Caborca; at the edge of desertscrub on the Caborca-Libertad road.

30°31′00″, 112°18′50″, 250 m.

Carrizal, Rancho el. A Phillips and Amadon (1952) collecting site about 39 km N of Altar. 31°03′00″, 111°42′00″, 820 m.

Carrizo, El. Large ranch in "Plains of Sonora"; "Masked" Bobwhite area. 30°02′50″, 111°13′45″, 720 m.

Carrizo, El. Ranch in thornscrub–deciduous forest on Arroyo el Carrizo 19 km ENE of Tonichi. 28°44′50″, 109°28′45″, 590 m.

Casita, La. Village on Highway 15 S of Nogales. 31°00′00″, 110°53′00″, 1030 m.

Cedros. Town SE of Rosario on upper Río Cedros. 27°45′35″, 109°17′20″, 360 m.

Charaibabi. Ranch in oak-grassland, SE of Cucurpe; also called Santa Ana on map. Van Rossem's "La Chumata" mine in the Sierra de San Antonio was presumably close to this location. 30°11′10″, 110°30′40″, 1180 m.

Chimeneas, Cajón las. Tributary of Arroyo Cajón Bonito in the Sierra San Luis, draining an area including a ranch called San Antonio. 31°13′30″, 108°53′00″, 1500 m.

Chinacas, Las. Village in Chihuahua. 27°14′30″, 108°38′50″, 1480 m.

Chinal, El Village in thornscrub–deciduous forest near Sinaloa border. 26°40′50″, 108°42′05″, 240 m.

Chinapa. Village on upper Río Sonora. 30°26′15″, 110°01′45″, 900 m.

Chino, Puerto el. A pass in hills about 13 km N of San Antonio (El Chino). Arroyo to S has fine deciduous forest. 26°40′20″, 108°58′10″, 260 m.

Chinobampo. Ranch SW of Alamos. 26°58′30″, 109°18′00″, 110 m.

Choya, Bahía la. Village 10 km WNW of Puerto Peñasco. 31°20′30″, 113°38′00″, 10 m. The bay to the N is also called Bahía la Choya; to the E is Estero la Choya.

Chuchupate, Sierra. See Metate.

Chueca, Punta. Seri village NW of Bahía Kino on Canal el Infiernillo. 29°00′50″, 112°09′40″, 2 m. There is also a small point here with the same name.

Chumata, La. See Charaibabi.

Cibuta (Cíbuta). Village on Highway 15 S of Nogales. 31°03′40″, 110°54′00″, 1060 m.

Ciénega, La. Village on Arroyo el Bamuri S of Pitiquito. 30°11′40″, 111°56′40″, 530 m.

Cirio, Punta. Desertscrub area with *Idria* (boojum trees), 8 km ESE of Puerto Libertad. 29°50′15″, 112°38′50″, 10 m.

Ciudad Obregon (Ciudad Obregón). Major city in agricultural area. Once called Cajeme. 27°29′30″, 109°56′00″, 35 m.

Claro, Río. Drainage in extreme N end of the Sierra los Ajos. A Marshall (1957), Phillips locality. Coordinates are of "El Seis" on map, which was the approximate location of the study site Marshall called "Fourth windmill" in his notes: 31°02′20″, 109°57′00″, 1680 m.

Club de Yates. Marina 5 km SE of Bahía Kino, in mangroves, on Estero Santa Cruz. 28°47′28″, 111°54′40″, 0 m.

Cobachi. Village in desert-thornscrub SW of Mazatán. 28°53′45″, 110°14′50″, 455 m.

Cobre, Arroyo el. Arroyo E of Güirocoba. Coordinates are 2 km ENE of Güirocoba: 26°54′40″, 108°40′25″, 320 m.

Cobre, El. Village 9 km N of Güirocoba. Also called Choquincahui. 26°59′00″, 108°40′30″, 580 m.

Cobre, Sierra el. Isolated mountain in desertscrub N of Caborca, near Arizona border. 31°31′00″, 112°05′00″, 1370 m.

Cochore, Estero. Small estero on E shore of Bahía Empalme, 4 km S of Empalme. 27°55′20″, 110°48′30″.

Cocorit (Cócorit). A van Rossem locality: a town 3 km W of Esperanza. The name Esperanza is used for this locality in this book.

Cocóspera. Mission ruin NE of Imuris. 30°54′20″, 110°37′00″, 1030 m.

Colas, Barranca las. See Sahuarivo.

Colonia, La. Village 2 km SE of Cumpas on Río Moctezuma. 29°58′30″, 109°46′00″, 735 m.

Colonia Independencia. See Independencia.

Colonia Oaxaca. Village on upper Río Bavispe. 30°43′50″, 109°03′20″, 900 m. Usually called simply Oaxaca.

Colorada, La. Village SE of Hermosillo on highway to Yécora. 28°48′10″, 110°34′40″, 380 m.

Coloradito, El. Fishing camp about 17 km SSE of Puerto Libertad. 29°46′10″, 112°36′20″, 2 m.

Colorado, Cajón. Coordinates are at stream about 5 km ESE of El Trigo de Cororepe: 28°35′20″, 108°55′30″, 1440 m.

Colorado, Volcan (Cerro Colorado). Crater in Sierra el Pinacate. 31°55′00″, 113°18′00″, 200 m.

Concepción, Río. The lower 40 km (approximately) of the Río Magdalena drainage. Coordinates at the mouth (where there are extensive mudflats when it is in flood): 30°32´22˝, 112°58´54˝, 1 m.

Consag, Roca. Small island SW of Golfo de Santa Clara. 31°07´00˝, 114°29´00˝.

Cortinez, Presa Adolfo Ruiz (Presa Adolfo Ruiz Cortínez). See Mocuzari, Presa.

Costa Rica, Rancho. A van Rossem locality N of Hermosillo-Kino road. 28°55´00˝, 111°34´30˝, 60 m.

Coyote, El. Ranch 17 km ENE of Huasabas, at junction of roads to Huasabas, Aribabi, Bacadehuachi; stream, cottonwoods, sycamores. 29°58´34˝, 109°07´52˝, 1050 m.

Coyote, El. Village WNW of Caborca. 30°50´00˝, 112°36´00˝, 140 m.

Coyote, Pozo. Ranch in desertscrub on the Río San Ignacio, about 13 km NNE of the Seri village of El Desemboque. 29°37´15˝, 112°21´50˝, 75 m.

Crisanto, Pozo. Ranch in desertscrub NW of El Oasis. Neff (1947) locality. 29°43´30˝, 111°10´45˝, 640 m.

Cruz del Diablo. Small, deep canyon (called Cañada Maimodochi on map) 8 km ENE of Huasabas. Coordinates are at head of canyon on road to Bacadehuachi: 29°56´30˝, 109°14´00˝, 1220 m.

Cruz, La. Village on Río Santa Cruz; Marshall locality (also used for Los Fresnos). 31°11´30˝, 110°36´00˝, 1320 m.

Cuauhtémoc. Village about 19 km WSW of Naco. Coordinates are for lake 3 km SW called Villa Verde: 31°08´40˝, 109°59´50˝, 1480 m.

Cuauhtémoc, Presa. Lake on Río Altar just W of Tabutama. 30°52´30˝, 111°31´00˝, 620 m.

Cuchujaqui, Río (El). A major river draining (into the Río Fuerte) the mountains E of Alamos. The most intensively visited site is at the road crossing about 10 km SE of Alamos: 26°56´30˝, 108°53´00˝, 260 m.

Cuchuta. A Lumholtz locality (Allen 1893), when it was spelled "Cachuta." N of Esqueda. 30°46´00˝, 109°35´30˝, 1180 m.

Cuchuverachi. Ranch on Río San Bernardino, N of Morelos. 31°07´30˝, 109°17´30˝, 980 m.

Cucurpe. Village on upper Río San Miguel. 30°19´50˝, 110°42´20˝, 860 m. There is a striking stand of palms in a canyon along the road from Magdalena. 30°27´30˝, 110°48´30˝, 1100 m.

Cuesta, Rancho La. Large ranch with much buffel grass. Coordinates are on road between San José de Pimas and Tecoripa: 28°38´00˝, 110°10´00˝, 420 m.

Cueva, La. Ranch in junipers 9 km N of Yécora. 28°27´10˝, 108°55´30˝, 1500 m.

Cuevitas, Las. An occasionally used fishing camp about 26 km NW of the Seri village of El Desemboque. 29°41´45˝, 112°33´20˝, 2 m.

Cumbre de los Pinos (= Quemado, Rancho). Site in pine-oak woodland, on an arroyo called Rancho Quemado. There was an abandoned stone house (Byerly Ranch house) here in the late 1980s. This location is used for van Rossem´s "Texas Canyon," which we never located on a map. On the basis of specimen labels covering several days, Texas Canyon is probably very close to Rancho Quemado. 27°34´40˝, 108°41´48˝, 1580 m.

Cumpas. Town on the Río Moctezuma. 29°59´30˝, 109°46´30˝, 720 m.

Cumuripa. Village on the lower Río Suaqui (Grande). 28°09´15˝, 109°54´30˝, 120 m.

Cuquiarichi. Village on Arroyo Cuquiarichi, about 12 km W of Fronteras. 30°52´30˝, 109°40´20˝, 1190 m. Presa Jacinto Lopez is about 1 km W.

Desemboque, El. Fishing village just N of the mouth of the Río Concepción. 30°34´00˝, 113°00´30˝, 2 m. There are extensive mudflats at the mouth of the Río Concepción when it is in flood. 30°32´22˝, 112°58´54˝, 1 m.

Desemboque, El. Seri village about 4 km SE of the mouth of the Río San Ignacio. 29°30´15˝, 112°23´30˝, 2 m.

Desierto, Gran. See Altar, Desierto de.

Diablo, El. Arroyo (locally called Rincon del Diablo) on extreme NW portion of Sierra San Luis. Drains into El Valle de Animas, New Mexico. Coordinates of "upper spring," as used in Marshall notes: 31°16´45˝, 108°47´30˝, 1800 m.

Diablo, El. Ranch in lower Cajón Bonito. 31°17´30˝, 109°02´45˝, 1240 m.

Divisadero. Town SE of Moctezuma. 29°36´50˝, 109°28´15˝, 675 m.

Doctor, El. Locality on railroad near mouth of Río Colorado. 31°57´30˝, 114°44´20˝, 5 m.

Dolar, El. Fishing camp at W end of mangroves at Estero Sargento. 29°19´35˝, 112°22´30˝, 1 m.

Elegante, Volcan el. Crater in Sierra el Pinacate. 31°50´40˝, 113°23´30˝, 280 m.

Empalme. Town E of Guaymas. Coordinates at old causeway on W edge of town: 27°57´15˝, 110°50´00˝, 1 m.

Encino, El. Locality about 1 km N of Puerto el Encino (which is on Moctezuma-Huasabas road); just below oaks on S end of Sierra la Madera. 29°52´30˝, 109°27´50˝, 1200 m.

Enmedio, Mesa de. Map name of mesa immediately to the S of high point on new Hermosillo-Yécora road. Usually called Mesa el Campanero (after small village on S end of mesa). Coordinates for La Lagunita on N end of mesa: 28°22´10˝, 109°02´10˝, 2140 m.

Esperanza. Town just N of Ciudad Obregon. 27°35´00˝, 109°55´45˝, 40 m. The Río Yaqui is only 3 km N.

Esperanza, La. Ranch in hilly desertscrub at edge of "Plains of Sonora." 29°27´05˝, 111°27´28˝, 320 m.

Esqueda. Town S of Agua Prieta. 30°43´30˝, 109°35´25˝, 1200 m.

Estiladero, El. Canyon SE of Divisadero, draining into the Río Bavispe. Coordinates are of a Russell site: 29°25´00˝, 109°20´50˝, 650 m.

Estrella, La. Village at bridge over Río Yaqui below Presa el Novillo on Hermosillo-Sahuaripa road. Coordinates at bridge: 28°56´50˝, 109°38´20˝, 240 m.

Estrellas, Las. Ranch in desertscrub on Río San Ignacio. 29°51´25˝, 112°05´30˝, 485 m.

Evans, Cañon. Drainage in NW portion of Sierra los Ajos. Called "Ajo Canyon" or "Molino Canyon" by Marshall (1957). Coordinates of El Cañon (Las Cabañas), where there is a small lake: 30°59´45˝, 110°00´20˝, 1780 m.

Ezcalqui. Village between agricultural region and desertscrub, NE of Bahía Kino. Also called Utah. 29°05´34˝, 111°43´50˝, 50 m.

Félix Gómez. Village on Río Bacoachito. Neff's (1947) "Pozo de Noriega" was nearby. 29°50´14˝, 111°28´40˝, 670 m.

Fraijo, Cañada el. Large arroyo S of Bacanuchi; oaks, mesquite. 30°32´20˝, 110°12´45˝, 1080 m.

Francisquito, Estero. A Mellink and Palacios (1993) site SE of Puerto Peñasco. 30°59´00˝, 113°05´00˝, 2 m.

Fresnos, Los. See Cruz, La.

Fronteras. Town S of Agua Prieta. 30°53´50˝, 109°33´30˝, 1140 m.

Fuerte, Río. The major river flowing from SW Chihuahua into Sinaloa; a bend flows for a few kilometers through Sonora.

Fundición. Town between Ciudad Obregon and Navojoa. 27°20´00˝, 109°44´00˝, 65 m.

Garrapata, La. Ranch W of El Sasabe. 31°25´30˝, 111°40´50˝, 880 m.

Gavilan, El. A Phillips site about 21 km WSW of Ures on the Río Sonora. 29°19´20˝, 110°32´30˝, 320 m.

Gavilan, Río. See Tres Rios.

Gohocopa, La Junta de. Village E of Sahuaripa in pine-oak woodland. 29°06´45˝, 108°58´00˝, 1480 m.

Golfo de Santa Clara. Village at mouth of the Río Colorado. 31°41´00˝, 114°30´00˝, 1 m.

Granados. Town on Río Bavispe. 29°51´40˝, 109°18´30˝, 530 m.

Grande, Mesa. Russell site in pines 11 km NW of Yécora. Coordinates near small pond: 28°26´40˝, 108°59´50˝, 1820 m.

Green, Mesa. Mesa in grasslands E of Cananea on road to Cañon Evans. 31°01´30˝, 110°05´00˝, 1540 m.

Guadalupe, Cajón. Arroyo arising in New Mexico and flowing through Arizona into the Río San Bernardino. Coordinates at bridge on Agua Prieta–Janos highway: 31°19´00˝, 109°08´10˝, 1210 m.

Guajaray (Guajaráy). Village in deciduous forest on the Río Guajaray (which flows SE into the Río Mayo). 27°36´15˝, 108°56´00˝, 300 m.

Guásimas. Coastal village on Bahía Guásimas SE of Empalme. 27°53´00˝, 110°35´00˝, 1 m.

Guásimas, Las. Ranch about 14 km E of Presa el Novillo, just N of road from Hermosillo to Sahuaripa. 28°58´30˝, 109°29´00˝, 900 m.

Guaymas. Coastal city. Early references to Guaymas included various localities in the vicinity. Coordinates are of the city center: 27°55´00˝, 110°54´00˝, 10 m.

Güirocoba. Village in deciduous forest in extreme SE Sonora. An important E. Harrison, W. Sheffler collecting locality. 26°54´15˝, 108°41´30˝, 400 m.

Guisamopa. Town on Río Sahuaripa. 28°38´45˝, 109°06´25˝, 680 m.

Hachita Hueca, Sierra. Small range NE of San Miguelito. 30°38´36˝, 110°50´45˝, 2490 m.

Hermosillo. Capital city. Coordinates for N end of dam at Presa Abelardo L. Rodriguez (Rodríguez): 29°05′00″, 110°55′00″, 225 m. There is an extensive irrigated agricultural area extending SW of Hermosillo, vaguely centered on a point about 60 km SW of the city. 28°42′00″, 111°22′00″, 115 m.

Higuera, La. Village about 10 km N of Alamos. The "Agua Marín" of van Rossem is assumed to be near this village. 27°06′30″, 108°57′30″, 270 m.

Higueritas, Las. Russell site in floodplain of Río Bavispe, 7 km S of Bacerac. 30°17′15″, 108°56′20″, 1080 m.

Hondo, Arroyo. Stream in thornscrub–deciduous forest. Coordinates 14 km N of Nuri where Nuri–San Nicolás road crosses: 28°14′10″, 109°18′00″, 360 m.

Hoyos, Los. Village on Río Moctezuma, 14 km N of Cumpas. 30°07′30″, 109°47′00″, 800 m.

Huachinera, Sierra. Not identified on 1:50000 maps. The mountains are between Huachineras and Mesa Tres Ríos and are W of the Río Bavispe. The highest point is perhaps Cerro Pico Guacamayas: 29°57′14″, 108°50′21″, 2630 m. Also see Apache, El.

Huachineras. Town on Río Huachinera, a tributary of the upper Río Bavispe. 30°12′40″, 108°57′30″, 1110 m.

Huivulai, Isla. Island in Estero Tóbari. 27°05′, 110°01′, 2 m.

Huasabas (Huásabas). Town on Río Bavispe. 29°54′30″, 109°18′00″, 550 m.

Huatabampito. Coastal village S of Huatabampo. 26°41′30″, 109°35′00″, 1 m.

Huatabampo. Agricultural town SW of Navojoa. 26°49′30″, 109°38′30″, 8 m.

Huatita, La. Russell site on Arroyo Huatita in thornscrub, hills, 11 km N of Bacadehuachi. 29°54′30″, 109°07′50″, 800 m.

Huepac (Huépac). Town on Río Sonora. 29°54′40″, 110°12′50″, 640 m.

Huicochic (Huicóchic). See Sahuarivo.

Humo, Sierra el. Isolated mountains SW of El Sasabe; oaks. The highest point is Cerro San Pedro: 31°13′30″, 111°55′30″, 1550 m.

Iglesia, Arroyo la. Stream, canyon, oaks. Coordinates are for site about 2 km SSW of village of La Iglesia: 28°43′20″, 108°58′00″, 1270 m.

Imuris. Town S of Nogales on the Río Magdalena. 30°46′45″, 110°51′10″, 820 m.

Independencia. Settlement on the road from San Luis Río Colorado to Golfo de Santa Clara, about 10 km E of the present channel of the Río Colorado. Once known as Colonia Independencia; locality is the "Las Carpas" of Mearns labels (van Rossem 1945). 32°15′00″, 114°55′00″, 30 m.

Infiernillo, Canal el. The straits between the mainland and Isla Tiburón. See Chueca, Punta.

Isleta, La. Village on the lower Río Cuchujaqui. 26°51′30″, 108°54′30″, 200 m.

Jacalón. Ranch 18 km NE of Rosario in deciduous forest on Arroyo Santa Rosa. On road to Sierra Oscura. 27°56′00″, 109°13′20″, 650 m.

Jacinto Lopez, Presa (Presa Jacinto López). Impoundment on Arroyo Cuquiarichi, W of Fronteras, dedicated in 1984. 30°52′30″, 109°40′45″, 1200 m.

Kino, Bahía. Old fishing village, now becoming resort oriented. Coordinates at center of Kino Viejo: 28°49′30″, 111°56′30″, 2 m. Coordinates for Cerro Prieto, a hill at the W end of the beach: 28°51′40″, 112°01′40″, 30 m.

Lanas. Large stock tank 12 km S of El Sasabe. 31°22′00″, 111°33′00″, 975 m.

Lattigura. Phillips collecting site about 10 km NNW of Bacoachi. Identified as "La Higura" on map. 30°43′00″, 110°00′00″, 1110 m.

Lavadero, El. Old mine; Russell site on Arroyo el Lavadero (which drains the N slope of the Sierra Aconchi), about 9 km WSW of Huepac. 29°52′20″, 110°16′45″, 650 m.

Libertad, Puerto. Coastal town. 29°54′20″, 112°41′00″, 1 m.

Liliba. Village (also called "Lobos") WSW of Ciudad Obregon on Estero Lobos. 27°21′00″, 110°27′00″, 3 m. Isla Lobos is 14 km WSW.

Llanos, Los. See San Pedro.

Lobos. See Liliba.

Lobos, Puerto. Fishing village. Bahía San Lorenzo is to the SE. 30°16′15″, 112°51′00″, 1 m.

Locos, Cerro los. Range NW of the Sierra Aconchi and contiguous with it. Visited by Phillips and Marshall. 29°51′10″, 110°24′30″, 1850 m.

Lo de Campa. Site repeatedly visited during this study, about 23 km ESE of Tecoripa. On Arroyo San Javier, in thornscrub–deciduous forest. 28°32′29″, 109°44′35″, 490 m.

Lopez, Presa Jacinto. See Jacinto Lopez, Presa.

Luna, Estero la. Estero about 14 km S of mouth of Río Yaqui. G. Bancroft collected here in May 1930. 27°31´15″, 110°34´30″.

Luna, Volcan la (Volcán la Luna). Crater in Sierra el Pinacate at E edge of Desierto de Altar. 31°45´20″, 113°41´10″, 65 m.

Mabibi. Ranch in oaks SW of Fronteras. 30°46´00″, 109°48´00″, 1390 m.

Macho, Puerto el. Pass in conifers 11 km WSW of Mesa Tres Ríos. 29°47´26″, 108°48´16″, 2180 m.

Madera, Sierra la. This is the "Sierra Oposura" of the ornithological literature. Coordinates for Cerro San Jose, the highest point: 29°56´40″, 109°31´30″, 2350 m. Coordinates of Marshall pine-oak woodland site: 29°54´00″, 109°29´00″, 2020 m.

Magdalena, Río. Major drainage arising S of Nogales and flowing SW. Also see Asunción, Río; Concepción, Río.

Magdalena de Kino. City on the Río Magdalena. 30°37´30″, 110°58´00″, 760 m.

Maimodochi, Cañada. See Cruz del Diablo.

Mariquita, Sierra. Mountains NW of Cananea. A road to a communications tower (microondas) from Highway 2 at Puerto Cananea provides access to oak-juniper woodland; pines are higher. 31°03´25″, 110°22´45″, 2500 m.

Mariscal, El. Ranch 8 km E of Mazocahui. 29°33´00″, 110°02´00″, 860 m.

Maromicon. See Baromico.

Martínez. Phillips collecting site on the Río Santa Cruz (not on 1:50000 map). 31°09´20″, 110°42´00″, 1250 m.

Mártir, Isla San Pedro. See San Pedro Mártir, Isla.

Mascareña. Village on Río Santa Cruz E of Nogales. 31°18´25″, 110°50´00″, 1150 m.

Masiaca. See San José Masiaca.

Masocarit, Isla. Large island about 10 km W of Agiabampo; in Estero Agiabampo. Called "Masocari Island" by van Rossem. 26°22´00″, 109°14´30″, 2 m.

Matape, Río (Río Mátape). Rises near Mazatán and flows SW to become the Río San Marcial near Punta de Agua.

Maycoba. Village E of Yécora on upper Río Maycoba. 28°23´30″, 108°39´30″, 1520 m.

Maycoba, Río. River in pine-oak woodland; it joins with the Arroyo (Río) Yécora to form the Río Mulatos (see also Río Aros). Coordinates are at bridge on road from Yécora to Maycoba: 28°22´36″, 108°45´50″, 1200 m.

Mayo, Río. A major river system arising in SW Chihuahua and flowing SW to the agricultural areas near Navojoa. Presa Mocuzari dams the Río Mayo.

Mazatán. Town E of Hermosillo in desert-thornscrub. 29°00´20″, 110°08´20″, 550 m.

Mazocahui. Village on Río Sonora. 29°32´20″, 110°07´10″, 490 m.

Mentidero, Arroyo. A major drainage on the E slope of the Sierra de Alamos. Coordinates where the Alamos–La Chinal road crosses it about 14 km S of Alamos, at a point about 1 km from the Río Cuchujaqui: 26°55´, 108°55´, 250 m.

Mesa Tres Ríos. Village in pine-oak woodland. 29°50´32″, 108°42´43″, 1880 m.

Metate. Russell site 17 km SE of Yécora, just E of Cerro el Metate; pine-oak woodland, large madrones, alders. At S end of Sierra el Chuchupate. 28°15´44″, 108°48´18″, 1900 m.

Mexiquillo. Village on W bank of Río Mayo. 27°21´00″, 108°55´20″, 160 m.

Miguel Hidalgo. Town SE of Nogales on the Río Santa Cruz. Also called San Lázaro. 31°09´00″, 110°38´30″, 1280 m.

Milpillas. Town in Chihuahua, NE of Alamos. 27°12´20″, 108°38´40″, 1380 m.

Miramar. The beach and older resort area of Guaymas, about 5 km W of the city center. 27°55´30″, 110°56´40″, 2 m.

Mirasol, El. Ranch 16 km NNE of Güirocoba. 27°01´10″, 108°36´30″, 1000 m.

Misa, La. Town on the Río San Marcial below Presa Alatorre. 28°22´30″, 110°31´55″, 170 m.

Mochos, Puerto los. Low pass in desertscrub in drainage of Río San Ignacio. 29°41´50″, 112°13´00″, 240 m.

Moctezuma. Town on the Río Moctezuma. The old name was Oposura. 29°48´10″, 109°40´50″, 615 m.

Moctezuma, Río. River arising near Nacozari and flowing S into Presa Novillo near San Pedro de la Cueva.

Mocuzari, Presa (Presa Mocúzari). On lower Río Mayo. 27°13´20″, 109°06´15″, 100 m. Also called Presa Adolfo Ruiz Cortinez.

Mojarit. Village E of Tepahui. 27°24′00″, 109°05′30″, 240 m.

Molino Canyon. See Evans, Cañon.

Monument 204. Boundary marker about 5 km E of present channel of Río Colorado, on NW edge of San Luis Río Colorado. Nineteenth-century collecting site (van Rossem 1945). 32°29′00″, 114°46′30″, 38 m.

Morelos. Town on upper Río Bavispe. 30°49′40″, 109°13′30″, 840 m.

Moreno, Estación. Village on railroad S of Hermosillo. 28°30′15″, 110°41′00″, 200 m.

Morúa, Estero. Estero 11 km ESE of Puerto Peñasco. 31°16′40″, 113°25′30″.

Movas. Village on Río Chico (a tributary of the Río Yaqui), NW of Nuri. 28°09′45″, 109°26′30″, 220 m.

Mututicache. Village SE of Cananea on upper Río Sonora. 30°44′40″, 110°01′00″, 1140 m.

Naco. Border town. 31°20′00″, 109°57′00″, 1410 m.

Nácori, Sierra. This mountain, not identified on the 1:50000 map series, is NE of the town of Nacori Chico and El Riito. Coordinates for a Marshall (1957) study site at Pinos Altos: 29°45′45″, 108°49′40″, 1480 m; the highest nearby peak is Cerro Obscuro: 29°44′56″, 108°47′18″, 2425 m.

Nacori Chico (Nácori Chico). Town on road between Huasabas and Mesa Tres Ríos. 29°41′20″, 108°58′45″, 820 m.

Nacozari. Mining town in NE. 30°22′30″, 109°41′15″, 1100 m.

Nahuibampo. Village on W bank of Río Mayo. 27°22′40″, 108°54′30″, 160 m.

Naranjo, El. Small ranch at road crossing of Río el Taymuco. 27°14′50″, 108°45′40″, 460 m.

Natora (Nátora). Village on Río Aros, E of Sahuaripa. 29°03′00″, 108°46′00″, 660 m.

Navojoa. Major city on Río Mayo in agricultural area. Coordinates at plaza: 27°05′00″, 109°27′00″, 45 m.

Negro, Río. See Tres Rios.

Nogales. Border city. Coordinates at border: 31°20′00″, 110°56′30″, 1200 m.

Nopalera, La. Ranch in desertscrub on Río Bacoachi S of Félix Gómez. 29°42′20″, 111°32′00″, 500 m.

Noria, La. Former border crossing, opposite Lochiel, Arizona, 2.5 km W of Río Santa Cruz. 31°20′00″, 110°37′15″, 1430 m.

Noria, La. Ranch headquarters in buffel grass area in the headwaters of the Río Agua Caliente. Coordinates are of a large pond N of the headquarters: 28°08′10″, 110°02′40″, 190 m.

Norteños, Los. Village on Río Sonoita between Sonoita and Puerto Peñasco. 31°40′40″, 113°18′30″, 155 m.

Novillo, Presa el. Dam and lake on Río Yaqui. Also known as Plutarco Elias Calles. Sometimes spelled "Novillero." Coordinates at dam: 28°58′40″, 110°38′30″, 320 m.

Nuevo, Rancho. See Bonito, Cajón.

Nuri. Town in deciduous forest, on Río Nuri. 28°07′00″, 109°19′40″, 360 m.

Oasis, El. In desertscrub on Nogales-Hermosillo highway, at Carbo junction. 29°41′20″, 111°02′30″, 545 m.

Obregon, Presa Alvaro (Presa Alvaro Obregón). Major dam and impoundment on lower Río Yaqui. See Oviachic.

Ocote, El. Ranch E of Sahuaripa in open pineland. 29°03′50″, 108°56′35″, 1475 m.

Ocuca. A Neff (1947) locality about 30 km WNW of Santa Ana. 30°36′00″, 111°25′30″, 600 m.

Ojos Azules, Sierra los. Mountains E of Colonia Oaxaca. 30°44′45″, 108°54′00″, 2450 m. Marshall had a site that he called "12 mi E of Colonia Oaxaca, Pacific side of crest," coordinates: 30°44′50″, 108°54′30″, 2380 m. Included in Sierra Xitahueca on some maps.

Onabas. Village on Río Yaqui. 28°27′40″, 109°31′40″, 185 m.

Opodepe. Town on Río San Miguel. 29°55′30″, 110°38′00″, 650 m.

Oposura. See Moctezuma.

Oposura, Sierra. See Sierra la Madera.

Oputo. See Villa Hidalgo.

Osa, El. Site of long-abandoned sawmill in Sierra Oscura. 27°59′40″, 108°59′00″, 2000 m.

Oscura, Sierra. Large mountain range E of Nuri mostly in Chihuahua, not named on 1:50000 maps. Coordinates of the sawmill village of Sierra Oscura (El Serruchito): 28°01′40″, 109°57′20″, 2030 m. Several parts of the Sonoran portion of the mountain are about 2100 m; the highest point in Sonora is probably close to a cemetery: 28°00′30″, 109°00′20″, 2210 m. Also see Aserraderos, Los; Osa, El; Rodriguez, Mesa; Tigre, El.

Oviachic (Oviáchic). Village at base of W end of dam at Presa Obregon, not marked on 1:50000 map. 27°49′20″, 109°53′47″, 60 m.

Pajaros, Isla (Isla Pájaros). Small island in Estero

Agiabampo; mangroves. E. Harrison collected here (called "Bird Island" on specimen labels). 26°17′15″, 109°14′00″, 1 m.

Palma, La. Ranch about 40 km S of Hermosillo on highway to Guaymas. 28°39′30″, 110°00′00″, 185 m.

Palmar, El. Russell site on Arroyo Palmar about 21 km SE of Tonichi on road to Yécora; palms, deciduous forest, clearings. 28°29′30″, 109°23′40″, 680 m.

Palmarito, El. Arroyo draining SW into the Río Cuchujaqui. Coordinates used for bird records are from its headwaters about 19 km NE of Alamos. 27°05′00″, 108°45′09″, 670 m.

Palmita, La. Small ranch in oaks W of Mesa de Enmedio and just N of new Hermosillo-Yécora road. Coordinates are on the road in arroyo just SE of the place named on the 1:50000 map: 28°22′00″, 109°04′00″, 1500 m.

Paredoncito, El. See Tóbari, Estero.

Peñasco, Puerto. Coastal town in NW. 31°18′00″, 113°32′00″, 2 m.

Pera, La. Ranch 11 km S of Esqueda. Includes riparian woodlands along Arroyo las Calabasas. 30°37′30″, 109°37′40″, 1270 m.

Pericos, Los. Russell site 3 km W of El Chinal in thornscrub–deciduous forest. 26°40′20″, 108°43′40″, 205 m.

Pesqueira. Village on railroad about 35 km NNE of Hermosillo. 29°23′00″, 110°54′00″, 330 m.

Piedra de Lumbre, Arroyo. A Russell site about 8 km N of Arizpe. 30°24′40″, 110°10′40″, 860 m.

Piedras, Arroyo las. Canyon on the NE slope of the Sierra de Alamos. 26°59′00″, 108°57′20″, 800 m.

Pilar, El. Russell site on Arroyo Agua Caliente branch of Presa Alvaro Obregon. 27°56′00″, 109°55′00″, 100 m.

Pilares. Van Rossem locality on Río Bavispe inundated by Presa Angostura. 30°26′15″, 109°22′40″, 800 m.

Pinacate, Sierra el. Extensive volcanic region NNW of Puerto Peñasco. 31°46′30″, 113°30′00″, 1190 m. Sometimes called Volcan Santa Clara.

Pinito, Arroyo el. Russell site in oaks on E side of Sierra Hachita Hueca. Coordinates on road from San Miguelito to Janos (Chihuahua), about 20 km NE of San Miguelito: 30°38′50″, 108°50′00″, 1640 m.

Pinito, Sierra el. Long north-south-oriented mountain mass SSE of Nogales. The highest point is probably an unnamed peak at the S end. 30°55′00″, 110°45′30″, 2090 m. Marshall had two pine-oak woodland study sites in this range. One he called "Cerro del Oso" (not on 1:50000 map), coordinates: 30°55′10″, 110°43′30″, 1300 m. The other he called "Los Pinitos," coordinates: 31°05′00″, 110°45′30″, 1500 m.

Pinos Altos. Phillips, Marshall site 14 km SW of Mesa Tres Ríos. 29°45′45″, 108°49′40″, 1480 m.

Piojillo. Village (= Porvenir Cumuripa) at N end of Presa Obregon, at confluence of the Río Suaqui and Río Yaqui. 28°07′20″, 109°53′30″, 100 m.

Pitiquito. Village E of Caborca on Río Asunción. 30°40′30″, 112°03′00″, 320 m.

Plomito, El. Mining site in desertscrub SW of Caborca. 30°15′00″, 112°22′10″, 380 m.

Plomo, El. Old mining town in desertscrub NW of Caborca. 31°16′00″, 112°07′30″, 610 m.

Plutarco Elias Calles, Presa. See Novillo, Presa el.

Porvenir Cumuripa. See Piojillo.

Potam (Pótam). Village once on Río Yaqui, now in area of irrigated agriculture. 27°37′30″, 110°25′00″, 7 m.

Potrero, El. Ranch or agricultural area 2 km S of Presa Obregon. Coordinates are for a Russell site on the E bank of the Río Yaqui, opposite the village of Buenavista: 27°47′55″, 109°53′45″, 60 m.

Potrero Alcantar. See Alcantar, Potrero.

Pozos, Rancho los. Ranch in thornscrub, 21 km SE of Mazatán. 28°53′20″, 109°57′20″, 660 m.

Prieto, Cerro. Isolated hill of volcanic origin in dunes, 12 km NNW of Puerto Peñasco. 31°23′20″, 113°36′00″, 125 m.

Prieto, Cerro. Small hill of volcanic origin at edge of dunes, 5 km ENE of Puerto Lobos. 30°18′20″, 112°49′50″, 175 m.

Promontorio, Mina. Abandoned mine in NW end of Sierra Alamos; specimens were taken here by M. Buchanan, L. Kiff, and G. Stiles. 27°01′10″, 109°01′50″, 600 m.

Puerto, El. Rancho (= Puerto Noche Bueno), in hills about 15 km NNE of Punta Chueca. 29°07′30″, 112°04′25″, 170 m.

Pulpito, El (El Púlpito). Canyon E of Morelos, draining the W slope of the Sierra Madre. The Sierra las Cuevas and Sierra la Cabellera are to the NW and the Sierra los

Azueles are to the SE. Coordinates are for Las Cuevas del Pulpito in the middle portion of the canyon: 30°49′50″, 108°58′30″, 1160 m.

Purinera, La. A Mellink and Palacios (1993) site SE of Puerto Peñasco. 31°04′30″, 113°07′30″, 2 m.

Quemado, Rancho. See Cumbre de los Pinos.

Querobabi. Town on railroad about 18 km SE of Benjamin Hill. 30°03′00″, 111°01′20″, 660 m.

Quintera, Mina. Abandoned mine in NW end of Sierra Alamos; specimens were taken here by M. Buchanan, L. Kiff, and G. Stiles. 27°02′, 109°13′, 800 m.

Quiriego. Town on Río Cedros E of Ciudad Obregon. 27°31′13″, 109°15′07″, 200 m.

Quitobaquito Springs. Spring-fed oasis W of Sonoita. It is in Arizona, just N of the border. 31°56′38″, 113°01′08″, 329 m.

Quitovac. Village SE of Sonoita in desertscrub. 31°31′00″, 112°45′00″, 425 m.

Rayon (Rayón). Town on Río San Miguel. 29°43′00″, 110°34′15″, 560 m.

Realito, El. Village on Río Yaqui. 28°20′25″, 109°36′50″, 160 m.

Rebeico. Village SE of Mazatán. 28°53′10″, 109°44′55″, 425 m.

Reparo, Arroyo el. Deep canyon with stream and riparian woodlands. Coordinates on stream about 6 km SE of Santa Ana de Yécora: 28°21′15″, 109°05′45″ 920 m.

Represito de Luján. Large ranch SSE of Hermosillo, largely planted in buffel grass. 28°35′55″, 110°45′00″, 280 m.

Riíto. Village in the Río Colorado delta (also known as Nuevo Michoacan). 32°10′00″, 114°57′30″, 12 m.

Riito, El. The crossing of El Riito (also known as Río Zatachi), 11 km NE of Nacori Chico on the road to Mesa Tres Ríos; riparian woodlands on permanent stream. 29°45′19″, 108°53′40″, 980 m.

Rodeo, El. Ranch W of Moctezuma. 29°38′15″, 109°56′40″, 920 m.

Rodriguez, Mesa (Mesa Rodríguez). Old sawmill site in the Sierra Oscura in pine-oak woodland; old fields, orchard. 28°00′40″, 109°00′50″, 2100 m.

Rodriguez, Presa Abelardo L. See Hermosillo.

Romo, Volcan. Crater (also called Sykes Crater) in Sierra el Pinacate. 31°56′20″, 113°34′00″, 340 m.

Rosario. Town in desertscrub–deciduous forest NE of Ciudad Obregon; also known as Tesopaco or Rosario de Tesopaco. 27°50′30″, 109°22′15″, 420 m.

Rosario, Sierra el. Small isolated granitic mountains in Desierto de Altar, about 70 km SE of San Luis Río Colorado. 32°05′00″, 114°12′30″, 600 m.

Sahuaribo. See Sahuarivo.

Sahuaripa. Town on Río Sahuaripa. 29°03′20″, 109°13′50″, 425 m.

Sahuarivo. Also spelled Sahuaribo (= Huicochic). Village in pine-oak woodland near Chihuahua border. 27°19′10″, 108°37′50″, 1500 m. Near head of Barranca las Colas, which drains into the Río Mayo.

Salcedo. Village on Río San Nicolás, in riparian and deciduous forest. 28°25′30″, 109°13′30″, 590 m.

Salina, La. Spring-fed saline lakes about 40 km ESE of Golfo de Santa Clara. 31°31′20″, 114°07′30″, 1 m.

Salinas, Las. Unmarked on 1:50000 map; coastal dune locality WNW of Caborca; estero. 30°57′15″, 113°05′45″, 6 m.

San Antonio. Once a sawmill and mine (Mina Abundancia); in pine-oak woodland. 27°08′40″, 108°38′30″, 1380 m.

San Antonio. Small ranch in pine-oak woodland on Arroyo las Chimeneas (a tributary of Arroyo Cajón Bonito) in the Sierra San Luis. 31°12′30″, 108°51′30″, 1620 m.

San Antonio. Village (= El Chino) in thornscrub at E edge of coastal plain. 26°34′15″, 108°58′30″, 140 m.

San Antonio, Sierra de. A van Rossem locality between the Ríos San Miguel and Sonora; not identified on 1:50000 maps. Charaibabi was used as the map location.

San Antonio de la Huerta. Village on W bank of Río Yaqui, NW of Tonichi. 28°37′50″, 109°36′40″, 195 m.

San Bernardino. Ranch at Arizona border. 31°19′30″, 109°16′00″, 1120 m.

San Bernardo. Town N of Alamos near Río Mayo. 27°24′05″, 108°50′45″, 210 m.

San Carlos. Resort area W of Guaymas. 27°57′00″, 111°03′30″, 3 m.

San Esteban, Isla. Island SW of Isla Tiburón. 28°42′, 112°34′.

San Francisco, Arroyo. Van Rossem used "San Francisco Canyon" for this locality, a tributary of the Arroyo los Limones, which flows S to the Río Fuerte. Coordinates are of a site in the canyon about 2.5 km SE of the village of Las Algarrobas: 26°55′00″, 108°29′00″, 600 m.

San Ignacio. Village NE of Magdalena de Kino on Río Magdalena. 30°41′40″, 110°55′15″, 780 m.

San Ignacio, Río. Small watershed (usually dry) draining desertscrub; enters El Golfo de California near the Seri village of El Desemboque. Coordinates at mouth: 29°31′20″, 112°25′40″.

San Javier. Mining town E of Tecoripa in deciduous forest. 28°35′42″, 109°44′22″, 760 m. It is in the headwaters of Arroyo San Javier, which drains S into the Río Yaqui.

San Jorge, Isla. Small rocky island SE of Puerto Peñasco. 31°00′30″, 113°14′30″.

San José. Small coastal village SW of Ciudad Obregon. There is an estero (Estero San José) to the north. 27°07′00″, 110°09′00″, 2 m.

San Jose, Sierra. Small isolated mountain WSW of Naco. 31°15′30″, 109°59′20″, 2540 m.

San José de Batuc. Village on W side of Río Moctezuma arm of Presa el Novillo, inundated by the lake. 29°12′50″, 109°45′20″, 380 m.

San José de Pimas. Village E of Hermosillo on highway to Yécora. Coordinates at crossing of Río Matape on W side of village; permanent stream here: 28°43′10″, 110°20′50″, 320 m.

San José Masiaca. Village in extreme S. Southernmost saguaros grow near here. 26°46′00″, 109°14′00″, 60 m.

San Lorenzo, Bahía. Bay immediately SE of Puerto Lobos.

San Luis, Sierra. Large mountain range in extreme NE. 31°14′50″, 108°48′35″, 2520 m.

San Luis Río Colorado. City on Arizona border, just E of Río Colorado. 32°28′00″, 114°46′00″, 45 m.

San Marcial. Village on Río Matape. 28°59′50″, 110°18′20″, 270 m.

San Marcial, Río. Arises near Mazatán and drains SW to Bahía Empalme. The upper portion is called the Río Matape.

San Miguel. Extensive agricultural area about 26 km SE of Estero Tastiota. 28°13′50″, 111°14′15″, 85 m.

San Miguel de Horcasitas, Río. Tributary of the Río Sonora arising SE of Magdalena, usually referred to as Río San Miguel.

San Miguelito. Village on upper Río Bavispe. 30°31′00″, 108°58′00″, 980 m.

San Nicolás. Village on Río San Nicolás. The river ultimately enters the Río Yaqui. 28°26′00″, 109°10′40″, 660 m.

San Pedro. Small ranch near gorge of Río Cuchujaqui, about 30 km E of Alamos by road. Another small ranch a kilometer NE is known as Los Llanos. 27°03′10″, 108°42′10″, 480 m.

San Pedro, Río. River flowing N into Arizona, NE of Cananea. Coordinates are at border: 31°20′00″, 110°09′00″, 1300 m.

San Pedro de la Cueva. Village on Presa Novillo. 29°17′15″, 109°44′10″, 460 m. The Río Moctezuma enters the lake nearby.

San Pedro Mártir, Isla. Small island S of Isla Tiburón. 28°23′, 112°20′.

San Rafael. Ranch in pine-oak woodland on edge of Barranca las Colas 3 km W of Sahuarivo. 27°19′30″, 108°39′30″, 1600 m.

San Rafael, Hacienda de. A major collecting site of Frazar and Wright (van Rossem, 1945). Not identified on 1:50000 map. In headwaters of Río Cuchujaqui, just inside Sonora. 27°42′00″, 108°38′20″, 620 m.

Santa Ana. Large town on Río Magdalena, at junction of highways from Hermosillo to Nogales and to San Luis Río Colorado. References in text to "Santa Ana" pertain to this locality. 30°33′00″, 111°07′00″, 660 m.

Santa Ana de Yécora. Village in eastern foothills. 28°23′03″, 109°08′54″, 880 m.

Santa Barbara, Rancho. A van Rossem locality where many specimens were collected. There is no Santa Barbara on the 1:50000 map at the coordinates given here; the map locality is called "Puerto Colorado." Data from specimen labels and from van Rossem (1945) were used to place the locality at Puerto Colorado. 27°14′25″, 108°40′45″, 1420 m. The "Santa Barbara" on the 1:50000 map at 27°07′20″ and 108°43′00″ does not fit the information available for van Rossem′s locality.

Santa Clara, Ciénega de. Marsh of 10 000 to 20 000 ha in the Río Colorado delta, fed by springs but more dependent upon saline groundwater from the Wellton-

Mohawk (Sánchez Taboada) drainage canal. Coordinates used for bird records in the ciénega: 32°03´45˝, 114°54´15˝, 2 m.

Santa Clara, Golfo de. See Golfo de Santa Clara.

Santa Cruz. Town on Río Santa Cruz. 31°14´00˝, 110°35´40˝, 1360 m.

Santa Rosa. Village on old road from San Nicolás to Yécora. 28°28´20˝, 109°06´30˝, 1030 m.

Santa Rosa, Punta. Named Punta San Miguel on map; NW of Bahía Kino. Mangroves (Estero Santa Rosa), sandspit. 28°58´20˝, 112°10´00˝, 2 m.

Santa Rosa Tesia. Major van Rossem collecting site on lower Río Mayo, called "Tesia" by him. 27°13´40˝, 109°19´50˝, 80 m.

Santo Domingo. A W. Schaldach locality W of El Sasabe. 31°28´00˝, 111°50´00˝ 820 m.

Santo Domingo, Arroyo. Small canyon draining S into Arroyo Saracachi. Coordinates used for bird records are for La Brisca, a ranch in the canyon: 30°24´40˝, 110°33´00˝, 980 m.

Santo Tomás. Village on Río Sahuaripa about 5 km N of Arivechi. Used by Clark (1984). 28°58´30˝, 109°11´40˝, 450 m.

Saracachi. Old ranch NE of Cucurpe on Arroyo Saracachi in the headwaters of the Río San Miguel. 30°21´20˝, 110°34´10˝, 950 m. There is a marshy area 3 km W: 30°21´35˝, 110°35´40˝, 930 m.

Sargento. Fishing camp on Estero Sargento, on mainland opposite N end of Isla Tiburón. The bay to the S is Bahía Sargento. The northernmost extensive stand of mangroves is adjacent to the Estero. 29°19´30˝, 112°20´30˝, 1 m.

Saric (Sáric). Town on upper Río Altar. 31°06´10˝, 111°22´45˝, 780 m.

Sasabe, El (El Sásabe). Town in desertscrub-oaks on Arizona border. 31°28´20˝, 111°32´40˝, 1060 m.

Saucito, El. Marshall site about 11 km SE of Bacadehuachi. 29°45´10˝, 109°03´10˝, 1070 m.

Sausa, Rancho la. Ranch SSE of Rancho el Carrizo; unnamed on map. "Masked" Bobwhite area. 29°56´00˝, 111°12´00˝, 775 m.

Sejaqui. Village W of Río Mayo in deciduous forest. 27°29´50˝, 108°55´40˝, 400 m.

Siari, El. Small fishing camp SW of Navojoa near S end of Estero Tóbari. 26°55´00˝, 109°53´10˝, 5 m.

Sierrita, Arroyo la. N flowing arroyo S of Trincheras. Coordinates of Russell site, with much of area cleared for buffel grass: 30°09´10˝, 111°33´00˝, 830 m.

Sinoquipe. Town on Río Sonora. 30°09´20˝, 110°14´40˝, 730 m.

Sonoita (usually spelled "Sonoyta," also Sonoíta). Town at Arizona border. 31°52´00˝, 112°51´00˝, 390 m. The Río Sonoita flows W through the town and ultimately disappears in the desert NE of Puerto Peñasco.

Sonora, Río. River arising near Cananea and flowing S and W to provide irrigation water for region SW of Hermosillo.

Sonoyta. See Sonoita.

Soyopa. Village on Río Yaqui. 28°45´45˝, 109°38´10˝, 235 m.

Suaqui. Small river arising N of Tecoripa and flowing S into the Río Yaqui at Piojillo. Sometimes called the Río Suaqui Grande. See Cumuripa.

Suaqui Grande. Village SE of Hermosillo on Río Suaqui. 28°23´14˝, 109°53´16˝, 230 m.

Sykes Crater. See Romo, Volcan.

Tábelo, El. Village on arroyo of same name, about 15 km N of Alamos. 27°09´50˝, 108°57´20˝, 190 m.

Tablas, Sierra las. Small mountain 12 km ESE of Güirocoba. 26°49´30˝, 108°39´200˝, 1240 m.

Tabutama (Tubutama on many maps). Village and mission site on the Río Altar. 30°53´00˝, 111°27´50˝, 650 m.

Tajo, El. Phillips collecting site 21 km S of Esqueda. 30°32´30˝, 109°39´30˝, 1370 m.

Tarachi. Village N of Yécora. 28°48´25˝, 108°56´20˝, 1180 m.

Tarahumara, Sierra. Small mountain 12 km N of Güirocoba. Coordinates for Cerro el Tarahumara: 27°00´00˝, 108°43´00˝, 1300 m.

Tastiota. Village SE of Bahía Kino on Estero Tastiota. 28°22´00˝, 111°27´25˝, 1 m.

Taymuco, El. Village NE of Alamos on arroyo of same name. 27°14´20˝, 108°44´15˝, 550 m.

Tecoripa. Town SE of Hermosillo on highway to Yécora, in desertscrub-thornscrub. 28°37´30˝, 109°57´20˝, 390 m. The Río Tecoripa flows S into Presa Obregon.

Tepache. Town SE of Moctezuma in thornscrub. 29°32´00˝, 109°31´45˝, 600 m.

Tepahui. Town on lower Río Cedros, above Presa Mocuzari. 27°23´45˝, 109°10´50˝, 160 m.

Tepoca. Town on Hermosillo-Yécora road on Arroyo la Quema (which becomes Río Tepoca downstream).

28°26′05″, 109°15′15″, 600 m. There is a short, deep canyon adjacent to the highway 3 km NW: 28°27′20″ 109°16′10″, 700 m.

Tepopa, Cerro. The small mountain on Punta Tepopa, on NW coast of Bahía Sargento. 29°21′30″, 112°24′00″, 630 m.

Terreito, El. A Russell campsite on road NNE of Los Algodones, in pine-oak woodland, upper deciduous forest. 27°32′40″, 108°42′40″, 1630 m.

Tesia. Ranch on Río Cedros 7 km ENE of Quiriego; sabinos. 27°34′40″, 109°12′50″, 240 m.

Tesia. See Santa Rosa Tesia.

Texas Canyon. See Cumbre de los Pinos.

Tiburón, Isla. The largest island in El Golfo de California, situated W of Bahía Kino. Approximately between 28°45′ and 29°15′, between 112°12′ and 112°45′.

Tierra Negra. Ranch on road to Sierra Oscura. 27°57′10″, 109°12′15″, 790 m.

Tigre, El. Ranch on escarpment at beginning of steep ascent to Sierra Oscura, in pine-oak woodland. 28°00′15″, 109°02′40″, 1370 m.

Tigre, Sierra el. Mountains E of Presa Angostura in loop of Río Bavispe. 30°32′40″, 109°09′00″, 2450 m. Marshall (1957) had a study site at a mining locality called "El Tigre": 30°35′20″, 109°13′00″, 1750 m.

Tinaja, La. Ranch 20 km SE of El Sasabe on Arroyo el Tesoro. 31°22′30″, 111°21′00″, 1040 m.

Tinaja, La. Small ranch in arroyo in deciduous forest; about 2 km from the Río Fuerte. 26°49′45″, 108°35′25″, 195 m.

Tiznado, Sierra el. Small rugged mountain ESE of Mazocahui. 29°30′45″, 110°00′15″, 1760 m.

Tóbari, Estero. Large estero (also known as Bahía Tóbari) W of Navojoa. Coordinates are of El Paredoncito (= El Tóbari), a village on E shore: 27°03′30″, 109°54′45″, 2 m.

Toledo. Abandoned mine site on W bank of the Río Yaqui, 3 km S of Tonichi. 28°33′43″, 109°33′36″, 180 m.

Tonichi (Tónichi). Town on E bank of Río Yaqui, just N of Hermosillo-Yécora highway. 28°35′55″, 109°33′30″, 195 m.

Topitzco, Arroyo. Coordinates are in hills about 10 km E of Pueblo de Alamos: 29°13′50″, 110°02′30″, 680 m.

Torimacuca. Ranch on Arroyo Torimacuca, which drains SE to the Río Suaqui. 28°19′50″, 110°03′00″, 400 m.

Torreoncito, El. Ranch E of Presa el Novillo on Hermosillo-Sahuaripa road; stream, palms. 28°56′45″, 109°30′40″, 940 m.

Tres Rios (Tres Ríos). Ranch at bridge over Río Negro (the upper Río Bavispe). The Río Gavilán enters the Río Negro about 1 km S. 29°53′18″, 108°37′48″, 1380 m.

Trigo, El. Ranch in thornscrub–deciduous forest 17 km E of Quiriego. 27°30′30″, 109°05′00″, 380 m.

Trigo, El. Village (= El Trigo de Cororepe) N of Yécora. 28°37′30″, 108°57′00″, 1380 m.

Trincheras. Town on Río Magdalena. 30°23′50″, 111°31′50″, 510 m.

Ures. City on Río Sonora. 29°25′40″, 110°23′10″, 350 m.

Utah. See Ezcalqui.

Vallecitos, Los. Ranch in oaks 12 km E of Yécora on road to Maycoba. 28°23′20″, 108°48′25″, 1400 m.

Venecia, Caleta. Tiny coastal bay between Bahía Kino and Guaymas. 28°07′50″, 111°17′35″, 1 m.

Vibora, La (La Víbora). Small pond in grasslands, on N side of Agua Prieta–Janos highway just W of Chihuahua border. 31°19′40″, 108°56′00″, 1575 m.

Viboras, Punta (Punta Víboras). Point about 18 km N of Punta Chueca; dunes, some mangroves. 29°09′20″, 112°11′50″, 1 m. Punta Arenas is 3 km NW.

Vicam (Vícam). Town between Guaymas and Ciudad Obregon in agricultural area. 27°38′30″, 110°17′45″, 15 m.

Viejo, Sierra el. Isolated mountain SW of Caborca. 30°17′50″, 111°58′40″, 1030 m.

Villa Hidalgo. Town on Río Bavispe. Called Oputo by van Rossem. 30°09′50″, 109°19′10″, 640 m.

Villa Pesquería (Villa Pesqueria). Village (= Mátape) NE of Mazatán. 29°07′00″, 109°58′00″, 730 m.

Villa Verde. See Cuauhtémoc.

Vinatería, La. Ranch on the Río Bacoachito 6 km SW of Félix Gómez; permanent stream for short distance. 29°47′30″, 111°31′20″, 605 m.

Yaqui, Río. Part of the largest river system in Sonora; it arises as the Ríos Bavispe and Aros and many tributaries, and it is impounded by Presa Obregon. Its channel in the lower portion N of Ciudad Obregon has little or no water; most has been diverted for agriculture.

Yates, Club de. See Club de Yates.

Yavaros. Village SW of Navojoa on W side of Bahía Yavaros. 26°42´10˝, 109°31´15˝, 1 m.

Yécora. Town in pines on Hermosillo-Chihuahua highway; much farmland, pastures. 28°22´20˝, 108°55´30˝, 1520 m.

Zapote, El. Town in deciduous forest and thornscrub. 26°46´40˝, 108°41´25˝, 340 m.

Zatachi, Río. See Riito, El.

Zonoqui. Russell campsite 4 km S of Aribabi in oaks at upper edge of deciduous forest–thornscrub. 30°02´00˝, 109°05´30˝, 1300 m.

Zoquete, El. Ranch about 25 km N of Caborca. 30°55´40˝, 112°07´40˝, 600 m.

Literature Cited

Alden P. 1969. Finding the birds in Western Mexico. Tucson: Univ Arizona Pr. 138 p.

Anderson DW. 1983. The sea birds. In: Case T, Cody M, editors. Island biogeography in the Sea of Cortez. Berkeley: Univ California Pr. 508 p.

Anderson DW, Mendoza J, Keith J. 1976. Seabirds in the Gulf of California: a vulnerable, international resource. Nat Resources J 16:483–505.

[AOU] American Ornithologists' Union. 1957. Check-list of North American birds. 5th ed. Baltimore: Port City Pr. 691 p.

[AOU] American Ornithologists' Union. 1983. Check-list of North American birds. 6th ed. Lawrence (KS): Allen Pr. 877 p.

[AOU] American Ornithologists' Union. In press. Check-list of North American birds. 7th ed. Lawrence (KS): Allen Pr.

[AOU] American Ornithologists' Union. 1985. 35th supplement to the American Ornithologists' Union Check-list of North American birds. Auk 102:680–686.

[AOU] American Ornithologists' Union. 1987. 36th supplement to the American Ornithologists' Union Check-list of North American birds. Auk 104:591–596.

[AOU] American Ornithologists' Union. 1989. 37th supplement to the American Ornithologists' Union Check-list of North American birds. Auk 106:532–538.

[AOU] American Ornithologists' Union. 1991. 38th supplement to the American Ornithologists' Union Check-list of North American birds. Auk 108:750–754.

[AOU] American Ornithologists' Union. 1993. 39th supplement to the American Ornithologists' Union

Check-list of North American birds. Auk 110:675–682.

[AOU] American Ornithologists' Union. 1995. 40th supplement to the American Ornithologists' Union Check-list of North American birds. Auk 112:819–830.

[AOU] American Ornithologists' Union. 1997. 41st supplement to the American Ornithologists' Union Check-list of North American birds. Auk 114:542–552.

Atwood JL. 1988. Speciation and geographic variation in Black-tailed Gnatcatchers. Ornith Monogr 42. 74 p.

Audubon JW. 1906. Audubon's western journal, 1849–1850. Cleveland: AH Clark. 249 p.

Bates JM. 1992a. Winter territorial behavior of Gray Vireos. Wilson Bull 104:425–433.

Bates JM. 1992b. Frugivory on *Bursera microphylla* (Burseraceae) by wintering gray vireos (*Vireo vicinior*, Vireonidae) in the coastal deserts of Sonora, Mexico. Southwestern Nat 37:252–258.

Bendire C. 1895. Life histories of North American birds. US Nat Mus Special Bull 3. Washington (DC): Smithsonian Inst. 518 p.

Bent AC, et al. 1938. Life histories of North American birds of prey. US Nat Mus Bull 170. Part 2. Washington (DC): Smithsonian Inst. 482 p.

Bent AC and collaborators. 1968. Life histories of North American cardinals, buntings, towhees, finches, sparrows, and allies. Austin OL Jr, editor. US Nat Mus Bull 237. Part 2. Washington (DC): Smithsonian Inst. p 603–1248.

Bowden C. 1993. The secret forest. Albuquerque: Univ New Mex Pr. 141 p.

Brown BT. 1988. Additional Bald Eagle nesting records from Sonora, Mexico. J Raptor Res 22:30–32.

Brown BT. 1990. Nesting Common Mergansers in Chihuahua, Mexico. Southwestern Nat 35:88–89.

Brown BT, Arenas J. 1988. Bald Eagles of northwest Mexico: 1988 monitoring and research findings. Final report submitted to US Bureau of Reclamation. Phoenix (AZ). 34 p.

Brown BT, Mesta R, Liebfried W, Arenas J. 1990. Bald Eagles of northwest Mexico: reproductive success of the 1990 nesting season. Final report submitted to US Bureau of Reclamation and the Salt River Project. Phoenix (AZ). 10 p.

Brown BT, Warren P. 1985. Wintering Bald Eagles along the Rio Yaqui, Sonora. Wilson Bull 97:224–226.

Brown BT, Warren P, Anderson LS. 1987. Bald Eagle surveys of the Rio Yaqui, Sonora: December 1986–July 1987. A final report submitted to the US Bureau of Reclamation. Phoenix (AZ). 21 p.

Brown DE. 1985. Arizona wetlands and waterfowl. Tucson: Univ Arizona Pr. 169 p.

Brown DE. 1989. Arizona game birds. Tucson: Univ Arizona Pr. 307 p.

Brown DE, editor. 1994. Biotic communities: southwestern United States and northwestern Mexico. Salt Lake City: Univ Utah Pr. 342 p.

Brown DE, Ellis D. 1984. Masked Bobwhite recovery plan. Revised by S Hoffman. Albuquerque: US Fish and Wildlife Service. 82 p.

Brown JL, Brown ER. 1985. Ecological correlates of group size in a communally breeding jay. Condor 87:309–315.

Brown L, Amadon D. 1968. Eagles, hawks, and falcons of the world. 2 vols. New York: McGraw-Hill. 945 p.

Clark TO. 1984. Notable records of birds from eastern Sonora, Mexico. W Birds 15:134–136.

Cowgill P. 1976. Big star dunes home of many hardy creatures. Arizona Daily Star [Tucson], Mar 7; F:6 (cols 5, 6).

Cowgill P. 1992. Redtail hawks' nest provided Baja bonus. Arizona Daily Star [Tucson], Mar 20; C:8 (cols 5, 6).

DeWeese LR, Anderson DW. 1976. Distribution and breeding biology of Craveri's Murrelet. Trans San Diego Soc Nat Hist 18:155–168.

Dunning JB Jr, editor. 1993. CRC handbook of avian body masses. Boca Raton (FL): CRC Pr. 370 p.

Eddleman WR. 1989. Biology of the Yuma Clapper Rail in the southwestern U.S. and northwestern Mexico. Final report, interagency agreement 4–AA–30–02060. Yuma (AZ): US Bureau of Reclamation, Yuma Projects Office. 189 p.

Escalante P, Sada AM, Gil JR. 1996. Listado de nombres comunes de las aves de México. Comisíon Nacional para el Conocimiento y Uso de la Biodiversidad. México: ASM.

Foust T. 1987. When bird hunting, don't overlook Hermosillo. Arizona Daily Star [Tucson] Jan 29; C:5.

Friedmann H, Griscom L, Moore RT. 1950. Distributional check-list of the birds of Mexico. Part 1. Pacific Coast Avifauna 29. Berkeley: Cooper Ornith Club. 202 p.

Gallucci T. 1981. Summer bird records from Sonora, Mexico. Am Birds 35:243–247.

Garrett KL, Dunn J. 1981. Birds of Southern California, status and distribution. Los Angeles Audubon Soc. 408 p.

Garza-Salazar F, Dobrott S, Haro-Rodriguez J. 1992. Observaciones recientes de las poblaciones de Codorniz Mascarita *(Colinus virginianus ridgwayi)* en Sonora, Mexico. Ecologica 2:1–6.

Gatz T, Base D, Moorhouse A, Tyler J, Thompson B. 1988. Black-shouldered Kite. In: Proc Southwest Raptor Management Symposium and Workshop, p 48–53. Inst Wildl Res, Natl Wildl Federation Sci and Tech Ser No. 11. Washington (DC): Natl Wildl Federation. 392 p.

Gentry HS. 1942. Rio Mayo plants. Publ 527. Carnegie Inst Washington. 328 p.

Grinnell J. 1928. A distributional summation of the ornithology of Lower California. Univ Calif Publ Zool. Vol 32. Berkeley. 300 p.

Groschupf K, Brown BT, Johnson RR. 1988. An annotated checklist of the birds of Organ Pipe Cactus National Monument, Arizona. Tucson (AZ): SW Parks and Mon Assoc. 45 p.

Harris J. 1977. Peregrine Falcon studies lead to explorations in tropical Mexico. Chihuahuan Desert Discovery, Newsletter No. 2, Dec 1977. Alpine (TX): Chihuahuan Desert Research Inst.

Harrison EN, Kiff LF. 1977. The nest and egg of the Black Solitary Eagle. Condor 79:132–133.

Henny CJ, Anderson DW. 1979. Osprey distribution, abundance, and status in Western North America: III. The Baja California and Gulf of California population. Bull So Calif Acad Sci 78:89–106.

Hilton JW. 1947. Sonora sketch book. New York: Macmillan. 333 p.

Howell SNG. 1993. Status of the Piping Plover in Mexico. Euphonia 2:51–54.

Howell SNG, Robbins MB. 1995. Species limits of the Least Pygmy-Owl *(Glaucidium minutissimum)* complex. Wilson Bull 107:7–25.

Howell SNG, Webb S. 1992. Noteworthy bird observations from Baja California, Mexico. W Birds 23:153–163.

Hunter WC. 1988. Status of the cactus ferruginous pygmy-owl *(Glaucidium brasilianum cactorum)* in the United States and Northern Mexico. Phoenix (AZ): US Fish and Wildlife Service. 13 p.

Kaufman K. 1990. A field guide to advanced birding. Boston: Houghton Mifflin. 299 p.

Kaufman K, Witzeman J. 1979. A Harlequin Duck reaches Sonora, Mexico. Continental Birdlife 1:16–17.

Knoder CE, Plaza P, Sprunt A IV. 1980. Status and distribution of the Jabiru Stork and other water birds in western Mexico. In: Proc Natl Audubon Soc Symposium, The birds of Mexico: their ecology and distribution. Tiburon (CA). 137 p.

Kramer GW, Euliss N Jr. 1986. Winter foods of Black-bellied Whistling-Ducks in northwestern Mexico. J Wildl Mgt 50:413–416.

Kuvlesky W Jr. 1994. Annual report: Masked Bobwhite reintroduction program—1993. Sasabe (AZ): Buenos Aires Nat Wildl Refuge. 16 p.

Lanning DV, Shiflet JT. 1983. Nesting ecology of Thick-billed Parrots. Condor 85:66–73.

Leopold Aldo S. 1949. A Sand County almanac and sketches here and there. New York: Oxford Univ Pr. 226 p.

Leopold A Starker. 1959. Wildlife of Mexico. Berkeley: Univ of Calif Pr. 568 p.

Lumholtz C. 1902. Unknown Mexico, vol. 1. New York: Scribner's. 530 p.

Marshall JT Jr. 1957. Birds of pine-oak woodland in southern Arizona and adjacent Mexico. Pacific Coast Avifauna 22. Berkeley: Cooper Ornith Soc. 125 p.

Marshall JT Jr. 1967. Parallel variation in North and Middle American Screech-Owls. Monogr No 1. Los Angeles: W Foundation Vert Zool. 72 p.

McMurry FB, Monson G. 1947. Least Grebe breeding in California. Condor 49:125–126.

Mellink E, Palacios E. 1993. Notes on breeding coastal waterbirds in northwestern Sonora. W Birds 24:29–37.

Mexico. 1971–1982. Estados Unidos Mexicanos: carta topographica [Col maps 58 x 72 cm, scale 1:50000].

Dirección de Estudios Economicos. Mexico City: Comisión de Estudios del Territorio Nacional.

Miller AH. 1948. A new subspecies of Eared Poor-will from Guerrero, Mexico. Condor 50:224–225.

Miller AH, Friedmann H, Griscom L, Moore RT, editors. 1957. Distributional check-list of the birds of Mexico. Part 2. Pacific Coast Avifauna 33. Berkeley: Cooper Ornith Soc. 436 p.

Monson G. 1986. Gray-collared Becard in Sonora. Am Birds 40:562–563.

Monson G, Phillips AR. 1981. Annotated checklist of the birds of Arizona. 2nd ed. Tucson: Univ Arizona Pr. 240 p.

Moore RT. 1938. Unusual birds and extensions of ranges in Sonora, Sinaloa, and Chihuahua. Condor 40:23–28.

Morlan J. 1981. Status and identification of forms of White Wagtail in western North America. Continental Birdlife 2:37–50.

Neff JA. 1947. Notes on some birds of Sonora. Condor 49:32–34.

Nelson JB. 1978. The Sulidae: gannets and boobies. New York: Oxford Univ Pr. 1012 p.

Palacios E, Mellink E. 1995. Breeding birds of Esteros Tóbari and San José. W Birds 26:99–103.

Palmer RS. 1988. Handbook of North American birds. Vol 5, Diurnal Raptors. Part 2. New Haven: Yale Univ Pr. 465 p.

Phillips AR. 1986. The known birds of North and Middle America. Part 1. Denver: Allan R. Phillips. 259 p.

Phillips AR. 1991. The known birds of North and Middle America. Part 2. Denver: Allan R. Phillips. 249 p.

Phillips AR, Amadon D. 1952. Some birds of northwestern Sonora, Mexico. Condor 54:163–168.

Phillips AR, Marshall JT, Monson G. 1964. The birds of Arizona. Tucson: Univ Arizona Pr. 212 p.

Pitelka FA. 1951. Generic placement of the Rufous-winged Sparrow. Wilson Bull 63:47–48.

Price W. 1899. Some winter birds of the Lower Colorado Valley. Bull Cooper Ornith Club 1:89–93.

Rising JD. 1988. Phenetic relationships among the warblers in the *Dendroica virens* complex and a record of *D. virens* from Sonora, Mexico. Wilson Bull 100:312–316.

Rosenberg KV, Ohmart RD, Hunter WC, Anderson BW. 1991. Birds of the lower Colorado River valley. Tucson: Univ Arizona Pr. 416 p.

Russell SM, Lamm DW. 1978. Notes on distribution of birds in Sonora, Mexico. Wilson Bull 90:123–131.

Saunders GB, Saunders DC. 1981. Waterfowl and their wintering grounds in Mexico, 1937–64. USFWS Resource Publ 138. Washington (DC). 1151 p.

Schaldach WJ Jr., Phillips AR. 1961. The Eared Poorwill. Auk 78:567–572.

Scott PE. 1994. Lucifer Hummingbird *(Calothorax lucifer)*. In: Poole A, Gill F, editors. The birds of North America. No 134. Philadelphia: Acad Nat Sciences, Am Ornith Union.

Short LL Jr. 1974. Nesting of southern Sonora birds during the summer rainy season. Condor 76:21–32.

Skutch AF. 1969. Life histories of Central American Birds III. Pacific Coast Avifauna 35. Berkeley: Cooper Ornith Soc. 580 p.

Smith R, Jensen G. 1970. Black Brant on the mainland coast of Mexico. Trans N Am Wildl Natural Resources Conf 35:227–241.

Terrill SB. 1981. Notes on the winter avifauna of two riparian sites in northern Sonora, Mexico. Continental Birdlife 2:11–18.

Terrill SB, Terrill L. 1986. Common Pauraque *(Nyctidromus albicollis)* record from Sonora, Mexico. Am Birds 40:430.

Thayer J, Bangs O. 1906. Breeding birds of the Sierra de Antonez, north central Sonora. Proc Biol Soc Wash 19:17–22.

Thompson CW. 1991. The sequence of molts and plumages in Painted Buntings and implications for theories of delayed plumage maturation. Condor 93:209–235.

Todd R. 1986. A saltwater marsh hen in Arizona. Completion Report, Federal Aid Proj. W–95–R. Phoenix: Arizona Game and Fish Dept. 290 p.

Tomlinson R. 1972a. Review of literature on the endangered Masked Bobwhite. Resource Publ 108. Washington (DC): US Bureau of Sport Fisheries and Wildlife. 28 p.

Tomlinson R. 1972b. Current status of the endangered Masked Bobwhite Quail. Trans of the Thirty-seventh North American Wildl and Natural Resources Conf.

Washington (DC): Wildl Mgt Inst. p. 294–310.

[USFWS] US Fish and Wildlife Service. 1947–1992. Winter waterfowl survey, Mexico. Annual reports written by various authors. Portland (OR), Columbia (MO).

[USFWS] US Fish and Wildlife Service. 1990. Endangered and threatened wildlife and plants. 50 CFR 17.11 and 17.12. April 15. Washington (DC).

van Rossem AJ. 1931. Report on a collection of land birds from Sonora, Mexico. Trans San Diego Soc Nat Hist 6:237–304.

van Rossem AJ. 1934. Critical notes on Middle American birds. Bull Mus Comp Zool 77:445.

van Rossem AJ. 1945. A distributional survey of the birds of Sonora, Mexico. Occasional Papers of the Mus Zool No 21. Baton Rouge: Louisiana St Univ Pr. 379 p.

van Rossem AJ, The Marquess Hachisuka. 1937. A further report on birds from Sonora, Mexico, with descriptions of two new races. Trans San Diego Soc Nat Hist 8:321–336.

Wilbur SR. 1987. Birds of Baja California. Berkeley: Univ Calif Pr. 253 p.

Williams SO III. 1987. The changing status of the Wood Duck *(Aix sponsa)* in Mexico. Am Birds 41:372–375.

Wilt R. 1976. Birds of Organ Pipe Cactus National Monument. Popular Series No 18. Globe (AZ): SW Parks and Mon Assoc. 82 p.

Zimmerman DA, Boettcher J. 1967. The Common Loon in Sonora, Mexico. Condor 69:527.

Index

This index contains the scientific names of bird species (in italics) and families (in capital letters), English and Spanish vernacular names, and some definitions. Plant and place names are not indexed; they may be identified in Appendixes A and B, respectively.

A

b

M

N

q

r

About the Authors

Stephen M. Russell became interested in birds while in high school in Abingdon, Virginia. He received a degree in biology from Virginia Polytechnic Institute and then earned a Ph.D. as a student of George H. Lowery, Jr., at Louisiana State University, where he did dissertation studies in British Honduras (Belize). He taught at the University of New Orleans for six years before joining the faculty at the University of Arizona. Recently he retired to emeritus status after 32 years at Arizona, where he taught, did research, and investigated the birds of Sonora. Russell has served as president of the Arizona-Sonora Desert Museum and the Cooper Ornithological Society and was secretary of the American Ornithologists' Union for several years. Currently he is chair of the North American Banding Council. He is the author of the *Birds of North America* account of the Anna's Hummingbird and coauthor of accounts of the Black-chinned Hummingbird and Rufous-winged Sparrow. He is author of a forthcoming paper on the birds of the tropical deciduous forest of southern Sonora. Current research interests focus upon the behavior and ecology of hummingbirds.

Gale Monson is a 1934 Bachelor of Science graduate of North Dakota State University in biology. His career has been in wildlife management and related work, mainly in Arizona but also in New Mexico and in Washington, D.C. He is one of the three authors of the classic *Birds of Arizona* (1964), the lead author of two editions of *An Annotated Checklist of the Birds of Arizona* (1964 and 1981), and the senior editor of *The*

Desert Bighorn (1980). He is a recipient of the Department of the Interior's Honor Award for Distinguished Service, the American Motors Conservation Award, and the Wildlife Society's Olé Timer Award in Wildlife Management in Arizona and New Mexico.